Leonardo Kapural
Editor

Chronic Abdominal Pain

An Evidence-Based, Comprehensive Guide to Clinical Management

Springer

Editor
Leonardo Kapural, M.D., Ph.D.
Professor of Anesthesiology
Wake Forest University, School of Medicine
Carolinas Pain Institute and Clinical Research Center
Winston-Salem, NC, USA

ISBN 978-1-4939-1991-8 ISBN 978-1-4939-1992-5 (eBook)
DOI 10.1007/978-1-4939-1992-5
Springer New York Heidelberg Dordrecht London

Library of Congress Control Number: 2014955051

Printed on acid-free paper

Springer is part of Springer Science+Business Media (www.springer.com)

This book is dedicated to my father Mile who was an internist and cardiologist, but above all, my teacher and inspiration in medicine early in my life. My mother Zlata, relentless in providing the best for her kids, with a smile on her face, even through difficult times.

My wife Miranda, my lifelong partner in medicine and at home, and my kids Daniella and Luka who showed incredible patience with me when taking a large chunk of quality time from them during weekends, evenings, and holidays in order to complete similar projects.

In Memoriam

James Crews, author of the chapter on regional anesthesia (Chap. 17) who left us early and unexpectedly before this textbook could be completed. His work at Wake Forest Regional Anesthesia and Acute Pain Program, his national and international work in pain medicine, and his presidency of Carolinas Pain Society will be long remembered

Preface

The greatest satisfaction for any physician, when treating a patient with chronic pain, is to achieve meaningful, and hopefully, long-lasting pain relief. When treating severe chronic abdominal pain, many obstacles are currently in our way to achieve just that. Those obstacles include (on occasion) elusive etiology, (frequently) lack of education of referring physicians on where to refer patient, presence of few long-lasting therapeutic options, and a strong affective response to the unrelenting pain.

The goal of this textbook was to direct an interested reader to a proper selection of various therapeutic approaches that currently exist in comprehensive treatment of chronic abdominal pain. However, in order to provide such information, accumulated knowledge on mechanisms of pain generation and adaptive mechanisms needed to be detailed first. In addition, various diagnostic approaches to investigate source of abdominal pain had to be presented.

In this textbook state-of-the-art therapeutic approaches for various causes of chronic abdominal pain were described by over 60 authors, most of them very busy clinicians, who invested in translational clinical research, from the bench to innovative therapies. They represent a wide range of specialties that include pain medicine, psychology, rehabilitation, gynecology, urology, abdominal surgery, neurology, anesthesiology, and neurosurgery.

Still, the core of this textbook is provided by interventional pain physicians. There are several reasons for this: a surge in various new minimally invasive approaches in treatment of abdominal pain that were mastered by this physician group, slow but steady departure from frequently controversial opioid management of abdominal syndromes, and unrelenting enthusiasm by this group to make a difference in treating serious chronic pain. Prolific growth of the Interventional Pain Management Centers and their central role in treatment of other chronic pain conditions, mainly chronic spinal issues, serves as a good base to tackle prevalent chronic abdominal pain.

This book, however, is a good reminder that the same problem should be treated by the multidisciplinary team having knowledge on proposed algorithms for the treatment of such maladies. A good example in this book is a treatment of chronic pancreatitis. From epidemiology, mechanisms, differential diagnosis, innovative new approaches to establish a diagnosis, to conservative and interventional treatment that includes blocks and radiofrequency ablation, to more advanced and invasive therapeutic approaches in neuromodulation, abdominal surgery and neurosurgical approaches were described through 12 different chapters (Chaps. 1, 2, 3, 4, 9, 14, 15, 16, 19, 22, 23, 24). In addition, psychological approaches were suggested in the last (but not least) chapter delivered by Dr. Sweis (Chap. 25). Readers will find such an approach very informative, however one will notice a lack of described step-by-step algorithm for any of the pain disorders described. The reason is simple, we are far from providing an accurate algorithm for any condition above, and suggesting an algorithm for various chronic abdominal problems may be a worthwhile task of the next issue editor. It is just too early for algorithms, mainly because of a lack of evidence-based literature in this area, and recent advances in the field without properly assessed risks and benefits.

I do hope that this textbook provides an original and necessary perspective from which to consider the challenge of treating abdominal pain: on how to select the right patients for the treatment, how to select the next proper step in treatment when the previous fails, and how to avoid unnecessary complications. Nevertheless, this book hopefully succeeds by asking the right questions, and providing a clinical snapshot from which future authors can take inspiration. More work is ahead of us to clearly determine if certain diagnostic tests, blocks, patient groups, or procedures will be predictive of long-term relief from severe chronic abdominal pain.

Winston-Salem, NC, USA Leonardo Kapural

Contents

Contributors

Jose De Andres, M.D., Ph.D., F.I.P.P., E.D.R.A. Anesthesia Division, Department of Surgery, Valencia School of Medicine, Valencia, Spain

Anesthesia Critical Care and Pain Management DepartmentGeneral University Hospital, Valencia, Spain

Juan Marcos Asensio-Samper Anesthesia Critical Care and Pain Management Department, General University Hospital, Valencia, Spain

Jaime Belkind-Gerson, M.D., M.Sc. Pediatric GI, Massachusetts General Hospital, Boston, MA, USA

Amit Bhatt, M.D. Gastroenterology and Hepatology, Cleveland Clinic Foundation, Cleveland, OH, USA

Daniel M. Birk, M.D. Department of Neurosurgery, University of Illinois at Chicago, Chicago, IL, USA

Robert Bolash, M.D. Pain Management Department, Cleveland Clinic, Cleveland, OH, USA

Brian M. Bruel, M.D. Department of Pain Medicine, University of Texas MD Anderson Cancer Center, Houston, TX, USA

Jorge F. Carillo, M.D. Department of Obstetrics and Gynecology, University of Rochester Medical Center, Rochester, NY, USA

Thomas Chai, M.D. Department of Pain Medicine, University of Texas MD Anderson Cancer Center, Houston, TX, USA

Jiande Chen, Ph.D. Division of Gastroenterology and Hepatology, Johns Hopkins Medicine, Baltimore, MD, USA

Ningbo Pace Translational Research Center, Zhejiang, Ningbo, China

Joseph P. Cravero, M.D. Anesthesiology, Perioperative and Pain Medicine, Boston Children's Hospital, Boston, MA, USA

James Crews, M.D. Department of Anesthesiology, Wake Forest University Baptist Medical Center, Winston-Salem, NC, USA

Roy Dekel, M.D. Department of Gastroenterology, Tel Aviv Medical Center, Tel Aviv, Israel

Mehul J. Desai, M.D., M.P.H. Pain Medicine and Research, International Spine, Pain and Performance Center, Silver Spring, MD, USA

Asbjørn Mohr Drewes, Ph.D., D.M.Sc. Mech-Sense, Department of Gastroenterology and Hepatology, Aalborg University Hospital, Aalborg, Denmark

Douglas A. Drossman, M.D. Division of Gastroenterology and Hepatology, University of North Carolina at Chapel Hill, Chapel Hill, NC, USA

Gustavo Fabregat Anesthesia Critical Care and Pain Management Department, General University Hospital, Valencia, Spain

Jens Brøndum Frøkjaer, M.D., Ph.D. Department of Radiology, Aalborg University Hospital, Aalborg, Denmark

Monica P. Garin-Laflam, M.D. Pediatric Gastroenterology, Virginia Tech-Carilion School of Medicine, Carilion Clinic Children's Hospital, Roanoke, VA, USA

Marcum Gillis, M.D. Section on Gastroenterology, North Carolina Baptist Medical Center, Winston-Salem, NC, USA

Teodor Goroszeniuk, F.F.A.R.C.S.I., F.F.P.M.C.A.I. Interventional Pain Management and Neuromodulation Practice, Pain Management and Neuromodulation Centre, London, UK

Kasper Grosen, M.H.Sc. Department of Cardiothoracic and Vascular Surgery, Aarhus University Hospital, Aarhus, Denmark

Amitabh Gulati, M.D. Department of Anesthesiology, Memorial Sloan Kettering Cancer Center, New York, NY, USA

Genaro J. Gutierrez, M.D. Anesthesiology and Critical Care Medicine, The George Washington University Hospital, Washington, DC, USA

Salim M. Hayek, M.D., Ph.D. Division of Pain Medicine, University Hospitals Case Medical Center, Cleveland, OH, USA

Daryl Steven Henshaw, M.D. Department of Anesthesiology, Wake Forest Baptist Health, Winston-Salem, NC, USA

Fred M. Howard, M.S., M.D. Department of Obstetrics and Gynecology, Strong Memorial Hospital, University of Rochester Medical Center, Rochester, NY, USA

Abdul Kanu, M.D. Cleveland Clinic Foundation, Cleveland, OH, USA

Leonardo Kapural, M.D., Ph.D. School of Medicine, Wake Forest University, Carolinas Pain Institute and Clinical Research Center, Winston-Salem, NC, USA

Kenneth L. Koch, M.D. Section on Gastroenterology, North Carolina Baptist Medical Center, Winston-Salem, NC, USA

Braden Kuo, M.D., M.Sc. GI Unit, Massachusetts General Hospital, Boston, MA, USA

Oscar A. de Leon-Casasola, M.D. Department of Anesthesiology, Roswell Park Cancer Institute, University at Buffalo, School of Medicine, Buffalo, NY, USA

Samer Narouze, M.D., Ph.D. Center for Pain Medicine, Western Reserve Hospital, Cuyahoga Falls, OH, USA

Anne Estrup Olesen, M.Sc. (Pharm.), Ph.D. Mech-Sense, Department of Gastroenterology and Hepatology, Aalborg University Hospital, Aalborg, Denmark

Søren Schou Olesen, M.D., Ph.D. Mech-Sense, Department of Gastroenterology and Hepatology, Aalborg University Hospital, Aalborg, Denmark

Stefano Palmisani, M.D. Pain Management and Neuromodulation Centre, Guy's and St Thomas' Hospital, London, UK

David Pang, M.B.Ch.B., F.R.C.A., F.F.P.M.R.C.A. Pain Management and Neuromodulation Centre, St Thomas' Hospital, Guy's & St. Thomas' NHS Foundation Trust Hospital, London, UK

Vicente Luis Villanueva Perez Anesthesia Critical Care and Pain Management Department, General University Hospital, Valencia, Spain

Luciano Perotti, M.D. Department of Anesthesia and Intensive Care, Civil Hospital of Vigevano, Pavia, Italy

Martine Puylaert, M.D., Ph.D., F.I.P.P. Ziekenhuis Oost-Limburg, Anesthesiology and Multidisciplinary Pain Center, Lanaken, Belgium

Maged K. Rizk, M.D. Cleveland Clinic Foundation, Cleveland, OH, USA

Justin Rodante, M.P.H. Department of Spine and Pain Medicine, The George Washington University Hospital, Washington, DC, USA

Rajiv K. Shah, M.D. Department of Anesthesiology and Pain Medicine, Weill Cornell School of Medicine, New York–Presbyterian Hospital, New York, NY, USA

Ravi Shah, M.D. Department of Pediatric Anesthesiology, Ann & Robert H. Lurie Children's Hospital of Chicago, Feinberg School of Medicine, Northwestern University, Chicago, IL, USA

Joshua Shroll, M.D., M.P.H. Cleveland Clinic Foundation, Cleveland, OH, USA

Konstantin V. Slavin, M.D. Department of Neurosurgery, University of Illinois at Chicago, Chicago, IL, USA

Dawn A. Sparks, D.O. Department of Anesthesiology, Dartmouth Hitchcock Medical Center, Children's Hospital at Dartmouth, Lebanon, NH, USA

Ami D. Sperber, M.D., M.S.P.H. Faculty of Health Sciences, Ben-Gurion University of the Negev, Beer-Sheva, Israel

Tyler Stevens, M.D. Gastroenterology and Hepatology, Cleveland Clinic Foundation, Cleveland, OH, USA

Yan Sun, M.D., Ph.D. Department of Internal Medicine, Texas Tech University Health Science Center, El Paso, TX, USA

Santhanam Suresh, M.D., F.A.A.P. Department of Pediatric Anesthesiology, Ann & Robert H. Lurie Children's Hospital of Chicago, Feinberg School of Medicine, Northwestern University, Chicago, IL, USA

Giries W. Sweis, Psy.D., M.H.S. Cleveland Clinic, Neurological Center for Pain, Cleveland, OH, USA

Matthew K. Tobin, B.S. Department of Neurosurgery, University of Illinois at Chicago, Chicago, IL, USA

Reda Tolba, M.D. Anesthesiology and Pain Management, Ochsner Health System, New Orleans, LA, USA

J. Gabriel Tsang, M.B.B.S. Department of Pain Medicine, University of Texas MD Anderson Cancer Center, Houston, TX, USA

I. Elias Veizi, M.D., Ph.D. Division of Pain Medicine, Department of Anesthesiology, University Hospitals Case Medical Center, Cleveland, OH, USA

Bruce Vrooman, M.D. Pain Management Department, Cleveland Clinic, Cleveland, OH, USA

R. Matthew Walsh, M.D. Department of General Surgery, Cleveland Clinic Foundation, Cleveland, OH, USA

Marisa Wynne, D.O. Division of Pain Medicine, University Hospitals Case Medical Center, Cleveland, OH, USA

Miya P. Yamamoto, M.D. Department of Obstetrics and Gynecology, Minimally Invasive Gynecologic Surgeon, Kaiser Permanente, Hayward, CA, USA

Etiology and Mechanisms of Chronic Abdominal Pain

1

Jaime Belkind-Gerson and Braden Kuo

Introduction

Nocioception is the detection of noxious stimuli [1] and acute nociceptive pain is produced when a noxious stimulus of enough intensity activates receptive pathways by damaging or threatening to cause tissue damage [2]. This mechanism is protective and helps prevent injury or further lesion by generating a reflex withdrawal and thus removal an offending stimulus. Not only is there a sub-conscious reflex elicited, but often the development of complex behavior or strategy in response to the unpleasant sensation with the principal goal of avoiding further damage and limiting injury. There are, however a variety of pain syndromes, some of which involve the viscera, where no significant gross tissue injury or structural disease is found even when using careful clinical methodology. Pain syndromes can also present despite the fact that inflammation or tissue damage has resolved. Such is often the case, for example, with functional gastrointestinal disorders and chronic abdominal pain. Functional chronic abdominal pain is formally defined by the ROME III Criteria as severe, usually generalized abdominal pain which is continuous and affects functioning and quality of life [3]. This is in contrast to functional dyspepsia (FD) and irritable bowel syndrome (IBS), which are intermittent. These abdominal pain syndromes such as FD and IBS are very common, so much so that in a study of consecutive outpatient visits to a university hospital, approximately 40 % of patients with a chief complaint that included abdominal symptoms were diagnosed as having a functional gastrointestinal disorder [4]. FD is a syndrome with symptoms centered in the upper abdominal region and include pain, heartburn, postprandial discomfort or bloating, and a heavy feeling in the stomach or fullness [5]. IBS is a syndrome that is characterized mostly by lower GI tract symptoms including abdominal pain or discomfort and altered bowel habits such as constipation and/or diarrhea, urgency and tenesmus [6]. The Rome III Diagnostic criterion for IBS is: Recurrent abdominal pain or discomfort at least 3 days/month in the last 3 months associated with two or more of the following: (1) Improvement with defecation. (2) Onset associated with a change in frequency of stool. (3) Onset associated with a change in form (appearance) of stool [7]. The Diagnostic criteria for FD, according to the Rome III criteria, must include one or more of the following: bothersome postprandial fullness, early satiation, epigastric pain, or epigastric burning and no evidence of structural disease (including at upper endoscopy) that is likely to explain the symptoms. This must be for the last 3 months with symptom onset and at least 6 months prior to diagnosis [8]. The different syndromes that comprise the functional gastrointestinal disorder spectrum are all made as diagnoses of exclusion, thus absence of an organic cause such as an ulcer, esophagitis, celiac disease, or cancer is important.

The cause and pathophysiology of FD is not completely defined, however several pathogenic factors have been proposed including motility abnormalities, visceral hypersensitivity, psychosocial factors, excessive gastric acid secretion, *Helicobacter pylori*, genetics, environment, diet, lifestyle, and post-infectious FD. It is likely that several factors may be involved even in the same individual. Many of these factors are also common to other functional gastrointestinal disorders and visceral pain syndromes [5].

Perception of pain caused by an innocuous peripheral stimulus, for example mechanical stimulation at a lower threshold than normal, in subsets of patients with IBS suggests abnormal processing of sensory information. This can occur by both the peripheral nervous system, and/or the central nervous system. A number of different potential mechanism have been proposed in IBS, including (1) a peripheral sensitization of sensory endings present in the gut wall;

J. Belkind-Gerson, M.D., M.Sc.
Pediatric GI, Massachusetts General Hospital, 175 Cambridge ST, Room 575, Boston, MA 02114, USA

B. Kuo, M.D., M.Sc. (✉)
GI Unit, Massachusetts General Hospital, GRJ 719, 55 Fruit St, Boston, MA 02114, USA

L. Kapural (ed.), *Chronic Abdominal Pain: An Evidence-Based, Comprehensive Guide to Clinical Management*, DOI 10.1007/978-1-4939-1992-5_1, © Springer Science+Business Media New York 2015

(2) increased flow of nociceptive information traveling through the sensory afferents at the level of the dorsal root ganglia or the nerve fibers of the exterior laminae of the spinal cord; (3) a reduced antinociceptive effect of descending inhibitory pathways acting in the spinal cord, and (4) a central amplification of afferent signals (anticipation and hypervigilance), possibly influenced by psychological factors as for example anxiety or depression [9]. We will review some of these, in particular peripheral and central sensitization, which appear to be key in the development of visceral pain.

Motility Abnormalities

The incidence and causality of motility abnormalities in FD has been investigated extensively and in different subsets of patients. A long list of abnormalities has been found including: impaired fundic accommodation, antral hypomotility, decreased antral distention, gastric dysrhythmias, and small bowel dysrhythmias [10], (Fig. 1.1). Some groups report a high prevalence of some or many of these specific abnormalities. In IBS, constipation or diarrhea can occur secondary to disordered motility from the small or large bowel (Fig. 1.2). Tack et al. found that 40–50 % of FD patients show impaired gastric accommodation, thought to cause early satiation [11]. The observations from different studies however have not been consistent and the strength of association of each of the motility disorders is not yet well defined. In a similar manner,

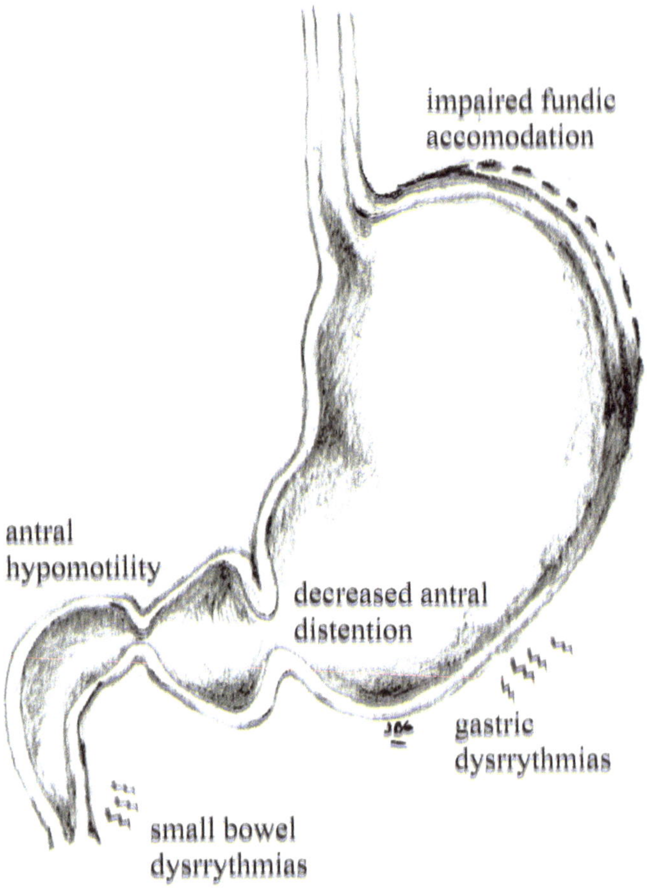

Fig. 1.1 Motor abnormalities in upper GI tract in FD

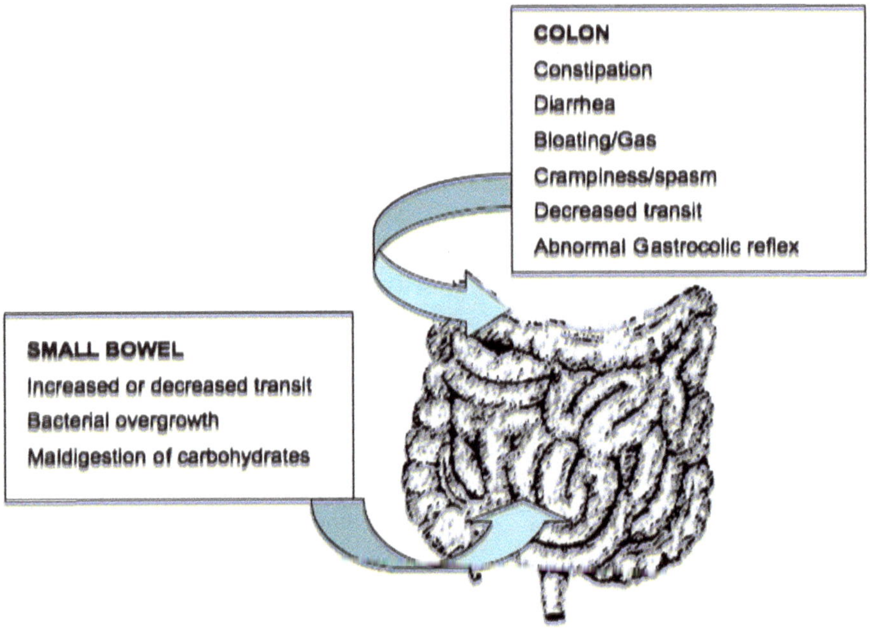

Fig. 1.2 Motor abnormalities and some associated conditions, in the lower GI tract in IBS

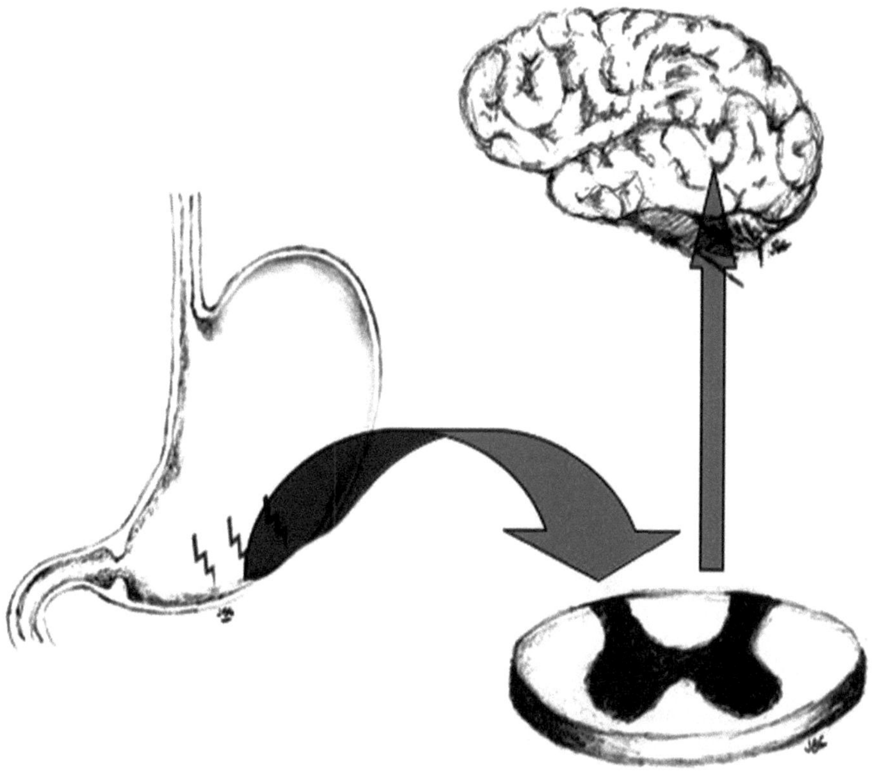

Fig. 1.3 Sensitization involves central nervous system synaptic, structural and chemical plasticity

delayed gastric emptying for solids has been found in approximately 40 % of FD patients [12], but the direct relationship between gastric emptying and dyspeptic symptoms is unclear [13]. Thus although motility disorders are common, their frequency and characteristics in each of the functional gastrointestinal pain syndromes vary and is still subject of debate. Additionally, it is unclear if motility abnormalities are a cause or an effect of the underlying pathogenic mechanism.

Sensitization

As a mechanism of added protection, sensitization is a physiological mechanism that enhances the nociceptive system [14]. Sensitization usually occurs immediately after exposure to an intense noxious event or after a repetitive damaging stimulus. When sensitization occurs, there is a reduction of the threshold needed for activation of the nociceptive response and in addition, an amplification of subsequent stimuli [15]. Sensitization is an adaptive process, which can occur in both normal and pathologic conditions and has the purpose of making the system hyper-alert to avoid ongoing damage where there is risk of further injury. It requires central nervous system synaptic, structural and chemical plasticity and although not necessarily permanent, some changes may be persistent [1]. In general, if the tissue injury or offending insult ceases, the state of heightened alertness returns over time to baseline and thus high-intensity stimuli are again required to initiate the response. In abnormal circumstances however, the state of heightened alertness becomes persistent despite an absence of ongoing tissue injury or of persistent nociceptive stimuli (Fig. 1.3).

Painful Syndromes that Involve Sensitization

Several painful syndromes including neuropathic pain [16], inflammatory pain [17, 18], migraine [19–21], and some types of headache [22], IBS [23, 24], fibromyalgia [25, 26], osteoarthritis [27], musculoskeletal disorders [28], generalized pain hypersensitivity [1, 29], temporomandibular joint disorders [30, 31], dental pain [32, 33], visceral pain hypersensitivity disorders [34–36], and postsurgical pain [29, 37–39] have been found to involve central sensitization. For example, visceral hypersensitivity thought to be determined by both central and peripheral mechanisms has been described in 20–90 % of patients with IBS [40]. Direct imaging of brain activity using functional magnetic resonance or positron emission tomography has demonstrated abnormal brain processing of peripheral sensory input [41]. In this scenario, the CNS malfunctions and the pain is no longer protective, on the contrary, it can be transformed to a new type of symptom often bothersome or worrisome to the patient and which may interfere with daily activity. In extreme cases, the pain or discomfort perceived may be severe. In these instances,

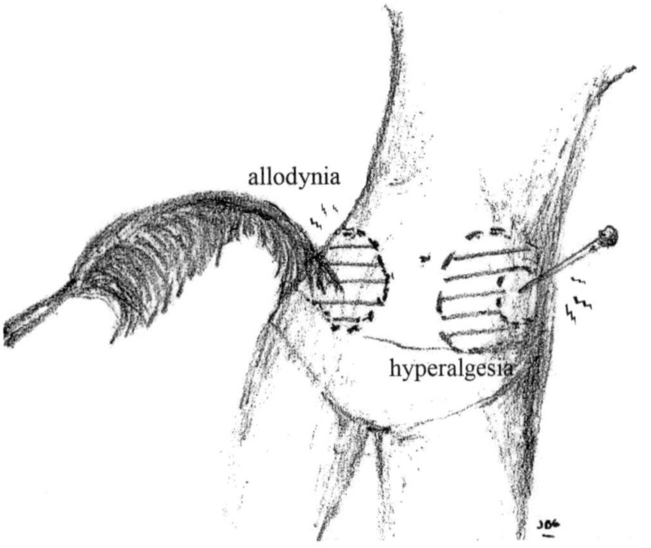

Fig. 1.4 Pain, which arises spontaneously or elicited by normally innocuous stimuli, is called allodynia. Hyperalgesia occurs when a painful sensation is exaggerated and/or prolonged

the pain arises spontaneously or can be elicited by normally innocuous stimuli, a process referred to as allodynia. When the painful sensation is exaggerated and/or prolonged in response to noxious stimuli it is known as hyperalgesia, and when it spreads beyond the site of injury as secondary hyperalgesia (Fig. 1.4).

Pathophysiology of Peripheral and Central Sensitization

Abnormal pain sensitivity is due to peripheral receptor and CNS changes a process known as peripheral and central sensitization respectively. As mentioned above, sensitization is believed to be an important mechanism explaining many acute and chronic pain syndromes. Peripheral and central sensitization differs both on the mechanisms that are involved in the pathogenesis as well as the manifestations that elicited. Peripheral sensitization involves mechanisms that lower the threshold of activation and amplify the response of pain signaling through a modification of the peripheral nocioception receptors [42, 43]. These receptors are high-threshold primary sensory neurons, which can undergo a change when exposed to inflammatory mediators and/or damaged tissue. This process is generally limited to the area that is injured or inflamed [43] and pain hypersensitivity or primary hyperalgesia at inflamed sites generally requires ongoing pathology for its maintenance. In certain cases, however the hypersensitivity may be longer lasting, for example in altered heat sensitivity where peripheral sensitization appears to play a major role. This is opposed to mechanical sensitivity which is a major feature of central sensitization [1].

In IBS the contribution of peripheral factors to pain perception has been increasingly recognized as being very important, since a subset of patients can develop IBS following and apparently as a result from an acute episode of infective gastroenteritis (i.e., post-infectious IBS). There is also additional evidence for gut immune, neural, endocrine and microbiological (i.e., intestinal microbiota) abnormalities in large subsets of patients. These factors likely influence each other and play an important contributing role in pain transmission from the periphery to the brain via sensory nerve pathways [9].

Inflammation in various areas of the GI tract as part of different disease states has been associated with the development of visceral hypersensitivity and pain as well. These diseases include: reflux esophagitis, Helicobacter pylori gastritis, celiac disease, acute infectious gastroenteritis, and inflammatory bowel disease [40, 44]. Using animal models, a wide array of mediators released by inflammatory cells have been found to induce or mediate the peripheral sensitization of mucosal neuronal afferents [44] and to recruit nociceptors that were previously silent [44]. These mediators include: cytokines, prostanoids, amines, neuropetides, and neurotrophins. Similar mechanisms may also cause altered motor and not only sensory function in post-infectious IBS [45, 46]. Stimulated by chemical and/or mechanical stimuli, terminal receptors of visceral afferent nerves respond by activation of ion channels which induces sensory transduction [47]. These receptors are located in the nerve terminals in the gut mucosa, muscle, and serosa and transmit sensory information from the viscera to the central nervous system through the vagal and spinal afferent nerves.

Recent studies have found evidence that help explain peripheral sensitization through changes in the enteric nervous system and/or afferent pathways in several visceral pain syndromes including functional gastrointestinal disorders like IBS. For example, an increased expression of transient receptor potential vanilloid type-1 (TRPV-1) has been described in animal models of intestinal inflammation [48]. TRPV-1 is an ion channel receptor activated by heat (>43 °C), acid (pH <5.9), inflammatory mediators, and capsaicin [49]. Increased levels of TRPV-1 were also found in patients with idiopathic rectal hypersensitivity and fecal urgency [48]. A correlation between increased expression of TRPV-1 and rectal distension sensory and health thresholds has been described [50].

In the past few years, new and exciting information has become available, shedding light into the mechanisms of bidirectional brain-gut interactions in both health and disease. Evidence is compelling for enteric flora including both commensal and pathogenic organisms to exert an important role in the brain-gut interaction in both directions: brain-gut and gut-brain [51, 52]. The brain influence on the intestinal microbiota can be through direct or indirect mechanisms. Indirect ways to influence bacterial composition and function include motility, secretion, and intestinal permeability [53]. Direct pathways can occur via signaling molecules secreted into the gut lumen by various types of cells in the mucosa, lamina propria, and even myenteric and neuronal layers of the gut. [54] Conversely, the influence of bacteria upon the intestine and eventually the brain occurs via signaling molecules which exert their actions through specific receptors expressed on gut epithelial cells or enterochromaffin cells. In cases of increased intestinal permeability which can occur with inflammatory states such as invasive infectious diarrhea, bacteria can directly stimulate host cells in the lamina propria.

A key element in bidirectional brain-gut communication is the enterochromaffin cell, functioning as a transducer for the communication between the gut lumen and the nervous system. Afferent stimuli involving pain and immune-response modulation and even background emotions as well as other functions are transmitted from enterochromaffin cells via their direct innervation by vagal branches. Disruption or abnormalities in the bidirectional interactions between the bacteria and the brain contribute to several acute and chronic gastrointestinal disease states which include, for example IBS and other functional as well as inflammatory bowel disorders [55]

The bidirectionality of brain-gut information may help explain the results of a prospective 12-year study on a random population of 41,775 who were surveyed in Australia. The investigators found that among people free of functional gastrointestinal disease (FGID) at baseline, higher levels of anxiety was a significant independent predictor of developing new onset FGID 12 years later. Conversely, among people who did not have elevated levels of anxiety and depression at baseline, those with a FGID at baseline had significantly higher levels of anxiety and depression at follow-up. In IBS, higher levels of anxiety and depression at baseline were predictive of IBS at follow-up, while only depression was predictive of FD at follow-up. [56]

In Central Sensitization, the pain occurs due to CNS dysfunction, with both brain and spinal cord undergoing changes that lead to an alteration of how it responds to sensory inputs. It does not depend on the presence of peripheral noxious stimuli, as in the case of Peripheral Sensitization. The hypersensitivity response in non-inflamed tissue occurs due to a change in the sensory response, which is elicited by normal inputs. This leads to increased pain sensitivity, often long after the initiating cause has disappeared and when no peripheral pathology is present. Thus Central sensitization is an abnormal state of increased or amplified responsiveness to peripheral signaling caused by increases in membrane excitability, synaptic efficacy and/or reduced inhibition.

The mechanics of this process involve the recruitment of inputs, which do not normally activate nociceptive pathways, like for example: large, low-threshold mechanoreceptor myelinated fibers which go on to transmit fiber-mediated pain. In this manner, mechanical sensitivity becomes a major feature of central sensitization. Since this process requires a change in the neuronal properties of the CNS, the sensation of pain is no longer proportional nor does it correspond to the intensity, or duration of the different peripheral stimuli, as normally occurs in acute nociceptive pain.

In central sensitization, somatosensory neurons that are not fixed or hard-wired, but are instead highly malleable go on to sensitization by synaptic enhancement or increased efficiency [1]. When this process becomes established pain is elicited, which is erroneously perceived as a peripheral nociceptive event, since often there is no abnormal peripheral event occurring. Rather there is an abnormal response to both innocuous and/or noxious stimuli with a spread of tenderness or pain beyond lesion sites. Therefore, when tissue inflammation or disease has been ruled out and the diagnosis is deemed to be central sensitization, the treatment for this altered sensory condition is aimed at the CNS and not at peripheral nociceptive receptors or peripheral tissue.

A growing body of evidence has shown that this process is not restricted to neurons, as glial activation plays an important role in the modulation of neuronal functions and affects the spinal processing of nociceptive signaling. Indeed, glia are also involved in central sensitization and chronic pain facilitation. For example, glial activation in the spinal cord is considered an important component in the development and maintenance of allodynia and hyperalgesia in various models of chronic pain, including: neuropathic pain, pain associated with peripheral inflammation, and some forms of visceral hyperalgesia [57].

For central sensitization to be induced, an intense, repetitive, or sustained noxious stimulus is required recruiting many neural fibers. This will not occur after a single stimulus. Important tissue injury almost always induces central sensitization, so that for example surgical trauma will often result in the development of central sensitization, but peripheral tissue injury is not always necessary. Nocioceptor afferents from different parts of the body may be involved, for example: those innervating muscles [18, 58], joints [17], skin [59, 60] or internal organs [61, 36]. Experimentally, central sensitization can be elicited by nocioceptor activation of the skin, as with topical or intradermal capsaicin or with repeated heat stimuli [62–64] and in the gastrointestinal tract by exposure to low pH solutions [65, 66] or with mustard oil instillation into the colon in anesthetized rats [67]. Exposure of the lower esophagus to acid, for example, induces central sensitization leading to viscero-visceral (pain hypersensitivity in the upper esophagus) and viscero-somatic hypersensitivity (allodynia on the chest wall) that can be captured by esophageal evoked potentials, and is associated with increased temporal summation [68]. Thermal and mechanical pain hypersensitivity in the rectum after esophageal stimulation using acid and capsaicin infusions demonstrates how widespread the effects of central sensitization are in the gastrointestinal tract [68]. Indeed, 35–50 % of FD patients are said to be hypersensitive to gastric distension stimuli [69]. In a similar manner in IBS, several studies have found that balloon inflation in the sigmoid colon evokes increased pain perception compared with healthy controls [70]. In one series, rectal hypersensitivity could be detected in 95 % of patients with IBS [71].

Once afferents are stimulated with enough intensity/duration, there are two temporal phases that are triggered in central sensitization, each of which involves two specific mechanistic stages: An early transcription-independent, phosphorylation-dependent phase and a later transcription-dependent phase. The early phase is mediated by rapid changes in glutamate receptor and ion channel properties. The later is longer-lasting, requiring synthesis of new proteins [15] and is the mechanism responsible for central sensitization in several pathological syndromes and conditions [15]. To be established, central sensitization involves primarily, but not exclusively the N-methyl-D-Aspartate receptors (NMDAR) expressed in postsynaptic spinal cord neurons from the dorsal horn. Besides NMDAR, other receptors include: ionotropic amino-3-hydroxy-5-methyl-4-isoxazole propionate (AMPA), Kainate (KA) receptors, and several metabotropic (G-protein coupled) glutamate receptor subtypes (mGluR). In the superficial laminae of the dorsal horn, AMPAR and NMDAR are present in virtually every synapse and these respond to the fast neurotransmitter glutamate, released by primary afferent neurons [72–74].

The activation of these receptors is very important at both initiating and also maintaining activity-dependent central sensitization. When using noncompetitive or competitive antagonists to experimentally block activation, the induced hyperexcitability of nociceptor conditioning is prevented [75]. Additionally, conditional deletion of NR1, the most common NMDA complexes in the dorsal horn, abolishes NMDA synaptic inputs and acute activity-dependent central sensitization [76], thus the NMDA receptor is both a trigger and an effector of central sensitization [1]. Stimulation of receptors: AMPAR and group I mGluRs [77–79] participate with NMDAR in the activation intracellular pathways, including the PLC/PKC pathway, the phosphatidylinositol-3-kinase (PI3 K) pathway and the mitogen-activated protein kinase pathway (MAPK) that involve the extracellular signal-regulated kinases (ERK1 and ERK2) and the cAMP response element binding protein (CREB). One way that ERK and CREB are activated is through an elevation in intracellular Ca2+ sufficient to drive a calmodulin-induced stimulation of adenylyl cyclases 1 and 8, whose cAMP production in turn activates PKA and subsequent cascade(s) [1]. Changes in the membrane receptors of the superficial lamina dorsal root neurons make previously Ca2+ impermeable to Ca2+permeable AMPARs and significantly contribute to the source of the [Ca2+]i increase in pain, specifically that which is inflammation-induced.

Together with Glutamine, the neurotransmitters: Substance P (SP) [80, 81] and Calcitonin Gene-Related Peptide (CGRP) [82–84] are also important in the induction of central somatization. Additionally CGRP enhances brain-derived neurotrophic factor (BDNF) release into the spinal cord from nociceptive neurons in an activity-dependent manner. BDNF binds to its high-affinity tyrosine-kinase receptor trkB and enhances NMDA-mediated C-fiber–evoked responses and causes activation of several signaling pathways in spinothalamic track neurons, including ERK [85, 86] and PKC, thus also contributing to central sensitization [87, 88]. Other important mediators that activate the CNS pathways and mediate central sensitization include: the inflammatory kinin bradykinin [89, 90] produced in the spinal cord as a response to peripheral noxious stimuli [91], nitric oxide synthesized by either neuronal or inducible NO synthase in the dorsal spinal cord [92–94], and serotonin (5-HT) [95] primarily through the 5-HT3 receptor.

Serotonin (5-HT) is a very important neurotransmitter and paracrine-signaling molecule between the brain and the gut [96]. Approximately 90 % of the production of 5-HT in the human body is in gastrointestinal tract, mainly synthesized by enterochromaffin cells, a subtype of enteroendocrine cells [96]. The remaining 10 % is produced in the brain in which it acts as a neurotransmitter. Peripherally produced 5-HT does not cross the blood–brain barrier. In the GI tract 5-HT activates multiple receptors (5-HT1 to 5-H7) mostly expressed by different intrinsic primary afferent neurons within the enteric nervous system and also extrinsic afferent sensory nerves [96–98]. 5-HT modulates visceral sensation via both 5-HT3 receptor and non 5-HT3 receptor-dependent mechanisms on vagal or spinal afferents [99]. The activity of 5-HT is terminated by the serotonin reuptake transporter (SERT), which is expressed in the enterocyte. 5-HT has received a great deal of attention in gastrointestinal pain syndromes and several reports have described an important role for this molecule. Alosetron a 5-HT3 receptor antagonist, for example, inhibits spinal cord c-fos expression in response to noxious colorectal distension [100]. This observation suggests that 5-HT plays a role in the transmission of noxious information within the spinal cord. 5-HT3 receptor antagonists and 5-HT4 receptor agonists in human trials have also been found to decrease multiple IBS symptoms as well [6, 101].

As previously mentioned, there are multiple different ways to initiate and precipitate the series of reactions that contribute to the establishment central sensitization. These different processes elicited as response to nociceptor input, can (1) increase membrane excitability, (2) facilitate syn aptic strength, or (3) decrease inhibitory influences in dorsal horn neurons [1]. Initial observations of central sensitization proved the plasticity of flexor motor neurons following peripheral nerve injury [102]. This led to finding similar changes in spinothalamic tract neurons [94, 103, 104], thalamus [105], spinal nucleus pars caudalis [106], anterior cingulate cortex [107], and amygdala [108, 109]. Thanks to modern imaging techniques such as functional magnetic resonance imaging, magnetoencephalography, and positron emission tomography, other brain structures involved in pain have demonstrated an increase in excitability and thus central sensitization. Amongst these are: the prefrontal cortex, superior colliculus, parabrachial nucleus, and periaqueductal gray area [110–116].

Central Sensitization in Pathological Settings

Phenotypic changes in myelinated fibers after inflammation and/or nerve injury can enable afferents to acquire the capacity to generate central sensitization. In this scenario, a normally protective mechanism may result in a pathological process. This most commonly occurs when inflammation or ongoing tissue injury persists. Additionally, there are abnormal situations where even in the absence of active peripheral pathology Central sensitization becomes autonomous and persistent and pain can be triggered not only by less intense inputs but also be maintained by different non-nociceptive stimuli. As an example: ongoing C-nerve fiber stimulation, even at levels that do not elicit central sensitization in basal conditions, is sufficient to maintain central sensitization once it has been induced for days [117]. When peripheral inflammation and/or nerve injury occur, transcription-dependent changes may ensue leading to longer-lasting effects [118] including, for example the expression of substance P and BDNF by neurons within the Dorsal Root Ganglia. Inflammation also exposes nerve terminals to nerve growth factor (NGF) [119, 120] which stimulates nociceptors expressing the tyrosine-kinase receptor TrkA a high-affinity receptor for NGF. These, when stimulated by NGF express higher levels of neuropeptides and other NGF-dependent proteins [121].

Once this process has occurred, low-intensity innocuous stimuli can now mediate the release of the aforementioned neuropeptides into the spinal cord [118]. This leads to an induction of cyclooxygenase-2 (Cox-2) in the dorsal horn neurons which then increases prostaglandin E2 (PGE2) production and release. When PGE2 binds to its dorsal horn neuron receptor, it elicits several changes including: potentiation of AMPAR and NMDAR currents, activation of non-selective cation channels, reduction in inhibitory glycinergic neurotransmission and an effect on EP4 receptors on presynaptic terminals resulting in an increase in neurotransmitter release [122]. In addition, spinal cord microglial cells also release pro-inflammatory cytokines including TNF and IL-1, which enhance excitatory and reduced inhibitory currents through COX-2 activation [123]. COX-2 has experimentally been suppressed in neurons, with a resulting near-complete loss of mechanical pain hypersensitivity but a retention of heat hyperalgesia in response to peripheral inflammation [122]. As mentioned above, mechanical sensitivity is a major feature of central sensitization [1], thus the COX-2 pathway's importance has been demonstrated.

Peripheral inflammation can also elicit other important changes, such as a shift from GluR2/3 to GluR1-containing AMPARs in the dorsal root neurons of the superficial lamina [124–126]. This leads to transforming previously $Ca2+$ impermeable to $Ca2+$-permeable AMPARs which then become a major source of the $[Ca2+]i$ increase in inflammatory pain, generating as much $Ca2+$ influx as with NMDAR activation [127]. Additionally, the NMDAR becomes phosphorylated by protein kinase C (pPKC) and extracellular signal-regulated kinase 2, which increase receptor activity [128]. Peng et al. investigated the participation of cyclin-dependent kinase-5 (Cdk5)-mediating NMDAR NR2B subunit phosphorylation in cross-organ reflex sensitization caused by colon irritation. They used external urethral sphincter electromyogram (EUSE) reflex activity evoked by the pelvic afferent nerve test stimulation (TS, 1 stimulation/30 s) and measured protein expression in the spinal cord and dorsal root ganglion tissue (T13-L2 and L6-S2 ipsilateral to the stimulation) in response to colon mustard oil instillation in anesthetized rats. They found that when compared with a baseline reflex activity with a single action potential evoked by the TS before the administration of test agents, mustard oil instillation into the descending colon sensitized the evoked activity characterized by elongated firing in the reflex activity in association with increased protein levels of Cdk5, PSD95, and phosphorylated NR2B (pNR2B) but not of total NR2B (tNR2B) in the spinal cord tissue. Both the cross-organ reflex sensitization and increments in protein expression were reversed by intra-colonic pretreatments with ruthenium red (a nonselective TRPV, antagonist), capsaizepine (a TRPV1-selective antagonist), lidocaine (a nerve conduction blocker) as well as by the intra-thecal pretreatment with APV (a NRMDR antagonist) Co-101244 (a NR2B-selective antagonist) and roscovitine (a Cdk5 antagonist). Moreover, compared with the control group, both the increase in pNR2B and the cross-organ reflex sensitization were attenuated in the si-RNA of NR2B rats. These results suggested that Cdk-dependent NMDAR NR2B subunit phosphorylation mediates the development of cross-organ pelvic-urethra reflex sensitization caused by acute colon irritation. The authors suggest that this mechanism could possibly underlie the high concurrence of pelvic pain syndrome with IBS [67]. Similar findings have been made in using other colitis models where inflammation up-regulates the activity of NMDARs in DRG neurons within ganglia innervating this tissue through mechanisms involving increased expression and persistent tyrosine phosphorylation [129]. Finally, peripheral inflammation also promotes group I mGluR insertion into the membrane (mGluR5) in a location closer to the synapse (mGluR1), thereby further increasing receptor expression at the level of the synaptic terminal [130].

In summary, pain is a complex and multidimensional process involving physical, emotional, and perceptual integration with the primary function of survival and safeguarding the individual from potential sources of tissue damage. In healthy condition, this process is adaptive, transient and has a protective role. Pathological process occurs in conditions where hyperalgesia and allodynia are present as a result of maladaptive neuroplastic changes lead to persistent increased perception and responsiveness to noxious stimuli, or response to normally non-noxious stimuli. Such neuroplastic changes can occur not only in primary afferent terminals (peripheral sensitization) but also in the spinal cord and in the brain (central sensitization) in both neurons and glia, thereby altering the processing of sensory information. These processes are very important in the development and persistence of visceral pain in several locations, predominantly gastrointestinal chronic pain syndromes.

References

1. Latremoliere A, Woolf CJ. Central sensitization: a generator of pain hypersensitivity by central neural plasticity. J Pain. 2009; 10(9):895–926.
2. Woolf CJ, Ma Q. Nociceptors–noxious stimulus detectors. Neuron. 2007;55(3):353–64.
3. Rome Foundation. Guidelines–Rome III diagnostic criteria for functional gastrointestinal disorders. J Gastrointestin Liver Dis. 2006;15(3):307–12.
4. Okumura T, Tanno S, Ohhira M. Prevalence of functional dyspepsia in an outpatient clinic with primary care physicians in Japan. J Gastroenterol. 2010;45(2):187–94.
5. Miwa H, Watari J, Fukui H, et al. Current understanding of pathogenesis of functional dyspepsia. J Gastroenterol Hepatol. 2011;26 Suppl 3:53–60.
6. Camilleri M, Northcutt AR, Kong S, et al. Efficacy and safety of alosetron in women with irritable bowel syndrome: a randomised, placebo-controlled trial. Lancet. 2000;355(9209):1035–40.
7. Sperber AD, Shvartzman P, Friger M, et al. A comparative reappraisal of the Rome II and Rome III diagnostic criteria: are we getting closer to the 'true' prevalence of irritable bowel syndrome? Eur J Gastroenterol Hepatol. 2007;19(6):441–7.
8. van Kerkhoven LA, Laheij RJ, Meineche-Schmidt V, et al. Functional dyspepsia: not all roads seem to lead to Rome. J Clin Gastroenterol. 2009;43(2):118–22.
9. Barbara G, Cremon C, De Giorgio R, et al. Mechanisms underlying visceral hypersensitivity in irritable bowel syndrome. Curr Gastroenterol Rep. 2011;13(4):308–15.
10. Waseem S, Moshiree B, Draganov PV. Gastroparesis: current diagnostic challenges and management considerations. World J Gastroenterol. 2009;15(1):25–37.
11. Tack J. Functional dyspepsia: impaired Fundic accommodation. Curr Treat Options Gastroenterol. 2000;3(4):287–94.
12. Quartero AO, de Wit NJ, Lodder AC, et al. Disturbed solid-phase gastric emptying in functional dyspepsia: a meta-analysis. Dig Dis Sci. 1998;43(9):2028–33.
13. Tack J, Bisschops R, Sarnelli G. Pathophysiology and treatment of functional dyspepsia. Gastroenterology. 2004;127(4):1239–55.
14. Ji RR, Kohno T, Moore KA, et al. Central sensitization and LTP: do pain and memory share similar mechanisms? Trends Neurosci. 2003;26(12):696–705.
15. Woolf CJ, Salter MW. Neuronal plasticity: increasing the gain in pain. Science. 2000;288(5472):1765–9.
16. Campbell JN, Meyer RA. Mechanisms of neuropathic pain. Neuron. 2006;52(1):77–92.
17. Neugebauer V, Schaible HG. Evidence for a central component in the sensitization of spinal neurons with joint input during development of acute arthritis in cat's knee. J Neurophysiol. 1990;64(1): 299–311.
18. Yunus MB. Role of central sensitization in symptoms beyond muscle pain, and the evaluation of a patient with widespread pain. Best Pract Res Clin Rheumatol. 2007;21(3):481–97.
19. Yarnitsky D, Goor-Aryeh I, Bajwa ZH, et al. 2003 Wolff Award: possible parasympathetic contributions to peripheral and central sensitization during migraine. Headache. 2003;43(7):704–14.
20. Malick A, Burstein R. Peripheral and central sensitization during migraine. Funct Neurol. 2000;15 Suppl 3:28–35.
21. Ghelardini C, Galeotti N, Grazioli I, et al. Indomethacin, alone and combined with prochlorperazine and caffeine, but not sumatriptan, abolishes peripheral and central sensitization in in vivo models of migraine. J Pain. 2004;5(8):413–9.
22. Fernandez-de-Las-Penas C, Ortega-Santiago R, Cuadrado ML, et al. Bilateral widespread mechanical pain hypersensitivity as sign of central sensitization in patients with cluster headache. Headache. 2011;51(3):384–91.
23. Verne GN, Price DD. Irritable bowel syndrome as a common precipitant of central sensitization. Curr Rheumatol Rep. 2002; 4(4):322–8.
24. Price DD, Craggs JG, Zhou Q, et al. Widespread hyperalgesia in irritable bowel syndrome is dynamically maintained by tonic visceral impulse input and placebo/nocebo factors: evidence from human psychophysics, animal models, and neuroimaging. Neuroimage. 2009;47(3):995–1001.
25. Staud R, Robinson ME, Price DD. Temporal summation of second pain and its maintenance are useful for characterizing widespread central sensitization of fibromyalgia patients. J Pain. 2007; 8(11):893–901.
26. Meeus M, Nijs J. Central sensitization: a biopsychosocial explanation for chronic widespread pain in patients with fibromyalgia and chronic fatigue syndrome. Clin Rheumatol. 2007; 26(4):465–73.
27. Gwilym SE, Keltner JR, Warnaby CE, et al. Psychophysical and functional imaging evidence supporting the presence of central sensitization in a cohort of osteoarthritis patients. Arthritis Rheum. 2009;61(9):1226–34.
28. Winkelstein BA. Mechanisms of central sensitization, neuroimmunology & injury biomechanics in persistent pain: implications for musculoskeletal disorders. J Electromyogr Kinesiol. 2004; 14(1):87–93.
29. Fernandez-Lao C, Cantarero-Villanueva I, Fernandez-de-las-Penas C, et al. Widespread mechanical pain hypersensitivity as a sign of central sensitization after breast cancer surgery: comparison between mastectomy and lumpectomy. Pain Med. 2011; 12(1):72–8.
30. Fernandez-de-las-Penas C, Galan-del-Rio F, Fernandez-Carnero J, et al. Bilateral widespread mechanical pain sensitivity in women with myofascial temporomandibular disorder: evidence of impairment in central nociceptive processing. J Pain. 2009;10(11): 1170–8.
31. Bonjardim LR, da Silva AP, Gameiro GH, et al. Nociceptive behavior induced by mustard oil injection into the temporomandibular joint is blocked by a peripheral non-opioid analgesic and a

central opioid analgesic. Pharmacol Biochem Behav. 2009; 91(3):321–6.
32. Zhang S, Chiang CY, Xie YF, et al. Central sensitization in thalamic nociceptive neurons induced by mustard oil application to rat molar tooth pulp. Neuroscience. 2006;142(3):833–42.
33. Park SJ, Zhang S, Chiang CY, et al. Central sensitization induced in thalamic nociceptive neurons by tooth pulp stimulation is dependent on the functional integrity of trigeminal brainstem subnucleus caudalis but not subnucleus oralis. Brain Res. 2006; 1112(1):134–45.
34. Brumovsky PR, Gebhart GF. Visceral organ cross-sensitization—an integrated perspective. Auton Neurosci. 2010;153(1–2):106–15.
35. Sipe WE, Brierley SM, Martin CM, et al. Transient receptor potential vanilloid 4 mediates protease activated receptor 2-induced sensitization of colonic afferent nerves and visceral hyperalgesia. Am J Physiol Gastrointest Liver Physiol. 2008; 294(5):G1288–98.
36. Anand P, Aziz Q, Willert R, et al. Peripheral and central mechanisms of visceral sensitization in man. Neurogastroenterol Motil. 2007;19(1 Suppl):29–46.
37. Juhl GI, Jensen TS, Norholt SE, et al. Central sensitization phenomena after third molar surgery: a quantitative sensory testing study. Eur J Pain. 2008;12(1):116–27.
38. Stubhaug A, Breivik H, Eide PK, et al. Mapping of punctuate hyperalgesia around a surgical incision demonstrates that ketamine is a powerful suppressor of central sensitization to pain following surgery. Acta Anaesthesiol Scand. 1997;41(9):1124–32.
39. Stubhaug A. A new method to evaluate central sensitization to pain following surgery. Effect of ketamine. Acta Anaesthesiol Scand Suppl. 1997;110:154–5.
40. Azpiroz F, Bouin M, Camilleri M, et al. Mechanisms of hypersensitivity in IBS and functional disorders. Neurogastroenterol Motil. 2007;19(1 Suppl):62–88.
41. Seminowicz DA, Labus JS, Bueller JA, et al. Regional gray matter density changes in brains of patients with irritable bowel syndrome. Gastroenterology. 2010;139(1):48–57.e2.
42. Guenther S, Reeh PW, Kress M. Rises in [Ca2+]i mediate capsaicin- and proton-induced heat sensitization of rat primary nociceptive neurons. Eur J Neurosci. 1999;11(9):3143–50.
43. Hucho T, Levine JD. Signaling pathways in sensitization: toward a nociceptor cell biology. Neuron. 2007;55(3):365–76.
44. Barbara G, Stanghellini V, Cremon C, et al. What is the effect of inflammation on intestinal function? Inflamm Bowel Dis. 2008;14 Suppl 2:S140–4.
45. Barbara G, Vallance BA, Collins SM. Persistent intestinal neuromuscular dysfunction after acute nematode infection in mice. Gastroenterology. 1997;113(4):1224–32.
46. Barbara G, De Giorgio R, Deng Y, et al. Role of immunologic factors and cyclooxygenase 2 in persistent postinfective enteric muscle dysfunction in mice. Gastroenterology. 2001;120(7): 1729–36.
47. Grundy D. 5-HT system in the gut: roles in the regulation of visceral sensitivity and motor functions. Eur Rev Med Pharmacol Sci. 2008;12 Suppl 1:63–7.
48. Jones 3rd RC, Xu L, Gebhart GF. The mechanosensitivity of mouse colon afferent fibers and their sensitization by inflammatory mediators require transient receptor potential vanilloid 1 and acid-sensing ion channel 3. J Neurosci. 2005;25(47): 10981–9.
49. Geppetti P, Trevisani M. Activation and sensitisation of the vanilloid receptor: role in gastrointestinal inflammation and function. Br J Pharmacol. 2004;141(8):1313–20.
50. Chan CL, Facer P, Davis JB, et al. Sensory fibres expressing capsaicin receptor TRPV1 in patients with rectal hypersensitivity and faecal urgency. Lancet. 2003;361(9355):385–91.
51. Bewick GA. Bowels control brain: gut hormones and obesity. Biochem Med (Zagreb). 2012;22(3):283–97.
52. Bonaz BL, Bernstein CN. Brain-gut interactions in inflammatory bowel disease. Gastroenterology. 2013;144(1):36–49.
53. Tillisch K, Labus JS. Advances in imaging the brain-gut axis: functional gastrointestinal disorders. Gastroenterology. 2011; 140(2):407–411.e1.
54. Keita AV, Soderholm JD. The intestinal barrier and its regulation by neuroimmune factors. Neurogastroenterol Motil. 2010;22(7): 718–33.
55. Rhee SH, Pothoulakis C, Mayer EA. Principles and clinical implications of the brain-gut-enteric microbiota axis. Nat Rev Gastroenterol Hepatol. 2009;6(5):306–14.
56. Koloski NA, Jones M, Kalantar J, et al. The brain–gut pathway in functional gastrointestinal disorders is bidirectional: a 12-year prospective population-based study. Gut. 2012;61(9):1284–90.
57. Bradesi S. Role of spinal cord glia in the central processing of peripheral pain perception. Neurogastroenterol Motil. 2010; 22(5):499–511.
58. Staud R. Is it all central sensitization? Role of peripheral tissue nociception in chronic musculoskeletal pain. Curr Rheumatol Rep. 2010;12(6):448–54.
59. Khodorova A, Strichartz GR. Contralateral paw sensitization following injection of endothelin-1: effects of local anesthetics differentiate peripheral and central processes. Neuroscience. 2010;165(2):553–60.
60. Wall PD, Woolf CJ. Muscle but not cutaneous C-afferent input produces prolonged increases in the excitability of the flexion reflex in the rat. J Physiol. 1984;356:443–58.
61. Cervero F. Visceral pain: mechanisms of peripheral and central sensitization. Ann Med. 1995;27(2):235–9.
62. Holst H, Arendt-Nielsen L, Mosbech H, et al. Increased capsaicin-induced secondary hyperalgesia in patients with multiple chemical sensitivity. Clin J Pain. 2011;27(2):156–62.
63. Modir JG, Wallace MS. Human experimental pain models 3: heat/capsaicin sensitization and intradermal capsaicin models. Methods Mol Biol. 2010;617:169–74.
64. Srbely JZ, Dickey JP, Bent LR, et al. Capsaicin-induced central sensitization evokes segmental increases in trigger point sensitivity in humans. J Pain. 2010;11(7):636–43.
65. van den Elzen BD, Tytgat GN, Boeckxstaens GE. Gastric hypersensitivity induced by oesophageal acid infusion in healthy volunteers. Neurogastroenterol Motil. 2009;21(2):160–9.
66. Graversen C, Brock C, Drewes AM, et al. Biomarkers for visceral hypersensitivity identified by classification of electroencephalographic frequency alterations. J Neural Eng. 2011; 8(5):056014.
67. Peng HY, Chen GD, Tung KC, et al. Colon mustard oil instillation induced cross-organ reflex sensitization on the pelvic-urethra reflex activity in rats. Pain. 2009;142(1–2):75–88.
68. Woolf CJ. Central sensitization: implications for the diagnosis and treatment of pain. Pain. 2011;152(3 Suppl):S2–15.
69. Tack J, Caenepeel P, Fischler B, et al. Symptoms associated with hypersensitivity to gastric distention in functional dyspepsia. Gastroenterology. 2001;121(3):526–35.
70. Whitehead WE, Holtkotter B, Enck P, et al. Tolerance for rectosigmoid distention in irritable bowel syndrome. Gastroenterology. 1990;98(5 Pt 1):1187–92.
71. Mertz H, Naliboff B, Munakata J, et al. Altered rectal perception is a biological marker of patients with irritable bowel syndrome. Gastroenterology. 1995;109(1):40–52.
72. Garry EM, Moss A, Delaney A, et al. Neuropathic sensitization of behavioral reflexes and spinal NMDA receptor/CaM kinase II interactions are disrupted in PSD-95 mutant mice. Curr Biol. 2003;13(4):321–8.
73. You HJ, Chen J, Morch CD, et al. Differential effect of peripheral glutamate (NMDA, non-NMDA) receptor antagonists on bee venom-induced spontaneous nociception and sensitization. Brain Res Bull. 2002;58(6):561–7.

74. Hudspith M, Munglani R. A role for presynaptic NMDA receptors in central sensitization in the spinal cord dorsal horn? Br J Anaesth. 1998;81(2):294–5.
75. Ma QP, Woolf CJ. Noxious stimuli induce an N-methyl-D-ASPARTATE receptor-dependent hypersensitivity of the flexion withdrawal reflex to touch: implications for the treatment of mechanical allodynia. Pain. 1995;61(3):383–90.
76. South SM, Kohno T, Kaspar BK, et al. A conditional deletion of the NR1 subunit of the NMDA receptor in adult spinal cord dorsal horn reduces NMDA currents and injury-induced pain. J Neurosci. 2003;23(12):5031–40.
77. Guo W, Wei F, Zou S, et al. Group I metabotropic glutamate receptor NMDA receptor coupling and signaling cascade mediate spinal dorsal horn NMDA receptor 2B tyrosine phosphorylation associated with inflammatory hyperalgesia. J Neurosci. 2004; 24(41):9161–73.
78. Ferguson AR, Bolding KA, Huie JR, et al. Group I metabotropic glutamate receptors control metaplasticity of spinal cord learning through a protein kinase C-dependent mechanism. J Neurosci. 2008;28(46):11939–49.
79. Sevostianova N, Danysz W. Analgesic effects of mGlu1 and mGlu5 receptor antagonists in the rat formalin test. Neuropharmacology. 2006;51(3):623–30.
80. Khasabov SG, Rogers SD, Ghilardi JR, et al. Spinal neurons that possess the substance P receptor are required for the development of central sensitization. J Neurosci. 2002;22(20):9086–98.
81. Xu XJ, Dalsgaard CJ, Wiesenfeld-Hallin Z. Spinal substance P and N-methyl-D-aspartate receptors are coactivated in the induction of central sensitization of the nociceptive flexor reflex. Neuroscience. 1992;51(3):641–8.
82. Cady RJ, Glenn JR, Smith KM, et al. Calcitonin gene-related peptide promotes cellular changes in trigeminal neurons and glia implicated in peripheral and central sensitization. Mol Pain. 2011;7(1):94.
83. Marquez de Prado B, Hammond DL, Russo AF. Genetic enhancement of calcitonin gene-related Peptide-induced central sensitization to mechanical stimuli in mice. J Pain. 2009;10(9):992–1000.
84. Sun RQ, Tu YJ, Lawand NB, et al. Calcitonin gene-related peptide receptor activation produces PKA- and PKC-dependent mechanical hyperalgesia and central sensitization. J Neurophysiol. 2004;92(5):2859–66.
85. Kawasaki Y, Kohno T, Zhuang ZY, et al. Ionotropic and metabotropic receptors, protein kinase A, protein kinase C, and Src contribute to C-fiber-induced ERK activation and cAMP response element-binding protein phosphorylation in dorsal horn neurons, leading to central sensitization. J Neurosci. 2004;24(38):8310–21.
86. Slack SE, Grist J, Mac Q, et al. TrkB expression and phospho-ERK activation by brain-derived neurotrophic factor in rat spinothalamic tract neurons. J Comp Neurol. 2005;489(1):59–68.
87. Kerr BJ, Bradbury EJ, Bennett DL, et al. Brain-derived neurotrophic factor modulates nociceptive sensory inputs and NMDA-evoked responses in the rat spinal cord. J Neurosci. 1999;19(12): 5138–48.
88. Lu VB, Ballanyi K, Colmers WF, et al. Neuron type-specific effects of brain-derived neurotrophic factor in rat superficial dorsal horn and their relevance to 'central sensitization'. J Physiol. 2007;584(Pt 2):543–63.
89. Liang YF, Haake B, Reeh PW. Sustained sensitization and recruitment of rat cutaneous nociceptors by bradykinin and a novel theory of its excitatory action. J Physiol. 2001;532(Pt 1):229–39.
90. Schuligoi R, Donnerer J, Amann R. Bradykinin-induced sensitization of afferent neurons in the rat paw. Neuroscience. 1994;59(1):211–5.
91. Giordano J, Gerstmann H. Patterns of serotonin- and 2-methylserotonin-induced pain may reflect 5-HT3 receptor sensitization. Eur J Pharmacol. 2004;483(2–3):267–9.
92. Ryu TH, Jung KY, Ha MJ, et al. Superoxide and nitric oxide involvement in enhancing of N-methyl-D-aspartate receptor-mediated central sensitization in the chronic post-ischemia pain model. Korean J Pain. 2010;23(1):1–10.
93. Wu J, Fang L, Lin Q, et al. Nitric oxide synthase in spinal cord central sensitization following intradermal injection of capsaicin. Pain. 2001;94(1):47–58.
94. Lin Q, Palecek J, Paleckova V, et al. Nitric oxide mediates the central sensitization of primate spinothalamic tract neurons. J Neurophysiol. 1999;81(3):1075–85.
95. Shevelkin AV, Nikitin VP, Kozyrev SA, et al. Serotonin imitates several of the neuronal effects of nociceptive sensitization in the common snail. Neurosci Behav Physiol. 1998;28(5):547–55.
96. Gershon MD, Tack J. The serotonin signaling system: from basic understanding to drug development for functional GI disorders. Gastroenterology. 2007;132(1):397–414.
97. Tonini M. 5-Hydroxytryptamine effects in the gut: the 3, 4, and 7 receptors. Neurogastroenterol Motil. 2005;17(5):637–42.
98. Spiller R. Recent advances in understanding the role of serotonin in gastrointestinal motility in functional bowel disorders: alterations in 5-HT signalling and metabolism in human disease. Neurogastroenterol Motil. 2007;19 Suppl 2:25–31.
99. De Giorgio R, Barbara G, Furness JB, et al. Novel therapeutic targets for enteric nervous system disorders. Trends Pharmacol Sci. 2007;28(9):473–81.
100. Kozlowski CM, Green A, Grundy D, et al. The 5-HT(3) receptor antagonist alosetron inhibits the colorectal distention induced depressor response and spinal c-fos expression in the anaesthetised rat. Gut. 2000;46(4):474–80.
101. Ford AC, Brandt LJ, Young C, et al. Efficacy of 5-HT3 antagonists and 5-HT4 agonists in irritable bowel syndrome: systematic review and meta-analysis. Am J Gastroenterol. 2009;104(7): 1831–43. quiz 1844.
102. Woolf CJ. Evidence for a central component of post-injury pain hypersensitivity. Nature. 1983;306(5944):686–8.
103. Lin Q, Peng YB, Wu J, et al. Involvement of cGMP in nociceptive processing by and sensitization of spinothalamic neurons in primates. J Neurosci. 1997;17(9):3293–302.
104. Simone DA, Sorkin LS, Oh U, et al. Neurogenic hyperalgesia: central neural correlates in responses of spinothalamic tract neurons. J Neurophysiol. 1991;66(1):228–46.
105. Guilbaud G, Benoist JM, Eschalier A, et al. Evidence for peripheral serotonergic mechanisms in the early sensitization after carrageenin-induced inflammation: electrophysiological studies in the ventrobasal complex of the rat thalamus using a potent specific antagonist of peripheral 5-HT receptors. Brain Res. 1989;502(1):187–97.
106. Burstein R, Yamamura H, Malick A, et al. Chemical stimulation of the intracranial dura induces enhanced responses to facial stimulation in brain stem trigeminal neurons. J Neurophysiol. 1998; 79(2):964–82.
107. Wei F, Zhuo M. Potentiation of sensory responses in the anterior cingulate cortex following digit amputation in the anaesthetised rat. J Physiol. 2001;532(Pt 3):823–33.
108. Kokkinidis L, Borowski TB. Sensitization of mesolimbic brain stimulation reward after electrical kindling of the amygdala. Brain Res Bull. 1991;27(6):791–6.
109. Hitchcock JM, Sananes CB, Davis M. Sensitization of the startle reflex by footshock: blockade by lesions of the central nucleus of the amygdala or its efferent pathway to the brainstem. Behav Neurosci. 1989;103(3):509–18.
110. Moylan Governo RJ, Morris PG, Prior MJ, et al. Capsaicin-evoked brain activation and central sensitization in anaesthetised rats: a functional magnetic resonance imaging study. Pain. 2006; 126(1–3):35–45.

111. Maihofner C, Jesberger F, Seifert F, et al. Cortical processing of mechanical hyperalgesia: a MEG study. Eur J Pain. 2010;14(1):64–70.
112. Mohr C, Leyendecker S, Mangels I, et al. Central representation of cold-evoked pain relief in capsaicin induced pain: an event-related fMRI study. Pain. 2008;139(2):416–30.
113. Shih YY, Chiang YC, Chen JC, et al. Brain nociceptive imaging in rats using (18)f-fluorodeoxyglucose small-animal positron emission tomography. Neuroscience. 2008;155(4):1221–6.
114. Zhao F, Welsh D, Williams M, et al. fMRI of pain processing in the brain: a within-animal comparative study of BOLD vs. CBV and noxious electrical vs. noxious mechanical stimulation in rat. Neuroimage. 2012;59(2):1168–79.
115. Mohr C, Leyendecker S, Petersen D, et al. Effects of perceived and exerted pain control on neural activity during pain relief in experimental heat hyperalgesia: A fMRI study. Eur J Pain. 2012;16(4):496–508.
116. Smith JK, Humes DJ, Head KE, et al. fMRI and MEG analysis of visceral pain in healthy volunteers. Neurogastroenterol Motil. 2011;23(7):648–e260.
117. Koltzenburg M, Lundberg LE, Torebjork HE. Dynamic and static components of mechanical hyperalgesia in human hairy skin. Pain. 1992;51(2):207–19.
118. Neumann S, Doubell TP, Leslie T, et al. Inflammatory pain hypersensitivity mediated by phenotypic switch in myelinated primary sensory neurons. Nature. 1996;384(6607):360–4.
119. Sevcik MA, Ghilardi JR, Peters CM, et al. Anti-NGF therapy profoundly reduces bone cancer pain and the accompanying increase in markers of peripheral and central sensitization. Pain. 2005;115(1–2):128–41.
120. Mannion RJ, Costigan M, Decosterd I, et al. Neurotrophins: peripherally and centrally acting modulators of tactile stimulus-induced inflammatory pain hypersensitivity. Proc Natl Acad Sci U S A. 1999;96(16):9385–90.
121. Woolf CJ. Phenotypic modification of primary sensory neurons: the role of nerve growth factor in the production of persistent pain. Philos Trans R Soc Lond B Biol Sci. 1996;351(1338):441–8.
122. Vasko MR, Campbell WB, Waite KJ. Prostaglandin E2 enhances bradykinin-stimulated release of neuropeptides from rat sensory neurons in culture. J Neurosci. 1994;14(8):4987–97.
123. Samad TA, Moore KA, Sapirstein A, et al. Interleukin-1beta-mediated induction of Cox-2 in the CNS contributes to inflammatory pain hypersensitivity. Nature. 2001;410(6827):471–5.
124. Larsson M, Broman J. Translocation of GluR1-containing AMPA receptors to a spinal nociceptive synapse during acute noxious stimulation. J Neurosci. 2008;28(28):7084–90.
125. Qu XX, Cai J, Li MJ, et al. Role of the spinal cord NR2B-containing NMDA receptors in the development of neuropathic pain. Exp Neurol. 2009;215(2):298–307.
126. Park JS, Voitenko N, Petralia RS, et al. Persistent inflammation induces GluR2 internalization via NMDA receptor-triggered PKC activation in dorsal horn neurons. J Neurosci. 2009;29(10):3206–19.
127. Luo C, Seeburg PH, Sprengel R, et al. Activity-dependent potentiation of calcium signals in spinal sensory networks in inflammatory pain states. Pain. 2008;140(2):358–67.
128. Weyerbacher AR, Xu Q, Tamasdan C, et al. N-Methyl-D-aspartate receptor (NMDAR) independent maintenance of inflammatory pain. Pain. 2010;148(2):237–46.
129. Li J, McRoberts JA, Ennes HS, et al. Experimental colitis modulates the functional properties of NMDA receptors in dorsal root ganglia neurons. Am J Physiol Gastrointest Liver Physiol. 2006;291(2):G219–28.
130. Pitcher MH, Ribeiro-da-Silva A, Coderre TJ. Effects of inflammation on the ultrastructural localization of spinal cord dorsal horn group I metabotropic glutamate receptors. J Comp Neurol. 2007;505(4):412–23.

The Epidemiology of Chronic Abdominal Pain

2

Reda Tolba, Joshua Shroll, Abdul Kanu, and Maged K. Rizk

Abbreviations

CD	Crohn's disease
CT	Computed tomography
ERCP	Endoscopic retrograde cholangiopancreatography
FAPS	Functional abdominal pain syndrome
GI	Gastrointestinal
IBD	Inflammatory bowel disease
IBS	Irritable bowel syndrome
UC	Ulcerative colitis
US	Ultrasound

Introduction

Chronic abdominal pain is a commonly seen complaint by primary care physicians, gastroenterologists, and pain physicians. Generally, it is defined as continuous or intermittent abdominal discomfort for at least 6 months, and can be caused by a wide variety of etiologies ranging from organic to functional. Organic causes can be anatomical, physiological, metabolical, or can arise from the abdominal wall musculature, fascia, or nerves. Functional abdominal pain is a more challenging problem and can be difficult to diagnose and manage. In patients with functional abdominal pain, frequently, there is no clear organic cause that can explain the underlying symptoms.

R. Tolba, M.D.
Anesthesiology and Pain Management, Ochsner Health System, 1514 Jefferson Highway, New Orleans, LA 70121, USA
e-mail: redatolba@hotmail.com

J. Shroll, M.D., M.P.H. • A. Kanu, M.D. • M.K. Rizk, M.D. (✉)
Cleveland Clinic Foundation, 9500 Euclid Avenue, A30, Cleveland, OH 44195, USA
e-mail: shroll3@gmail.com; kanua@ccf.org; RIZKM@ccf.org

Epidemiology

The prevalence of unspecified chronic abdominal pain is suggested by the epidemiological data to be around 22.9 per 1,000 person-years. Abdominal pain is a common complaint with cross-sectional data suggesting that up to 25 % of adult populations have abdominal pain at any one time [1–3]. The prevalence is equal across different age groups, ethnicities, and geographic regions [4–9]. In a national, cross-sectional, telephone survey of US households, Sandler et al. [3] suggest that the prevalence of abdominal pain and discomfort was 22 % overall, and 16 % in individuals of age 60 and older. The same study suggests that women are more likely to report abdominal pain than men. Other studies found that the overall frequency of abdominal pain and discomfort of more than six times per year was 21 % in healthy individuals [10] and 24 % in people of age 65 and older [11]. There is a wide range of variation in the reported prevalence of upper abdominal symptoms (mostly upper abdominal pain or discomfort) ranging from 8 to 54 % [4]. The most likely explanation of the broad range in reported prevalence is variation in the definition of symptoms.

Visceral Chronic Abdominal Pain

A. *Inflammatory Bowel Disease*: Of the chronic abdominal pain etiologies which are of a primarily visceral origin, the most common and most costly to our healthcare resources is that of inflammatory bowel disease (IBD), specifically Ulcerative Colitis (UC) and Crohn's Disease (CD). Unfortunately, very limited data exist on tools that could help identify those patients with IBD that may go on to develop a chronic pain syndrome. Therefore, it is of particular importance to understand the current epidemiologic trends of the disease process itself, using a wealth of data currently available to researchers… IBD is an ongoing area of needed research as the past several decades

L. Kapural (ed.), *Chronic Abdominal Pain: An Evidence-Based, Comprehensive Guide to Clinical Management*, DOI 10.1007/978-1-4939-1992-5_2, © Springer Science+Business Media New York 2015

have seen a rapid rise in incidence and shift in susceptible populations. Large disparities are seen globally as changing worldwide demographics have seen a rise in areas previously believed to be resistant to disease.

The incidence of IBD worldwide is generally thought to be in the range of 0.6–20.3/100,000 for Crohn's disease and 0.1–15.6/100,000 for Ulcerative colitis [12]. Such wide incidence range and high variability are due to a large disparity between geographical areas. Historically, believed to be a "westernized disease" or "urbanized disease," the highest IBD incidence rates are seen in North America, the United Kingdom, and northern Europe as compared to southern Europe, Asia, and Africa. In North America, incidence rates for IBD ranges from 2.2 to 14.3/100,000 with 3.1–14.6/100,000 cases of CD and 2.2–14.3 cases of UC diagnosed annually [12–17]. Similar incidence data are recorded in Europe with ranges of 1.5–20.3/100,000 for UC and 0.7–9.8/100,000 for CD. A large scale study out of Europe in the 1990s found the mean annual incidence to be in the range of 7.6–13.1 for UC and 2.8–8.3 for CD. Of particular interest is that this study showed a predominant north hemisphere versus south distribution with rates in northern Europe to be 40 % higher for UC and 80 % higher for CD [14]. This trend appears to be shifting slightly, however, with data suggesting a disproportionate rise in incidence in areas such as Japan, Korea, and northern India [18–20]. IBD is generally considered to be a disease of younger adults and adolescents. Peak incidence for CD is from 15 to 25 years of age, and for UC 25–35 years of age with about 10 % of cases diagnosed before the age of 18. A second and rather modest peak in incidence for both diseases is seen between the ages of 50–60 [16, 17, 21]. Recent data suggested a rise in the incidence of pediatric IBD with a greater proportion of these cases being CD. A recent statewide epidemiologic survey from Wisconsin demonstrated the highest rate of pediatric IBD in the world to date, with an overall incidence of 7.05/100,000 in children <18 years of age with 4.56/100,000 newly diagnosed cases of CD and 2.14/100,000 cases of UC [21]. Similar studies out of Europe, including Sweden and Finland have seen incidence rates of pediatric IBD almost double, with the majority of this being in cases of CD while UC has remained relatively stable [22, 23] (Fig. 2.1). In regards to gender prevalence, in UC, there is a male predominance; while in CD female, those gender differences appear to be decreasing [12, 14]. Breakdowns of racial and ethnic predispositions are another area that is continually changing. Historically, IBD was thought to be more common in Caucasians and people of Jewish descent, with lower rates in African-Americans and Asian-Americans were documented. A recent data suggested that this gap is closing [16, 17, 24, 25]. Data from urbanized areas of the United States have shown that disease rates amongst African-Americans and Caucasian populations are similar. Studies of migrant populations suggest that the ethnic and racial differences may be more related to lifestyle and environmental influences than true genetic differences [12]. In regards to potential risk factors identified for IBD, a thorough review is outside the scope of this discussion, however, briefly those factors which have been identified, and are under current investigation include cigarette smoking/tobacco use, diet, high stress occupations, sanitation and exposure to infection, gut flora, and oral contraceptives [12, 15, 25].

TABLE 2. IBD in Children in Southern Finland, 1987–2003

Year	Incidence (95% CIs)*			
	IBD	CD	UC	IC
1987	3.9 (2.5–5.8)	1.7 (0.8–3.1)	2.2 (1.2–3.7)	0.0 (0.0–0.6)
1988	3.0 (1.8–4.8)	1.0 (0.4–2.2)	1.3 (0.6–2.7)	0.7 (0.2–1.7)
1989	3.4 (2.1–5.2)	0.8 (0.3–2.0)	2.0 (1.0–3.5)	0.5 (0.1–1.5)
1990	4.7 (3.1–6.7)	1.2 (0.5–2.4)	2.8 (1.6–4.5)	0.2 (0.2–1.7)
1991	6.4 (4.6–8.8)	2.0 (1.0–3.4)	3.9 (2.5–5.9)	0.5 (0.1–1.4)
1992	4.1 (2.6–6.0)	1.0 (0.4–2.1)	2.6 (1.5–4.2)	0.5 (0.1–1.4)
1993	5.0 (3.4–7.1)	1.8 (0.9–3.2)	2.9 (1.7–4.6)	0.3 (0.0–1.2)
1994	5.0 (3.4–7.1)	1.5 (0.7–2.8)	2.6 (1.5–4.2)	0.9 (0.4–2.1)
1995	4.7 (3.1–6.7)	1.3 (0.6–2.5)	1.9 (1.0–3.4)	1.4 (0.7–2.7)
1996	6.1 (4.3–8.4)	2.2 (1.2–3.8)	2.7 (1.6–4.4)	1.1 (0.5–2.3)
1997	4.6 (3.1–6.7)	1.8 (0.9–3.1)	2.2 (1.2–3.8)	0.6 (0.2–1.6)
1998	9.7 (7.5–12.5)	3.4 (2.1–5.1)	5.1 (3.5–7.2)	1.2 (0.6–2.5)
1999	7.4 (5.4–9.8)	2.7 (1.6–4.4)	3.4 (2.1–5.1)	1.2 (0.6–2.5)
2000	6.6 (4.7–8.9)	1.9 (1.0–3.4)	3.5 (2.2–5.3)	1.1 (0.5–2.3)
2001	7.7 (5.7–10.3)	2.6 (1.5–4.2)	4.0 (2.6–5.9)	1.1 (0.5–2.3)
2002	8.7 (6.6–11.4)	3.6 (2.2–5.4)	4.8 (3.3–6.9)	0.3 (0.0–1.2)
2003	7.0† (5.0–9.4)	2.6 (1.5–4.2)	3.2 (2.0–5.0)	1.0 (0.4–2.1)

*Expressed as n/100,000.

†In 1 case, diagnosis unsettled because of missing data and therefore included only in total incidence figure.

Fig. 2.1 Incidence of Pediatric IBD over 17-year period in southern Finland. The mean annual incidence rate increased from 3.9/100,000 (95 % confidence interval [CI] 2.5–5.8) in 1987 to 7.0/100,000 (CI 5.0–9.4) in 2003 (P<0.001). Reprinted with permission from Elsevier - Askling J, Grahnquist L, Ekbom A, Finkel Y. Incidence of paediatric Crohn's disease in Stockholm, Sweden. Lancet. 1999 Oct 2;354(9185):1179

With a better understanding of the scope and makeup of the IBD patient population, we can now shift our focus to a specific subset of this population, those patients who experience chronic abdominal pain. As IBD is a disease of chronic inflammation, it is not surprising that 50–70 % of patients cite pain as their initial symptom, or as a prevalent symptom during exacerbations of their disease. What is surprising, however, is that up to 20 % of IBD patients will report chronic pain, despite a clinical

Annual incidence	Crohn's disease (n=50)	Ulcerative colitis (n=27)	Unspecified colitis (n=14)
1990	0·0 (0·0–3·3)	0·9 (0·0–5·0)	0·9 (0·0–5·0)
1991	2·7 (0·6–7·8)	3·6 (1·0–9·1)	1·8 (0·2–6·4)
1992	4·4 (1·4–10)	4·4 (1·4–10·2)	0·0 (0·0–3·2)
1993	1·7 (0·2–6·2)	0·9 (0·0–4·8)	1·7 (0·2–6·2)
1994	4·2 (1·4–9·8)	2·5 (0·5–7·4)	1·7 (0·2–6·1)
1995	2·7 (0·9–6·3)	1·6 (0·3–4·8)	0·0 (0·0–2·0)
1996	4·3 (1(9–8·5)	2·7 (0·9–6·3)	1·6 (0·3–4·7)
1997	5·9 (2·9–11)	2·1 (0·6–5·5)	1·6 (0·3–4·7)
1998	5·9 (2·9–11)	0·5 (0·0–3·0)	0·5 (0·0–3·0)

Incidence (per 100 000, 95% CI) of inflammatory bowel disease among individuals below 17 years of age in northern Stockholm, Sweden, 1990–98

Fig. 2.2 In the TREAT registry, patients using narcotic analgesics had increased mortality rates (OR 1.84, $P = 0.044$). Also the use of narcotic analgesics was an independent predictor of serious infection (OR 2.38, $P < 0.001$). Reprinted with permission from Elsevier – Lichtenstein GR, Feagan BG, Cohen RD, Salzberg BA, Diamond RH, Chen DM, et al. Serious infections and mortality in association with therapies for Crohn's disease: TREAT registry. Clin Gastroenterol Hepatol. 2006 May;4(5):621–30

diagnosis of remission and negative endoscopic findings [26]. Up to 15 % of them continue with opioid use for treatment of their chronic abdominal pain [13, 26–29]. This is of particular importance, as studies have shown an increase in the morbidity and mortality of those patients which require chronic opioid use [28].

Analysis from the Therapy Resource, Evaluation and Assessment Tool (TREAT) registry showed that chronic use of opioids increased the risk of serious infections, possibly by decreasing gut transit, increasing bacterial growth within the gut, or masking early symptoms. Also, of concern with regards to chronic opioid use is the risk of narcotic bowel syndrome (NBS), risk of toxic megacolon, narcotic dependence, and masking of more serious complications, such as bowel perforation [27, 28] (Fig. 2.2). There is a limited data to identify those risk factors, or patients at a proportionally higher risk for chronic pain, or with need for ongoing opioid therapy. Edwards et al. found a high rate of preexisting psychiatric illness amongst IBD patients on chronic opiates (up to 67 %) [27]. In a retrospective study of 291 CD patients over a 5-year period, Cross et al. found that patients using chronic opioids were more likely to be female, at the higher rates of disability, a longer duration of disease, and were more likely to be active smokers [29]. Finally, in a case–control study of 100 IBD patients, Hanson et al. found significant associations between chronic opioid use and female gender, two or more previous surgeries, higher rates of depression/anxiety, and a history of sexual, emotional, or physical abuse [13]. Again, as limited epidemiologic data are available it is difficult to make generalizations or truly make cause-effect relationships but it does identify a growing need for more data in this patient population.

Key Points

- There is a rapid rise in IBD incidence over the past several decades, with CD becoming as equally apparent as UC.
- Up to 1.8-fold increase in incidence in pediatric IBD with the majority of cases being CD.
- Racial/ethnic disparities are becoming less evident, as population differences appear to be more geographically versus genetically dependent.
- Up to 20 % of IBD patients report chronic abdominal pain despite clinical remission and the majority of these patients require chronic opioid use.
- Limited epidemiologic data in IBD patients that report chronic pain, however, associations that have been drawn include preexisting psychiatric illness, female gender, smoking, and longer duration of disease.

B. *Chronic Pancreatitis*: While not as common as IBD, chronic pancreatitis is an inflammatory condition that leads to progressive and irreversible destruction of tissue and has a significant impact on the quality of life of patients with ever increasing healthcare costs everywhere in the world.

Epidemiologic studies regarding prevalence of chronic pancreatitis over the past several decades are few and not consistent. Given the natural history of the disease process, and constantly changing disease classification, clear comparison amongst patient groups is very difficult. In regards to the incidence and prevalence of the disease, we do know that it is rising [30–32]. Most studies suggest the incidence of chronic pancreatitis to be around 3–14/100,000 in Europe, 6–7/100,000 [31] in the United States [30, 33], and 5–6/100,000 in Japan [32, 34]. The overall prevalence of the disease has been on the rise, with an increased incidence worldwide at 13–35 cases per 100,000 [30–34]. There is a great variability in peak age of onset of disease amongst the studies, but in general the mean age range when diagnosis is established is between 39.7 and 57 years of age [31]. There is a marked disparity in disease prevalence between men and women, mostly related to a higher incidence of chronic alcoholism in men, with male to female ratios of 3:1. That may be direct consequence of majority, 68.5–91 % of patients with chronic pancreatitis being men [31, 32, 34]. In regards to etiology of disease, chronic alcohol use is the most common cause as 60–90 % of cases can be related to alcohol [35, 36]. This number, however, appears to be declining, with a higher incidence of idiopathic cases recently mostly in women [32, 34, 35]. A more recent study by Cote et al. suggested that only 44.5 % of causative etiology could be attributed to alcohol, and only 59.4 % occurred in men. Smoking in particular has been identified as a potential major risk factor for developing chronic pancreatitis, as well as advancing the rate of progression [35]. Such increased incidence may be also related to use

	Pain pattern			Pain pattern		
Variable	Intermittent	Constant	p Value	Mild to moderate	Severe	p Value
Number (%)	186 (44.9)	228 (55.1)		96 (23.2)	318 (76.8)	
Gender						
Male	98 (52.7)	107 (46.9)	0.28	52 (54.2)	153 (48.1)	0.35
Age at enrolment (mean±SD)	50.6±17.1	47.6±13.8	0.05	50.1±14.6	48.6±15.7	0.38
Race (%) (n=413)						
White	160 (86.5)	184 (80.7)	0.26	76 (80.0)	268 (84.3)	0.22
Black	19 (10.3)	31 (13.6)		16 (16.8)	34 (10.7)	
Others or mixed	6 (3.2)	13 (5.7)		3 (3.2)	16 (5.0)	
Body mass index (n=406)						
Normal or low	109 (59.6)	130 (58.3)	0.95	50 (53.2)	189 (60.6)	0.45
Overweight	52 (28.4)	68 (30.5)		34 (36.2)	86 (27.6)	
Obese	22 (12.0)	25 (11.2)		10 (10.6)	37 (11.9)	
Drinking category (%) (n = 413)						
Abstainer	48 (25.9)	47 (20.6)	0.01	16 (16.7)	79 (24.9)	0.50
Light	38 (20.5)	41 (18.0)		24 (25.0)	55 (17.4)	
Moderate	43 (23.2)	41 (18.0)		20 (20.8)	64 (20.2)	
Heavy	20 (10.8)	28 (12.3)		9 (9.4)	39 (12.3)	
Very heavy	36 (19.5)	71 (31.1)		27 (28.1)	80 (25.2)	
Smoking (%) (n = 412)						
Never	65 (35.3)	55 (24.1)	0.02	24 (25.3)	96 (30.3)	0.25
Past	44 (23.9)	52 (22.8)		28 (29.5)	68 (21.5)	
Current	75 (40.8)	121 (53.1)		43 (45.3)	153 (48.3)	
Amount of smoking (%) (n=401)						
Never	65 (36.3)	55 (24.8)	0.05	24 (25.8)	96 (31.2)	0.52
<1 packs/day	45 (25.1)	71 (32.0)		30 (32.3)	86 (27.9)	
≥1 packs/day	69 (38.5)	96 (43.2)		39 (41.9)	126 (40.9)	
Acute pancreatitis ever (%) (n=411)						
Yes	122 (66.3)	146 (64.3)	0.62	58 (61.7)	210 (66.2)	0.62
No	29 (15.8)	44 (19.4)		17 (18.1)	56 (17.7)	
Unclear	33 (17.9)	37 (16.3)		19 (20.2)	51 (16.1)	
Regular use of pain medication (%) (n=330)						
Yes	37 (22.3)	119 (72.6)	<0.001	36 (45.0)	120 (48.0)	0.73
No	129 (77.7)	45 (27.4)		44 (55.0)	130 (52.0)	
Disability (%) (n=399)						
Yes	32 (17.5)	91 (42.1)	<0.001	23 (25.0)	100 (32.6)	0.21
No	151 (82.5)	125 (57.9)		69 (75.0)	207 (67.4)	

Fig. 2.3 Survey of 540 patients with chronic pancreatitis. Fifty-five percent of patients identified a chronic (versus intermittent) pain pattern and in this subgroup, 72.6 % required the use of daily analgesics, 75.9 % were current or ex-smokers, and 42.1 % identified themselves as currently disabled. Reprinted with permission from BMJ Publishing Group Ltd.—Mullady DK, Yadav D, Amann ST, O'Connell MR, Barmada MM, Elta GH, et al. Type of pain, pain-associated complications, quality of life, disability and resource utilisation in chronic pancreatitis: A prospective cohort study. Gut. 2011 Jan;60(1):77–84

of more sensitive diagnostic modalities such as US, CT, and ERCP [32, 34].

The socioeconomic impact of chronic pancreatitis is of obvious concern. These patients who have a significantly poorer quality of life, require extended hospitalizations, and typically require chronic analgesia. According to current literature, 27–67 % of patients with chronic alcoholic pancreatitis experience chronic pain, and as high as 80–90 % of those patients report either chronic, or recurrent pain during the course of their illness [37, 38]. In some of these cases the source of their pain is apparent, such as bile duct or duodenal stenosis, pancreatic fibrosis/inflammation, or intra-pancreatic causes, however, in the majority of the cases a definitive source of pain cannot be identified [38]. There is a limited data on the epidemiology of chronic pain in the setting of chronic pancreatitis. Probably the largest study comes from Mullady et al. [39] in which 540 chronic pancreatitis patients were identified of which 414 self-identified a particular pain pattern A-E which were defined by the temporal nature (intermittent/chronic) and pain severity (mild, moderate, severe) (Fig. 2.3). This study revealed that 55 % of patients identified a chronic pain pattern as opposed to intermittent. Of particular importance, 72.6 % of patients that identified a chronic pain pattern required the use of daily pain medications. This is in comparison to those patients which identified as intermittent in which only 22.3 % required chronic pain medications. Of this specific patient population, 46.9 % of these patients were men, 75.9 % of patients were either current or ex-smokers, and a much higher percentage, 42.1 % versus 17.5 %, reported themselves as disabled. There was no difference in frequency of intermittent or constant pain based on race, BMI, family history of pancreatic cancer, or personal history of acute pancreatitis [39]. Somewhat higher proportion (63 %) of patients with ongoing alcohol use were also reporting constant pain patterns.

Key Points

- Incidence/prevalence of chronic pancreatitis is on the rise worldwide, and potential causes include increased use of alcohol, but also use of more sensitive diagnostic modalities.
- There is a subset of rising CP cases in which alcohol is not identified as the primary cause, and it appears to be higher in women.
- As high as 80–90 % of patients identify either chronic or intermittent, recurrent pain during the course of their illness and a large majority of these patients require chronic pain medications.
- Limited epidemiologic data are available regarding prevalence of pain from chronic pancreatitis; however there appears to be association between ongoing alcohol use, smoking, and patients with chronic pain syndrome.

C. *Adhesions/Postsurgical*: Intra-abdominal adhesion-related diseases include postsurgical chronic abdominal pain syndromes, most common of which are post- cholecystectomy, herniorrhaphy, and lysis of adhesions. Very limited epidemiological data exist on this subset of patients, as the operative management remains controversial, and true causal relationships have been difficult to prove.

In many patients who initially present with chronic abdominal pain, no immediate source of intra-abdominal adhesions can be identified [40]. Postsurgical adhesions incidence varied in the literature, from 45 to 90–100 % [40–45]. Post-mortem studies done out of the UK, suggested that in as high as 28 % of autopsies in patients without prior abdominal surgery, found intra-abdominal adhesive disease believed to be related to intra-abdominal infections. Indeed, a true causal relationship between presence of abdominal adhesions and chronic abdominal pain could not be consistently found in the literature [40, 43, 44]. Despite the fact that the link between presence of adhesions and chronic abdominal pain is difficult to make, several studies provided evidence that diagnostic laparoscopy benefits this patient population, providing postsurgical pain relief rates as high as 80 %, even regardless of whether adhesiolysis was performed [40, 43]. Risk factors for development of adhesions post-surgery predominantly revolve around surgical approach, patient age, type of procedure, use of foreign bodies such as peritoneal mesh, and presence or absence of a contaminated field (i.e., gallbladder debris) [40–45].

Other risk factors for development of chronic postsurgical pain included type of surgery, duration, open versus minimally invasive, intraoperative nerve damage and gender [46–52]. Of these, most striking association is type of surgery, specifically cholecystectomy, herniorrhaphy, pelvic procedures, and adhesiolysis. Rates for post cholecystectomy chronic abdominal pain have been reported to be in the range of 3–56 % [46, 47, 49, 53]. More specifically, risk factors include preexisting psychiatric illness, female gender, long duration of symptom prior to surgery, and pain at 6 weeks post-surgery [53]. Post-herniorraphy chronic pain incidence rate was also high, from 0 to 63 % [46–49, 51, 53]. Conclusions from these studies suggest that recurrent disease, presence of preoperative pain, severity acute postoperative pain, higher body mass index, and younger age correlate with higher rates of chronic pain development. On the other hand it seems to be consistent throughout the studies that older patients have a reduced risk for the development of chronic pain.

Key Points

- Rate of painful post abdominal surgery adhesion development is between 45 and 90 %. Risk of adhesion formation is higher in open procedures (versus minimally invasive), use of foreign bodies (i.e., mesh), and presence of a contaminated surgical field (i.e., gallbladder/bowel contamination).
- Most common surgical procedures linked to chronic abdominal pain include cholecystectomy, herniorraphy, and adhesiolysis.
- Postsurgical chronic pain risk factors identified include type and duration of surgical procedure, preexisting psychiatric illness, female gender, and younger age at operation.

Chronic Somatosensory Pain of the Abdomen

Abdominal pain is one of the frequent patient complaints in the United States. Acute abdominal pain differential diagnosis is extensive, with predominance of abdominal organ functional disorders. Possible defined sources of chronic abdominal pain (defined as an interval between 3 and 12 months) may remain elusive for the treating clinician. Chronic abdominal pain can be roughly divided into visceral, somatosensory, and functional. While visceral pain typically originates from deep, internal abdominal structures, somatic pain originates from nociceptors in superficial tissues (i.e., skin), or the musculoskeletal system (i.e., bones, ligaments, muscles, etc.). In addition, the nerves that innervate these superficial structures can incur injury leading to neuropathic causalgia. The most common causes of chronic abdominal pain in these superficial structures will be the focus of the next several sections.

A. *Postsurgical Pain*: It is common to experience pain immediately after major abdominal surgery. However, chronic pain (defined as pain >6 months following surgery) varies in incidence from 3 to 80 %, depending on the type of surgery performed [53]. Chronic postsurgical pain can be associated with limb amputation (i.e., phantom limb pain), or post-thoracotomy [54, 55]. Chronic abdominal pain after major abdominal surgery incidence varies from 3 to 50 % [56–58]. Various peri-operative risk factors in patients undergoing specific surgery have been linked to chronic abdominal pain. For example, psychological vulnerability, male gender, and long-standing symptoms before cholecystectomy were preoperative risk factors for chronic postoperative pain [59]. Interestingly, surgical approach (laparoscopic versus open cholecystectomy) made no difference in the incidence of abdominal pain 1 year after the surgery [60]. One of the strongest predictors of persistent pain at 1 year is presence of the pain at 6 weeks after procedure [61]. Incidence of chronic pain after inguinal hernia repair varies from 0 to 37 %, on average, around 11 % [53]. Patients with recurrent hernias and those with occupation-related injuries had a higher incidence of pain at 1 year [62]. Various studies that compared open versus laparoscopic inguinal hernia repair and, mesh versus non-mesh repair showed no statistically significant difference in the incidence of chronic pain between those surgical approaches [53]. Again, the incidence of pain at 4 weeks postoperatively was a strong predictor of future chronic abdominal pain [63]. For abdominal surgery, patients with psychological vulnerability, preexisting chronic pain, and persistent pain at 4–6 weeks after surgery seem to have highest likelihood for developing chronic abdominal pain.

   Organic causes such as nerve damage, adhesions, or continued bowel dysfunction have been suggested as possible etiologies for continued pain at these postoperative intervals (3–12 months). Laparoscopic adhesiolysis has been attempted in chronic abdominal pain patients who have had previous abdominal surgery. Of the patients who underwent laparoscopic adhesiolysis, 47 % became pain free, 36 % reported relief, and 16 % had no change in symptoms [64]. Similar results have been found in patients female patients reporting chronic pelvic pain [65]. These studies suggest that non-obstructive adhesions, which are a common postoperative complication, could significantly contribute to the development of chronic abdominal pain. Perhaps, further studies looking at quality rather than severity of pain would help differentiate the origin (visceral versus somatic) of pain. Hyperalgesia after nociceptors activation (i.e., surgery) has been suggested as another mechanism for chronic pain [66, 67]. One study looked at the effect of systemic lidocaine in blunting the sensitization of nociceptive afferents. Peri-operative lidocaine was shown to decrease total peri-operative morphine consumption [66]. In addition, patients in the lidocaine treatment group reported less pain with movement (i.e., walking, deep inspiration, etc.). It has been suggested that lidocaine may inhibit peripheral neuropeptides which activate nociceptors and produce a central hyperalgesic effect.

B. *Chronic Abdominal Wall Pain*: Chronic abdominal wall pain is a common, yet elusive, clinical entity that may account for up to 10 % of patients seen in gastroenterologists' practice [68]. The difficulty in identifying chronic abdominal wall pain is that clinicians often search for etiologies affecting visceral organs. The clinical work-up for the organic causes has huge economic implications (roughly $1,300 dollars per patient in 1993). The incidence of chronic abdominal wall pain varies greatly depending on the study with ranges from 11 to 74 %. However, most notably Rubio et al. [69] reported an 11 % incidence of abdominal wall pain in patients with abdominal pain of obscure etiology over a 2-year interval. Multiple etiologies have been suggested as the cause of chronic abdominal wall pain. Commonly this is related to nerve injuries of the anterior abdominal wall such as entrapment of the anterior abdominal cutaneous nerve [70] caused by increased abdominal pressure, impingement by a surgical scar, or perhaps a painful rib. Other less commonly suggested etiologies include rectus sheath hematomas, incisional hernia, or radicular pain (T7-12). Various studies have investigated the utility of anesthetic injections as a therapeutic and diagnostic maneuver for these patients. Abdominal wall pain was defined as superficial tenderness localized to a distinct point with abdominal wall tensing (positive Carnett test). Roughly 60–90 % percent of patients reported pain relief with anesthetic (local anesthetics and/or steroids) injections to the anterior cutaneous nerve block [70].

C. *Radiculopathy*: The abdominal wall is innervated by spinal nerves exiting T7-T12. Irritation of a nerve root due to disc herniation or degeneration will produce neurogenic pain in a radicular pattern. The patient may also have associated sensory deficits or decrease in elicited reflexes. Although L5-S1 and C5-C7 are the most commonly involved roots for radiculopathy, thoracic nerves can also produce radicular symptoms due to spinal pathology. The difficulty of specifically isolating abdominal pain in the setting of radiculopathy has perhaps contributed to the difficulty in the studying the prevalence of this clinical entity. Despite radiculopathy being a relatively common occurrence, there is a paucity of literature regarding the epidemiology of radicular chronic abdominal pain.

D. *Diabetic Neuropathy*: Patients with long-standing diabetes mellitus can experience pain and/or weakness in the distal aspects of their extremities. Less frequently described is an abdominal radiculopathy originating from the thoracic nerves. Although no prospective clinical or epidemiological studies have investigated abdominal diabetic neuropathy, a case series form the Mayo Clinic [71] illustrated four patients, which after extensive clinical investigation for other etiologies ultimately were found to have diffuse neuropathy in the abdomen by electromyography. This was presumed to be related to diabetic neuropathy. Fortunately, all patients in this case series had spontaneous resolution with conservative medical management. Again, due to lack of epidemiologic data in this area, the extent of this disease process is unclear.

E. *Post herpetic Neuralgia*: Acute herpes zoster (shingles) is the reactivation of dormant varicella-zoster virus in ganglionic neurons. The incidence has been shown to be in the range of 0.4–1.6 cases per 1,000 in patients under the age of 20 years old. The incidence in patients over the age of 80 is much higher at 4.5–11 cases per 1,000 [72]. Immunocompromised patients (leukemia and transplant recipients) also experience a much higher incidence with rates as high as 45 per 1,000 in this susceptible population. [73]. Thoracic nerves (T7-12) are most commonly affected followed by trigeminal, cervical, and sacral spinal roots. Pain typically follows a unilateral dermatome progression with a rash that spontaneously resolves over the course of several weeks. Of the patients with acute herpes zoster 10–70 % progress to post herpetic neuralgia [73, 74].

Functional Abdominal Pain

Abdominal pain of unknown origin represents a frustrating topic among gastroenterologists and pain physicians due to difficulties in diagnoses and management. Extensive gastrointestinal evaluations often fail to show any correctible pathology. This is due to the complexities of visceral innervation and the fact that some chronic abdominal pain may not be visceral in origin. Epidemiological studies suggest many patients with chronic abdominal pain have a functional GI disorder such as irritable bowel syndrome (IBS) or functional dyspepsia [1–9, 75]. Often the pain associated with functional GI disorders coexists with other organic disorders. Psychological risk factors such as fatigue, psychological distress, health anxiety, and illness behavior are predictors of the development of new onset abdominal pain [8]. The results of this latter study suggest that functional abdominal pain is consistent with other non-organic pain syndromes. Many patients with abdominal pain have no obvious cause of their symptoms and receive an inconclusive diagnosis or no diagnosis at all. One study [1] suggests that patients consulting for the first time with a diagnosis of unspecified abdominal pain were 16–27 times more likely than controls to receive a new diagnosis of gallbladder disease, diverticular disease, pancreatitis, or appendicitis. US Householder Survey of Functional Gastrointestinal Disorders by Drossman et al. [6] suggests that females reported a higher incidence of IBS, functional abdominal pain, and functional biliary pain than males. Symptoms reported tended to decline with age. The survey also showed that patients in lower socioeconomic categories had an increased incidence. The rate of school / work absenteeism and physician visits is increased for those having a functional gastrointestinal disorder.

Common causes of functional abdominal pain include: IBS, functional dyspepsia, and functional abdominal pain syndrome (FAPS).

A. *Irritable bowel syndrome*: IBS is a symptom-based condition in which affected individuals report recurrent episodes of abdominal pain or discomfort associated with altered bowel habits [10]. Population-based studies report that the IBS prevalence is 7–15 % and that IBS occurs more commonly in women than men [76–78] (Fig. 2.4). Most healthcare providers consider IBS a diagnosis of exclusion after first ruling out organic causes and other functional causes of abdominal pain. Patients with IBS often undergo an extensive work up to rule out organic causes such as IBD or colorectal cancer. Community based surveys indicate that half of IBS patients undergo colonoscopy as part of an evaluation of their symptoms [79]. One national database analysis found that up to 25 % of all colonoscopies performed in the United States are for symptoms related to IBS [80]. One prospective controlled US trial by Chey et al. [81] demonstrates the yield of colonoscopy in patients with non-constipated IBS. The study showed the prevalence of structural abnormalities of the colon is equal between suspected IBS patients as opposed to healthy controls. Microscopic colitis was identified in a small portion (1.5 %) of patients with IBS symptoms. Recent research on the pathophysiology of painful functional gastrointestinal disorders, including IBS, has focused on visceral hypersensitivity and dysregulation of brain-gut interaction [82, 83]. One prospective controlled study [84] shows that women undergoing gynecological surgeries for non-pain indications may develop abdominal pain (17 %), and that IBS might be a consequence as well, although the previous observation did not reach statistical significance. Psychological factors and adverse life events are often implicated in the etiology of IBS [85, 86]. There is also evidence for the overlap

Demographic characteristics	Demographic composition (%) within IBS subtype				Prevalence (%) within each demographic stratum			
	Diarrhoea ($n = 901$)	Alternator ($n = 453$)	Constipation ($n = 333$)	Overall ($n = 1713$)	Diarrhoea ($n = 901$)	Alternating ($n = 453$)	Constipation ($n = 333$)	Overall ($n = 1713$)
Overall	NA	NA	NA	NA	3.5	1.7	1.3	6.6 (6.3–6.9)
Gender								
Male	39.0	32.9	29.1	34.9	2.8	1.2	0.8	4.7 (4.4–5.1)
Female	61.0	67.1	70.9	63.6	4.1	2.3	1.8	8.2 (7.7–8.7)
Education level								
Less than HS	5.0	7.9	9.3	6.5	3.0	2.4	2.1	7.4 (6.2–8.9)
HS	23.9	25.6	29.4	25.0	3.3	1.8	1.5	6.7 (6.1–7.3)
Some college	46.4	42.8	39.6	43.4	4.0	1.9	1.3	7.2 (6.7–7.7)
Bachelor or higher	24.8	23.6	21.6	23.5	2.9	1.4	0.9	5.2 (4.7–5.8)
Race/ethnicity								
White	85.5	78.8	79.9	81.3	3.6	1.6	1.2	6.4 (6.1–6.8)
Black/African-American	8.1	13.2	13.2	10.3	2.8	2.3	1.7	6.8 (5.9–7.9)
Other combined	6.2	6.2	6.0	6.1	3.5	1.8	1.3	6.6 (5.4–7.9)
Latino/Hispanic	5.4	7.5	9.6	6.7	2.6	1.8	1.7	6.2 (5.2–7.4)
Age								
<35	21.8	24.7	23.4	22.5	3.1	1.8	1.2	6.2 (5.6–6.8)
35–44	31.1	29.6	30.3	30.1	3.7	1.8	1.4	6.9 (6.3–7.5)
45–54	28.6	29.4	26.4	28.0	3.6	1.8	1.2	6.6 (6.0–7.2)
55+	18.5	16.3	19.8	17.9	3.4	1.5	1.3	6.2 (5.5–6.9)
Income								
<$20 000	14.7	23.2	21.0	17.9	3.9	3.1	2.0	9.0 (8.1–10.0)
$20 000–$34 999	18.6	17.0	18.9	18.0	4.0	1.8	1.5	7.4 (6.6–8.2)
$35 000–$49 999	22.2	19.9	21.3	21.1	3.7	1.6	1.3	6.6 (6.0–7.3)
$50 000–$74 999	23.6	20.5	24.0	22.5	3.2	1.4	1.2	5.8 (5.2–6.4)
$75 000+	20.5	19.2	14.7	18.7	3.0	1.4	0.8	5.2 (4.7–5.8)
Marital status								
Not married	37.8	44.4	37.8	39.0	3.9	2.3	1.4	7.7 (7.2–8.3)
Married	62.0	55.4	62.2	59.4	3.24	1.4	1.2	5.9 (5.6–6.2)
Employment status								
Working	81.5	78.6	64.0	76.1	3.4	1.2	1.0	5.9 (5.6–6.3)
Not working	15.5	18.1	21.3	17.1	4.8	2.8	2.4	10.0 (8.9–11.1)
Disabled	–	–	14.7	2.9	0.0	0.0	4.5	4.5 (3.4–5.9)
Head of household								
No	15.9	20.3	15.6	16.8	3.8	2.5	1.4	7.7 (6.9–8.6)
Yes	83.6	79.2	84.4	81.3	3.4	1.6	1.3	6.3 (6.0–6.6)

Fig. 2.4 Web-based survey sent to 31,829 individuals of which 25,986 responded. Prevalence of IBS was similar across all race/ethnicity groups, was the highest in persons without a high school education, and increased as income decreased. Higher rates were also seen in unemployed or unmarried individuals. Reprinted with permission from John Wiley and Sons—Andrews EB, Eaton SC, Hollis KA, Hopkins JS, Ameen V, Hamm LR, et al. Prevalence and demographics of irritable bowel syndrome: Results from a large web-based survey. Aliment Pharmacol Ther. 2005 Nov 15;22(10):935–42

of symptoms in different functional disorders. In one study of patients with chronic fatigue, there was a point prevalence of IBS symptoms of 63 % [87]. It is difficult to prove whether these factors are a predictor of onset, or merely a consequence, of symptoms. Furthermore, research into IBS has mainly taken place in primary or secondary care settings. The findings, therefore, cannot be easily extrapolated to other populations as consultation and referral behavior as well as recall bias of adverse events render the subjects highly selected and unrepresentative of the general population [88, 89].

B. *Functional dyspepsia*: Dyspepsia refers to a constellation of upper gastrointestinal symptoms that commonly occurs in adults. Dyspepsia can occur as a result of organic causes; however the majority of patients suffer from non-ulcer or functional dyspepsia. Functional dyspepsia is defined as the presence of recurrent pain or discomfort centered in the upper abdomen in the absence of any known structural cause and without any features of IBS [90]. Mahadeva and Goh [91] have reviewed epidemiological data from population-based studies of various geographical locations. The summary of the data collected supports the notion that dyspepsia is common in most populations in the world. The varying prevalence of uninvestigated dyspepsia in different populations appears to be related to the different definitions of dyspepsia used by investigators of different surveys. The true prevalence of functional dyspepsia amongst the general population has not been evaluated, due to the difficulties in excluding organic disease in large cohorts. However, several studies [91–94] have been able to examine this in some detail. The estimated prevalence of functional dyspepsia globally is between 11.5 and 29.2 %. Epidemiologically, according to the Mahadeva and Goh review [95], it appears that risk factors for functional dyspepsia are different than that of organic dyspepsia or uninvestigated dyspepsia. Female gender and underlying psychological disturbances have been shown to be the important factors in functional dyspepsia [82, 92, 94, 96, 97]. In contrast, environmental and life style habits such as poor socioeconomic status, smoking, increased caffeine intake, and NSAID ingestion appear to be more relevant to uninvestigated dyspepsia. That might be the result of a greater rate of organic disease in these populations. Dyspepsia has a peak prevalence between the ages of 40–50 years [93, 98]. There does not appear to be a significant difference in the incidence amongst varying ethnic groups [91].

C. *Functional abdominal pain syndrome*: FAPS represents a pain syndrome attributed to the abdomen that is poorly related to gut function, is associated with some loss of daily activities, and has been present for at least 6 months. The pain is constant or very frequent. The principal criterion differentiating FAPS from other functional gastrointestinal disorders, such as IBS and functional dyspepsia, is the lack of symptom relationship to food intake or defecation. The epidemiology of FAPS is limited due to a lack of available data as well as difficulties in establishing a diagnosis that can be differentiated from other more common functional gastrointestinal disorders, such as IBS and functional dyspepsia. Reported prevalence figures in North America range from 0.5 to 2 % and do not differ from those reported in other countries [97, 99, 100]. The disorder is more common in women, with a female to male ratio of 3:2 [6]. The prevalence peaks in the fourth decade of life [6, 101]. Patients with FAPS have a high rate of work absenteeism and healthcare utilization and impose a significant economic burden [6, 99, 101]. FAPS shows a close relationship with a variety of psychiatric and psychological conditions. Clinical evidence suggests that there is strong association of adverse life events and psychological stressors with increased pain reports in functional gastrointestinal disorders [102, 103]. The combination of genetic factors, vulnerabilities factors, and adult stress may determine in part the effectiveness of endogenous pain modulation systems and thereby influence the development of FAPS [75]. Studies have confirmed a significant association between chronic abdominal pain and affective disorders, most notably anxiety and depression [104]. FAPS may be seen with other somatoform disorders such as somatization disorder, conversion disorder, and hypochondriasis [105]. Patients with FAPS may exhibit ineffective coping strategies or have poor social and family support [106–110]. Histories of sexual and physical abuse are common [111, 112].

D. *Opioid bowel dysfunction and NBS*: NBS is a recognized subset of opioid bowel dysfunction that is characterized by chronic or frequently recurring abdominal pain that worsens with continued or escalating dosages of narcotics [113, 114]. This syndrome is thought to be underrecognized, but probably is becoming more prevalent due to an increase in the use of opiate analgesia. Opioid bowel dysfunction is manifested by symptoms of constipation, nausea, bloating, ileus, and sometimes worsening abdominal pain [115, 116]. The effects of opioids on bowel function have been best studied in patients with cancer pain [117]. A population-based study by Choung et al. [118] demonstrated that NBS is a relatively rare disorder with a prevalence of 0.17 %. Those patients using prescription narcotics, however, were much more likely to report increased gastrointestinal symptoms and use more laxatives. In a case series, four patients with NBS were identified over a 20-year period. The authors [119] suggested that NBS may now become more prevalent because of the use of narcotics for chronic nonmalignant painful disorders.

References

1. Wallander MA, Johansson S, Ruigomez A, Garcia Rodriguez LA. Unspecified abdominal pain in primary care: the role of gastrointestinal morbidity. Int J Clin Pract. 2007;61(10):1663–70.
2. Penston JG, Pounder RE. A survey of dyspepsia in Great Britain. Aliment Pharmacol Ther. 1996;10(1):83–9.
3. Sandler RS, Stewart WF, Liberman JN, Ricci JA, Zorich NL. Abdominal pain, bloating, and diarrhea in the United States: prevalence and impact. Dig Dis Sci. 2000;45(6):1166–71.
4. Heading RC. Prevalence of upper gastrointestinal symptoms in the general population: a systematic review. Scand J Gastroenterol Suppl. 1999;231:3–8.
5. Jones R, Lydeard S. Irritable bowel syndrome in the general population. BMJ. 1992;304(6819):87–90.
6. Drossman DA, Li Z, Andruzzi E, Temple RD, Talley NJ, Thompson WG, et al. U.S. householder survey of functional gastrointestinal disorders. Prevalence, sociodemography, and health impact. Dig Dis Sci. 1993;38(9):1569–80.
7. Talley NJ. Irritable bowel syndrome: definition, diagnosis and epidemiology. Baillieres Best Pract Res Clin Gastroenterol. 1999;13(3):371–84.
8. Halder SL, McBeth J, Silman AJ, Thompson DG, Macfarlane GJ. Psychosocial risk factors for the onset of abdominal pain. Results from a large prospective population-based study. Int J Epidemiol. 2002;31(6):1219–25; discussion 1225–6.
9. Hotopf M, Carr S, Mayou R, Wadsworth M, Wessely S. Why do children have chronic abdominal pain, and what happens to them when they grow up? Population based cohort study. BMJ. 1998;316(7139):1196–200.
10. Longstreth GF, Thompson WG, Chey WD, Houghton LA, Mearin F, Spiller RC. Functional bowel disorders. Gastroenterology. 2006; 130(5):1480–91.

11. Talley NJ, O'Keefe EA, Zinsmeister AR, Melton III LJ. Prevalence of gastrointestinal symptoms in the elderly: a population-based study. Gastroenterology. 1992;102(3):895–901.
12. Loftus Jr EV. Clinical epidemiology of inflammatory bowel disease: incidence, prevalence, and environmental influences. Gastroenterology. 2004;126(6):1504–17.
13. Hanson KA, Loftus Jr EV, Harmsen WS, Diehl NN, Zinsmeister AR, Sandborn WJ. Clinical features and outcome of patients with inflammatory bowel disease who use narcotics: a case-control study. Inflamm Bowel Dis. 2009;15(5):772–7.
14. Binder V. Epidemiology of IBD during the twentieth century: an integrated view. Best Pract Res Clin Gastroenterol. 2004;18(3): 463–79.
15. Bernstein CN. New insights into IBD epidemiology: are there any lessons for treatment? Dig Dis. 2010;28(3):406–10.
16. Calkins BM, Mendeloff AI. Epidemiology of inflammatory bowel disease. Epidemiol Rev. 1986;8:60–91.
17. Calkins BM, Lilienfeld AM, Garland CF, Mendeloff AI. Trends in incidence rates of ulcerative colitis and Crohn's disease. Dig Dis Sci. 1984;29(10):913–20.
18. Morita N, Toki S, Hirohashi T, Minoda T, Ogawa K, Kono S, et al. Incidence and prevalence of inflammatory bowel disease in Japan: nationwide epidemiological survey during the year 1991. J Gastroenterol. 1995;30 Suppl 8:1–4.
19. Yang SK, Hong WS, Min YI, Kim HY, Yoo JY, Rhee PL, et al. Incidence and prevalence of ulcerative colitis in the Songpa-Kangdong District, Seoul, Korea, 1986–1997. J Gastroenterol Hepatol. 2000;15(9):1037–42.
20. Sood A, Midha V, Sood N, Bhatia AS, Avasthi G. Incidence and prevalence of ulcerative colitis in Punjab, North India. Gut. 2003;52(11):1587–90.
21. Kugathasan S, Judd RH, Hoffmann RG, Heikenen J, Telega G, Khan F, et al. Epidemiologic and clinical characteristics of children with newly diagnosed inflammatory bowel disease in Wisconsin: a statewide population-based study. J Pediatr. 2003;143(4):525–31.
22. Askling J, Grahnquist L, Ekbom A, Finkel Y. Incidence of paediatric Crohn's disease in Stockholm, Sweden. Lancet. 1999; 354(9185):1179.
23. Turunen P, Kolho KL, Auvinen A, Iltanen S, Huhtala H, Ashorn M. Incidence of inflammatory bowel disease in Finnish children, 1987–2003. Inflamm Bowel Dis. 2006;12(8):677–83.
24. Kurata JH, Kantor-Fish S, Frankl H, Godby P, Vadheim CM. Crohn's disease among ethnic groups in a large health maintenance organization. Gastroenterology. 1992;102(6):1940–8.
25. Hanauer SB. Inflammatory bowel disease: epidemiology, pathogenesis, and therapeutic opportunities. Inflamm Bowel Dis. 2006;12 Suppl 1:S3–9.
26. Bielefeldt K, Davis B, Binion DG. Pain and inflammatory bowel disease. Inflamm Bowel Dis. 2009;15(5):778–88.
27. Edwards JT, Radford-Smith GL, Florin TH. Chronic narcotic use in inflammatory bowel disease patients: prevalence and clinical characteristics. J Gastroenterol Hepatol. 2001;16(11):1235–8.
28. Lichtenstein GR, Feagan BG, Cohen RD, Salzberg BA, Diamond RH, Chen DM, et al. Serious infections and mortality in association with therapies for Crohn's disease: TREAT registry. Clin Gastroenterol Hepatol. 2006;4(5):621–30.
29. Cross RK, Wilson KT, Binion DG. Narcotic use in patients with Crohn's disease. Am J Gastroenterol. 2005;100(10):2225–9.
30. Etemad B, Whitcomb DC. Chronic pancreatitis: diagnosis, classification, and new genetic developments. Gastroenterology. 2001;120(3):682–707.
31. Jupp J, Fine D, Johnson CD. The epidemiology and socioeconomic impact of chronic pancreatitis. Best Pract Res Clin Gastroenterol. 2010;24(3):219–31.
32. Lin Y, Tamakoshi A, Matsuno S, Takeda K, Hayakawa T, Kitagawa M, et al. Nationwide epidemiological survey of chronic pancreatitis in Japan. J Gastroenterol. 2000;35(2):136–41.
33. Spanier BW, Dijkgraaf MG, Bruno MJ. Epidemiology, aetiology and outcome of acute and chronic pancreatitis: an update. Best Pract Res Clin Gastroenterol. 2008;22(1):45–63.
34. Otsuki M. Chronic pancreatitis in Japan: epidemiology, prognosis, diagnostic criteria, and future problems. J Gastroenterol. 2003;38(4):315–26.
35. Cote GA, Yadav D, Slivka A, Hawes RH, Anderson MA, Burton FR, et al. Alcohol and smoking as risk factors in an epidemiology study of patients with chronic pancreatitis. Clin Gastroenterol Hepatol. 2011;9(3):266–73; quiz e27.
36. Dani R, Mott CB, Guarita DR, Nogueira CE. Epidemiology and etiology of chronic pancreatitis in Brazil: a tale of two cities. Pancreas. 1990;5(4):474–8.
37. Ammann RW, Muellhaupt B. The natural history of pain in alcoholic chronic pancreatitis. Gastroenterology. 1999;116(5): 1132–40.
38. Drewes AM, Krarup AL, Detlefsen S, Malmstrom ML, Dimcevski G, Funch-Jensen P. Pain in chronic pancreatitis: the role of neuropathic pain mechanisms. Gut. 2008;57(11):1616–27.
39. Mullady DK, Yadav D, Amann ST, O'Connell MR, Barmada MM, Elta GH, et al. Type of pain, pain-associated complications, quality of life, disability and resource utilisation in chronic pancreatitis: a prospective cohort study. Gut. 2011;60(1):77–84.
40. Swank DJ, Swank-Bordewijk SC, Hop WC, van Erp WF, Janssen IM, Bonjer HJ, et al. Laparoscopic adhesiolysis in patients with chronic abdominal pain: a blinded randomised controlled multicentre trial. Lancet. 2003;361(9365):1247–51.
41. Dijkstra FR, Nieuwenhuijzen M, Reijnen MM, van Goor H. Recent clinical developments in pathophysiology, epidemiology, diagnosis and treatment of intra-abdominal adhesions. Scand J Gastroenterol Suppl. 2000;232:52–9.
42. Attard JA, MacLean AR. Adhesive small bowel obstruction: epidemiology, biology and prevention. Can J Surg. 2007;50(4): 291–300.
43. Swank DJ, Jeekel H. Laparoscopic adhesiolysis in patients with chronic abdominal pain. Curr Opin Obstet Gynecol. 2004;16(4): 313–8.
44. Swank DJ, van Erp WF, Repelaer van Driel OJ, Hop WC, Bonjer HJ, Jeekel J. Complications and feasibility of laparoscopic adhesiolysis in patients with chronic abdominal pain. A retrospective study. Surg Endosc. 2002;16(10):1468–73.
45. Peters AA, Van den Tillaart SA. The difficult patient in gastroenterology: chronic pelvic pain, adhesions, and sub occlusive episodes. Best Pract Res Clin Gastroenterol. 2007;21(3):445–63.
46. Katz J, Seltzer Z. Transition from acute to chronic postsurgical pain: risk factors and protective factors. Expert Rev Neurother. 2009;9(5):723–44.
47. Visser E. Chronic post-surgical pain: epidemiology and clinical implications for acute pain management. Acute Pain. 2006;8(2): 73–81.
48. Macrae WA. Chronic post-surgical pain: 10 years on. Br J Anaesth. 2008;101(1):77–86.
49. Kehlet H, Jensen TS, Woolf CJ. Persistent postsurgical pain: risk factors and prevention. Lancet. 2006;367(9522):1618–25.
50. Crombie IK, Davies HT, Macrae WA. Cut and thrust: antecedent surgery and trauma among patients attending a chronic pain clinic. Pain. 1998;76(1–2):167–71.
51. Wilder-Smith OH, Schreyer T, Scheffer GJ, Arendt-Nielsen L. Patients with chronic pain after abdominal surgery show less preoperative endogenous pain inhibition and more postoperative hyperalgesia: a pilot study. J Pain Palliat Care Pharmacother. 2010;24(2):119–28.
52. Poobalan AS, Bruce J, Smith WC, King PM, Krukowski ZH, Chambers WA. A review of chronic pain after inguinal herniorrhaphy. Clin J Pain. 2003;19(1):48–54.
53. Perkins FM, Kehlet H. Chronic pain as an outcome of surgery. A review of predictive factors. Anesthesiology. 2000;93(4):1123–33.

54. Nikolajsen L, Ilkjaer S, Jensen TS. Effect of preoperative extradural bupivacaine and morphine on stump sensation in lower limb amputees. Br J Anaesth. 1998;81(3):348–54.
55. Dajczman E, Gordon A, Kreisman H, Wolkove N. Long-term postthoracotomy pain. Chest. 1991;99(2):270–4.
56. Stiff G, Rhodes M, Kelly A, Telford K, Armstrong CP, Rees BI. Long-term pain: less common after laparoscopic than open cholecystectomy. Br J Surg. 1994;81(9):1368–70.
57. Hay JM, Boudet MJ, Fingerhut A, Poucher J, Hennet H, Habib E, et al. Shouldice inguinal hernia repair in the male adult: the gold standard? A multicenter controlled trial in 1578 patients. Ann Surg. 1995;222(6):719–27.
58. de Pouvourville G, Ribet-Reinhart N, Fendrick M, Houry S, Testas P, Huguier M. A prospective comparison of costs and morbidity of laparoscopic versus open cholecystectomy. Hepatogastroenterology. 1997;44(13):35–9.
59. Fenster LF, Lonborg R, Thirlby RC, Traverso LW. What symptoms does cholecystectomy cure? Insights from an outcomes measurement project and review of the literature. Am J Surg. 1995; 169(5):533–8.
60. Nicholl JP, Brazier JE, Milner PC, Westlake L, Kohler B, Williams BT, et al. Randomised controlled trial of cost-effectiveness of lithotripsy and open cholecystectomy as treatments for gallbladder stones. Lancet. 1992;340(8823):801–7.
61. Borly L, Anderson IB, Bardram L, Christensen E, Sehested A, Kehlet H, et al. Preoperative prediction model of outcome after cholecystectomy for symptomatic gallstones. Scand J Gastroenterol. 1999;34(11):1144–52.
62. Salcedo-Wasicek MC, Thirlby RC. Postoperative course after inguinal herniorrhaphy. A case-controlled comparison of patients receiving workers' compensation vs patients with commercial insurance. Arch Surg. 1995;130(1):29–32.
63. Callesen T, Bech K, Kehlet H. Prospective study of chronic pain after groin hernia repair. Br J Surg. 1999;86(12):1528–31.
64. Mueller MD, Tschudi J, Herrmann U, Klaiber C. An evaluation of laparoscopic adhesiolysis in patients with chronic abdominal pain. Surg Endosc. 1995;9(7):802–4.
65. Kresch AJ, Seifer DB, Sachs LB, Barrese I. Laparoscopy in 100 women with chronic pelvic pain. Obstet Gynecol. 1984;64(5): 672–4.
66. Koppert W, Weigand M, Neumann F, Sittl R, Schuettler J, Schmelz M, et al. Perioperative intravenous lidocaine has preventive effects on postoperative pain and morphine consumption after major abdominal surgery. Anesth Analg. 2004;98(4):1050–5, table of contents.
67. Koppert W, Ostermeier N, Sittl R, Weidner C, Schmelz M. Low-dose lidocaine reduces secondary hyperalgesia by a central mode of action. Pain. 2000;85(1–2):217–24.
68. Srinivasan R, Greenbaum DS. Chronic abdominal wall pain: a frequently overlooked problem. Practical approach to diagnosis and management. Am J Gastroenterol. 2002;97(4):824–30.
69. Rubio M, Gitlin MC, Lambaise LR. Abdominal wall pain as a cause of abdominal pain of obscure etiology. Am J Gastroenterol. 1998;93:1740.
70. Applegate W. Abdominal cutaneous nerve entrapment syndrome (ACNES): a commonly overlooked cause of abdominal pain. Perm J. 2002;6(3):20–7.
71. Longstreth GF, Newcomer AD. Abdominal pain caused by diabetic radiculopathy. Ann Intern Med. 1977;86(2):166–8.
72. Kost RG, Straus SE. Postherpetic neuralgia—pathogenesis, treatment, and prevention. N Engl J Med. 1996;335(1):32–42.
73. Balfour Jr HH. Varicella zoster virus infections in immunocompromised hosts. A review of the natural history and management. Am J Med. 1988;85(2A):68–73.
74. Ragozzino MW, Melton III LJ, Kurland LT, Chu CP, Perry HO. Population-based study of herpes zoster and its sequelae. Medicine (Baltimore). 1982;61(5):310–6.
75. Clouse RE, Mayer EA, Aziz Q, Drossman DA, Dumitrascu DL, Monnikes H, et al. Functional abdominal pain syndrome. Gastroenterology. 2006;130(5):1492–7.
76. Andrews EB, Eaton SC, Hollis KA, Hopkins JS, Ameen V, Hamm LR, et al. Prevalence and demographics of irritable bowel syndrome: results from a large web-based survey. Aliment Pharmacol Ther. 2005;22(10):935–42.
77. Hungin AP, Chang L, Locke GR, Dennis EH, Barghout V. Irritable bowel syndrome in the United States: prevalence, symptom patterns and impact. Aliment Pharmacol Ther. 2005;21(11): 1365–75.
78. Saito YA, Talley NJ, Melton L J, Fett S, Zinsmeister AR, Locke GR. The effect of new diagnostic criteria for irritable bowel syndrome on community prevalence estimates. Neurogastroenterol Motil. 2003;15(6):687–94.
79. Talley NJ, Gabriel SE, Harmsen WS, Zinsmeister AR, Evans RW. Medical costs in community subjects with irritable bowel syndrome. Gastroenterology. 1995;109(6):1736–41.
80. Lieberman DA, Holub J, Eisen G, Kraemer D, Morris CD. Utilization of colonoscopy in the United States: results from a national consortium. Gastrointest Endosc. 2005;62(6):875–83.
81. Chey WD, Nojkov B, Rubenstein JH, Dobhan RR, Greenson JK, Cash BD. The yield of colonoscopy in patients with non-constipated irritable bowel syndrome: results from a prospective, controlled US trial. Am J Gastroenterol. 2010;105(4):859–65.
82. Kellow JE, Delvaux M, Azpiroz F, Camilleri M, Quigley EM, Thompson DG. Principles of applied neurogastroenterology: physiology/motility-sensation. Gut. 1999;45 Suppl 2:II17–24.
83. Drossman DA. The functional gastrointestinal disorders and the Rome III process. Gastroenterology. 2006;130(5):1377–90.
84. Sperber AD, Morris CB, Greemberg L, Bangdiwala SI, Goldstein D, Sheiner E, et al. Development of abdominal pain and IBS following gynecological surgery: a prospective, controlled study. Gastroenterology. 2008;134(1):75–84.
85. Kay L, Jorgensen T, Jensen KH. The epidemiology of irritable bowel syndrome in a random population: prevalence, incidence, natural history and risk factors. J Intern Med. 1994;236(1): 23–30.
86. Talley NJ, Fett SL, Zinsmeister AR. Self-reported abuse and gastrointestinal disease in outpatients: association with irritable bowel-type symptoms. Am J Gastroenterol. 1995;90(3):366–71.
87. Gomborone JE, Gorard DA, Dewsnap PA, Libby GW, Farthing MJ. Prevalence of irritable bowel syndrome in chronic fatigue. J R Coll Physicians Lond. 1996;30(6):512–3.
88. Smith RC, Greenbaum DS, Vancouver JB, Henry RC, Reinhart MA, Greenbaum RB, et al. Psychosocial factors are associated with health care seeking rather than diagnosis in irritable bowel syndrome. Gastroenterology. 1990;98(2):293–301.
89. Drossman DA, McKee DC, Sandler RS, Mitchell CM, Cramer EM, Lowman BC, et al. Psychosocial factors in the irritable bowel syndrome. A multivariate study of patients and nonpatients with irritable bowel syndrome. Gastroenterology. 1988;95(3):701–8.
90. Talley NJ, Stanghellini V, Heading RC, Koch KL, Malagelada JR, Tytgat GN. Functional gastroduodenal disorders. Gut. 1999;45 Suppl 2:II37–42.
91. Shaib Y, El-Serag HB. The prevalence and risk factors of functional dyspepsia in a multiethnic population in the United States. Am J Gastroenterol. 2004;99(11):2210–6.
92. Bernersen B, Johnsen R, Straume B. Non-ulcer dyspepsia and peptic ulcer: the distribution in a population and their relation to risk factors. Gut. 1996;38(6):822–5.
93. Hirakawa K, Adachi K, Amano K, Katsube T, Ishihara S, Fukuda R, et al. Prevalence of non-ulcer dyspepsia in the Japanese population. J Gastroenterol Hepatol. 1999;14(11):1083–7.
94. Lu CL, Lang HC, Chang FY, Chen CY, Luo JC, Wang SS, et al. Prevalence and health/social impacts of functional dyspepsia in Taiwan: a study based on the Rome criteria questionnaire survey

assisted by endoscopic exclusion among a physical check-up population. Scand J Gastroenterol. 2005;40(4):402–11.
95. Mahadeva S, Goh KL. Epidemiology of functional dyspepsia: a global perspective. World J Gastroenterol. 2006;12(17):2661–6.
96. Li Y, Nie Y, Sha W, Su H. The link between psychosocial factors and functional dyspepsia: an epidemiological study. Chin Med J (Engl). 2002;115(7):1082–4.
97. Koloski NA, Talley NJ, Boyce PM. Epidemiology and health care seeking in the functional GI disorders: a population-based study. Am J Gastroenterol. 2002;97(9):2290–9.
98. Caballero-Plasencia AM, Sofos-Kontoyannis S, Valenzuela-Barranco M, Martin-Ruiz JL, Casado-Caballero FJ, Lopez-Manas JG. Irritable bowel syndrome in patients with dyspepsia: a community-based study in southern Europe. Eur J Gastroenterol Hepatol. 1999;11(5):517–22.
99. Kwan AC, Bao TN, Chakkaphak S, Chang FY, Ke MY, Law NM, et al. Validation of Rome II criteria for functional gastrointestinal disorders by factor analysis of symptoms in Asian patient sample. J Gastroenterol Hepatol. 2003;18(7):796–802.
100. Thompson WG, Heaton KW, Smyth GT, Smyth C. Irritable bowel syndrome in general practice: prevalence, characteristics, and referral. Gut. 2000;46(1):78–82.
101. Maxton DG, Whorwell PJ. Use of medical resources and attitudes to health care of patients with 'chronic abdominal pain'. Br J Med Econ. 1992;2:75–9.
102. Mayer EA, Derbyshire S, Naliboff BD. Cerebral activation in irritable bowel syndrome. Gastroenterology. 2000;119(5):1418–20.
103. Mayer EA, Collins SM. Evolving pathophysiologic models of functional gastrointestinal disorders. Gastroenterology. 2002; 122(7):2032–48.
104. Bixo M, Sundstrom-Poromaa I, Bjorn I, astrom M. Patients with psychiatric disorders in gynecologic practice. Am J Obstet Gynecol. 2001;185(2):396–402.
105. Gmitrowicz A, Kucharska A. Developmental disorders in the fourth edition of the American classification: diagnostic and statistical manual of mental disorders (DSM IV–optional book). Psychiatr Pol. 1994;28(5):509–21.
106. Sarason IG, Sarason BR, Potter III EH, Antoni MH. Life events, social support, and illness. Psychosom Med. 1985;47(2):156–63.
107. Jamison RN, Virts KL. The influence of family support on chronic pain. Behav Res Ther. 1990;28(4):283–7.
108. Scarinci IC, McDonald-Haile J, Bradley LA, Richter JE. Altered pain perception and psychosocial features among women with gastrointestinal disorders and history of abuse: a preliminary model. Am J Med. 1994;97(2):108–18.
109. Drossman DA, Whitehead WE, Toner BB, Diamant N, Hu YJ, Bangdiwala SI, et al. What determines severity among patients with painful functional bowel disorders? Am J Gastroenterol. 2000;95(4):974–80.
110. Drossman DA, Leserman J, Li Z, Keefe F, Hu YJ, Toomey TC. Effects of coping on health outcome among women with gastrointestinal disorders. Psychosom Med. 2000;62(3):309–17.
111. Drossman DA, Talley NJ, Leserman J, Olden KW, Barreiro MA. Sexual and physical abuse and gastrointestinal illness. Review and recommendations. Ann Intern Med. 1995;123(10):782–94.
112. Wurtele SK, Kaplan GM, Keairnes M. Childhood sexual abuse among chronic pain patients. Clin J Pain. 1990;6(2):110–3.
113. Sandgren JE, McPhee MS, Greenberger NJ. Narcotic bowel syndrome treated with clonidine. Resolution of abdominal pain and intestinal pseudo-obstruction. Ann Intern Med. 1984;101(3):331–4.
114. Rogers M, Cerda JJ. The narcotic bowel syndrome. J Clin Gastroenterol. 1989;11(2):132–5.
115. Pappagallo M. Incidence, prevalence, and management of opioid bowel dysfunction. Am J Surg. 2001;182(5A Suppl):11S–8.
116. Panchal SJ, Muller-Schwefe P, Wurzelmann JI. Opioid-induced bowel dysfunction: prevalence, pathophysiology and burden. Int J Clin Pract. 2007;61(7):1181–7.
117. Grond S, Zech D, Diefenbach C, Bischoff A. Prevalence and pattern of symptoms in patients with cancer pain: a prospective evaluation of 1635 cancer patients referred to a pain clinic. J Pain Symptom Manage. 1994;9(6):372–82.
118. Choung RS, Locke III GR, Zinsmeister AR, Schleck CD, Talley NJ. Opioid bowel dysfunction and narcotic bowel syndrome: a population-based study. Am J Gastroenterol. 2009;104(5):1199–204.
119. Grunkemeier DM, Cassara JE, Dalton CB, Drossman DA. The narcotic bowel syndrome: clinical features, pathophysiology, and management. Clin Gastroenterol Hepatol. 2007;5(10):1126–39. quiz 1121–2.

Establishing Diagnosis of Chronic Abdominal Pain: Gastroenterologist View

3

Amit Bhatt and Tyler Stevens

Introduction

Abdominal pain is a common complaint encountered in gastroenterology practices [1]. Most patients with abdominal pain have a functional disorder (e.g., irritable bowel disorder) or a benign and self-limited condition. However, abdominal pain may sometimes indicate a serious or life-threatening illness. The primary role of the gastroenterologist is to differentiate organic from functional disease and to provide directed treatment for the underlying cause of pain. The clinical challenge and expense of evaluating abdominal pain arises from the concern over missing a structural or organic disease.

Many gastrointestinal and systemic disorders may cause abdominal pain (Table 3.1). The gastroenterologist must consider these myriad possibilities and carry out a rationale evaluation based on plausible causes. Functional disorders should be considered once organic pathology has been confidently excluded. This chapter will focus on the diagnostic tools which gastroenterologists and internists utilize in the evaluation of abdominal pain, ranging from a careful history to sophisticated invasive testing.

History

The clinician must initially adopt a broad differential diagnosis that becomes more focused as the investigation progresses. The history should inquire about the characteristics of abdominal pain including the onset, duration, location, nature, radiation, associated features, and relieving and aggravating factors. Establishing the duration of pain is very useful in narrowing the differential diagnosis. Chronic abdominal pain is defined as constant or intermittent pain occurring for greater than 6 months. Acute abdominal pain is when pain has been occurring for up to several days, and sub-acute abdominal pain is from several days to 6 months. After establishing chronicity, the location, nature, and radiation of pain should be determined to help focus attention to certain pathologies. Upper abdominal pain can arise from biliary, pancreatic, gastric, and duodenal pathology. Mid-abdominal pain likely originates from the small bowel (e.g., Crohn's disease, celiac disease, bacterial overgrowth, partial small bowel obstruction, chronic mesenteric ischemia). Lower abdominal pain arises from the colon (e.g., irritable bowel syndrome, colitis), bladder, or reproductive organs. It is important to differentiate between constant and intermittent chronic abdominal pain. While intermittent pain can have many causes, constant abdominal pain results from only a few gastrointestinal etiologies (Table 3.2). The presence of aggravating and relieving factors can be quite informative. Pain that is positional in nature is likely to be of musculoskeletal origin. Worsening of pain with eating is typical in peptic ulcer disease, chronic mesenteric ischemia, and in the presence of biliary and pancreatic pathologies. Relief with bowel movements is expected with constipation and irritable bowel syndrome (IBS). Pain related to menstruation may signify a gynecological cause. The clinician should probe for coexisting symptoms such as nausea, vomiting, diarrhea, blood in stools, and systemic symptoms like fever or rash. The presence of diarrhea suggests IBS, chronic pancreatitis, inflammatory bowel disease, celiac disease, and bacterial overgrowth. "Alarm" symptoms of fever, weight loss, night sweats, appetite, or nocturnal awakening often indicate organic pathology.

Rare medical causes of abdominal pain should be considered when structural etiologies are ruled out. Recurrent attacks of fever, joint pain, and abdominal pain suggest familial Mediterranean fever [2]. Recurrent attacks of abdominal pain, tachycardia, constipation, and dark urine suggest acute intermittent porphyria. The presence of hyponatremia, hyperkalemia, and hyperpigmentation should raise suspicion for

A. Bhatt, M.D. • T. Stevens, M.D. (✉)
Gastroenterology and Hepatology, Cleveland Clinic Foundation, 9500 Euclid Avenue, Desk Q3, Cleveland, OH 44112, USA
e-mail: Bhatta3@ccf.org; Stevent@ccf.org

L. Kapural (ed.), *Chronic Abdominal Pain: An Evidence-Based, Comprehensive Guide to Clinical Management*,
DOI 10.1007/978-1-4939-1992-5_3, © Springer Science+Business Media New York 2015

Table 3.1 Etiology of chronic abdominal pain

Etiology	Typical diagnostic tests	Treatment
Organic		
Gallstones	US, HIDA scan	Cholecystectomy
Cholangitis	RUQ US, ERCP	ERCP
Appendicitis	CT scan	Appendectomy
Peptic ulcer disease	Upper endoscopy, *H. pylori* testing	Proton pump inhibitor treatment of *H. pylori*
Chronic pancreatitis	EUS, CT scan, MRI,	Life style modifications Pancreatic enzymes Celiac plexus block
Inflammatory bowel disease	CTE, colonoscopy, EGD	5-ASA, Budesonide, prednisone, Imuran, 6-MP, cyclosporine, Anti-TF agents
Mesenteric ischemia	Mesenteric ultrasound, CT angiography	Endovascular or surgical revascularization
Hernias	CT scan	Hernia repair
Intestinal obstruction	CT scan, small bowel series	Surgical repair
Abdominal adhesions	Ct scan, small bowel series	Lysis of adhesions Symptomatic management
Abdominal neoplasms	CT scan, MRI, EUS	Surgical resection Endoscopic resection
Lactulose intolerance	Breath testing Trial of withdrawal	Lactulose avoidance
Small bowel bacterial overgrowth	Breath testing	Antibiotics
Gastroparesis	Gastric emptying study	Promotility agents
Pelvic inflammatory disease	Laboratory testing Gram stain and microscopic examination of vaginal discharge Ultrasound	Antibiotics
Mittelschmertz	History	Symptomatic management
Diabetic neuropathy	History	Symptomatic management
Eosinophilic gastroenteritis	Upper and lower endoscopy	Budesonide Prednisone Oral cromolyn
Familial Mediterranean fever	History Genetic testing	Colchicine
Hereditary angioedema	C4 esterase levels	Avoid triggers C1 esterase inhibitor replacement protein Ecallantide Icatibant
Porphyria	Porphyria screen	Avoid triggers Intravenous hemin
Celiac artery syndrome	Mesenteric ultrasound CT angiogram MR angiogram	Surgery
Superior mesenteric artery syndrome	Mesenteric ultrasound CT angiogram MR angiogram	Surgery
Abdominal migraine	History	Anti-migraine medications
Herpes Zoster	Physical examination PCR Viral culture DFA test	Nucleoside analogues
Lead poisoning	Blood lead level	Reduce lead exposure Chelation therapy
Neuromuscular		

(continued)

Table 3.1 (continued)

Etiology	Typical diagnostic tests	Treatment
Anterior cutaneous nerve entrapment syndrome	History and physical examination	Local anesthetic injection
Myofascial pain syndrome	History and physical examination	Physical therapy Anti-depressants Sedatives
Slipping rib syndrome	History and physical examination	Local anesthetic injection
Thoracic nerve radiculopathy	X-ray MRI	Treatment based on underlying process
Functional gastrointestinal disorders		
Gallbladder dyskinesia	HIDA scan	Cholecystectomy
Sphincter of oddi dysfunction	Timed HIDA scan ERCP with manometry	ERCP with sphincterotomy
Functional abdominal pain syndrome	History and physical examination Exclusion of other etiology	Tricyclic antidepressants
Functional dyspepsia	History and physical examination Upper endoscopy *H. pylori* testing	Acid suppressive drugs Eradication of *H. pylori* Antidepressants
Irritable bowel syndrome	History and physical examination Exclusion of other etiology	High-fiber diet Antispasmodics Lubiprostone SSRI TCA
Levator ani syndrome	History and physical examination	Sitz baths Perineal strengthening exercises

Table 3.2 Etiology of chronic constant abdominal pain

Chronic pancreatitis
Malignancy
Abscess
Psychiatric
Inexplicable

adrenal insufficiency. Hereditary angioedema should be considered in patients with intermittent abdominal pain who have a history of recurrent angioedema without urticaria. History of exposure, metallic taste in mouth, and cognitive impairment should direct attention to heavy metal poisoning. The presence of coexisting medical illnesses may also suggest a cause of abdominal pain. A history of vasculopathy raises suspicion of chronic mesenteric ischemia. A history of physical or sexual abuse is common in patients with functional gastrointestinal disorders [3]. A family history of gastrointestinal malignancy, pancreatic disorders, or inflammatory bowel disease should be elicited.

Physical Examination

A complete abdominal examination includes inspection, auscultation, percussion, and palpation. Surgical scars on inspection should be noted. Identification of a bruit on auscultation may indicate chronic mesenteric ischemia. Light and deep palpation should be performed to check for masses, ascites, hernias, and organomegaly. Observing the patient's response to palpation can be helpful in differentiating functional from organic disease. A closed eye sign and stethoscope sign are seen more in functional gastrointestinal disorders. A closed eye sign is when patients close their eyes during examination [4], in contrast to patients with acute abdominal pain whose eyes open in fearful anticipation. The stethoscope sign is the detection of less tenderness during pressure with a stethoscope than with palpation [5]. Hover sign and Carnett's sign are seen in abdominal wall pain. Hover sign is when lightly touching the area of pain and patient guards the area with his hand or grabs the examining hand [6]. Carnett's sign is increased abdominal tenderness when the patient tenses their abdominal muscles [7]. Patients with chronic abdominal pain may still present with an acute abdomen and care should be taken to look for peritoneal signs of rebounding and guarding.

It is important to also perform a complete physical examination looking for systemic disease. Signs of malnutrition, vitamin deficiency, and skin changes can signify organic illness. Skin rashes can be helpful in narrowing the diagnosis. Dermatitis herpetiformis is associated with celiac disease (Fig. 3.1). Erythema nodosum, pyoderma gangernosum, and sweets syndrome may be seen in inflammatory bowel disease (Fig. 3.2). Acanthosis nigricans, Leser–Trélat sign,

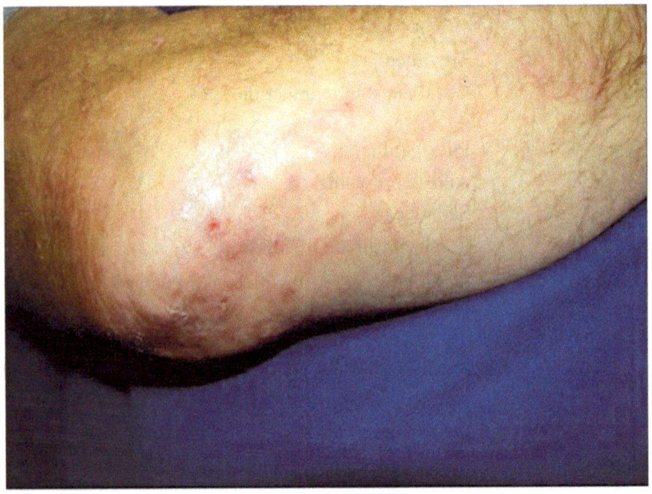

Fig. 3.1 Dermatitis herpetiformis (Thank you to Dr. Pooja Kheera for the picture)

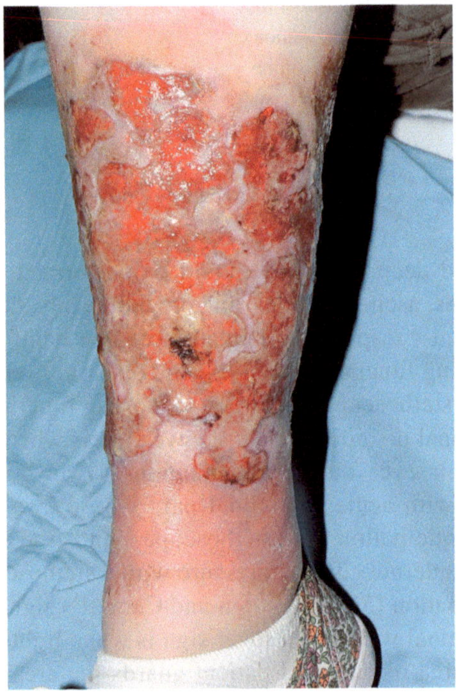

Fig. 3.2 Pyoderma gangernosum (Thank you to Dr. Pooja Kheera for the picture)

hypertrichosis lanuginosa, Tylosis, and Tripe palm can signify underlying malignancy.

Laboratory Testing

Laboratory test abnormalities are common in patients with organic pathology, while normal lab tests are expected in patients with functional bowel disorders. Routine laboratory evaluation includes complete blood cell count (CBC). Anemia can raise suspicion of IBD, celiac disease, or gastrointestinal malignancies. Elevated platelet counts and white blood cell count can be seen in inflammatory diseases. Additional laboratory testing should be based on history and physical examination. Testing for *Helicobacter pylori* antibody should be considered in patients with upper abdominal pain. Celiac serology testing should be considered in those with suspicion of celiac disease. Liver function tests should be checked in those with suspicion of biliary pathology. If recurrent pancreatitis is considered, amylase and lipase should be checked.

Specialized laboratory testing for "rare" medical conditions should be obtained based on clinical suspicion. C4 esterase levels can be checked for hereditary angioedema, cortisol stimulation test for adrenal insufficiency, heavy metal screen for heavy metal poisoning, porphyria screen for acute intermittent porphyria, and genetic testing for familial Mediterranean fever.

Radiologic Imaging

Radiographic tests identify structural abnormalities of the gastrointestinal system. Radiologic evaluation should be tailored based on presenting symptoms, physical examination, and laboratory findings. An abdominal X-ray is a reasonable "screen" for various causes of chronic abdominal pain. It may still detect excessive stool in constipation, calcifications in chronic pancreatitis, appendicolith, partial bowel obstruction from adhesions, and foreign bodies.

Trans-abdominal ultrasound (US) is another safe and noninvasive imaging test, which is fast, portable, and uses no ionizing radiation. The most useful roles of US are in evaluating the hepatobiliary tract and assessing the patency of mesenteric vessels. Right upper quadrant ultrasound evaluates the gallbladder, biliary tree, and adjacent structures. Stones or sludge in the gallbladder may suggest a source of biliary colic or recurrent pancreatitis. Biliary dilation may indicate biliary obstruction arising from a pancreatic mass or bile duct stone. Since US does not easily visualize the distal (periampullary) bile duct, it has relatively poor sensitivity (22–55 %) for detecting common bile duct stones [8, 9]. However, it is able to detect common bile duct dilation that is associated with choledocholithiasis (sensitivity 77–87 %) [9, 10]. In the presence of an intact gallbladder the normal bile duct can range from 3 to 6 mm, while a common bile duct greater than 8 mm is indicative of biliary obstruction [11, 12].

CT scan should be considered based on clinical suspicion of structural pathology and in those with "alarm symptoms." Intravenous contrast (IV) during the CT scan helps establish vascular patency, organ perfusion, and differentiate hypervascular from hypovascular lesions. Oral contrast helps differentiate collapsed bowel from a collection/mass, identifies leaks and fistulae, detects intestinal obstruction, and assesses wall thickness and enhancement. CT protocols can be modified and tailored based on the clinical suspicion. For example, a CT angiogram is performed by timing image capture when the IV contrast is within the arterial system, allowing detection of aneurysms, artery stenosis, arteriovenous malformation, and thrombosis. A CT enterography uses a neutral oral contrast like Volumen, which provides a more detailed evaluation of inflammation, thickening, and luminal patency of the small bowel [13].

MRI is another cross-sectional imaging modality that is especially useful in the evaluation of biliary and pancreatic disease. MRI imaging uses T1 imaging which highlights fat and T2 imaging which highlights fluid. Although more costly than CT, MRI imparts much less radiation exposure and more detailed imaging in pancreatic and biliary diseases. Magnetic resonance cholangiopancreatography (MRCP) relies on the strong T_2 signal from stationary liquid (bile, pancreatic fluid, etc.) to generate images. The resulting images show fluid-filled structures as bright, producing excellent imaging of the biliary tree and pancreatic duct [14]. MRCP has good sensitivity (85–92 %) and specificity (93–97 %) for detecting choledocholithiasis in two recent systemic reviews [15, 16]. MRCP is able to assess the pancreatic parenchyma, main pancreatic duct, and side branches allowing evaluation of early chronic pancreatitis [17]. Pancreatic duct imaging can be enhanced by the use of secretin. By observing duodenal filling from the pancreatic duct after secretin administration pancreatic exocrine function can be assessed [18]. In one "head to head" study MRCP and EUS were compared to a composite gold standard of endoscopic retrograde cholangiopancreatography (ERCP), pathology, and long-term follow-up. The investigators found EUS was more sensitive than MRCP (93 % vs. 63 %) and equally specific (93 % vs. 90 %) [18]. MR enterography is an alternative to CT enterography in the evaluation of Crohn's disease, especially if there is concern of radiation exposure.

Endoscopy

Endoscopy evaluates the mucosal surfaces of the digestive disease tract. The common diagnostic functions are visual inspection and obtaining mucosal biopsies. Upper endoscopy provides inspection of the esophagus, stomach, and proximal duodenum and detects peptic ulcers, tumors, and inflammation. Colonoscopy examines the rectum, entire colon, and small part of the terminal ileum and detects colitis, tumors, and diverticulae. Upper and lower endoscopies are safe procedures with a low complication rate. Upper endoscopy should be considered in any patient with persistent upper abdominal symptoms with "alarm symptoms," over the age of 55, or persistent symptoms despite an appropriate trial of therapy (e.g., acid-suppressing medications or treatment for *H. pylori*) [19]. Colonoscopy should be considered if there is a suspicion of inflammatory bowel disease, diarrhea, iron deficiency anemia or if the patient is older than 50.

ERCP

During an ERCP, a side-viewing endoscope is passed into the duodenum to identify the major papilla. The bile and/or pancreatic ducts are cannulated and injected with contrast to provide detailed fluoroscopic imaging of the ducts. ERCP is not a benign procedure and carries risks of pancreatitis, cholangitis, bleeding, and perforation. Due to the risks careful patient selection is needed when considering ERCP. With the advent of MRCP imaging, the use of ERCP as a purely diagnostic procedure has waned. However, ERCP is a vital part of the therapeutic armamentarium for patients with pancreaticobiliary diseases. A number of tools (catheters, balloons, stents, etc.) can be passed into the duct to allow therapeutic interventions such as removing stones, dilating strictures, and placing stents for drainage. In the setting of chronic abdominal pain, ERCP aids in the evaluation and management of sphincter of Oddi dysfunction and chronic pancreatitis.

The sphincter of Oddi is a smooth muscle sphincter which surrounds the opening of the bile and pancreatic ducts at their entry into the duodenum. It consists of three components (biliary, pancreatic, and common). Impaired drainage through the sphincter due to spasm or stenosis is termed sphincter of Oddi dysfunction (SOD). Clinically this can present as either recurrent biliary type pain or recurrent pancreatitis. Patients with biliary sphincter of Oddi dysfunction typically present with episodic epigastric or right upper quadrant pain that may radiate to the right shoulder blade following cholecystectomy. The reason biliary SOD is mostly recognized in patients who have undergone cholecystectomy may be related to the removal of the gallbladder that was functioning as a reservoir to accommodate increased pressure in the biliary tree. The gold standard for diagnosing SOD is ERCP with manometry. Manometry involves placing a thin catheter in the biliary or pancreatic sphincter and directly measuring the pressure. The first report of ERCP in the management of SOD was published in 1989 [20]. In a double-blinded study 47 patients with suspected SOD were randomized to endoscopic sphincterotomy or sham sphincterotomy. Manometry identified 23 patients with increased sphincter pressure who were eligible for randomization. Sphincterotomy improved pain scores in 10 of 11 patients with elevated sphincter pressure. In patients who received the sham procedure only 3 of the 12 patients showed

Table 3.3 Modified Milwaukee classification

SOD Type	Typical biliary pain	Abnormal liver enzymes[a]	Dilated bile duct greater than 8 mm	Response rate[b] to sphincterotomy (%)
1	+	+	+	65–95
2	+	Either abnormal liver enzymes or dilated bile duct		50–63
3	+	–	–	12–28

[a]Abnormal liver enzymes—AST, ALT, or AP > 2 times normal values documented on two or more occasions
[b]Reviewed in Corazziari E. Sphincter of Oddi Dysfunction. Dig Liver Dis 2003;35 Supple 3:S26–9
SOD Sphincter of Oddi Dysfunction

improvement in pain scores. Patients with suspected biliary SOD are classified according to the revised Milwaukee classification as this helps predict outcomes after biliary sphincterotomy [21, 22] (Table 3.3).

In SOD patients, ERCP has a high rate of pancreatitis occurring up to 25 % of time [23–25]. Several noninvasive tests have been studied in SOD, but they have not gained widespread use due to poor sensitivity and specificity. The mainstay of treatment is endoscopic sphincterotomy. Prior to referring a patient for ERCP and sphincter of Oddi manometry for SOD, it is important to adequately exclude other causes of pain. A careful history and review of lab and imaging tests helps to verify that the pain is truly biliary or pancreatic in nature, and that the clinical features support the diagnosis (e.g., elevated liver function tests, dilated ducts). Finally patients must be carefully counseled as to their chances of benefit based on the Milwaukee classification, and the significant risk of pancreatitis which may rarely be life-threatening.

Abdominal pain is a common and debilitating symptom of chronic pancreatitis. The pathogenesis of pain in CP is multifactorial. However, a major component of the pain may relate to increased intraductal pressure as a result of pancreatic strictures and/or calculi [26]. ERCP has been frequently used to assess and treat ductal pathology arising in chronic pancreatitis. Main pancreatic duct strictures are treated with dilation and placement of stents. Management of obstructing pancreatic ductal calculi may entail extracorpeal shock wave lithotripsy to fragment the stone followed by endoscopic removal of fragmented stones. There is an increased risk of pancreatic cancer in CP and should be excluded in any patient with a ductal stricture.

Endoscopic Ultrasound

Ultrasound imaging has distinct advantages including detailed soft tissue imaging and the ability to provide real-time guidance for tissue sampling. However, US is limited by its inability to image beyond air filled or extremely dense structures (e.g., calcifications). Transabdominal ultrasound does not usually adequately image all of the intra-abdominal structures when there is a large amount of intervening fat or air artifact. Endoscopic ultrasound (EUS) overcomes these limitations by endoscopically placing the ultrasound probe in the stomach and duodenum next to the organs of interest. For example, an ultrasound probe in contact with the duodenal wall will be within 5 mm of the intrapancreatic portion of the common bile duct [27]. EUS is a minimally invasive procedure with a low risk profile similar to upper endoscopy [28]. While EUS has many diagnostic and therapeutic indications, its main role in chronic abdominal pain is evaluation of biliary disorders and chronic pancreatitis.

Chronic pancreatitis (CP) is an inflammatory disease of the pancreas that results in fibrosis and scarring of the pancreas. These changes result in both pancreatic exocrine and endocrine insufficiency. While severe chronic pancreatitis can be seen on radiologic imaging, early CP is harder to detect. This was demonstrated by Walsh in a report titled "Minimal Change Chronic Pancreatitis" [29]. Walsh described 16 patients with typical pancreatic pain but with negative or equivocal imaging work-up. The patients eventually underwent pancreatic resection due to the strong suspicion of CP. Histologic specimens demonstrated subtle, but definite histologic changes of chronic pancreatitis in 15 of the 16 patients. EUS strength lies in being able to detect mild parenchymal and ductal abnormalities not seen on CT scan making it a good test in evaluating patients with typical pancreatic pain, but with non-diagnostic imaging.

The normal endosonographic appearance of the pancreas include homogenous and granular echotexture ("salt and pepper"), smooth gland borders, and a smoothly tapering pancreatic duct. The main pancreatic duct wall and side branches are hard to visualize. The upper limit of normal of the pancreatic duct is 3.6 mm in the head, 3.0 mm in the body, and 2.0 mm in the tail [30]. The EUS changes seen in CP include ductal and parenchymal changes. The parenchymal changes include hyperechoic foci and stranding, lobularity, cysts, and calcifications [31, 32]. The ductal changes include a dilated and irregular main pancreatic duct with hyperechoic walls, ductal calculi, and dilated side branches [31, 32]. Not all features are needed for the diagnosis of CP, and some features may be seen in patients without CP. To help organize the EUS changes of chronic pancreatitis scoring systems have been developed. The traditional EUS scoring system is an unweighted scoring system that has been shown to help diagnose CP [33]. A limitation of this scoring system was that each criterion was weighted equally, though some criteria have more diagnostic importance. To address these issues the consensus-based Rosemont classification was formed. The Rosemont classification differentiates EUS findings into minor and major categories and established four diagnostic categories (Normal, indeterminate for chronic pancreatitis, suggestive of chronic pancreatitis, consistent with chronic pancreatitis) [34]. The Rosemont criteria look promising but require further validation in multicenter studies. Limitations of EUS include inter- and intra-observer variability, operator dependence, and an incomplete understanding of its true accuracy. Despite these limitations EUS is the best test for evaluating early chronic pancreatitis.

Pancreatic Function Test in Chronic Pancreatitis

Pancreatic function tests (PFT) assess pancreatic exocrine function. Exocrine function may decline in the initial stages of chronic pancreatitis, making PFT a good test for early CP. PFT involves administration of either secretin or CCK which stimulate the exocrine pancreas, followed by collection and analysis of pancreatic secretion.

Historically PFTs were performed by using double lumen gastroduodenal collection tubes, which are cumbersome and time-consuming to use limiting their clinical application. Recently, endoscopic PFTs have been shown to be useful with the direct collection of fluid aspirated through endoscopes suction channel.

By performing EUS at the same time of endoscopic PFT both the structure and function of the pancreas can be assessed. A recent retrospective study comparing EUS and endoscopic PFT to histology suggested combined EUS and PFT may increase sensitivity in detecting CP [35].

Conclusion

Chronic abdominal pain may be a challenging diagnosis, as the concern of missing a serious medical condition needs to be balanced with the expense of excessive diagnostic testing in a patient with a functional disorder. Certain testing can also have significant risk of harm as in ERCP for sphincter of Oddi dysfunction and must be considered cautiously. A careful history and physical examination helps guide effective diagnostic testing and limits unnecessary testing. While it is important to confidently exclude organic disease prior to diagnosing a functional disorder, it is also important to prevent the "endless" loop of diagnostic testing many of these patients undergo. Knowing when to stop diagnostic testing is an important and challenging part of managing chronic abdominal pain.

References

1. Mitchell CM, Drossman DA. Survey of the AGA membership relating to patients with functional gastrointestinal disorders. Gastroenterology. 1987;92(5 Pt 1):1282–4.
2. Livneh A, Langevitz P, Zemer D, Zaks N, Kees S, Lidar T, et al. Criteria for the diagnosis of familial Mediterranean fever. Arthritis Rheum. 1997;40(10):1879–85.
3. Drossman DA, Talley NJ, Leserman J, Olden KW, Barreiro MA. Sexual and physical abuse and gastrointestinal illness. Review and recommendations. Ann Intern Med. 1995;123(10):782–94.
4. Gray DW, Dixon JM, Collin J. The closed eyes sign: an aid to diagnosing non-specific abdominal pain. BMJ. 1988;297(6652):837.
5. Drossman DA. Functional abdominal pain syndrome. Clin Gastroenterol Hepatol. 2004;2(5):353–65.
6. Hershfield NB. The abdominal wall. A frequently overlooked source of abdominal pain. J Clin Gastroenterol. 1992;14(3):199–202.
7. Thomson H, Francis DM. Abdominal-wall tenderness: a useful sign in the acute abdomen. Lancet. 1977;2(8047):1053–4.
8. Einstein DM, Lapin SA, Ralls PW, et al. The insensitivity of sonography in the detection of choledocholithiasis. AJR Am J Roentgenol. 1984;142:725–8.
9. Maple JT, Ben-Menachem T, Anderson MA. The role of endoscopy in the evaluation of suspected choledocholithiasis. Gastrointest Endosc. 2010;71(1):1–9.
10. Lapis JL, Orlando RC, Mittelstaedt CA. Ultrasonography in the diagnosis of obstructive jaundice. Ann Intern Med. 1978;89(1):61–3.
11. Parulekar SG. Ultrasound evaluation of common bile duct size. Radiology. 1979;133(3 Pt 1):703–7.
12. Baron RL, Stanley RJ, Lee JK, Koehler RE, Melson GL, Balfe DM, et al. A prospective comparison of the evaluation of biliary obstruction using computed tomography and ultrasonography. Radiology. 1982;145(1):91–8.
13. Paulsen SR, Huprich JE, Fletcher JG, Booya F, Young BM, Fidler JL, et al. CT enterography as a diagnostic tool in evaluating small bowel disorders: review of clinical experience with over 700 cases. Radiographics. 2006;26(3):641–57. Discussion 657–62.
14. Mehta SN, Reinhold C, Barkun AN. Magnetic resonance cholangiopancreatography. Gastrointest Endosc Clin N Am. 1997;7(2):247–70.
15. Romagnuolo J, Bardou M, Rahme E, Joseph L, Reinhold C, Barkun AN. Magnetic resonance cholangiopancreatography: a meta-analysis of test performance in suspected biliary disease. Ann Intern Med. 2003;139(7):547–57.
16. Verma D, Kapadia A, Eisen GM, Adler DG. EUS vs MRCP for detection of choledocholithiasis. Gastrointest Endosc. 2006;64(2):248–54.
17. Czako L, Endes J, Takacs T, Boda K, Lonovics J. Evaluation of pancreatic exocrine function by secretin-enhanced magnetic resonance cholangiopancreatography. Pancreas. 2001;23(3):323–8.
18. Pungpapong S, Wallace MB, Woodward TA, Noh KW, Raimondo M. Accuracy of endoscopic ultrasonography and magnetic resonance cholangiopancreatography for the diagnosis of chronic pancreatitis: a prospective comparison study. J Clin Gastroenterol. 2007;41(1):88–93.
19. Talley NJ, Vakil N, Practice Parameters Committee of the American College of Gastroenterology. Guidelines for the management of dyspepsia. Am J Gastroenterol. 2005;100(10):2324–37.
20. Geenen JE, Hogan WJ, Dodds WJ, Toouli J, Venu RP. The efficacy of endoscopic sphincterotomy after cholecystectomy in patients with sphincter-of-Oddi dysfunction. N Engl J Med. 1989;320(2):82–7.
21. Hogan WJ, Geenen JE. Biliary dyskinesia. Endoscopy. 1988;20 Suppl 1:179–83.

22. Behar J, Corazziari E, Guelrud M, Hogan W, Sherman S, Toouli J. Functional gallbladder and sphincter of oddi disorders. Gastroenterology. 2006;130(5):1498–509.
23. Freeman ML. Role of pancreatic stents in prevention of post-ERCP pancreatitis. JOP. 2004;5(5):322–7.
24. Singh P, Gurudu SR, Davidoff S, et al. Sphincter of Oddi manometry does not predispose to post-ERCP acute pancreatitis. Gastrointest Endosc. 2004;59:499–505.
25. Guda NM, Freeman ML. True culprit or guilt by association? Is sphincter of Oddi manometry the cause of post-ERCP pancreatitis in patients with suspected sphincter of Oddi dysfunction, or is it the patients' susceptibility? Rev Gastroenterol Disord. 2004;4(4): 211–3.
26. Di Sebastiano P, di Mola FF, Buchler MW, Friess H. Pathogenesis of pain in chronic pancreatitis. Dig Dis. 2004;22(3):267–72.
27. Shami VM, Kahaleh M. Clinical gastroenterology: endoscopic ultrasound. Springer Science LLC: New York; 2010.
28. Rathod V, Maydeo A. How safe in endoscope ultrasound? A retrospective analysis of complications encountered during diagnosis and interventional Endosonography in large individual series of 3006 patients from India [abstract]. Gastrointest Endosc. 2002;56(Suppl):S144.
29. Walsh TN, Rode J, Theis BA, Russell RC. Minimal change chronic pancreatitis. Gut. 1992;33(11):1566–71.
30. Wiersema MJ, Hawes RH, Lehman GA, Kochman ML, Sherman S, Kopecky KK. Prospective evaluation of endoscopic ultrasonography and endoscopic retrograde cholangiopancreatography in patients with chronic abdominal pain of suspected pancreatic origin. Endoscopy. 1993;25(9):555–64.
31. Dancygier H, Classen M. Endosonographic diagnosis of benign pancreatic and biliary lesions. Scand J Gastroenterol Suppl. 1986;123:119–22.
32. Lees WR. Endoscopic ultrasonography of chronic pancreatitis and pancreatic pseudocysts. Scand J Gastroenterol Suppl. 1986;123: 123–9.
33. Sahai AV, Zimmerman M, Aabakken L, Tarnasky PR, Cunningham JT, van Velse A, et al. Prospective assessment of the ability of endoscopic ultrasound to diagnose, exclude, or establish the severity of chronic pancreatitis found by endoscopic retrograde cholangiopancreatography. Gastrointest Endosc. 1998;48(1):18–25.
34. Catalano MF, Sahai A, Levy M, Romagnuolo J, Wiersema M, Brugge W, et al. EUS-based criteria for the diagnosis of chronic pancreatitis: the Rosemont classification. Gastrointest Endosc. 2009;69(7):1251–61.
35. Albashir S, Bronner MP, Parsi MA, Walsh RM, Stevens T. Endoscopic ultrasound, secretin endoscopic pancreatic function test, and histology: correlation in chronic pancreatitis. Am J Gastroenterol. 2010;105(11):2498–503.

Establishing Diagnosis of Chronic Abdominal Pain: Pain Medicine view

4

I. Elias Veizi, Marisa Wynne, and Salim M. Hayek

Introduction

Abdominal pain is a common patient complaint with significant social and healthcare system burden. It can manifest in acute, recurrent, and chronic forms, which are due to organic or dysfunctional causes. Abdominal pain originating from viscera may result from: (a) acute or chronic inflammation of the visceral organ (pancreatitis, cholecystitis, inflammatory bowel disease); (b) obstruction of flow and distension of the organ (kidney stones, bile duct stones); (c) ischemia due to atherosclerosis or vasoconstriction (mesenteric ischemia); and (d) organ dysfunction (irritable bowel syndrome; IBS). Pain caused by interstitial cystitis, gastro esophageal reflux, and endometriosis/dysmenorrhea illustrates visceral pain's widespread impact. Furthermore the presence of concurrent painful conditions in more than one organ, abdominal wall pain and psychogenic factors could complicate the clinical evaluation of visceral pain. The paradigm shift from the notion of considering viscera as insensitive to pain, to examining the role of visceral sensory neurons [visceral nociceptors] has peaked the interest of clinicians and researchers in exploring visceral pain mechanisms. Crosstalk between the visceral nociceptors results in diffused abdominal pain perception which makes the diagnosis and treatment of chronic abdominal pain challenging. Apparent diffuse pain perception of the patients with abdominal pain conditions, due to recently discovered cross-talk between different organ visceral nociceptors, makes it difficult to diagnose and treat. In recent years, nerve bocks have been used with increasing frequency in the diagnosis and therapy of visceral disease. Their effectiveness depends largely on proper application. Appropriate use of nerve blocks requires understanding of the clinical complexity of chronic abdominal pain (CAP), neurophysiology, neuroanatomy, and segmental supply of viscera, good technical skills, and understanding of the limitations of nerve blocks as diagnostic tools. This chapter will provide a review of various concepts on the most common etiologies of chronic abdominal pain as well as techniques employed from pain physicians in improving diagnosis and prognosis of CAP.

Clinical Characteristics of Abdominal Pain

Chronic pain is recognized as a multidimensional experience including several coexisting components. The sensory-discriminative component represents the ability to localize a stimulus in space and time and assess its intensity. The affective-motivational component refers to the experience of the unpleasant and emotional aspects of the pain. Finally, the cognitive-evaluative component consists of evaluation and interpretation of the meaning of the pain experience. CAP due to true visceral pain can have an insidious course, and is characterized by a vague and poorly defined sensation. Regardless of the organ origin, it is often perceived as a dull, aching sensation and can be associated with autonomic phenomena such as nausea, vomiting, sweating, and GI disturbances as well as affective responses such as anxiety and catastrophizing (Table 4.1.). The most common hallmark is "referred" pain at sites of the abdominal wall whose innervation is at the same [or adjacent] spinal cord level from the involved organ. Convergence of somatic and visceral nociceptive pathways in the same second order neuron leads to brain "misinterpretation" of pain perception from the referred site. Referred pain can often be characterized by deep muscle hyperalgesia in the affected region; however it is better localized and similar in quality to somatic pain. Clinical impression of CAP is sometimes more confusing due to visceral cross-organ

I.E. Veizi, M.D., Ph.D.
Division of Pain Medicine, Department of Anesthesiology, University Hospitals Case Medical Center, 11100 Euclid Avenue, Cleveland, OH 44106, USA

M. Wynne, D.O. • S.M. Hayek, M.D., Ph.D. (✉)
Division of Pain Medicine, University Hospitals Case Medical Center, 11100 Euclid Avenue, Cleveland, OH 44106, USA
e-mail: Marisa.Wynne@uhhospitals.org; Salim.Hayek@uhhospitals.org

L. Kapural (ed.), *Chronic Abdominal Pain: An Evidence-Based, Comprehensive Guide to Clinical Management*, DOI 10.1007/978-1-4939-1992-5_4,

Table 4.1 Clinical and neurophysiological characteristics of visceral and somatic pain

Component	Visceral	Somatic
Nerve fiber	Ad, C	Ab [allodynia], Ad, C
End organ silent nociceptors	Yes (~80 %)	Yes (~20 %)
DH laminar synapse	I, V, X	I-to-V and X
Trigger/insult	Distension Ischemia	Direct trauma Ischemia
Character/location	Dull, poorly localized	Sharp, precisely localized,
Relation to stimulus	Referred site	Radiation along the nerve
Associated symptoms	Nausea, vomiting, sick feeling	Tenderness, soreness
Stimulus dependent summation	Intensity dependent summation	No
Sensitization	Mechanical	No
1st Hyperalgesia	Yes	Yes
2nd Hyperalgesia	Yes [at referred site]	Yes
Sensitization	Mechanical-common	Thermal-common Mechanical-uncommon

convergence phenomenon. Viscero-visceral hyperalgesia is most likely produced by sensitization of neurons in the CNS. Augmentation of pain is due to sensory interaction between two different internal organs that share afferent circuitry or converge at the same second order neuron in the spinal cord. For example patients with pancreatitis have more upper GI symptoms whereas patients with interstitial cystitis (IC)/Painful Bladder Syndrome (PBS) experience more genital pain, or pain with defecation due to lumbar hyperalgesia [T10-L1 convergence]. This phenomenon is crucial in treatment decisions since effective treatment of one cause may improve symptoms from a different organ system [e.g., treatment of IC/PBS improves pain with constipation].

In summary, characteristics of visceral pain include:

- **Referred** to body wall
- **Diffuse & poorly localized**
- Accompanied by increased **motor & autonomic reflexes**
- Severity of pain **doesn't always reflect disease severity** (e.g., mild/no pain in colon cancer, severe pain when passing a stool in IBS)
- **Not evoked from all organs** (solid < hollow)
- **Not always linked to injury [functional disorders]**

Neurophysiology and Neuroanatomy of Abdominal Pain

Most of the knowledge on chronic pain has derived from research on somatic nociceptors. As a result, the unique visceral sensory components are typically less understood which in turn results in less efficacy of the treatment of visceral pain in comparison to that of somatic pain.

Visceral Neurophysiology

Visceral afferent nerves differ functionally from non-visceral afferents. Visceral nociceptors (Table 4.1) are:

- Mechanically insensitive
- *Not* sensitive to physiological/supraphysiological stimuli
- Recruited *only* in pathological situations
- Sensitive to inflammatory mediators
- Activated by distension, pressure, and ischemia

The principal visceral afferents are anatomically associated with sympathetic and parasympathetic fibers [splanchnic afferents, pelvic nerve afferents]. There are two types of peripheral afferent nerve endings: (a) low threshold; and (b) the largest population—high threshold [reserve, inactive, nonfunctional usually]. Low threshold peripheral organ mechanoreceptors [also called "wide dynamic range"] encode organ distension from low physiologic pressures to high noxious pressures. They are insensitive to touch, cut and generally any form of mechanosensation. The high threshold endings become active upon development of end organ pathological processes such as local inflammation and stimulation by chemokines and interleukins. This capacity for neuroplasticity is the basis of peripheral end organ sensitization, which results in central sensitization and expansion of receptive fields. Furthermore, the ability to localize the source of pain (spatial resolution) in visceral pain is poor due to two factors: (1) Viscera are relatively sparsely innervated in comparison to somatic structures (nearly 1:10 ration in comparison to somatic structures); and (2) Secondary spinal neurons receive convergent input from both viscera and skin. Pain pathways involved in visceral nociception are maintained by both efferent mechanisms [sympathetically maintained] and afferent visceral sensory mechanisms. A study by AK Houghton and colleagues concluded that dorsal column pathways play an important role in the processing and relaying of pancreatic visceral nociceptive information [1].

Morphological Consideration

Neuronal cell types are largely defined by size. Large "light" neurons have A fibers and small "dark" neurons have C-fibers. Fiber size is directly related to the degree of fiber myelination, which is, in turn, directly related to conduction velocity. Thus the large, myelinated A fibers have the fastest conduction velocity whereas the small, unmyelinated C fibers have the slowest conduction velocity [2, 3]. Comparing nociceptive to non-nociceptive afferent neurons [4], nociceptive neurons were found to have longer action potential duration and slower maximum firing rate. These properties appeared to be graded according to the conduction velocity with the slowest (C) fibers having the longest action potential and the least rate of fiber firing (C>A-δ>A-β/α). Up to 80 % of visceral dorsal root ganglia cell somata are C-fibers, and the rest Aδ fibers. Aβ fibers are rarely encountered on viscera. In comparison, only 7–17 % of C-fibers are found at the L4 DRG part of the sciatic nerve. Spinal visceral afferents terminate in the superficial dorsal horn, lamina V and around the central canal (lamina X). Somatic cutaneous afferent neurons terminate in the superficial lamina of the dorsal horn and lamina V which explains the convergence of visceral and somatic stimuli at the level of second order neurons.

Neuroanatomy of Abdominal Visceral Sensory System

As discussed earlier, abdominal viscera receive dual innervation from the afferent and efferent system.

Sympathetic Nerves

Preganglionic sympathetic fibers, which control the function of the abdominal viscera, have their cell bodies in the spinal cord segments T5-to-L2. Their axons pass via the white rami communicates and through the sympathetic trunk without synapsing to become the splanchnic nerves. The splanchnic nerves end in the prevertebral ganglia where they synapse with postganglionic neurons. The postganglionic neurons leave the celiac and related ganglia, and together with the sensory and vagal fibers, reach the end organs. Each organ is supplied by sympathetic nerves, which originate in specific spinal cord segments.

Parasympathetic Nerves

The upper abdominal viscera are supplied by parasympathetic fibers from the cell bodies in the dorsal motor nucleus of the vagus, which pass via the prevertebral visceral plexuses to end in the terminal ganglia in the walls of the viscera. Parasympathetic fibers do not transmit pain sensations.

Sensory Nerves

Sensory nerves supplying the viscera accompany both the sympathetic and parasympathetic fibers. Sensory fibers, which are components of the splanchnic nerves, enter the spinal cord via the posterior roots of the T5 to L2 nerves. They serve as the primary pain pathways from the abdominal viscera.

Diagnostic and Prognostic Nerve Blocks in Chronic Abdominal Pain

According to the International Association for the Study of Pain (IASP) taxonomy, CAP can be classified into:

I. Abdominal wall pain
II. Abdominal pain of visceral origin
III. Abdominal pain syndromes of generalized diseases
IV. Pain of psychological origin
V. Chronic pelvic pain syndromes
VI. Diseases of the bladder, uterus, ovaries, testes, and prostate, and their adnexa
VII. Pain Perceived in the rectum, perineum, and external genitalia of nociceptive or neuropathic cause

Given the difficulty introduced by neurophysiological, neuroanatomical, and behavioral factors, and in order to increase the diagnostic accuracy as well as improve the prognosis of any subsequent invasive treatments, a pain physician may employ nerve blocks (Fig. 4.1.). Many techniques derived from regional anesthesia have consistently stood the test of time and preserved their clinical utility as diagnostic tools. However, these techniques are not to be used as sole approaches to arrive at clinical diagnoses or clinical decision-making. Rather, the information obtained from nerve blocks should be used in conjunction with the patient's history, physical examination and labs, imaging and GI evaluation to guide future management interventions (Fig. 4.2.). On their own and without taking into account the whole clinical picture, nerve blocks may be of limited value in chronic abdominal pain. In addition, choosing the appropriate nerve block in the proper patient is critical for the usefulness of these interventions.

In general, nerve blocks may be of therapeutic, diagnostic, or prognostic value. Often, initial nerve block interventions in CAP are done for diagnostic or prognostic purposes. These can utilize an anatomic or pharmacologic approach to delineate likely processes involved in CAP maintenance. Two types of nerve blocks have a significant role in the differential diagnosis and prognosis of CAP:

(a) Neuraxial diagnostic blocks;
(b) Peripheral nerve blocks: splanchnic nerve block (SNB), celiac nerve block, superior hypogastric nerve block, ganglion impar blocks, paravertebral nerve blocks, and intercostal nerve blocks.
Transversus Abdominis Plane (TAP) blocks and blocks of the ilioinguinal and genitofemoral nerves may be performed for diagnostic or therapeutic reasons in postsurgical pain or abdominal nerve entrapment situations. They are discussed elsewhere.

Neuraxial Diagnostic Differential Neural Blockade

Nearly 80 years ago Gaser & Erlanger demonstrated that less cocaine is needed to stop conduction in thin myelinated A nerve fibers than in thicker ones. Since then, decades of research have documented differential local anesthetic susceptibility of A-, C-, and B fibers. There are two approaches to achieving this differential block of nerve fibers: an anatomic approach and a pharmacologic approach. The anatomic approach is based upon the notion that there is enough anatomic separation of the nerve fibers (afferent and efferent) to allow selective blockade of particular painful abdominal

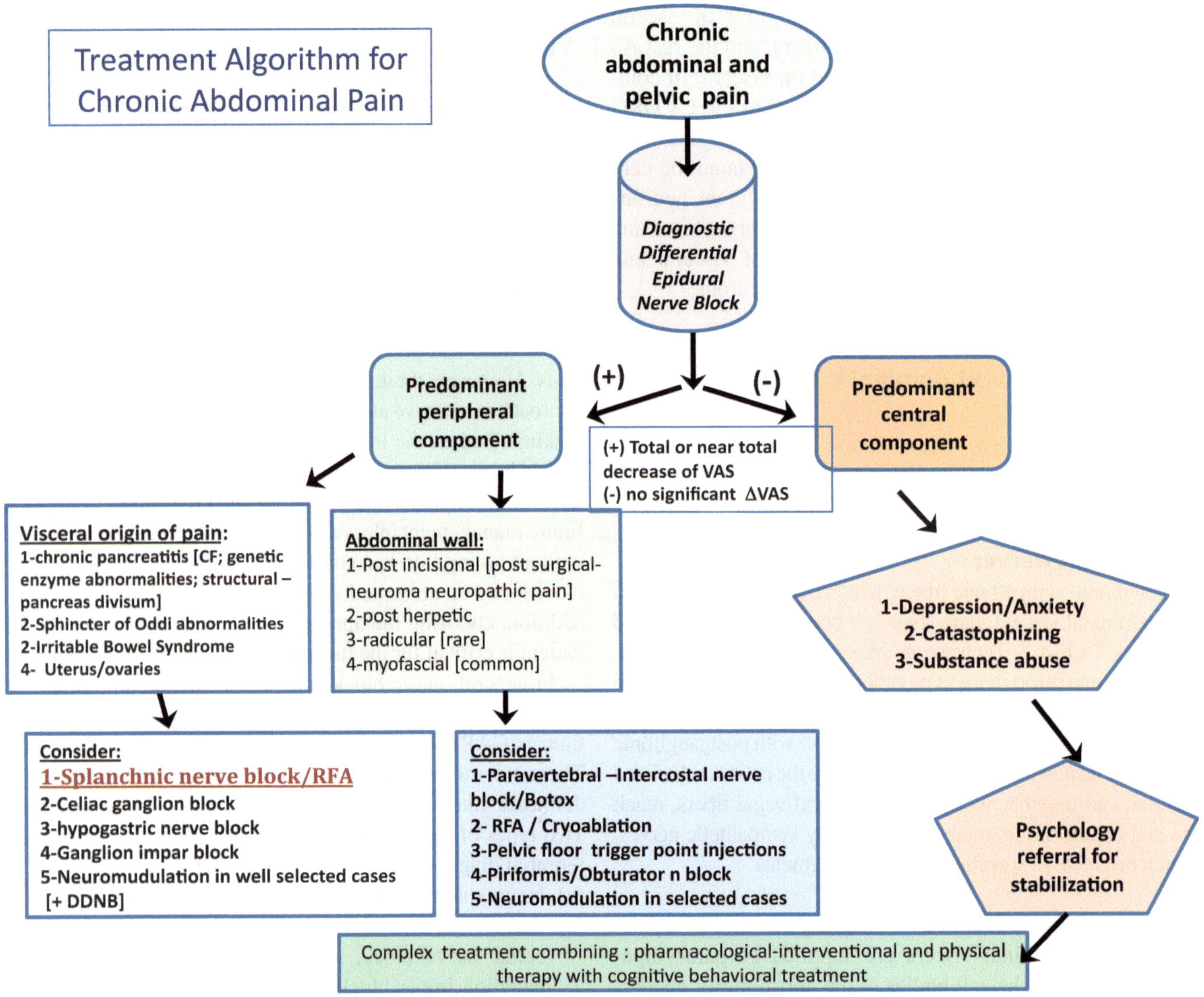

Fig. 4.1 Common diagnostic nerve blocks. Neuraxial and peripheral nerve block sites of action. Neuroanatomical and neurophysiological specificity is hypothesized to make diagnostic blocks valuable tools in aiding pain physicians in the diagnosis and treatment of visceral abdominal pain patients

structures. The pharmacologic approach takes advantage of the fact that sympathetic and somatic nerve fibers exhibit different sensitivities to local anesthetics, thus the injection of various concentrations of local anesthetics would allow for a differential blockade of the fibers [5]. By observing the analgesic and anesthetic responses to injections of normal saline (placebo) and different amounts of local anesthetics, pain may be distinguished as visceral or non-visceral in origin.

Diagnostic Differential Nerve Block

While differential neural blockade has been observed using both subarachnoid and epidural anesthesia, the epidural technique was found to be more advantageous and controllable in the differential diagnosis of CAP. ***Epidural diagnostic differential nerve block (DDNB) is the most useful tool in the armamentarium of the Pain Medicine physician for diagnostic purposes in CAP.***

Purpose: The goal of DDNB is to delineate underlying pain mechanisms involved in maintenance of the chronic pain state [6]. Epidural differential nerve blockade is performed with injections of placebo, and differing amounts of local anesthetics through an epidural catheter to achieve surgical anesthesia in the afferent neuronal distribution that overlaps the patient's site of pain

The Role of Local Anesthetics

The duration of nerve conduction block is dependent upon the protein binding affinity of the local anesthetic while lipid solubility determines potency of the agent. Thus, lipophilic

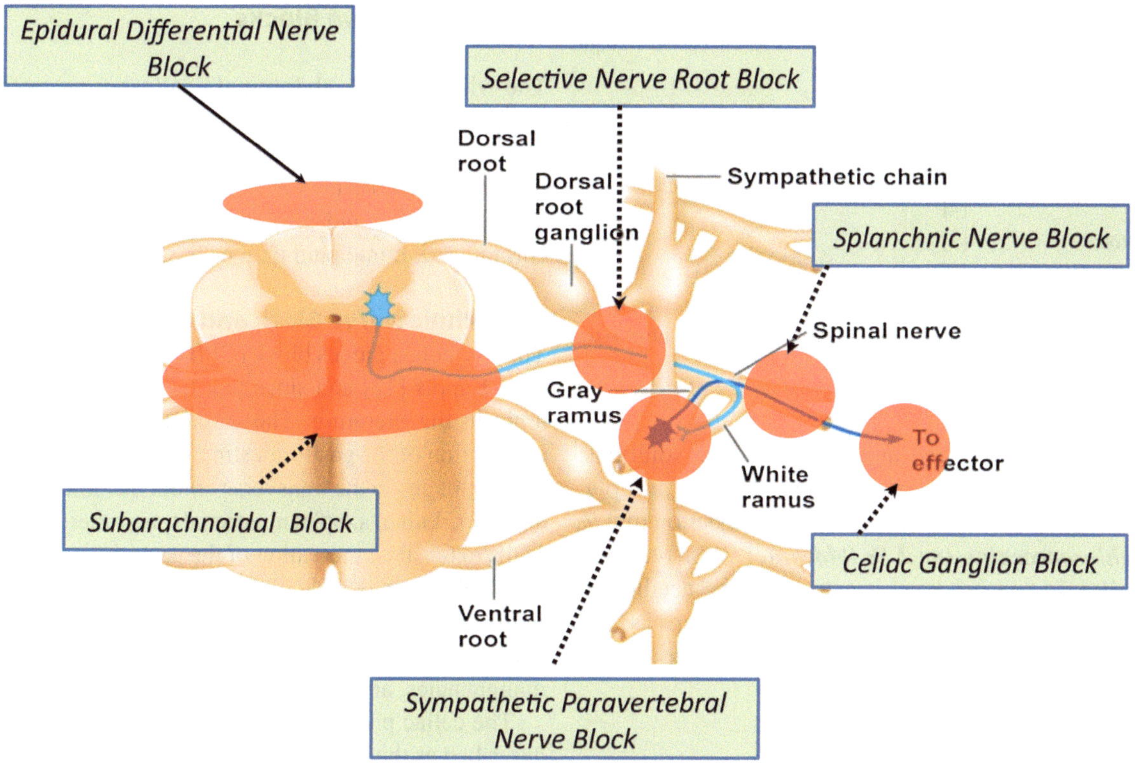

Fig. 4.2 Descriptive algorithm for diagnosis and treatment of chronic abdominal pain

amide-based local anesthetics with high protein binding properties, such as bupivacaine, may produce longer neuronal blockade and at lower concentrations than their less lipophilic ester-based counterparts such as 2-chloroprocaine. An amide local anesthetic with lower lipophilicity and less protein binding affinity than bupivacaine, such as lidocaine, may also be useful. Because of these reasons, 2 % lidocaine or 2-chloroprocaine are often used for epidural diagnostic differential neural blockade. In vivo studies have shown that a differential susceptibility exists based upon fiber size regardless of the type of local anesthetic used [7]. A-α fibers have been shown to be less sensitive to local anesthetics, regardless of type, in comparison with the smaller C or A-δ fibers. The order of susceptibility to local anesthetic blockade from most to least susceptible is: B<C<Aδ<Aγ<Aβ<Aα [5, 8–10].

Differential Neural Blockade: Epidural Approach

The classic spinal approach involved placing the patient in the lateral decubitus position and administering saline and various concentrations of local anesthetics through a spinal needle. Given drawbacks to this conventional subarachnoid differential technique, the epidural approach was suggested by Raj in 1977 [5] in an attempt to avoid positional issues and the possibility of a post-dural puncture headache following differential spinal (intrathecal) blockade [11]. Raj's technique was identical to that of the subarachnoid classic approach in that a placebo solution and increasing concentrations of local anesthetic were injected into the epidural space and the patient was observed for the onset of pain relief with the onset of neural blockade of the various fibers. However, because the injections occurred within the epidural space as opposed to the subarachnoid space, the local anesthetic utilized was lidocaine in concentrations considered be the mean concentrations leading to blockade of the various neural fibers within the epidural space.

Raj's epidural technique brought about two problems: (1) the onset of neural blockade is slower in the epidural space, thus Raj's technique is more time-consuming than the classic subarachnoid approach; and (2) local anesthetic injected into the epidural space gives even less discreet endpoints than injection into the spinal space, leading to possible misinterpretation of results. The two disadvantages of Raj's technique can be circumvented in the same way the modified spinal approach circumvented the same drawbacks to the classic spinal approach as described below.

Modified Epidural Procedure

The patient is placed in the supine, lateral, or sitting position and the back is prepped and draped in the usual sterile fashion. An epidural needle is utilized to gain access to the epidural space utilizing a loss of resistance technique. At this time, an epidural catheter is placed

Table 4.2 Transient Differential: Time Sequence of functional impairment at onset of Differential epidural block and expected change in VAS score

Effect observed	Saline solution	Chloroprocaine 3 %
Vasomotor	[–]	+/–sympathetolytic
Warmth and pinprick	[–]	
Cold	[–]	
Touch/Motor	[–]	
Change in pain intensity [DVAS]	[–]	[+] or [–]

*Differential epidural block related to agent diffusion and fiber size/conduction velocity

for the administration of the inactive and active solutions. Preservative free normal saline is injected initially as a placebo and is followed 10–15 min later by the injection of ***2 % or 3 % chloroprocaine (or 2 % lidocaine)*** in incremental doses until a T4 sensory level is achieved. The needle/and catheter are removed. Prior to injecting the solutions as well as after each injection, the same observations mentioned in Table 4.2 are assessed.

Interpretation

If the patient experiences pain relief after injection of saline, the pain relief is considered to be influenced by suggestion and reflects a place. If the patient does not experience pain relief following the injection of the anesthetic with a resulting appropriate sensory level, the pain is considered to be central in origin. However, when pain relief does occur following injection of the local anesthetic, the pain is considered to be organic in nature. Whether the pain is sympathetic or somatic in origin cannot be delineated until neural function begins to return. If the pain reappears with the return of the perception to pinprick, the pain is deemed to be somatic. If the return of pain occurs well after the return of sensation the pain is consider sympathetically mediated or visceral in origin.

Limitations of Epidural DDNB

- Does not measure precisely the extent of nerve fiber blocked.
- More than one fiber type is simultaneously blocked leading to misinterpretation of the results.
- It is impossible to blind the patient, as such a placebo effect is significant especially if it occurs on repeat differential block procedure.
- Neuroplasticity changes could lead to block of nerve conduction at subanesthetic concentrations altering result interpretation.
- Significant behavioral components of chronic abdominal pain patients could influence patient response.

In conclusion: in spite of its shortcomings, epidural DDNB is as a critical part of the treatment algorithm (Fig. 4.2.) remains a clinically useful tool to aid in the diagnosis of chronic abdominal pain.

Peripheral Nerve Blocks

The visceral and abdominal wall sensory afferent and sympathetic efferent nerve supply can be interrupted at peripheral neural structures such as peripheral nerves and peripheral ganglia/plexuses. Below are the most commonly used diagnostic and prognostic peripheral nerve blocks in chronic abdominal pain patients.

Splanchnic Nerve Block and Celiac Plexus Block

The first celiac plexus block (CPB) was performed in 1914 by Max Kappis [12] and since that time has become a frequent topic in the literature. The celiac plexus is responsible for innervation of the gastrointestinal tract from the distal third of the esophagus to the transverse colon, including the liver, biliary tract, kidneys, spleen, adrenals, and mesentery. Because of the extensive innervation of the gastrointestinal tract by the celiac plexus, blockade of the celiac plexus has been utilized in the diagnosis and treatment of visceral abdominal pain in patients diagnosed with pancreatic cancer, upper abdominal malignancies, and chronic pancreatitis [13, 14].

The celiac plexus is found retroperitoneally and has been described as the largest autonomic plexus. It unites two large celiac ganglia and intimately surrounds the celiac artery as well as the superior mesenteric artery. These celiac ganglia can be found medial to the adrenal glands and anterior to the crura of the diaphragm [15]. There is variation in both the morphology and location of the celiac ganglia; an examination of 20 adult cadavers reported diameters to vary between 0.5 and 4.5 cm, number to vary from 1 to 5, and location to vary from the middle of L2 to the intervertebral disk between T12 and L1 [16]. Recently, Zhang et al. [17] have found 94 % of 65 cadavers to have celiac ganglia at the level of T12 or L1. The celiac plexus is comprised of a network of sympathetic and parasympathetic nerve fibers and receives sympathetic fibers from the **three splanchnic nerves (greater, lesser, and least) deriving from T5-T12 segmental levels.** In addition, the celiac plexus receives parasympathetic fibers from the vagus nerve. The celiac plexus is the site of origination of nerves supplying the liver, pancreas, gallbladder, stomach, spleen, kidneys, intestines, adrenal glands, and blood vessels [18].

There are three approaches [19] to blocking visceral nociception of the upper abdomen which include:

(a) **Retrocrural technique**
(b) **Anterocrural technique**
(c) **Splanchnic nerve block**

Irrespective of the approach, the needle(s) are inserted at the level of the first lumbar vertebra, approximately 5–7 cm from midline. When utilizing the retrocrural technique the needle tip is directed toward the upper third of the body of L1 and when utilizing the anterocrural technique the needle tip is directed toward the lower third of the body of L1. In order to maintain the needle tip in a retrocrural position, when utilizing the retrocrural technique the needle tip is not to be advanced more than 0.5 cm anterior to the anterior border of L1. However, when utilizing the anterocrural technique on the left side, the tip of the needle is advanced through the aorta until blood flow through the needle ceases. Because of this, the anterocrural technique is also known as the transaortic approach. When utilizing the splanchnic nerve block, the needle is directed toward the body of T12, with final position of the needle tip at the junction of the anterior third and lower third of the T12 vertebral body in the lateral fluoroscopic view [18].

In addition to techniques guided by fluoroscopy, CT- and ultrasound-guided techniques have also emerged. These techniques have enabled patients who are unable to assume the prone positions to undergo neurolysis of the celiac plexus from a transabdominal approach. Additionally, CT-guided techniques have enabled the performance of the anterocrural technique without penetration or the aorta, thus decreasing the risk associated with this technique [18].

Complications

Blockade of the celiac plexus is associated with several potential complications, which include accidental epidural or dural puncture [19], orthostatic hypotension due to sympathetic chain neurolysis, backache due to local trauma during needle placement, retroperitoneal hemorrhage secondary to vascular injury from needle damage, diarrhea as a result of sympathetic blockade of the bowel [18] chylothorax [20], pneumothorax, renal trauma [21], and abdominal aortic dissection all from direct needle damage [18]. The complications associated with blockade seem to be related to which technique is utilized. Ischia and colleagues [22] compared the efficacy and complications of three techniques (retrocrural, transcrural, and transaortic) in 61 patients with pancreatic cancer. It was found that orthostatic hypotension occurred most frequently when the retrocrural or splanchnic techniques were utilized with an incidence of 50 % and 52 %, respectively. However, the anterocrural technique yielded a 10 % incidence of hypotension. Alternatively, diarrhea occurred more often following the anterocrural technique with an incidence of 65 % as compared to the splanchnic (5 %) and retrocrural (25 %) techniques. When assessing the incidence of dysesthesia, hematuria, hiccups, interscapular back pain, and reactive pleurisy, no difference was found between the three techniques.

Superior Hypogastric Plexus Block

Like the ganglion Impar, the superior hypogastric plexus is a retroperitoneal structure. It is a bilateral structure and is located at the lower third of the fifth lumbar vertebra and the upper third of the first sacral vertebra at the sacral promontory. Its location is in proximity to the bifurcation of the common iliac vessels. The superior hypogastric plexus, by way of the hypogastric nerves, innervates the viscera of the pelvis.

The patient is placed in the prone position with a pillow under the pelvis to flatten the lumbar lordosis. The lumbosacral spine is prepped and draped in the usual sterile fashion. Under AP fluoroscopic guidance the L4-5 interspace is identified. Bilaterally, 5–7 cm from midline, the skin and subcutaneous tissue is anesthetized with 0.5 % lidocaine at the level of the L4-5 interspace. Following local anesthesia, two 22-gauge 7 in. needles are inserted into each of the anesthetized areas with the bevel directed medially. The needles are then directed 45° medial and 30° caudal, toward the anterolateral aspect of the L5 vertebral body. The L5 transverse process as well as the iliac crest can, at times, serve as barriers to the passage of the needle, necessitating slight adjustments to the needle entry site or needle direction. Fluoroscopy is utilized intermittently during advancement of the needle. The needle is advanced until bony contact with the L5 vertebral body is made or until the tip is noted to be located at the anterolateral aspect of the L5 vertebral body on fluoroscopic imaging. If bony contact is made with the L5 vertebral body, the needle tip is walked off and advanced 1 cm anterior to the vertebral body. At this time a loss of resistance can be encountered, indicating the needle tip has been advanced through the psoas fascia and into the retroperitoneal space. Anteroposterior (AP) fluoroscopic imaging at this time should indicate the needle tip lies at the L5-S1 vertebral junction and lateral imaging should indicate the needle tip is just anterior to the anterolateral border of the L5 vertebral body. Following negative aspiration, injection of 3–4 ml of radiopaque contrast dye is then carried out under live fluoroscopy and should indicate in the lateral view a smooth posterior border, formed by the psoas fascia. AP imaging should display the dye confined to the midline.

Because the iliac crest and the L5 transverse process, as well as the L5 nerve root can serve as potential barriers to the performance of the classic approach, with the potential for resulting damage in the case of the nerve root, the transdiscal approach was described by Turker et al. in 2005 [23]. With the patient in the lateral position, the L5/S1 interspace is identified with fluoroscopic guidance. After the skin is prepped and draped in the usual sterile fashion, the skin and subcutaneous tissues are locally anesthetized. A 15 cm 20-gauge Chiba needle is then inserted and advanced under fluoroscopic guidance toward the intervertebral disc. The needle should traverse the intrathecal sac under lateral fluoroscopy. The avoidance of nerve injury is confirmed by a lack of paresthesia. The needle is advanced through the intervertebral disc until it penetrates the anterior portion of the disc. Contrast is injected to verify correct needle tip placement in the AP and lateral fluoroscopic views.

Complications

Of the studies and case reports regarding superior hypogastric plexus block and neurolysis reviewed in the literature, all were without complication with the exception of two cases.

One complication was published in a report of 6 cases of CT-guided superior hypogastric plexus block from the posterior approach and 1 case of CT-guided superior hypogastric plexus block from the anterior approach in radiology [24]. It was noted that one of the cases was terminated because of injection of local anesthetic into the peritoneal cavity, however, it was not specified whether this was during the anterior or posterior approach. Additionally, a case was reported of injury to a somatic nerve resulting in a sensory deficit in a severely kyphoscoliotic patient following CT-guided superior hypogastric plexus neurolysis [25]. The studies during which no complications occurred include de Leon Casassola's 1993 report of 26 patients, Plancarte et al.'s 1997 report of 226 patients undergoing both diagnostic as well as neurolytic block utilizing the classic as well as Nabil and Eissa's report of 26 patients undergoing a posteromedial transdiscal approach after failure of the classic approach.

There are several potential complications related to the adjacent anatomy of the superior hypogastric plexus. Potential complications to the block include vascular puncture with resulting intravascular injection and/or hemorrhage and hematoma formation because of the close proximity of the bifurcation of the common iliac vessels. Additionally, intramuscular or intraperitoneal injection could occur as a result of inappropriate estimation of needle depth. Less likely complications include epidural or intrathecal injection, L5 nerve root injury, and renal or ureteral puncture [26].

Not only are there potential complications related to the adjacent anatomy of the superior hypogastric plexus but inherent to the approach undertaken in the performance of the block. A potential complication specific to the transdiscal approach is a small chance of discitis and thus prophylactic antibiotics are recommended any time the integrity of the disc is disrupted [27, 28]. Additionally, penetration of the disc may predispose to subsequent disc degeneration [29]. Complications inherent to the anterior approach include damage to the bladder, common iliac artery, and bowel, with an increased risk of infection related to traversing the bowel [23].

Ganglion Impar

The ganglion Impar is also referred to as the ganglion of Walther or sacrococcygeal ganglion [30]. The structure is located at the sacrococcygeal junction at the anterior aspect of the sacrococcygeal joint (SCJ) and is a solitary retroperitoneal structure. It marks the caudal termination of the paired paravertebral sympathetic chains [30, 31]. The innervation of the ganglion Impar may include the perineum, distal rectum, anal region, distal urethra, vulva/scrotum, and the distal third of the vagina [32].

Block of the ganglion Impar has been utilized for the diagnosis and treatment of visceral or sympathetically maintained pain of the above-mentioned regions [31]. Trans-sacrococcygeal approach, which was described by Wemm and Semerski in 1995, is currently widely used [33]. In this procedure, a straight needle is inserted via the sacrococcygeal junction into the retroperitoneal space. Because of the difficulty with anteroposterior fluoroscopic view of the SCJ when there is associated bowel gas or impacted stool, Lin et al. have described the successful use of ultrasound guidance as an adjuvant to fluoroscopy during this technique [30]. Ultrasound is utilized for the identification of the SCJ and both anteroposterior (AP) and lateral fluoroscopy are utilized, the latter for identification of needle depth and observation of spread of contrast dye. Another modification of the block was introduced in order to overcome difficulty traversing the sacrococcygeal ligament due to calcification or because of difficulty performing the trans-sacrococcygeal approach [34]. During this procedure, under fluoroscopic guidance the needle is passed in a trajectory toward the inferior portion of the transverse process of the coccyx. Once bony contact is made with the transverse process, the needle is walked inferior and is passed along the anterior surface of the coccyx to the level of the SCJ. CT-guided ganglion Impar block has also been described in the radiological literature [35].

During the trans-sacrococcygeal approach the patient is placed in the prone position and location of the SCJ is identified utilizing AP fluoroscopy. The skin and subcutaneous tissue along the needle trajectory to the SCJ are anesthetized with 0.5 % lidocaine. A 22-gauge spinal needle is then advanced under AP followed by lateral fluoroscopic guidance into the retroperitoneal space. Confirmation of needle position is obtained with the injection of 0.5 ml of non-ionic contrast solution, which demonstrates spread along the anterior portion of the coccyx revealing a "comma sign" [36–38].

Complications

Complications appear to be specific to a ganglion Impar block as well as specific to how the ganglion Impar block was performed. A review of current literature reveals that all formal studies of the block indicate that no complications occurring during the trial period, however, authors have reported theoretical complications or those that have in their past practice experienced without discussing particular rates of complications [35, 38, 39]. Complications specific to the curved/bent needle technique include needle breakage, tissue trauma along the plane of needle angulation, rectal perforation leading to needle contamination and increased risk of needle puncture to the operator's finger, vascular damage leading to bleeding, periosteal injection, and a high rate of failure (20–30 %) [36, 38]. The trans-coccygeal approach is more comfortable for patients and theoretically causes less tissue trauma since that a straight needle is utilized, however, the approach may be difficult at times given that the sacrococcygeal disc can ossify later in life [40] which renders it difficult to visualize utilizing fluoroscopy [38]. Potential complications specific to the trans-coccygeal approach related to penetration of the disc include discitis, fistula formation, and bleeding [36, 41]. Munir et al. have described a needle-through-needle technique to overcome this issue [36].

Paravertebral Block

Introduction and Anatomy

Analgesia by way of a paravertebral block (PVB) was first described by Sellheim and Leipzig in 1905 as an alternative to neuraxial anesthesia in patients undergoing cesarean section. Since that time, PVB has been utilized to provide anesthesia for patients undergoing surgical procedures of the thoracic, abdominal, and pelvic regions as well as those suffering from pain secondary to trauma or chronic pain syndromes. The paravertebral space (PVS) exists as a potential space. The boundaries of the PVS are the following: posterior—the superior costotransverse ligament, more laterally the posterior intercostal membrane; anterior—parietal pleura; medial—posterolateral aspect of the vertebra, intervertebral disc, intervertebral foramen; superior—heads and necks of ribs; inferior—limited by the crura of the diaphragm, (although this is not agreed upon in the literature); and lateral—no limit, contiguous with the intercostal space. Several neural structures are located within the wedge-shaped PVS and include the anterior and posterior ramus of the intercostal nerve, sympathetic chain, rami communicantes, and the sinu-vertebral nerve. Anesthesia secondary to PVB occurs from direct penetration of these neural structures by local anesthetic. During the performance of a single PVB, local anesthetic is capable of traveling anterior to the transverse process providing anesthesia to adjacent levels. In addition to this multilevel spread, local anesthetic is capable of spread to the intercostal space, the intervertebral foramen with associated epidural spread, as well as spread to the contralateral PVS. In fact, the incidence of epidural spread in chronic pain patients as found to be as high as 70 %, however, other studies have stated that epidural spread is a rare occurrence.

The patient is placed in the prone position and the appropriate thoracic levels are identified utilizing fluoroscopic guidance. The back is prepped and draped in the usual sterile fashion. A under fluroscopic guidance a 22-gauge spinal needle is inserted 3 cm lateral to the spinous process, toward the most superior aspect of the transverse process. Once bony contact is made with the superior aspect of the transverse process, the needle is withdrawn slightly, redirected cranial, and walked off the transverse process. Once the needle is walked off, it is advanced 1–1.5 cm to enter the PVS. A loss of resistance technique can be utilized (instead of advancing a set distance of 1–1.5 cm), the subtle loss being felt when the tip of the needle traverses the superior costotransverse ligament (the posterior border of the PVS). After negative aspiration, needle tip location is confirmed with the injection of contrast medium. A bolus dose of 2 mg/kg of ropivacaine with 1/200,000 epinephrine has been found to be safe by Karmakar et al. [42].

Complications

Studies regarding the complications secondary to PVB have demonstrated the incidence of such to be low, with a high success rate for the block (unilateral PVB success 94 %). The three main risks associated with PVB include high blood levels of local anesthetic, hypotension, and pneumothorax. A case has also been reported of total spinal anesthesia. In reviewing two studies which investigated cardiovascular parameters in the absence of fluid loading following unilateral PVB, one study demonstrating somatic and sympathetic block of 5 and 8 dermatomes, respectively, the other demonstrating 12 and 6, respectively, neither found significant postural changes in the measured cardiovascular. The risk of accidental pleural puncture during unilateral PVB has been described as 0.8 % with the risk of subsequent development of pneumothorax to be 0.5 % of the total patient population. Given these results PVB can be considered both safe and effective.

Intercostal Nerve Block

Introduction and Anatomy

Intercostal nerve block (INB) and continuous intercostal nerve infusion have been utilized for the diagnosis and treatment of chronic pain syndromes [43, 44]. Anatomically, the anterior divisions of the thoracic spinal nerves are the intercostal nerves. The lower intercostal nerves (7 through 11) continue anteriorly and supply the parietal pleura of the thorax as well as the abdominal wall. The xyphoid process serves as the termination point of the seventh intercostal nerve, the umbilicus as the termination of the tenth intercostal nerve, and the lower abdominal wall and groin as the termination of the twelfth thoracic (also known as the subcostal nerve). In the case of abdominal pain, these landmarks are utilized in determining which intercostal nerve(s) to block for the diagnosis and treatment of chronic pain [45].

With the patient in the prone position, fluoroscopic guidance is utilized to locate the corresponding ribs at a location that is the midpoint between the thoracic spine and the posterior axillary line. The skin and the subcutaneous tissues slightly inferior to the inferior border of the rib are then anesthetized. Following local anesthesia, a 22-gauge 2 in. spinal needle is directed under fluoroscopic guidance cephalad toward the inferior border of the corresponding rib. Once bony contact is made, the needle tip is withdrawn slightly and walked off the inferior border of the rib, ensuring to advance only a few millimeters. A stimulating needle can also be utilized in place of a spinal needle to allow for the use of a peripheral nerve stimulator in confirmation of needle tip location. The needle is then aspirated to confirm an absence of blood and air. After negative aspiration, contrast medium is injected under live fluoroscopy and should indicate spread of the dye along the inferior border of the rib in a medial-to-lateral fashion, as well as a lack of an intravascular uptake or a dispersion of the dye, indicating intrapleural injection. Five milliliter of 0.5 % ropivacaine is then injected to perform the neural blockade. The procedure can be performed simultaneously at multiple levels.

Contraindications and Complications

Complications specific to ICB include intravascular injection, as the intercostal nerve is located in a highly vascular area; and pneumothorax, which leads to a contraindication in the performance of an ICB in the presence of a contralateral pneumothorax. The incidence of pneumothorax with ICB has been shown to be 1.4 % for each intercostal nerve that is blocked [46].

References

1. Houghton AK, Wang CC, Westlund KN. Do nociceptive signals from the pancreas travel in the dorsal column? Pain. 2001;89(2–3):207–20. Research Support, U.S. Gov't, P.H.S.
2. Djouhri L, Lawson SN. Abeta-fiber nociceptive primary afferent neurons: a review of incidence and properties in relation to other afferent A-fiber neurons in mammals. Brain Res Brain Res Rev. 2004;46(2):131–45. Comparative Study Research Support, Non--U.S. Gov't Review.
3. Djouhri L, Lawson SN. Differences in the size of the somatic action potential overshoot between nociceptive and non-nociceptive dorsal root ganglion neurones in the guinea-pig. Neuroscience. 2001;108(3):479–91. Research Support, Non-U.S. Gov't.
4. Lawson SN. Phenotype and function of somatic primary afferent nociceptive neurones with C-, Adelta- or Aalpha/beta-fibres. Exp Physiol. 2002;87(2):239–44. Research Support, Non-U.S. Gov't Review.
5. Winnie A, Candido K. Differential neural blockade for the diagnosis of pain. In: Waldman S, editor. Pain management. Philadelphia: Saunders; 2007. p. 155–66.
6. Hogan QH, Abram SE. Neural blockade for diagnosis and prognosis. A review. Anesthesiology. 1997;86(1):216–41. Review.
7. Ford DJ, Raj PP, Singh P, Regan KM, Ohlweiler D. Differential peripheral nerve block by local anesthetics in the cat. Anesthesiology. 1984;60(1):28–33. Comparative Study.
8. Molloy R, Candido K. Diagnostic nerve blocks. In: Benzon H, Raja SN, Molloy R, editors. Essentials of Pain Medicine and Regional Anesthesia. 2nd ed. Philadelphia: Elsevier; 2005. p. 181–9.
9. Rosenberg PH, Heinonen E, Jansson SE, Gripenberg J. Differential nerve block by bupivacaine and 2-chloroprocaine. An experimental study. Br J Anaesth. 1980;52(12):1183–9. Comparative Study In Vitro Research Support, Non-U.S. Gov't.
10. Fink BR. Mechanisms of differential axial blockade in epidural and subarachnoid anesthesia. Anesthesiology. 1989;70(5):851–8. Research Support, U.S. Gov't, P.H.S.
11. Cherry DA, Gourlay GK, McLachlan M, Cousins MJ. Diagnostic epidural opioid blockade and chronic pain: preliminary report. Pain. 1985;21(2):143–52. Case Reports.
12. Kappis M. Erfahungen mit localanasthesie bie bauchoperationen. Verh Dtsch Gesellsch Chir. 1914;43:97–9.
13. Eisenberg E, Carr DB, Chalmers TC. Neurolytic celiac plexus block for treatment of cancer pain: a meta-analysis. Anesth Analg. 1995;80(2):290–5. Meta-Analysis Research Support, U.S. Gov't, P.H.S.
14. Brown DL. A retrospective analysis of neurolytic celiac plexus block for nonpancreatic intra-abdominal cancer pain. Reg Anesth. 1989;14(2):63–5. Research Support, Non-U.S. Gov't.
15. Loukas M, Klaassen Z, Merbs W, Tubbs RS, Gielecki J, Zurada A. A review of the thoracic splanchnic nerves and celiac ganglia. Clin Anat. 2010;23(5):512–22. Review.
16. Ward EM, Rorie DK, Nauss LA, Bahn RC. The celiac ganglia in man: normal anatomic variations. Anesth Analg. 1979;58(6):461–5.
17. Zhang XM, Zhao QH, Zeng NL, Cai CP, Xie XG, Li CJ, et al. The celiac ganglia: anatomic study using MRI in cadavers. AJR Am J Roentgenol. 2006;186(6):1520–3. Research Support, Non-U.S. Gov't.
18. de Leon-Casasola O. Neurolysis of the sympathetic axis for cancer pain management. In: Benzon H, Rathmell J, Wu C, Turk D, Argoff C, editors. Raj's Practical Management of Pain. Philadelphia: Mosby Elsevier; 2008.
19. Molloy RE, Benzon HT. Neurolytic blocking agents: Uses and complications. In: Benzon H, Rathmell J, Wu C, Turk D, Argoff C, editors. Raj's Practical Management of Pain. Philadelphia: Mosby Elsevier; 2008.
20. Fine PG, Bubela C. Chylothorax following celiac plexus block. Anesthesiology. 1985;63(4):454–6. Case Reports.
21. Leung JW, Bowen-Wright M, Aveling W, Shorvon PJ, Cotton PB. Coeliac plexus block for pain in pancreatic cancer and chronic pancreatitis. Br J Surg. 1983;70(12):730–2.
22. Ischia S, Ischia A, Polati E, Finco G. Three posterior percutaneous celiac plexus block techniques. A prospective, randomized study in 61 patients with pancreatic cancer pain. Anesthesiology. 1992;76(4):534–40. Clinical Trial Randomized Controlled Trial.
23. Turker G, Basagan-Mogol E, Gurbet A, Ozturk C, Uckunkaya N, Sahin S. A new technique for superior hypogastric plexus block: the posteromedian transdiscal approach. Tohoku J Exp Med. 2005;206(3):277–81. Case Reports.
24. Wechsler RJ, Maurer PM, Halpern EJ, Frank ED. Superior hypogastric plexus block for chronic pelvic pain in the presence of endometriosis: CT techniques and results. Radiology. 1995;196(1):103–6.
25. Chan WS, Peh WC, Ng KF, Tsui SL, Yang JC. Computed tomography scan-guided neurolytic superior hypogastric block complicated by somatic nerve damage in a severely kyphoscoliotic patient. Anesthesiology. 1997;86(6):1429–30. Case Reports Letter.
26. Plancarte R, Amescua C, Patt RB, Aldrete JA. Superior hypogastric plexus block for pelvic cancer pain. Anesthesiology. 1990;73(2):236–9.
27. Bajwa ZH, Ho C, Grush A, Kleefield J, Warfield CA. Discitis associated with pregnancy and spinal anesthesia. Anesthesia Analg. 2002;94(2):415–6. Case Reports, Table of contents.
28. Osti OL, Fraser RD, Vernon-Roberts B. Discitis after discography. The role of prophylactic antibiotics. J Bone Joint Surg Br. 1990;72(2):271–4. Research Support, Non-U.S. Gov't.
29. Carragee EJ, Don AS, Hurwitz EL, Cuellar JM, Carrino JA, Herzog R. 2009 ISSLS Prize Winner: Does discography cause accelerated progression of degeneration changes in the lumbar disc: a ten-year matched cohort study. Spine (Phila Pa 1976). 2009;34(21):2338–45. Comparative Study.

30. Lin CS, Cheng JK, Hsu YW, Chen CC, Lao HC, Huang CJ, et al. Ultrasound-guided ganglion impar block: a technical report. Pain Med. 2010;11(3):390–4.
31. Waldman S. Hypogastric plexus block and impar ganglion block. Philadelphia: Saunders Elsevier; 2007.
32. Reig E, Abejon D, del Pozo C, Insausti J, Contreras R. Thermocoagulation of the ganglion impar or ganglion of Walther: description of a modified approach. Preliminary results in chronic, nononcological pain. Pain Pract. 2005;5(2):103–10.
33. Wemm Jr K, Saberski L. Modified approach to block the ganglion impar (ganglion of Walther). Reg Anesth. 1995;20(6):544–5. Letter.
34. Nebab EG, Florence IM. An alternative needle geometry for interruption of the ganglion impar. Anesthesiology. 1997;86(5):1213–4. Letter.
35. Datir A, Connell D. CT-guided injection for ganglion impar blockade: a radiological approach to the management of coccydynia. Clin Radiol. 2010;65(1):21–5. Evaluation Studies.
36. Munir MA, Zhang J, Ahmad M. A modified needle-inside-needle technique for the ganglion impar block. Can J Anaesth. 2004;51(9):915–7.
37. R P, R V, RB P. Neurolytic blocks of the sympathetic axis. RB P, editor. Philadelphia: Lippincott; 1993.
38. Toshniwal GR, Dureja GP, Prashanth SM. Transsacrococcygeal approach to ganglion impar block for management of chronic perineal pain: a prospective observational study. Pain physician. 2007;10(5):661–6. Clinical Trial.
39. Bhatnagar S, Khanna S, Roshni S, Goyal GN, Mishra S, Rana SP, et al. Early ultrasound-guided neurolysis for pain management in gastrointestinal and pelvic malignancies: an observational study in a tertiary care center of urban India. Pain Pract. 2011;12(1): 23–32.
40. Cramer GD, Darby SA, editors. Basics and clinical anatomy of the spine, spinal cord, and ANS. St. Louis: Mosby-Year Book; 1995.
41. Loev MA, Varklet VL, Wilsey BL, Ferrante FM. Cryoablation: a novel approach to neurolysis of the ganglion impar. Anesthesiology. 1998;88(5):1391–3. Case Reports.
42. Karmakar MK, Ho AM, Law BK, Wong AS, Shafer SL, Gin T. Arterial and venous pharmacokinetics of ropivacaine with and without epinephrine after thoracic paravertebral block. Anesthesiology. 2005;103(4):704–11. Randomized Controlled Trial.
43. Doi K, Nikai T, Sakura S, Saito Y. Intercostal nerve block with 5% tetracaine for chronic pain syndromes. J Clin Anesth. 2002; 14(1):39–41.
44. Shir Y, Shapira S, Magora F. Continuous intercostal nerve block for chronic pain. Harefuah. 1989;116(9):461–3.
45. Benzon HT, Raj PP. Raj's practical management of pain. 4th ed. Philadelphia: Mosby-Elsevier; 2008.
46. Shanti CM, Carlin AM, Tyburski JG. Incidence of pneumothorax from intercostal nerve block for analgesia in rib fractures. J Trauma. 2001;51(3):536–9.

5 Lessons Learned from Visceral Sensory Stimulation: Implications for Treatment of Chronic Abdominal Pain

Søren Schou Olesen, Anne Estrup Olesen, Jens Brøndum Frøkjaer, Kasper Grosen, and Asbjørn Mohr Drewes

Introduction

Abdominal pain is of frequent occurrence, even in the normal population, and it is probably the most prevalent symptom in the gastroenterology clinic. Consequently, characterization of gut pain is fundamental in the diagnosis and assessment of organ dysfunction, and optimal treatments will only be achieved on the basis of a better understanding of underlying pathology and pain mechanisms. In the clinical setting, many patients with chronic abdominal pain suffer from comorbidity such as nausea, narcotic addiction, physical and emotional disability, and malnutrition. Therefore, a detailed characterization of pain symptoms is often difficult to obtain and is often blurred by symptoms from the associated comorbidities as well as medication. This is particularly problematic when underlying pain mechanisms are under investigation. In order to bypass this problem, experimental pain models based on quantitative sensory testing (QST) can be used [1–3]. QST provides information on sensory function at the peripheral and central level of the nervous system by recording subjects' responses (subjective or objective) to different external stimuli of controlled intensity. The primary advantages of QST are that a pain stimulus can be controlled, delivered repeatedly, and modulated, and that the responses can be assessed qualitatively and quantitatively with psychophysical, neurophysiological, or different imaging methods—Fig. 5.1. The methods have proven to be an important instrument to characterize basic physiology as well as mechanisms underlying pathological pain disorders [1–3]. The interest in human *visceral* QST has increased rapidly during the last decade, and also in gastroenterology the focus has been on developing methods for experimental induction and assessment of pain.

Experimental Visceral Pain Stimuli

The natural origin of visceral pain is not fully understood, although a variety of innate stimuli are clearly associated with pain from the viscera. Naturally occurring visceral stimuli are distention of hollow organs, ischemia, inflammation, spasms, and traction of the gut. Also, thermal stimuli (heat and cold) may provoke pain from the viscera although (apart from the esophagus) this seems not to occur under normal physiological conditions [4].

The ideal experimental stimulus to elicit gut pain in humans should mimic innate visceral stimuli, be minimally invasive, reliable in test–retest experiments, and quantifiable. The response to the stimulus should increase with increasing stimulus intensity and preferably the pain should reflect observations in diseased organs by evoking phenomena such as allodynia and hyperalgesia [5, 6]. The different methods currently available for visceral sensory stimulation are:

- Electrical stimuli
- Mechanical stimuli
- Thermal stimuli
- Chemical stimuli and models evoking visceral hyperalgesia

Ischemic stimuli are difficult to quantify in human and is normally not used as a direct stimulus. In the following sections, the individual pain stimuli most widely used for visceral QST and experimental-evoked hyperalgesia are briefly discussed.

S.S. Olesen, M.D., Ph.D. (✉) • A.E. Olesen, M.Sc. (Pharm), Ph.D.
A.M. Drewes, Ph.D., D.M.Sc.
Mech-Sense, Department of Gastroenterology and Hepatology, Aalborg University Hospital, Mølleparkvej 4, Aalborg, Denmark 9000
e-mail: sso@mech-sense.com; aneso@rn.dk; amd@rn.dk

J.B. Frøkjaer, M.D., Ph.D.
Department of Radiology, Aalborg University Hospital, PO Box 365, Aalborg, Denmark
e-mail: frokjaer@mail.tele.dk; jf@mech-sense.com

K. Grosen, M.H.Sc.
Department of Cardiothoracic and Vascular Surgery, Aarhus University Hospital, Brendstrupgaardsvej 100, Aarhus N, Denmark 8200
e-mail: kasper.grosen@ki.au.dk

L. Kapural (ed.), *Chronic Abdominal Pain: An Evidence-Based, Comprehensive Guide to Clinical Management*, DOI 10.1007/978-1-4939-1992-5_5, © Springer Science+Business Media New York 2015

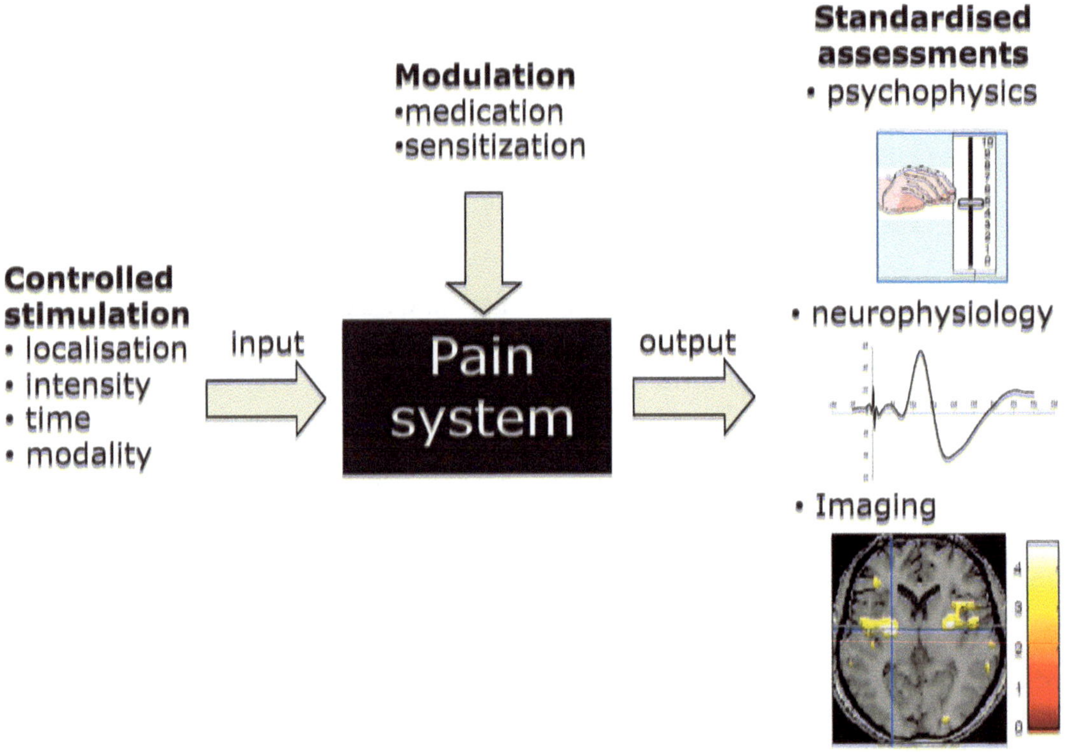

Fig. 5.1 The concept of experimental pain. The pain system can be considered as a "black box" between the experimental stimulation (input) and the response (output). When input and output are reproducible, it is possible to reveal differences in pain processing between, e.g., healthy volunteers and patients. Furthermore, modulation of the pain system is possible through various mechanisms (e.g., medication, modulation, or sensitization) and may provide additional information

Electrical Stimulation

Depolarization of visceral nerve afferents by electrical current has been widely used as an experimental stimulus of the human gut. The electrical stimuli have proved to be safe in all parts of the gastrointestinal (GI) tract and are easily controlled over time. As the stimulus by-pass peripheral receptors in the gut wall, the method is used to characterize afferent transmission and central processing of visceral stimuli [1–3, 7]. A major challenge of visceral electrical stimulation is varying electrode contact with gut mucosa. Integration of the electrical stimulation device and a biopsy forceps of an endoscope provide an elegant solution to this problem and allows application of stimuli in well-defined areas throughout the GI tract with high spatial precision.

Mechanical Stimulation

The mechanical properties of the gut are important for its function as a digestive organ and it contains numerous mechanoreceptors distributed mainly in the muscle layers of the gut [8]. Mechanical stimulation of the gut is typically done by distension of a balloon positioned in the segment under investigation. Widely used methods are computerized systems such as the "Barostat," where balloon pressure and volume can be strictly controlled during distension and thereby transmit a controlled mechanical stimulus [9]. The major advantage of the Barostat system and similar pressure–volume-based methods are the relatively low costs and reliability, making it useful for routine purposes. However, the accuracy of these systems has been questioned mainly due to uncontrolled elongation of the balloon during distension in nonspheric organs such as the rectum. Accordingly, elongation and deformation of the balloon during distention may not reliably reflect mechanoreceptor activation. These problems may be overcome by calculation of the balloon radius and tissue strain using impedance planimetry or imaging-based methods such as ultrasound or magnetic resonance. In accordance with recent studies, strain of the gut is probably the most consistently mechanical parameter relating to mechanoreceptors activation and possibly the subjective sensory response [10]. However, the technical complexity of such systems has so far limited their use to advanced experimental GI research and they are not widely used in the clinical setting.

Thermal Stimulation

Short-lasting thermal stimuli of the human GI tract are believed to activate unmyelinated afferents in the mucosa through the transient receptor potential cation channel subfamily V member 1 (TRPV1) receptor. This is opposed to mechanical and electrical stimuli activating afferents in both superficial and deeper layers in the viscera [8]. Although thermal stimuli of the gut have been used to some extent in animal studies, temperature stimuli in the human GI tract have only been used in a limited number of studies. This has mainly been due to difficulties in controlling the temperature rate (being essential for control of nociceptor activation). However, new technologies for thermal stimulation of the GI tract have been developed. These are based on continuously recirculating of water inside a balloon with concomitant measurement of balloon temperature [11]. The model has been used in many studies unraveling pain mechanisms in patients and was recently modified for use in the lower gut (rectosigmoid) [7]. Based on this method, the temperature stimuli show a linear stimulus–response relationship, thus demonstrating validity of the model. However, uncertainty in pain assessments due to fast increase in temperature (2 °C/min) has been demonstrated and recently it was proposed that individual differences in reaction time could affect the accuracy of rating. Consequently, in future studies a slower temperature increase (0.2 °C/min) is recommended [12, 13].

Chemical Stimulation and Models Evoking Visceral Allodynia and Hyperalgesia

Inflammation of the gut generally leads to altered sensations including pain. This has been investigated experimentally in patients with, e.g., esophagitis [4]. Chemical stimulation of the GI tract more closely resembles clinical diseases and is believed to approach the ideal experimental visceral pain stimulus [14]. Most chemical stimuli are assumed to activate unmyelinated C-fibers. Following chemical stimulation, tissue injury generates release of multiple molecules acting synergistically to produce inflammatory responses and hyperalgesia. Acid stimulation is the most common used method to evoke such visceral hyperalgesia, although chemical stimulation with alcohol, bradykinin, glycerol, capsaicin, and hypertonic saline has also been used to induce gut sensitization [1–3]. To mimic the clinical situation experimentally, it may be necessary to use a mixture of chemical substances with diverse tissue effects. An example of such mixed chemical stimulation is seen in the combination of acid and capsaicin working through different cellular interaction sites. Accordingly, acid targets the TRPV1 receptor extracellularly, whereas the capsaicin targets the TRPV1 receptor predominantly intracellularly. The method has been applied in, e.g., the esophagus of healthy volunteers and provides a human model to study visceral hyperalgesia [12, 13]. Most studies on visceral hyperalgesia have demonstrated increased pain to one or more stimulation modalities after experimentally induced sensitization by chemicals. Also, the duration and magnitude of hypersensitivity has been shown to be related to exposure area and dose of the chemicals [15]. Although chemical stimulation and experimental induced hyperalgesia generally posses high reproducibility in test–retest experiments, it has been demonstrated that the hyperalgesic response to acid is variable comparing the first time a subject is exposed to esophageal acid perfusion with the second time [12, 13]. Therefore, it is recommended to have a training session, where the subjects are introduced and exposed to chemical perfusion in order to familiarize them with the stimulus.

Multimodal Visceral Stimulation

The major limitation of the existing human models for visceral pain stimulation is that they may not mimic clinical pain because they are based on single, short-lasting stimuli only partly involving the many mechanisms typically activated during diseases. Therefore, the basic neurobiological mechanisms in clinical pain may be different from those relating to an experimental stimulation and experimental visceral models mimicking more closely the clinical situation are needed. Such a model should be based on multimodal testing regimens in which different receptors and central nervous system mechanisms are activated. Hence, a test battery where different stimuli are used will increase the probability for activation of a range of relevant nervous mechanisms. Especially if the stimulation is relatively long lasting and includes modalities known to evoke peripheral as well as central sensitization of the nervous system, the likelihood that part of the model will mimic clinical pain is high despite the nonharmful nature of the stimulation. To fulfill these requirements, a multimodal testing approach has been developed for experimental stimulation of the gut—Fig. 5.2 [11].

Experimental Studies and Pathophysiology of Chronic Abdominal Pain

QST has been used as an attempt to explain the pathophysiology of both functional and organic disorders of the gut. It is generally the belief that the central component of the pain system plays a major role in functional disorders such as functional chest pain, functional dyspepsia, and irritable bowel syndrome (IBS). On the other hand, in organic diseases such as inflammatory bowel disease and chronic pancreatitis, the pain regulatory systems are intact and the

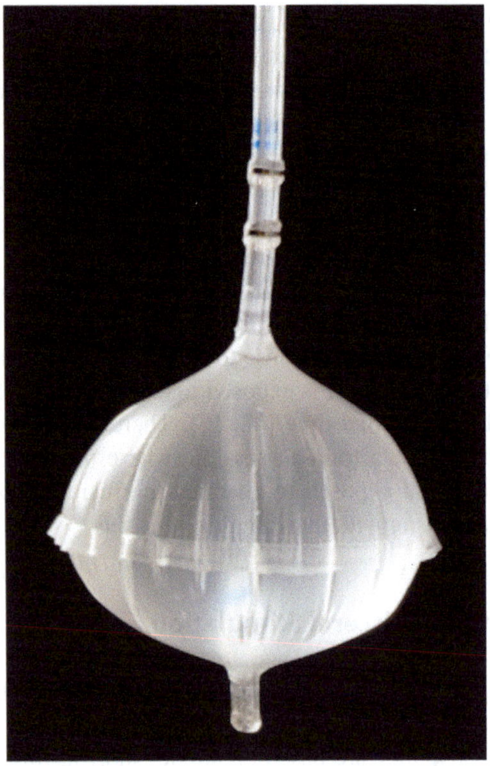

Fig. 5.2 The multimodal probe for electrical, mechanical, cold, warmth, and chemical stimuli. The probe has a bag for mechanical and thermal stimuli, the latter given by recirculating water. Electrodes for electrical stimuli are mounted on the probe above the bag. A side hole in the tube proximal to the bag allows perfusion with acid and other chemicals

balance between afferent activity and local/central pain inhibition is functioning differently. In the following section, selected methods to stimulate and assess the pain system in different examples of functional and organic diseases are highlighted.

In order to unravel abnormal pain processing, several pain assessment methods are available being more or less directly associated with the stimulus. These are for example:

- The subjective response using different rating scales
- The size and localization of the referred pain area
- Detection of viscero-visceral hyperalgesia
- The response to repeated stimulation (a proxy to wind-up or central integration)

The *subjective sensory response to QST* may reflect abnormal processing of the pain. Patients with functional pain disorders, such as functional chest pain and IBS, typically have hyperalgesia and allodynia to experimental stimuli of the organs thought to be diseases [16]. Only 40–60 % of the IBS patients show lowered rectal discomfort thresholds to mechanical stimulation, but when other perceptual abnormalities (altered referral pattern and increased intensity of sensations) were considered, 94 % of IBS patients had at least one abnormality [17]. In order to unravel disease pathogenesis, more advanced methods, such as the multimodal probe, can be used to detect sensory abnormalities. This approach has been used in patients with functional and organic diseases. As stated previously can the TRPV1 receptor be activated by a variety of stimuli, including acid (protons) and increases in temperature that reach the noxious range. Hence, patients with organic diseases, such as nonerosive and erosive esophagitis, were shown to have specific hyperalgesia to heat reflecting activation of the receptor by the natural acid reflux. On the other hand in patients with functional chest pain acid, there was a pathological response to experimental acid perfusion likely reflecting activation of central pain mechanisms [4]. In general, chronic tissue injury and pain has been associated with higher thresholds to mechanical stimulation in different regions of the GI tract. For example, chronic inflammation of the small bowel in patients with inflammatory bowel disease is associated with mechanical hypoalgesia of the rectum [18]. However, the pain response can vary according to the tissues that are stimulated such as seen in patients with chronic pancreatitis. This may reflect the complex pain mechanisms and interaction between sensitization and descending control systems [19, 20]—see also section about viscera-visceral hyperalgesia.

Referred pain is a normal phenomenon seen in clinical practice where pain originating from the viscera is also felt in somatic areas remote from the organ. Convergence between visceral and somatic afferents in the spinal cord seems to be of importance (Fig. 5.3). In organic diseases, it is believed that referred hyperalgesia of somatic tissues is caused by a normal process of central sensitization triggered by massive afferent visceral barrage [21]. However, in functional disorders, abnormal central processing of the afferent stimulation is likely of more importance. Hence, if the patients are properly instructed, the referred pain can be used as a proxy for the central changes. Experimentally, we have found that the referred pain area in healthy volunteers typically changed location after acid perfusion of the esophagus [1–3]. In patients with organic diseases such as those with gastro-esophageal reflux disease (GORD) and chronic pancreatitis, the referred pain area to stimulation of the esophagus and duodenum is increased in size and this is likely reflecting the increased afferent visceral barrage and subsequent activation of second-order neurons [4, 19]. In functional disorders, however, there seems to be a change in localization as well as an increased size of the referred pain area such as seen to experimental visceral stimulation in patients with functional chest pain, functional dyspepsia, and IBS [22].

Changes in the sensitivity and skin temperature in the referred pain area have also been shown in experimental studies of healthy volunteers [23, 24]. Correspondingly, abnormal superficial and deep sensations have been demonstrated in patients with renal stones, appendicitis, and cholecystolithiasis [25–27]. In patients with chronic pancreatitis

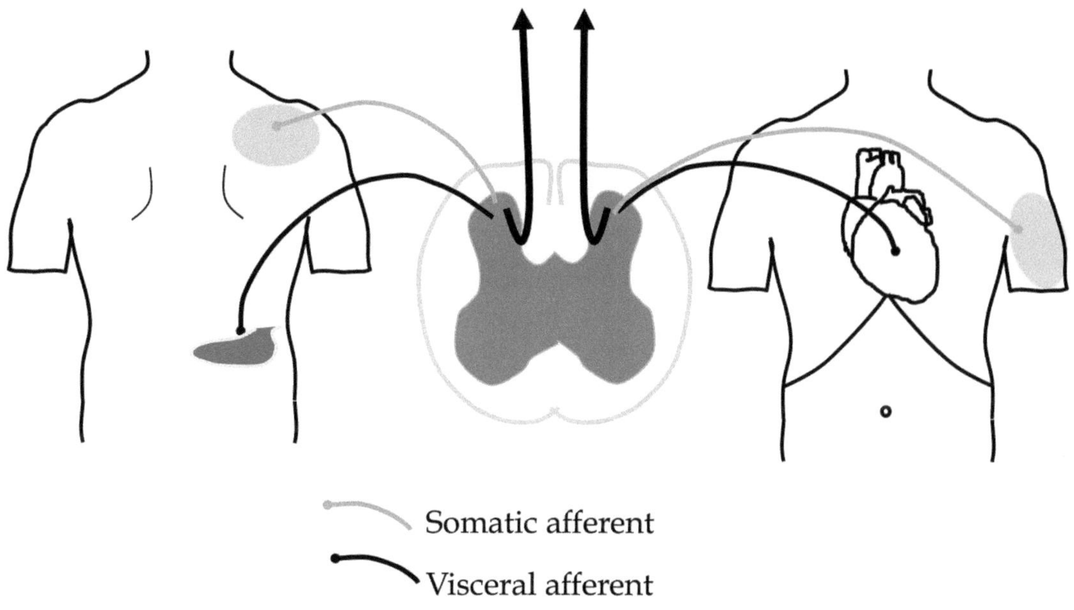

Fig. 5.3 Pain referral to somatic areas remote from the visceral organs is a common finding in GI diseases and known as "referred pain," e.g., pain referral to the right shoulder in acute cholecystitis. The underlying mechanism is related to convergence between visceral and somatic afferents in the spinal cord although in principal more complicated

sensory changes have also been seen in the corresponding "viscerotome" [28]. Such changes in localization and sensitivity of the referred pain areas may be a hallmark of diseased organs and if the experimental methods are improved they may serve as a biomarker of the disease.

Viscero-visceral hyperalgesia is a complex form of hypersensitivity probably explained by more than one mechanism. Since this phenomenon takes place between visceral organs which share their central afferent termination, it is plausible that central sensitization plays an important role [29]. Recently, human experimental studies support the role of viscero-visceral hyperalgesia in GI diseases. Acidification of the distal esophagus resulted in hyperalgesia in the proximal esophagus, and duodenal acidification was shown to induce esophageal hypersensitivity [30]. Recently, we showed that acidification of the esophagus in healthy volunteers involve widespread changes in the perception of experimental pain from remote organs such as the rectum [31]. The widespread visceral hypersensitivity in functional GI disorders (IBS, functional dyspepsia, etc.) may be due to this mechanism. As an example a marked reduction in colonic perception thresholds and alternation in the viscero-somatic referral pattern were seen in patients with IBS after lipid administration in the duodenum [32]. Viscero-visceral hyperalgesia may also explain the epidemiological findings in several clinical conditions with organic diseases such as an increased number of anginal attacks in patients with gallbladder calcinosis, and increased number of colics in dysmenorrheic patients with urinary calculosis [33]. Evidence for viscero-visceral hyperalgesia has also been provided in experimental studies of organic diseases, e.g., in patients with gastro esophageal reflux disease (GERD) where increased sensitivity to gastric distension was shown. Therefore, the frequent airway symptoms in GERD (often refractory to treatment with proton pump inhibitors) may not only be related to direct aspiration of the gastric refluxate, but vasovagal reflex mechanisms evoked by acid-related hyperalgesia may also be important [34].

Repeated stimulations: Sensitization of the spinal neurons is known to occur with prolonged or repeated stimulation ("wind-up" or temporal summation) of the peripheral afferents. Thus, temporal summation results in a short-lasting spinal cord sensitization that persists after discontinuing the peripheral stimulation. In the laboratory, this is perceived as increased pain to a series of stimuli with the same intensity.

Repeated electrical or mechanical stimuli to the small and large intestine in volunteers may cause increased sensation to subsequent stimuli, and this may be used as a model for enhanced central gain [1–3]. In functional pain Munakata et al. showed the importance of central mechanisms. In their study, patients with IBS developed rectal hyperalgesia following repetitive sigmoid distensions [35]. Paterson et al. [36] as well as studies from our group [4] also showed that repeated distensions conditioned the esophagus in functional chest pain patients resulted in higher pain scores. In organic diseases, repeated stimuli were also used to show the central amplification of pain in patients with chronic pancreatitis [37]. This stimulation paradigm can also be used to understand the changes in referred pain. If electrical stimuli are repeated over time, the pain and the area of referred pain increase progressively [23].

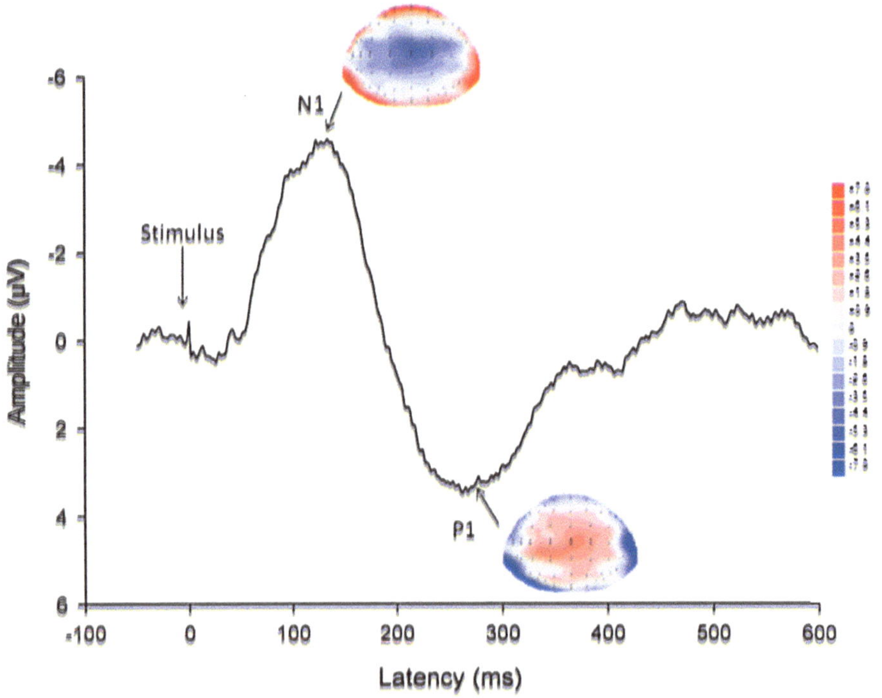

Fig. 5.4 A typical evoked potential (vertex-electrode) recorded after stimulation of the rectosigmoid junction in a healthy volunteer. Note the different peaks denoted N1 and P1, each defined by latency (ms), amplitude (µV), and a corresponding topographic map made from the 64 electrodes covering the head

QST can also be used to unravel pain mechanism at higher centers using electrophysiological and imaging methods. There are several possibilities, but the most used methods are as follows:

- The nociceptive reflex
- Electroencephalography (EEG) and magnetoencephalography (MEG)
- Imaging

The nociceptive (RIII) reflex is a spinal reflex that is elicited by painful stimulation of a sensory nerve. For example, stimulation of the sural nerve at the ankle evokes a flexion reflex that can be measured by quantification of the electromyographic response in the biceps femoris muscle. The connection from the primary afferents to the motor neurons is a polysynaptic pathway, which can be modulated by other afferent input, spinal neuronal excitability, and activity in descending control systems. Bouhassira et al. showed that tonic distension of the stomach and rectum resulted in inhibition of the reflex, whereas phasic mechanical stimuli of the rectum resulted in more complex modulation [38, 39]. Sensitization of the esophagus with acid resulted in a significant increase in the baseline reflex excitability, followed by a gradual inhibition during continuous distension of the organ [1–3]. Analgesics can modify the reflex and hence it may indirectly be used for basic and pharmacological studies of pain pathways in the GI tract [40].

The EEG monitors the brain activity to external stimuli directly in real time. The resting EEG has been used to unravel pain mechanisms in visceral diseases [41]. However, when a repetitive stimulus is applied and the cortical electrical activity is averaged (time-locked to the stimulus), the stimulus-evoked cortical potential (EP) can be extracted from the background electrical activity and is shown in shape of a waveform with different peaks (Fig. 5.4). Each peak in the EP represents a synaptic event associated with the transmission of afferent information from one group of neurons to another. The early peaks are supposed only to be influenced by the stimulation rate, intensity, and localization, and they reflect to a major degree the brain loci that process the pain intensity and localization [42]. EPs have been used to explain abnormal central pain processing in patients with functional disorders such as functional chest pain and IBS, suggesting an increased central nervous system response to visceral stimuli and reorganization of brain activation in the cingulate gyrus among others [43–45]. Studies have also suggested that different subgroups of patients with IBS exist such as those with short latency of the early EP components having sensitization of GI afferent pathways, and those with long latencies and enhanced late responses reflecting hypervigilance and increased affective processing [46].

Inverse modeling of the EPs can be used to identify the original electrical sources in the brain—for details see [47].

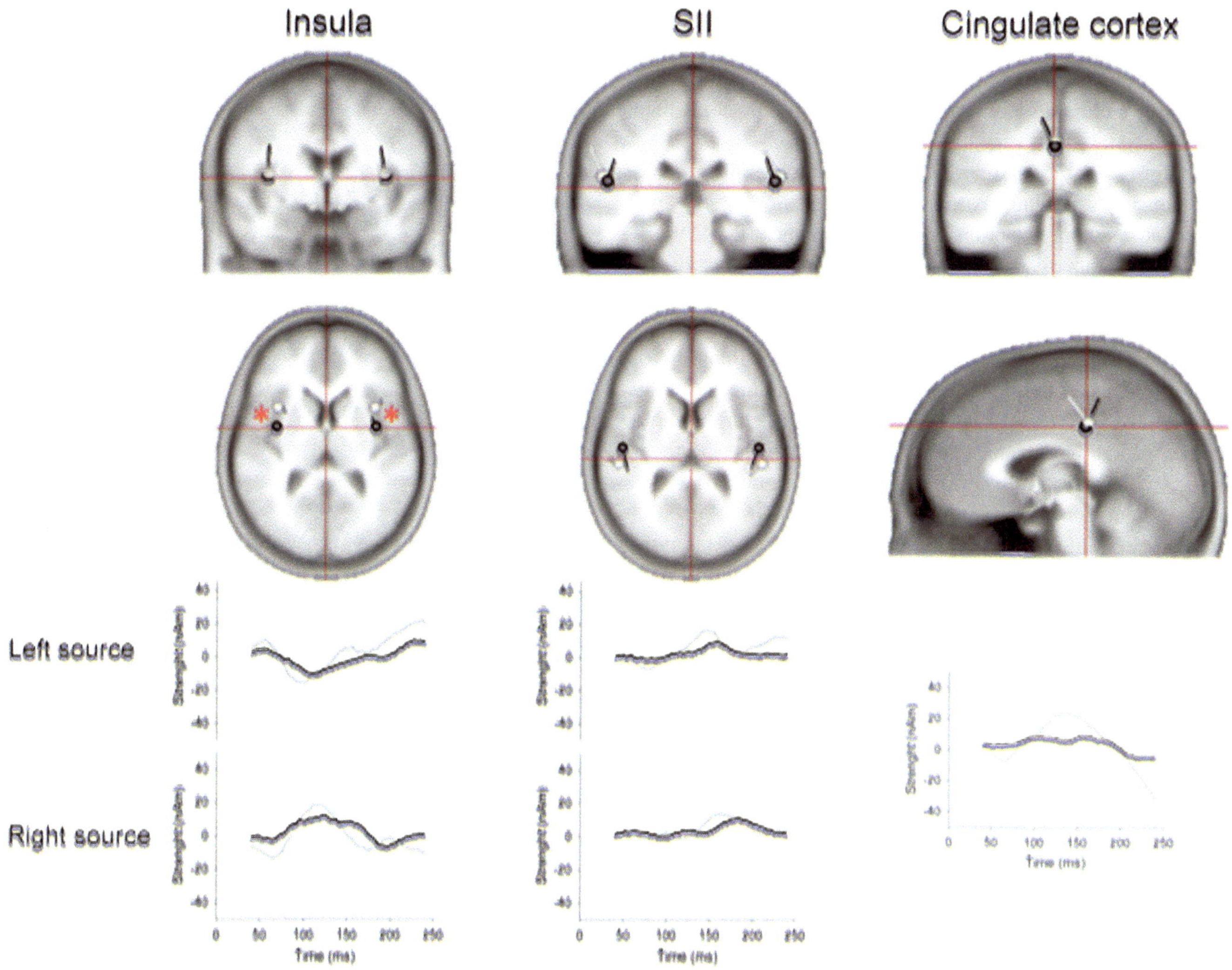

Fig. 5.5 *Top panels*: locations of brain sources evoked by painful stimulation of the sigmoid in patients with chronic pancreatitis (*black*) and healthy volunteers (*white*). The locations of insular sources differed between the groups. *Lower panel*: sequential activity of the brain sources throughout the time window of analysis (40–240 ms) in chronic pancreatitis patients (*black*) and healthy volunteers (*grey*). Modified from Olesen et al. (2011a) [91]

In organic diseases such chronic pancreatitis analysis of the EP topography has revealed a shift in insular dipole localization which was correlated with the patients' clinical pain scores (Fig. 5.5) [48]. Comparable findings have been reported in experimentally induced visceral hypersensitivity in healthy volunteers and may reflect the neurophysiologic correlate of functional reorganization. Insula has an important function for integrating the visceral sensory and motor activity together with limbic integration and is particularly important in pain perception from the gut. Experimental and clinical studies of somatic pain conditions, such as phantom limb pain, have also showed a correlation between clinical pain scores and reorganization, with the most suffering patients showing the most pronounced reorganization (i.e., a maladaptive pain response). Hence, the reorganization in chronic pancreatitis may be due to an "overactivation" of pain areas in the brains pain matrix, inducing a functional reorganization of the insular cortex. Such analysis may increase our understanding of the pain pathogenesis where the pain processing in the brain is of major importance, and there is preliminary evidence that these abnormalities may serve as predictors of treatment response.

MEG is a noninvasive technique for mapping brain activity by recording magnetic fields produced by electrical currents in the brain. MEG is a technically demanding technique and is only available in few specialist centers. Furthermore, it is limited by its incapability to resolve radial currents generated by deep brain sources, e.g., in the cingulate cortex. However, the spatial resolution of more superficial cortical activity is in the mm range which is better than the EEG (for review see [47]). The methods have been used to follow the brain activation following esophageal electrical stimulation in healthy volunteers, but otherwise studies of visceral pain has until today been very limited [49].

Imaging methods may also be used to explore pain mechanisms following experimental stimulation of the gut. Improved methods for brain imaging techniques (fMRI, PET, and SPECT) have vastly increased our understanding of the central processing of GI sensation and pain in both healthy volunteers as well as in patients suffering from GI disorders.

Magnetic resonance imaging (MRI) allows imaging of both brain structure and activity. Brain activity measured by functional MRI (fMRI) has most commonly been acquired by the blood oxygenation level dependent (BOLD) technique, which is based on different paramagnetic properties of oxy- and deoxyhemoglobin in the blood. fMRI has an excellent spatial resolution (2–5 mm) and operates in a noninvasive and nonradioactive environment allowing subjects to be studied repetitively. The BOLD signal reflects simultaneously changes in local blood flow, volume, and deoxyhemoglobin content, which derive from changes in neuronal activity [50]. Regions of activation are identified by subtracting regional BOLD signal during a control/resting condition from the signal during a stimulus condition—Fig. 5.6. Recently, other techniques such as arterial spin labeling which allows the measurement of whole brain cerebral blood flow in absolute units through the use of magnetically labeled endogenous water in blood allowing assessment of the temporal dynamics of the neural activation induced by pain. This has been used to detect changes in regional cerebral blood flow associated with a standard cutaneous heat pain [49] and infusion of hypertonic saline [51]. Arterial spin labeling is particularly suited to studies of prolonged pain since it becomes increasingly more sensitive than BOLD to changes in neural activation as the stimulus duration exceeds one minute [52]. A new technique called signal enhancement by extravascular water protons has been used in fMRI of the spinal cord, which is essential in the complete mapping of the pain system, and spinal cord and brain stem sensory-related neural activity has been consistently observed in a number of studies. Recently also, resting state fMRI has been applied in pain research including connectivity analysis between multiple brain networks [53]. Additionally, structural information obtained by other MRI techniques can been superimposed on the functional data: diffusion tensor imaging with assessment of microstructural integrity in sensory-related brain areas, tractography with tracing of nerve fibers, volumetry of cortical regions with assessment of the neuroplastic response to long-standing pain, and spectroscopy assessing the concentration of metabolites [54–57]. This allows more explanatory information on the neural structures, function, and connections between the centers involved in pain processing.

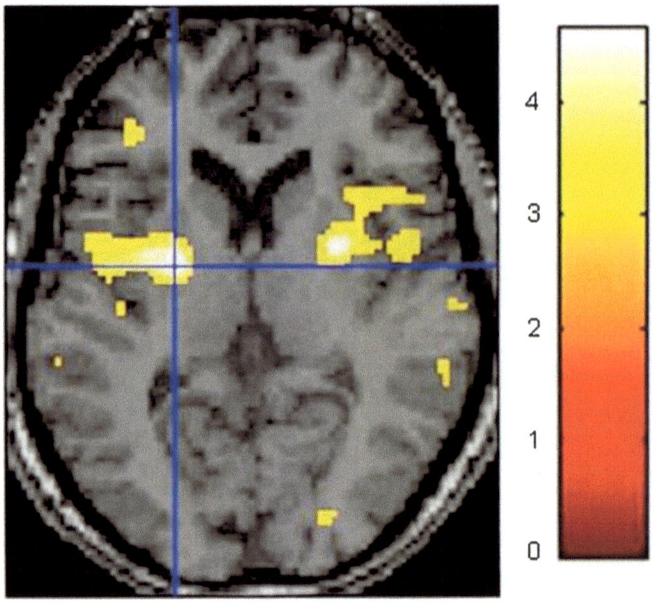

Fig. 5.6 Functional magnetic imaging (fMRI) with illustration of the brain activity induced by painful thermal stimulation of the right forearm in a single subject. This is based on the BOLD technique, which is based on different paramagnetic properties of oxy- and deoxyhemoglobin in the blood where the color code shows signal intensity. Regions of activation (here in the insular regions) are identified by subtracting regional BOLD signal during a resting condition from the signal during the painful stimulus

fMRI has been used in several studies for demonstrating abnormal brain processing in particular functional GI disorders. Few studies have also been conducted in organic diseases such as inflammatory bowel disease. Kwan et al. identified abnormal event-related sensations in five brain regions following rectal distensions in IBS [58]. In the primary somatosensory cortex, urge-related responses in the IBS group were seen compared to the control group. This could be interpreted as upregulated afferent input underlying visceral hypersensitivity or "visceral allodynia." In the IBS group, pain-related responses were seen in the medial thalamus and hippocampus, but not in the control group. However, pronounced urge- and pain-related activations were present in the right anterior insula and the right anterior cingulate cortex in the control group, but not the IBS group. Finally, lack of activation in right anterior insula was found in IBS patients, interpreted by the authors as either a ceiling effect or a dysfunction in interoceptive processing or control of visceromotor responses. In controls, patients with inflammatory bowel disease and IBS patients, Bernstein et al. performed rectal balloon distention to a sensation of stool and to a sensation of pain while undergoing fMRI [59]. All three groups share similar loci of activations to visceral sensations of stool and pain, but both activation and deactivation of particular regions of interest was differentiated between the groups. Finally, fMRI has been used to evaluate the effect of the tricyclic antidepressant amitriptyline, which is believed to be of clinical benefit in IBS patients [60]. Amitriptyline reduced pain-related cerebral activations in the pACC and the left posterior parietal cortex compared to placebo, but only during mental stress [61].

Single photon emission computed tomography (SPECT) and positron emission tomography (PET) are nuclear imaging techniques that can trace radiolabeled molecules injected into the blood stream, whereby the distribution, density and activity of receptors in the brain can be visualized. This provides an insight into the organization of functional networks in the brain, which cannot be achieved by morphologic investigations or imaging of blood flow and metabolism [62]. Using this molecular imaging technique, pharmaceutical compounds can be used as radiolabeled tracers combined with kinetic models allowing quantification of receptor sites and enzyme function [63]. PET is superior in imaging radiopharmaceuticals and/or other ligands as it offers the ability to study receptor distribution and explore the site of action.

Both SPECT and PET have been used in studies investigating which brain areas are activated during painful stimuli [64]. Nevertheless, it has not been used very widely in clinical pain studies. A study by Fukumoto et al. assessed regional cerebral blood flow of the contralateral thalamus in ten patients with reflex sympathetic dystrophy syndrome [65], but has so far not been used in the investigation of visceral pain. Several studies have used PET for investigating brain activation during visceral pain [66–68], but to our knowledge no studies of specific receptor systems have been conducted.

Assessment of Analgesic Effects by Visceral QST

The effect of analgesics on visceral pain is difficult to evaluate in the clinic, due to the deep and diffuse nature of the pain and the accompanying autonomic symptoms [1–3]. Application of experimental pain models in a crossover study design with appropriate baseline recordings offers a unique opportunity of revealing analgesic effects [5]. It has been recommended to include pain models in various tissues as, e.g., opioid analgesia can exhibit tissue dissimilarities [69]. Moreover, different modalities activate distinct pain mechanisms and make it possible to investigate on a mechanistic basis how analgesics work. The effect on pain from deeper structures (muscle and viscera) is considered most important as, e.g., opioid analgesia is more robust in deep pain [69].

To induce deep pain, experimental pain has been evoked in different parts of the GI tract [1–3, 12, 13]. Sensitization of the nervous system is also possible by, e.g., perfusion of the gut with chemical substances. Thus, peripheral and central mechanisms relating to the clinical situation can be evoked, and the effect on pharmacological modulation evaluated. Experimental pain studies can be conducted in healthy volunteers or in patients to evaluate analgesic effects. In the next section, some examples are discussed, for more comprehensive review the reader is referred to [69, 70].

Healthy Volunteers

Experimental pain in healthy volunteers appears to be suited to investigate the analgesic effects, especially when deep pain and hyperalgesia is evoked to mimic the clinical situation [5, 6, 69, 70].

Ketamine is an NMDA-antagonist that has been widely studied in experimental pain in healthy volunteers. Animal studies that investigated the analgesic role of NMDA receptors proposed that NMDA-receptor-related transmission is more important in acute nociceptive responses involving visceral tissues, whereas involvement in somatic nociception may be more dependent on mechanisms active in inflammation and hyperalgesia [71–73]. Therefore, visceral stimulations should be included in the experimental pain model when investigating effects of NMDA-antagonists. However, only one study investigated the effect of ketamine in a model including visceral sensory stimulations. It was found that pain from visceral distension was decreased by ketamine [74]. The findings on analgesic effects of ketamine in acute visceral pain in humans are in agreement with these animal data since the ketamine-related attenuation of pain intensity appeared more pronounced for noxious visceral than for cutaneous stimulation [74].

Morphine and oxycodone are opioids and have both been tested in experimental visceral pain studies in healthy volunteers. They were both effective against mechanical and electrical esophageal pain, but only oxycodone attenuated thermal pain [5, 6, 75]. Moreover, oxycodone and morphine have been tested in esophageal hyperalgesia induced by a combination of acid and capsaicin. In visceral hyperalgesia, only oxycodone showed effect on pain to electrical stimulation and the referred pain area to heat stimuli [76]. Morphine and oxycodone also showed different effects comparing somatic and visceral pain. This reflects the clinical situation where visceral pain in contrast to somatic pain can be difficult to treat with traditional μ-opioid agonists, and oxycodone has in a few clinical studies been found more effective than morphine [77, 78].

New drugs: Human experimental pain models in healthy volunteers have also been used to evaluate analgesic effects of new drugs. For example, the effect of a new TRPV1 antagonist (AZD1386) was assessed by our group in experimentally induced esophageal pain. It was found that it increased pain thresholds to heat stimuli of the esophagus, whereas pain thresholds to other stimuli were unaffected. AZD1386 treatment also attenuated, but did not prevent, acid-induced hyperalgesia [79].

Patients

Pain experienced and reported by healthy volunteers is different from clinical pain, and in the laboratory it is not possible to reproduce the pathophysiology and the full complexity of the pain experience in patients [80, 81]. As described previously, pain in patients is accompanied by several factors such as fear, emotions, anxiety, etc. influencing the overall sensory experience [82]. Hence, improvement in, e.g., depression during treatment with a new drug can result in less pain ratings. It can therefore be difficult to evaluate analgesic effects and specific mechanisms in patients with pain, and even studies with well-known analgesics, such as NSAIDs, are frequently inconclusive [83]. However, experimental pain models can be applied in patient groups to investigate analgesic effect in the actual patient group providing controlled stimuli and quantitative assessments. Below some examples are provided to give insight into this testing.

Gabapentin and pregabalin decreases hyperalgesia and allodynia and are widely used in treating neuropathic pain. Gabapentin and pregabalin also exert antinociceptive effects in animal models of neuropathic, surgical, inflammatory, acute, and chronic pain. This was supported by positive findings in the described human experimental pain models in patients [84, 85]. The mechanism of action is not fully known, but part of the therapeutic action on neuropathic pain is thought to involve voltage-gated calcium ion channels [86, 87]. Gabapentin has been investigated in experimental visceral pain in patients with diarrhea-predominant IBS where pain was evoked by rectal distensions. The distending pressure triggering a first sensation of defecation was not altered, but threshold pressures for bloating, discomfort, and pain were increased [85]. Pregabalin was also studied in patients with IBS. Rectal sensitivity was assessed using a Barostat technique and pregabalin significantly increased the sensory thresholds, desire to defecate and pain [84]. In patients with chronic pancreatitis thought to have a strong neuropathic pain component [88], pregabalin was also tested. Here, the experimental measures were translated into a clinical efficacy, confirmed by traditional questionnaire endpoints [89]. In these patients, perceptual thresholds to electrical stimulation of the sigmoid with recording of corresponding evoked brain potentials were also obtained. Pregabalin increased pain threshold to electrical gut stimulation, whereas no differences in evoked brain potential characteristics or corresponding brain sources were seen. It was concluded that the antinociceptive effects of pregabalin is mediated primarily through subcortical mechanisms [90].

Opioids: In an experimental pain study in patients with chronic pancreatitis, it was found that mechanical, heat, and electrical pain in skin and mechanical and electrical muscle pain was unaffected by morphine. However, morphine increased esophageal mechanical pain-tolerance threshold, whereas esophageal heat and electrical pain thresholds were unaffected [91]. Another study investigated the effect of morphine in patients with chronic pancreatitis and found no effect on rectal distension thresholds [92]. In patients undergoing abdominal hysterectomies, morphine increased pain tolerance to rectal distension, whereas no effect on transcutaneous electric sensation or skin electric pain-tolerance thresholds was found [93]. The effect of oxycodone was only investigated in one experimental pain study in patients with chronic pancreatitis. Oxycodone showed more pronounced effects than morphine on skin, muscle, and visceral stimulations [91] again demonstrating a differential effect on opioids.

QST in Prediction of Response to Analgesics

It has recently been shown that QST has the potential to stratify patients into responders and nonresponders to analgesic treatment. Such results are promising and indicate that the methods may be useful as a clinical tool in tailoring individualized therapy. For example, heat pain threshold was correlated to the effect of oxycodone on pain following cold pressor testing in healthy volunteers [94]. Likewise, electrical, heat, and pressure-evoked pain have been shown predict postoperative analgesic consumption in surgical patients. Hence, *electrical* pain stimulation was correlated to postoperative consumption of acetaminophen and morphine after caesarean section and percutaneous nephrolithotomy [95, 96]. *Pressure* pain was correlated to morphine consumption following hysterectomy [97]. Finally, preoperative *heat* stimulations predicted morphine use following knee arthroplasty and caesarean section [98, 99], as well as ibuprofen requirement within the first ten postoperative days following laparoscopic tubal ligation [100]. In contrast, three studies have been unable to find a relationship between electrical pain thresholds and subsequent analgesic consumption [101–103]. These apparently conflicting results regarding electrical stimulation are most likely related to differences in study methodology across studies.

In *patients with neuralgia*, Edwards et al. [104] found that heat pain sensitivity predicted the effect of morphine, but not the responses to nortriptylin or placebo. Likewise, Attal et al. [105] reported a correlation between baseline heat pain and the effect of lidocaine and mexiletine. Recently, Yarnitsky et al. [106] suggested that in patients with painful diabetic neuropathy those with less efficient conditioned pain modulation were most likely to benefit from duloxetine. Finally, in patients with *chronic pancreatitis*, Olesen et al. [107] showed that the effect of pregabalin was associated with increased sensitivity to electrical stimulation in the pancreatic viscerotome compared to a control area. In summary, the evidence remains insufficiently robust to suggest any specific QST measure to discriminate between patients who are likely to respond to analgesic treatment. However, results are promising and call for future well designed and sufficiently powered studies focusing on different modalities of experimental pain modulation rather than a single static pain paradigm.

Conclusion

Painful sensations from the gut tract are very common in the clinic, but underlying diseases can be difficult to diagnose and treat successfully. Findings from basic, experimental, and clinical research have gained new insight about the GI pain system, and evidence for sensitization at both the peripheral and the central level seems to be of major importance in the explanation and treatment. The methods have also been used to test the effect and mechanisms of existing and new drugs and in prediction of the responses to treatment. This information and knowledge should be implemented in the clinic leading to the right diagnosis and directing future treatment approaches against underlying visceral pain mechanism.

References

1. Drewes AM, Gregersen H, Arendt-Nielsen L. Experimental pain in gastroenterology: a reappraisal of human studies. Scand J Gastroenterol. 2003;38:1115–30.
2. Drewes AM, Pedersen J, Liu W, Arendt-Nielsen L, Gregersen H. Controlled mechanical distension of the human oesophagus: sensory and biomechanical findings. Scand J Gastroenterol. 2003;38:27–35.
3. Drewes AM, Schipper KP, Dimcevski G, Petersen P, Andersen OK, Gregersen H, Arendt-Nielsen L. Multi-modal induction and assessment of allodynia and hyperalgesia in the human oesophagus. Eur J Pain. 2003;7:539–49.
4. Drewes AM, Reddy H, Pedersen J, Funch-Jensen P, Gregersen H, Arendt-Nielsen L. Multimodal pain stimulations in patients with grade B oesophagitis. Gut. 2006;55:926–32.
5. Staahl C, Christrup LL, Andersen SD, Arendt-Nielsen L, Drewes AM. A comparative study of oxycodone and morphine in a multi-modal, tissue-differentiated experimental pain model. Pain. 2006;123:28–36.
6. Staahl C, Reddy H, Andersen SD, Arendt-Nielsen L, Drewes AM. Multi-modal and tissue-differentiated experimental pain assessment: reproducibility of a new concept for assessment of analgesics. Basic Clin Pharmacol Toxicol. 2006;98:201–11.
7. Brock C, Nissen TD, Gravesen FH, Frokjaer JB, Omar H, Gale J, Gregersen H, Svendsen O, Drewes AM. Multimodal sensory testing of the rectum and rectosigmoid: development and reproducibility of a new method. Neurogastroenterol Motil. 2008;20:908–18.
8. Sengupta JN, Gebhart GF. Gastrointestinal afferent fibers and sensation. In: Johnson LR, editor. Physiology of the gastrointestinal tract. New York, NY: Raven; 1994. p. 483–519.
9. van der Schaar PJ, Lamers CB, Masclee AA. The role of the barostat in human research and clinical practice. Scand J Gastroenterol Suppl. 1999;230:52–63.
10. Gregersen H, Kwiatek MA, Schwizer W, Tutuian R. Contribution of sensitivity, volume and tone to visceral perception in the upper gastrointestinal tract in man: emphasis on testing. Neurogastroenterol Motil. 2007;19:47–61.
11. Drewes AM, Schipper KP, Dimcevski G, Petersen P, Andersen OK, Gregersen H, Arendt-Nielsen L. Multimodal assessment of pain in the esophagus: a new experimental model. Am J Physiol Gastrointest Liver Physiol. 2002;283:G95–103.
12. Olesen AE, Staahl C, Brock C, Arendt-Nielsen L, Drewes AM. Evoked human oesophageal hyperalgesia: a potential tool for analgesic evaluation? Basic Clin Pharmacol Toxicol. 2009;105:126–36.
13. Olesen SS, Olesen AE, Gravesen F, Poulsen JK, Funch-Jensen P, Gregersen H, Drewes AM. An endoscopic method for thermal and chemical stimulation of the human oesophagus. Neurogastroenterol Motil. 2009;21:1250–e116.
14. Ness TJ, Gebhart GF. Visceral pain: a review of experimental studies. Pain. 1990;41:167–234.
15. Sarkar S, Hobson AR, Furlong PL, Woolf CJ, Thompson DG, Aziz Q. Central neural mechanisms mediating human visceral hypersensitivity. Am J Physiol Gastrointest Liver Physiol. 2001;281:G1196–202.
16. Rossel P, Drewes AM, Petersen P, Nielsen J, Arendt-Nielsen L. Pain produced by electric stimulation of the rectum in patients with irritable bowel syndrome: further evidence of visceral hyperalgesia. Scand J Gastroenterol. 1999;34:1001–6.
17. Mertz H, Naliboff B, Munakata J, Niazi N, Mayer EA. Altered rectal perception is a biological marker of patients with irritable bowel syndrome. Gastroenterology. 1995;109:40–52.
18. Bernstein CN, Niazi N, Robert M, Mertz H, Kodner A, Munakata J, Naliboff B, Mayer EA. Rectal afferent function in patients with inflammatory and functional intestinal disorders. Pain. 1996;66:151–61.
19. Dimcevski G, Schipper KP, Tage-Jensen U, Funch-Jensen P, Krarup AL, Toft E, Thorsgaard N, Arendt-Nielsen L, Drewes AM. Hypoalgesia to experimental visceral and somatic stimulation in painful chronic pancreatitis. Eur J Gastroenterol Hepatol. 2006;18:755–64.
20. Olesen SS, Brock C, Krarup AL, Funch-Jensen P, Arendt-Nielsen L, Wilder-Smith OH, Drewes AM. Descending inhibitory pain modulation is impaired in patients with chronic pancreatitis. Clin Gastroenterol Hepatol. 2010;8:724–30.
21. Arendt-Nielsen L, Laursen RJ, Drewes AM. Referred pain as an indicator for neural plasticity. Prog Brain Res. 2000;129:343–56.
22. Drewes AM, Pedersen J, Reddy H, Rasmussen K, Funch-Jensen P, Arendt-Nielsen L, Gregersen H. Central sensitization in patients with non-cardiac chest pain: a clinical experimental study. Scand J Gastroenterol. 2006;41:640–9.
23. Arendt-Nielsen L, Drewes AM, Hansen JB, Tage-Jensen U. Gut pain reactions in man: an experimental investigation using short and long duration transmucosal electrical stimulation. Pain. 1997;69:255–62.
24. Arendt-Nielsen L, Schipper KP, Dimcevski G, Sumikura H, Krarup AL, Giamberardino MA, Drewes AM. Viscero-somatic reflexes in referred pain areas evoked by capsaicin stimulation of the human gut. Eur J Pain. 2008;12:544–51.
25. Vecchiet L, Giamberardino MA, Dragani L, Albe-Fessard D. Pain from renal/ureteral calculosis: evaluation of sensory thresholds in the lumbar area. Pain. 1989;36:289–95.
26. Stawowy M, Funch-Jensen P, Arendt-Nielsen L, Drewes AM. Somatosensory changes in the referred pain area in patients with cholecystolithiasis. Eur J Gastroenterol Hepatol. 2005;17:865–70.
27. Stawowy M, Rossel P, Bluhme C, Funch-Jensen P, Arendt-Nielsen L, Drewes AM. Somatosensory changes in the referred pain area following acute inflammation of the appendix. Eur J Gastroenterol Hepatol. 2002;14:1079–84.

28. Buscher HC, Wilder-Smith OH, van Goor H. Chronic pancreatitis patients show hyperalgesia of central origin: a pilot study. Eur J Pain. 2006;10:363–70.
29. Qin C, Malykhina AP, Akbarali HI, Foreman RD. Cross-organ sensitization of lumbosacral spinal neurons receiving urinary bladder input in rats with inflamed colon. Gastroenterology. 2005;129:1967–78.
30. Sarkar S, Aziz Q, Woolf CJ, Hobson AR, Thompson DG. Contribution of central sensitisation to the development of non-cardiac chest pain. Lancet. 2000;356:1154–9.
31. Frokjaer JB, Andersen SD, Gale J, Arendt-Nielsen L, Gregersen H, Drewes AM. An experimental study of viscero-visceral hyperalgesia using an ultrasound-based multimodal sensory testing approach. Pain. 2005;119:191–200.
32. Simren M, Abrahamsson H, Bjornsson ES. An exaggerated sensory component of the gastrocolonic response in patients with irritable bowel syndrome. Gut. 2001;48:20–7.
33. Giamberardino MA. Recent and forgotten aspects of visceral pain. Eur J Pain. 1999;3:77–92.
34. Fass R, Achem SR, Harding S, Mittal RK, Quigley E. Review article: supra-oesophageal manifestations of gastro-oesophageal reflux disease and the role of night-time gastro-oesophageal reflux. Aliment Pharmacol Ther. 2004;20 Suppl 9:26–38.
35. Munakata J, Naliboff B, Harraf F, Kodner A, Lembo T, Chang L, Silverman DH, Mayer EA. Repetitive sigmoid stimulation induces rectal hyperalgesia in patients with irritable bowel syndrome. Gastroenterology. 1997;112:55–63.
36. Paterson WG, Wang H, Vanner SJ. Increasing pain sensation to repeated esophageal balloon distension in patients with chest pain of undetermined etiology. Dig Dis Sci. 1995;40:1325–31.
37. Dimcevski G, Staahl C, Andersen SD, Thorsgaard N, Funch-Jensen P, Arendt-Nielsen L, Drewes AM. Assessment of experimental pain from skin, muscle, and esophagus in patients with chronic pancreatitis. Pancreas. 2007;35:22–9.
38. Bouhassira D, Chollet R, Coffin B, Lemann M, Le Bars D, Willer JC, Jian R. Inhibition of a somatic nociceptive reflex by gastric distention in humans. Gastroenterology. 1994;107:985–92.
39. Bouhassira D, Sabate JM, Coffin B, Le Bars D, Willer JC, Jian R. Effects of rectal distensions on nociceptive flexion reflexes in humans. Am J Physiol. 1998;275:G410–7.
40. Curatolo M, Petersen-Felix S, Arendt-Nielsen L. Sensory assessment of regional analgesia in humans: a review of methods and applications. Anesthesiology. 2000;93:1517–30.
41. Olesen SS, Hansen TM, Graversen C, Steimle K, Wilder-Smith OH, Drewes AM. Slowed EEG rhythmicity in patients with chronic pancreatitis: evidence of abnormal cerebral pain processing? Eur J Gastroenterol Hepatol. 2011;23:418–24.
42. Valeriani M, Le Pera D, Tonali P. Characterizing somatosensory evoked potential sources with dipole models: advantages and limitations. Muscle Nerve. 2001;24:325–39.
43. Frobert O, Arendt-Nielsen L, Bak P, Funch-Jensen P, Peder BJ. Pain perception and brain evoked potentials in patients with angina despite normal coronary angiograms. Heart. 1996;75: 436–41.
44. Hollerbach S, Bulat R, May A, Kamath MV, Upton AR, Fallen EL, Tougas G. Abnormal cerebral processing of oesophageal stimuli in patients with noncardiac chest pain (NCCP). Neurogastroenterol Motil. 2000;12:555–65.
45. Rossel P, Pedersen P, Niddam D, Arendt-Nielsen L, Chen AC, Drewes AM. Cerebral response to electric stimulation of the colon and abdominal skin in healthy subjects and patients with irritable bowel syndrome. Scand J Gastroenterol. 2001;36:1259–66.
46. Arebi N, Bullas DC, Dukes GE, Gurmany S, Hicks KJ, Kamm MA, Hobson AR. Distinct neurophysiological profiles in irritable bowel syndrome. Am J Physiol Gastrointest Liver Physiol. 2011;300:G1086–93.
47. Frokjaer JB, Olesen SS, Graversen C, Andresen T, Lelic D, Drewes AM. Neuroimaging of the human visceral pain system—a methodological review. Scand J Pain. 2011;2:95–104.
48. Olesen SS, Frokjaer JB, Lelic D, Valeriani M, Drewes AM. Pain-associated adaptive cortical reorganisation in chronic pancreatitis. Pancreatology. 2010;10:742–51.
49. Hobson AR, Furlong PL, Worthen SF, Hillebrand A, Barnes GR, Singh KD, Aziz Q. Real-time imaging of human cortical activity evoked by painful esophageal stimulation. Gastroenterology. 2005;128:610–9.
50. Turner R. Magnetic resonance imaging of brain function. Am J Physiol Imaging. 1992;7:136–45.
51. Owen DG, Clarke CF, Ganapathy S, Prato FS, St Lawrence KS. Using perfusion MRI to measure the dynamic changes in neural activation associated with tonic muscular pain. Pain. 2010;148:375–86.
52. Wang J, Aguirre GK, Kimberg DY, Roc AC, Li L, Detre JA. Arterial spin labeling perfusion fMRI with very low task frequency. Magn Reson Med. 2003;49:796–802.
53. Napadow V, LaCount L, Park K, As-Sanie S, Clauw DJ, Harris RE. Intrinsic brain connectivity in fibromyalgia is associated with chronic pain intensity. Arthritis Rheum. 2010;62:2545–55.
54. Nucifora PG, Verma R, Lee SK, Melhem ER. Diffusion-tensor MR imaging and tractography: exploring brain microstructure and connectivity. Radiology. 2007;245:367–84.
55. Frokjaer JB, Bouwense SA, Olesen SS, Lundager FH, Eskildsen SF, van Goor H, Wilder-Smith OH, Drewes AM. Reduced cortical thickness of brain areas involved in pain processing in patients with chronic pancreatitis. Clin Gastroenterol Hepatol. 2012;10: 434–8.e1.
56. Frokjaer JB, Olesen SS, Gram M, Yavarian Y, Bouwense SA, Wilder-Smith OH, Drewes AM. Altered brain microstructure assessed by diffusion tensor imaging in patients with chronic pancreatitis. Gut. 2011;60:1554–62.
57. Fregni F, Potvin K, Dasilva D, Wang X, Lenkinski RE, Freedman SD, Pascual-Leone A. Clinical effects and brain metabolic correlates in non-invasive cortical neuromodulation for visceral pain. Eur J Pain. 2011;15:53–60.
58. Kwan CL, Diamant NE, Pope G, Mikula K, Mikulis DJ, Davis KD. Abnormal forebrain activity in functional bowel disorder patients with chronic pain. Neurology. 2005;65:1268–77.
59. Bernstein CN, Frankenstein UN, Rawsthorne P, Pitz M, Summers R, McIntyre MC. Cortical mapping of visceral pain in patients with GI disorders using functional magnetic resonance imaging. Am J Gastroenterol. 2002;97:319–27.
60. Clouse RE, Lustman PJ. Use of psychopharmacological agents for functional gastrointestinal disorders. Gut. 2005;54:1332–41.
61. Morgan V, Pickens D, Gautam S, Kessler R, Mertz H. Amitriptyline reduces rectal pain related activation of the anterior cingulate cortex in patients with irritable bowel syndrome. Gut. 2005;54:601–7.
62. Heiss WD, Herholz K. Brain receptor imaging. J Nucl Med. 2006;47:302–12.
63. Wirrwar A, Schramm N, Vosberg H, Muller-Gartner HW. High resolution SPECT in small animal research. Rev Neurosci. 2001;12:187–93.
64. Talbot JD, Marrett S, Evans AC, Meyer E, Bushnell MC, Duncan GH. Multiple representations of pain in human cerebral cortex. Science. 1991;251:1355–8.
65. Fukumoto M, Ushida T, Zinchuk VS, Yamamoto H, Yoshida S. Contralateral thalamic perfusion in patients with reflex sympathetic dystrophy syndrome. Lancet. 1999;354:1790–1.
66. Ladabaum U, Minoshima S, Hasler WL, Cross D, Chey WD, Owyang C. Gastric distention correlates with activation of multiple cortical and subcortical regions. Gastroenterology. 2001;120: 369–76.

67. Vandenbergh J, Dupont P, Fischler B, Bormans G, Persoons P, Janssens J, Tack J. Regional brain activation during proximal stomach distention in humans: a positron emission tomography study. Gastroenterology. 2005;128:564–73.
68. Stephan E, Pardo JV, Faris PL, Hartman BK, Kim SW, Ivanov EH, Daughters RS, Costello PA, Goodale RL. Functional neuroimaging of gastric distention. J Gastrointest Surg. 2003;7:740–9.
69. Staahl C, Olesen AE, Andresen T, Arendt-Nielsen L, Drewes AM. Assessing analgesic actions of opioids by experimental pain models in healthy volunteers—an updated review. Br J Clin Pharmacol. 2009;68:149–68.
70. Staahl C, Olesen AE, Andresen T, Arendt-Nielsen L, Drewes AM. Assessing efficacy of non-opioid analgesics in experimental pain models in healthy volunteers: an updated review. Br J Clin Pharmacol. 2009;68:322–41.
71. Ji Y, Traub RJ. Spinal NMDA receptors contribute to neuronal processing of acute noxious and nonnoxious colorectal stimulation in the rat. J Neurophysiol. 2001;86:1783–91.
72. McRoberts JA, Coutinho SV, Marvizon JC, Grady EF, Tognetto M, Sengupta JN, Ennes HS, Chaban VV, Amadesi S, Creminon C, Lanthorn T, Geppetti P, Bunnett NW, Mayer EA. Role of peripheral N-methyl-D-aspartate (NMDA) receptors in visceral nociception in rats. Gastroenterology. 2001;120:1737–48.
73. Olivar T, Laird JM. Differential effects of N-methyl-D-aspartate receptor blockade on nociceptive somatic and visceral reflexes. Pain. 1999;79:67–73.
74. Strigo IA, Duncan GH, Bushnell MC, Boivin M, Wainer I, Rodriguez Rosas ME, Persson J. The effects of racemic ketamine on painful stimulation of skin and viscera in human subjects. Pain. 2005;113:255–64.
75. Arendt-Nielsen L, Olesen AE, Staahl C, Menzaghi F, Kell S, Wong GY, Drewes AM. Analgesic efficacy of peripheral kappa opioid receptor agonist CR665 compared to oxycodone in a multimodal, multi-tissue experimental human pain model: selective effect on visceral pain. Anesthesiology. 2009;111:616–24.
76. Olesen AE, Staahl C, Arendt-Nielsen L, Drewes AM. Different effects of morphine and oxycodone in experimentally evoked hyperalgesia: a human translational study. Br J Clin Pharmacol. 2010;70:189–200.
77. Lenz H, Sandvik L, Qvigstad E, Bjerkelund CE, Raeder J. A comparison of intravenous oxycodone and intravenous morphine in patient-controlled postoperative analgesia after laparoscopic hysterectomy. Anesth Analg. 2009;109:1279–83.
78. De Schepper HU, Cremonini F, Park MI, Camilleri M. Opioids and the gut: pharmacology and current clinical experience. Neurogastroenterol Motil. 2004;16:383–94.
79. Krarup AL, Ny L, Astrand M, Bajor A, Hvid-Jensen F, Hansen MB, Simren M, Funch-Jensen P, Drewes AM. Randomised clinical trial: the efficacy of a transient receptor potential vanilloid 1 antagonist AZD1386 in human oesophageal pain. Aliment Pharmacol Ther. 2011;33:1113–22.
80. Curatolo M, Arendt-Nielsen L, Petersen-Felix S. Central hypersensitivity in chronic pain: mechanisms and clinical implications. Phys Med Rehabil Clin N Am. 2006;17:287–302.
81. Arendt-Nielsen L, Curatolo M, Drewes A. Human experimental pain models in drug development: translational pain research. Curr Opin Investig Drugs. 2007;8:41–53.
82. Melzack R. The McGill Pain Questionnaire: major properties and scoring methods. Pain. 1975;1:277–99.
83. Bjordal JM, Ljunggren AE, Klovning A, Slordal L. Non-steroidal anti-inflammatory drugs, including cyclo-oxygenase-2 inhibitors, in osteoarthritic knee pain: meta-analysis of randomised placebo controlled trials. BMJ. 2004;329:1317.
84. Houghton LA, Fell C, Whorwell PJ, Jones I, Sudworth DP, Gale JD. Effect of a second-generation alpha2delta ligand (pregabalin) on visceral sensation in hypersensitive patients with irritable bowel syndrome. Gut. 2007;56:1218–25.
85. Lee KJ, Kim JH, Cho SW. Gabapentin reduces rectal mechanosensitivity and increases rectal compliance in patients with diarrhoea-predominant irritable bowel syndrome. Aliment Pharmacol Ther. 2005;22:981–8.
86. Field MJ, Cox PJ, Stott E, Melrose H, Offord J, Su TZ, Bramwell S, Corradini L, England S, Winks J, Kinloch RA, Hendrich J, Dolphin AC, Webb T, Williams D. Identification of the alpha2-delta-1 subunit of voltage-dependent calcium channels as a molecular target for pain mediating the analgesic actions of pregabalin. Proc Natl Acad Sci U S A. 2006;103:17537–42.
87. Bauer CS, Rahman W, Tran-van-Minh A, Lujan R, Dickenson AH, Dolphin AC. The anti-allodynic alpha(2)delta ligand pregabalin inhibits the trafficking of the calcium channel alpha(2) delta-1 subunit to presynaptic terminals in vivo. Biochem Soc Trans. 2010;38:525–8.
88. Drewes AM, Krarup AL, Detlefsen S, Malmstrom ML, Dimcevski G, Funch-Jensen P. Pain in chronic pancreatitis: the role of neuropathic pain mechanisms. Gut. 2008;57:1616–27.
89. Olesen SS, Bouwense SA, Wilder-Smith OH, van Goor H, Drewes AM. Pregabalin reduces pain in patients with chronic pancreatitis in a randomized, controlled trial. Gastroenterology. 2011;141:536–43.
90. Olesen SS, Graversen C, Olesen AE, Frokjaer JB, Wilder-Smith O, van Goor H, Valeriani M, Drewes AM. Randomised clinical trial: pregabalin attenuates experimental visceral pain through sub-cortical mechanisms in patients with painful chronic pancreatitis. Aliment Pharmacol Ther. 2011;34:878–87.
91. Staahl C, Dimcevski G, Andersen SD, Thorsgaard N, Christrup LL, Arendt-Nielsen L, Drewes AM. Differential effect of opioids in patients with chronic pancreatitis: an experimental pain study. Scand J Gastroenterol. 2007;42:383–90.
92. Wilder-Smith CH, Hill L, Osler W, O'Keefe S. Effect of tramadol and morphine on pain and gastrointestinal motor function in patients with chronic pancreatitis. Dig Dis Sci. 1999;44:1107–16.
93. Wilder-Smith CH, Hill L, Wilkins J, Denny L. Effects of morphine and tramadol on somatic and visceral sensory function and gastrointestinal motility after abdominal surgery. Anesthesiology. 1999;91:639–47.
94. Eisenberg E, Midbari A, Haddad M, Pud D. Predicting the analgesic effect to oxycodone by 'static' and 'dynamic' quantitative sensory testing in healthy subjects. Pain. 2010;151:104–9.
95. Buhagiar L, Cassar OA, Brincat MP, Buttigieg GG, Inglott AS, Adami MZ, Azzopardi LM. Predictors of post-caesarean section pain and analgesic consumption. J Anaesthesiol Clin Pharmacol. 2011;27:185–91.
96. Pedersen K, Olesen A, Osther P, Arendt-Nielsen L, Drewes A. Prediction of postoperative pain after percutaneous nephrolithotomy: can preoperative experimental pain assessment identify patients at risk? Urolithiasis. 2013;41(2):169–77.
97. Hsu YW, Somma J, Hung YC, Tsai PS, Yang CH, Chen CC. Predicting postoperative pain by preoperative pressure pain assessment. Anesthesiology. 2005;103:613–8.
98. Martinez V, Fletcher D, Bouhassira D, Sessler DI, Chauvin M. The evolution of primary hyperalgesia in orthopedic surgery: quantitative sensory testing and clinical evaluation before and after total knee arthroplasty. Anesth Analg. 2007;105:815–21.
99. Pan PH, Coghill R, Houle TT, Seid MH, Lindel WM, Parker RL, Washburn SA, Harris L, Eisenach JC. Multifactorial preoperative predictors for postcesarean section pain and analgesic requirement. Anesthesiology. 2006;104:417–25.
100. Rudin A, Wolner-Hanssen P, Hellbom M, Werner MU. Prediction of post-operative pain after a laparoscopic tubal ligation procedure. Acta Anaesthesiol Scand. 2008;52:938–45.
101. Nielsen PR, Norgaard L, Rasmussen LS, Kehlet H. Prediction of post-operative pain by an electrical pain stimulus. Acta Anaesthesiol Scand. 2007;51:582–6.

102. Wilder-Smith OH, Tassonyi E, Crul BJ, Arendt-Nielsen L. Quantitative sensory testing and human surgery: effects of analgesic management on postoperative neuroplasticity. Anesthesiology. 2003;98:1214–22.
103. Aasvang EK, Hansen JB, Kehlet H. Can preoperative electrical nociceptive stimulation predict acute pain after groin herniotomy? J Pain. 2008;9:940–4.
104. Edwards RR, Sarlani E, Wesselmann U, Fillingim RB. Quantitative assessment of experimental pain perception: multiple domains of clinical relevance. Pain. 2005;114:315–9.
105. Attal N, Rouaud J, Brasseur L, Chauvin M, Bouhassira D. Systemic lidocaine in pain due to peripheral nerve injury and predictors of response. Neurology. 2004;62:218–25.
106. Yarnitsky D, Granot M, Nahman-Averbuch H, Khamaisi M, Granovsky Y. Conditioned pain modulation predicts duloxetine efficacy in painful diabetic neuropathy. Pain. 2012;153:1193–9.
107. Olesen SS, Graversen C, Bouwense SAW, van Goor H, Wilder-Smith OHG, Drewes AM. Quantitative sensory testing predicts pregabalin efficacy in painful chronic pancreatitis. PLoS One. 2013;8(3):e57963.

Abdominal Pain in Irritable Bowel Syndrome (IBS)

6

Roy Dekel, Douglas A. Drossman, and Ami D. Sperber

Key Points

- Irritable bowel syndrome (IBS) is the most common functional gastrointestinal disorder (FGID), affecting up to 15 % of the general population.
- It is characterized by chronic abdominal pain that can be mild and intermittent, or severe, constant, and debilitating. Pain in IBS, as in other chronic pain disorders, is a complex symptom resulting from the interplay between peripheral (visceral) stimulation (enteric nervous system) and central modulation (central nervous system).
- As the severity of pain increases central processing plays an increasingly important role compared to peripheral input. In IBS, the normal adaptive central inhibitory response to painful visceral stimuli is diminished. This change is modulated by psychosocial factors such as anxiety, depression, poor social support, and impaired coping skills.
- Successful treatment begins with a therapeutic doctor–patient partnership. Medical treatment of IBS includes peripherally acting and centrally acting agents with antidepressants playing a central role. Cognitive behavioral therapy (CBT), interpersonal (psychodynamic) therapy, hypnosis, stress reduction, and mindfulness meditation have been shown to be effective in the treatment of IBS.

R. Dekel, M.D. (✉)
Department of Gastroenterology, Tel Aviv Medical Center, Waitzman 6, Tel Aviv, Israel 64239
e-mail: royd@tasmc.health.gov.il

D.A. Drossman, M.D.
Center for Education and Practice of Biopsychosocial Care and Center for Functional GI and Motility Disorders, Drossman Gastroenterology and University of North Carolina at Chapel Hill, 55 Vilcom Center Drive Suite 110, Chapel Hill, NC 27514, USA
e-mail: drossman@med.unc.edu

A.D. Sperber, M.D., M.S.P.H.
Faculty of Health Sciences, Ben-Gurion University of the Negev, Beer-Sheva, Israel
e-mail: amiroie@me.com

Introduction

Most patients who present with gastrointestinal symptoms have no clear organic cause even after an extensive investigation and are diagnosed with a functional gastrointestinal disorder (FGID). Among the FGIDs, irritable bowel syndrome (IBS) is the most common, affecting up to 15 % of the general population. The hallmark of IBS is chronic abdominal pain associated with irregular bowel movements. The pain can be mild and intermittent or severe, constant, and debilitating. IBS patients are major healthcare utilizers and are seen and treated not only by primary care physicians and gastroenterologists but also by surgeons, gynecologists, pain specialists, and rheumatologists. Thus, it is important for physicians in diverse subspecialties to be familiar with the diagnosis and management of this disorder.

The purpose of this chapter is to review the epidemiology and diagnosis of IBS and provide an in-depth look into the pathogenesis and treatment of pain in IBS patients.

Epidemiology

IBS is a common functional disorder with a symptom-based diagnosis (Rome III diagnostic criteria, Table 6.1) [1]. The reported prevalence of IBS varies from study to study depending on diagnostic criteria used as well as other methodological differences among studies [2]. However, some findings on the epidemiology of IBS appear to hold true and are as follows:

1. IBS is a global problem that affects individuals all over the world [3]. The reported worldwide prevalence rates for IBS range from 5 % to 20 %.
2. In most countries IBS affects women (60–70 %) more than men [4, 5]. The East is unique in that there are reports from China, Taiwan, and Singapore of a similar prevalence between males and females [6, 7]. There are conflicting reports from India with community-based surveys reporting

L. Kapural (ed.), *Chronic Abdominal Pain: An Evidence-Based, Comprehensive Guide to Clinical Management*,
DOI 10.1007/978-1-4939-1992-5_6, © Springer Science+Business Media New York 2015

Table 6.1 Rome III diagnostic criteria[a] for IBS

Recurrent abdominal pain or discomfort[b] at least 3 days/month in the last 3 months associated with *two or more* of the following:
1. Improvement with defecation
2. Onset associated with a change in frequency of stool
3. Onset associated with a change in form (appearance) of stool

[a]Criterion fulfilled for the last 3 months with symptom onset at least 6 months prior to diagnosis
[b]"Discomfort" means an uncomfortable sensation not described as pain

higher prevalence of IBS among females in the general population and hospital-based surveys reporting higher proportion of males among patients in gastroenterology clinics [8, 9]. The latter observation might reflect cultural aspects of healthcare-seeking behaviors in Indian society.

3. Although IBS can appear at any age, it is more common in young and middle-aged patients and tends to be less common in the elderly [10, 11].
4. Socioeconomic status may play a role in the epidemiology of IBS, which has been reported in some countries to be more prevalent in lower socioeconomic classes [4, 12, 13], although the data on this factor are not consistent.

As a prevalent chronic disorder, IBS places a major economic burden on health care. A meta-analysis of 18 studies from the USA and the UK estimated the annual direct cost of an IBS patient (drugs, procedures, and doctor visits) at $348–8,750 and the annual indirect costs (loss of work days and deceased productivity) at $355–3,344 [14, 15]. Another US study estimated the overall annual direct cost of IBS to be $228 million in doctor visits and $80 million in drugs [15].

Diagnosis

There is no specific diagnostic finding or biomarker for IBS, so the diagnosis is based on patients' reports of their symptoms. In the past, IBS was considered a diagnosis of exclusion, but inherent to this approach is an exhaustive diagnostic work-up that involves unpleasant and potentially risky tests for the patient and is not cost effective. Thus, a symptom-based diagnostic system, known as the Rome criteria, was developed. The main concept introduced by the Rome criteria is that the diagnostic process of a functional disorder should be based on two components. The first is the presence of a typical cluster of symptoms and the second is the absence of "red flags" including initial presentation of symptoms at an age over 50, unexplained weight loss, fever, nocturnal symptoms, blood in the stool, a family history of gastrointestinal malignancy or disease (e.g., celiac or inflammatory bowel disease), or an abnormal finding on physical examination. Basic laboratory tests, such as a complete blood count and celiac serology, are usually enough to complete the diagnostic process and establish a firm diagnosis. Patients who fulfill the criteria and do not have red flags need a minimal diagnostic work-up after which the diagnosis of IBS can be made with confidence [16, 17].

The latest update of the Rome diagnostic criteria for IBS is Rome III, in which the diagnosis of IBS requires the presence of abdominal pain or discomfort for at least 10 % of the time over the previous three months with symptom onset at least six months earlier [18]. Additionally, pain should be relieved by defecation and associated with a change in the frequency of bowel movements or a change in the form of the stool. Accompanying symptoms, although not essential for the diagnosis, are a feeling of incomplete evacuation, abnormal stool frequency (less than three times a week or more than three times a day) or consistency, straining at defecation, urgency, mucus discharge, and bloating. IBS can be further divided into three main subgroups according to bowel habit as constipation predominant (IBS-C), diarrhea predominant (IBS-D), and those exhibiting an alternating bowel pattern [19]. Patients may switch from one subclass to another during the course of their illness. It has been demonstrated repeatedly that the use of positive symptom-based diagnostic criteria in conjunction with the use of red flags to guide further investigation in selected cases is a reliable and cost-effective approach. After establishing the diagnosis of IBS, based on the Rome criteria, it is rarely necessary to change the diagnosis [20–22].

The Pathophysiology of Pain in IBS

Abdominal pain is a hallmark of IBS and is essential for its diagnosis. In IBS, as in many other chronic pain syndromes, pain is a complex experience resulting from the interplay between peripheral (visceral) stimulation (enteric nervous system) and central modulation (central nervous system [CNS]).

Afferent stimulation from the colon is transmitted to second-order neurons in the spinal cord and then ascends to the brain through the spinothalamic, spinoreticular, and spinomesencephalic tracts. These tracts connect to the somatosensory cortex responsible for registration and localization of painful visceral and somatic stimuli. They also connect to structures in the limbic system that are involved in the reflexive, affective, and motivational responses to pain [23]. The afferent pathways project to the perigenual anterior cingulate cortex (pACC), which is involved in affective modification, and to the midcingulate cortex (MCC), which is involved in the behavioral response.

The amplification of afferent visceral stimulation can result from increased excitability of peripheral receptors or impaired spinal and/or central pain regulatory systems. Increased excitability can produce the two related phenomena of hyperalgesia (increased pain response to painful stimuli) and allodynia (increased pain response to nonpainful stimuli) [24]. Thus, afferent visceral stimulation can be experienced as painful not only as a result of peripheral intensity but also as a result of central processing that may be modulated by psychosocial factors such as anxiety, depression, poor social support, and impaired coping skills [25]. As the severity of pain increases central processing plays an increasingly important role compared to peripheral input. Once a pattern of central sensitization has taken hold, patients may even experience severe pain without ongoing peripheral nociceptive stimulation [26, 27]. This is the extreme end of the IBS severity spectrum.

While we do not have full knowledge of all the causes of excessive peripheral stimulation, there is good evidence that eating, infection, inflammation, physical injury, hormones (e.g., menses), or colonic motility may play a role.

Up to 15 % of IBS patients attribute the beginning of their symptoms to an acute episode of gastrointestinal infection. A meta-analysis of eight papers including almost 600,000 patients over a follow-up up to one year found that the odds ratio for developing IBS after such an episode is seven [28]. IBS that follows acute intestinal infection has been shown to be associated with a persistent or chronic state of inflammation that cannot be identified by routine clinical tests and procedures [29, 30].

Risk factors for postinfectious IBS are related to not only to the severity of the acute infectious episode (fever, bloody stools, and need for hospitalization) but also to patient characteristics such as female gender, stress, anxiety, and depression [31]. This is a good example of how excessive afferent stimulation, induced in this case by a microinflammatory state, can develop into a chronic condition such as IBS-D after central sensitization occurs in a susceptible person with psychological comorbidity.

Peripheral stimulation and its interplay with central amplification are also reflected in the development of chronic abdominal pain following abdominal or pelvic surgery. IBS patients reported up to twice the number of appendectomies and hysterectomies and up to three times the number of cholecystectomies compared with those without IBS [32]. Surgery may cause visceral afferent sensitization that eventually results in allodynia and chronic pain even in the presence of normal gut function.

This contention is supported by a study that evaluated the development of abdominal pain after elective gynecologic surgery for nonpainful indications [32]. Patients with no prior history of chronic abdominal pain undergoing gynecological surgery for nonpainful indications were followed for the development of de novo abdominal pain following surgery. They were compared with a control group comprised of nonsurgical patients who came to a gynecologic clinic for nonpain-related reasons. At one-year follow-up significantly more patients in the surgery group complained of chronic abdominal pain (15.3 %) than in the control group (3.6 %, $p=0.003$). There was no association between any surgery-related variables and the subsequent development of chronic abdominal pain. The only predictors of chronic abdominal pain at one-year follow-up were associated with the patients' preoperative psychological profile. Patients anticipating difficulty with surgery or recovery from it and those with lower scores on the Sense of Coherence questionnaire (an index of coping skills) were more likely to develop chronic postoperative abdominal pain. In these cases, the interplay of peripheral visceral stimulus together with central sensitization related to psychosocial variables affected the de novo development of chronic abdominal pain.

Studies using functional MRI and PET CT have demonstrated that the ACC, which is responsible for descending pain inhibition, is less active in IBS patients. This phenomenon is also found in other chronic pain syndromes such as fibromyalgia [33–35]. In contrast, the MCC, which is associated with unpleasantness and fear, is overactive. Therefore, in IBS patients the normal adaptive inhibitory response to painful visceral stimuli is diminished and replaced by a maladaptive, presumably even aggravating, response [33, 34, 36]. The factors that ultimately lead to this shift into a maladaptive pattern are psychosocial in nature. This connection was elegantly demonstrated in the case report of a patient with a severe functional gastrointestinal pain syndrome and a history of abuse [37]. Her baseline brain scan demonstrated marked activation of the MCC and the somatosensory cortex. Following successful treatment with antidepressants and psychotherapy a repeated scan demonstrated diminished MCC activity and increased insular activation. Thus, maladaptive brain responses are reversible and so is the patient's clinical situation.

Treatment of Abdominal Pain in IBS

As in other fields of medicine, in particular in patients with chronic painful conditions, the healing process for IBS patients begins when the patient enters the doctor's office before any medicine has been prescribed. It is of the outmost importance to establish a good doctor–patient relationship in order to succeed in the therapeutic process [38, 39]. Some of the essentials of a salutary doctor–patient relationship are discussed below:

1. Allow enough time especially for the first meeting. The patient should feel that the doctor is listening to and him/her and that their symptoms are considered legitimate and are being taken seriously.
2. Take a full detailed history and perform a physical examination: These basic measures of good clinical practice help to foster the doctor–patient relationship.

3. It is very helpful to remember four key questions that patients should be asked:
   a. What brings you here at this time? IBS is a chronic condition and many patients have their symptoms for years before consulting a specialist. Consultation is often driven by a specific anxiety or a stressful situation that should be addressed.
   b. What do you think is the cause of your symptoms? Many IBS patients attribute their symptoms to undiagnosed cancer, infection, inflammatory bowel disease, or food allergy.
   c. What are your concerns or worries? It is important to understand the patient's agenda and to address their primary concerns such as "What exactly do I have?" or "Do I have cancer," or alternatively related to the symptoms like "I can't deal with this pain anymore."
   d. What are your expectations from me? Some patients have the unrealistic expectation of a "quick fix" for their situation that can lead to mutual frustration and treatment failure [40]. It should be emphasized that treating IBS is a process rather than an isolated consultation and that the goal of treatment is to reduce their suffering and to improve their quality of life rather than to "cure" them.

Many IBS patients have never received a comprehensive explanation about the nature of their problem. This may be the basis for the unwarranted fears ("I might have cancer") and feelings of frustration ("why can't they figure out what I have"). A detailed explanation about the nature of functional disorders and their natural history is very important to deal with these issues.

Treating IBS patients is an ongoing process that takes time. Throughout this process patients are likely to encounter difficulties, setbacks, and frustration. Patients should not feel that they are left alone to deal with their setbacks. Scheduling a follow-up phone call, for example, is a simple measure that is often sufficient to allay patients' new concerns [41]. Physicians should inquire about comorbid gastrointestinal and nongastrointestinal functional disorders. IBS patients have a high prevalence of other functional disorders [42], leading some patients to feel that they are very ill. By providing patients with a unifying paradigm that connects different, apparently unrelated, symptoms to one disorder (i.e., central sensitization), we can alleviate much of their fears and concerns.

For some patients with mild symptoms, these steps may be enough to alleviate fears and concerns regarding their symptoms. These patients often continue to cope successfully with their symptoms and need no further treatment. However, the majority of patients will require more specific treatment.

The treatment options for IBS can be divided into pharmacological and nonpharmacological treatment modalities (Fig. 6.1).

Medical Treatment

Medical treatment of IBS includes peripherally acting agents and centrally acting agents.

Peripherally Acting Agents

These drugs act on the gut itself and are targeted against specific IBS symptoms such altered bowel movements, bloating, and cramps. Because they are not key agents in

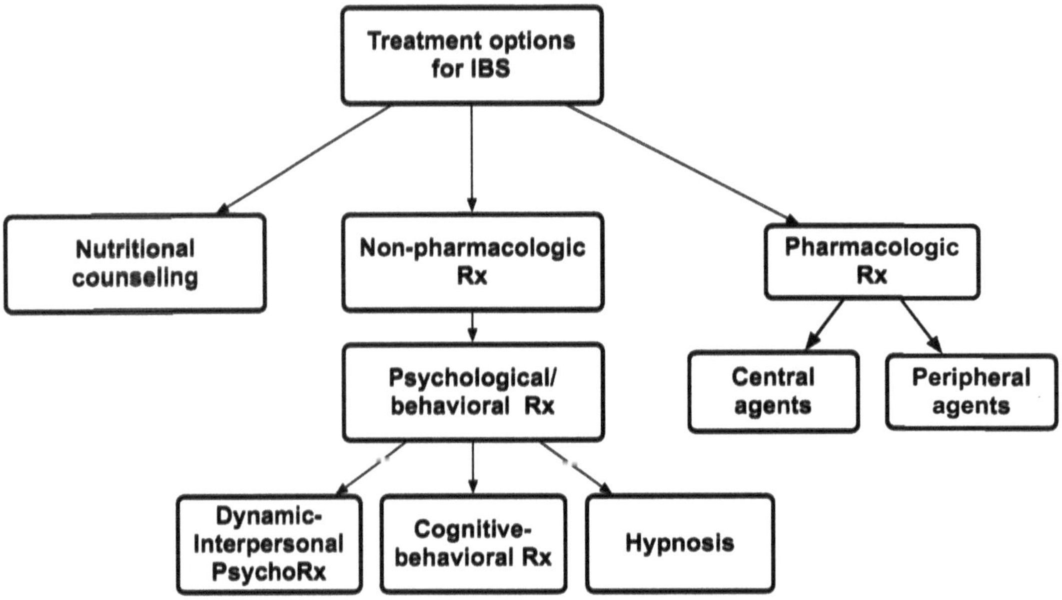

Fig. 6.1 Treatment options for IBS in addition to a therapeutic doctor–patient therapeutic partnership. Although there is not cure for IBS, a large number of treatment options are available to reduce suffering and improve quality of life. Doctors need not feel "empty handed" when coming to treat these patients

Table 6.2 Peripheral agents used most commonly in the treatment of IBS. Peripheral agents, although not primarily directed against pain, have an important role in IBS treatment. In mild IBS cases, they might suffice but in more severe IBS cases and, where pain is a cardinal symptom, central agents are preferred

Class	Drug	Mechanism of action	Comments
• Antispasmodics	• Pinaverium • Mebeverin	• Direct visceral smooth muscle relaxants	• Modest effect on IBS spastic pain
	• Colpermin (peppermint oil) • Hyoscamine dicyclomine	• Anticholinergic/antimuscarinic	• Otilinium bromide, hyoscine, and colpermin; best evidence for effectiveness
• Serotonergic and other agents	• Alosetron	• 5HT3 receptor antagonist	• Available only through a restricted access program; increased incidence of ischemic colitis
	• Tegaserode	• 5HT4 receptor agonist	• Withdrawn from the US market; an increased incidence of cardiovascular adverse events
	• Linaclotide	• Guanylate cyclase-C agonist	• Recently approved in Europe and the US for IBS-C
	• Lubiprostone	• Chloride channel activator	• In phase 3 studies, lubiprostone was almost twice as effective for IBS symptoms as placebo

IBS pain management only some of them are discussed in detail and the rest is mentioned briefly. Table 6.2 summarizes the main facts about the different peripheral agents. Serotonin (5HT) is an important neurotransmitter that coordinates gut function and has played a key role in research and drug development. It is secreted from enterochromaffin cells in the mucosa and is involved in almost every aspect of gut function including motility, sensation, and secretion. Alosetron is a 5HT3 receptor antagonist that was shown to improve global IBS symptoms and pain in women with IBS-D. A meta-analysis comparing 12 randomized controlled trials that evaluated the efficacy of alosetron compared to placebo found an odds ratio of 1.85 for improvement in the alosetron group [43]. Unfortunately, after initial FDA approval, safety issues and in particular ischemic colitis and severe constipation led to its withdrawal from the market. It was reintroduced in 2002 under a restricted access program. Under this program, alosetron can be prescribed (under some restrictions) to women with severe IBS-D who have failed to respond to traditional medical therapies.

Lubiprostone is a chloride channel activator that has been approved by the FDA for chronic constipation and IBC-C. In phase 3 studies, patients receiving lubiprostone were almost twice as likely to gain relief from overall IBS symptoms compared to patients who received placebo [44]. The main side effect of lubiprostone, nausea, is reported in 8 % of IBS-C patients who receive 8 mcg twice daily.

Centrally Acting Agents

Centrally acting agents should be the cornerstone of treatment in moderate-to-severe cases of IBS [45]. The main classes of drugs that are being used are the selective serotonin reuptake inhibitors (SSRIs), selective serotonin-norepinephrine reuptake inhibitors (SNRIs), and tricyclic antidepressants (TCAs). Other drugs, such as Mirtazapine, Buspiron, and the atypical antipsychotic Quetiapine, can also be used. These drugs were developed for the treatment of anxiety and depression, but can and should be used in IBS as discussed below. The different drugs and dosages are summarized in Table 6.3.

Antidepressants play a central role in medical therapy for IBS for two main reasons. First, they have a direct analgesic effect and are used in various pain syndromes, with or without concomitant depression, to elevate pain thresholds via central and peripheral effects. Second, since many IBS patient have psychological comorbidity, they can gain direct benefit from these drugs. Whether the main effect of antidepressants stems from central mechanisms (modulation of central pain processing) or from peripheral effects (effects on motility and secretion and reduction of afferent pain signals) or just from reducing depression and anxiety is still uncertain. The actual mechanism is probably a combination of all three. A recent meta-analysis found all classes of antidepressants to be effective in IBS with a number needed to treat as low as four [46].

Antidepressants in IBS (especially TCAs) are given at much lower doses then those used for the treatment of depression. The usual starting dose in 25–50 mg and can be increased as needed. SSRIs and SNRIs are usually given in the lower range of the "regular" psychiatric doses, for example, 10–20 mg of escitalopram or 30 mg of duloxetine.

Since TCAs and SNRIs have an independent indication in other pain syndromes, such as neuropathic pain and fibromyalgia, they are the drugs of choice for painful IBS. The choice between them is often based on the therapeutic profile of the drugs including potential adverse effects.

Table 6.3 Common interventions used in IBS. For optimal results these interventions can be used in combination ("augmentation" therapy). The use of more than one drug at a low dose can augment the therapeutic response and minimize the side effects

Drug	Drug (daily dose range [mg])	Comments
TCA	• Desipramine (25–150) • Nortriptyline (25–150) • Amitriptyline (25–150)	• Begin with low dose and titrate by response • Allow 4–8 weeks for maximal response
SSRIs	• Paroxetine (20–60) • Escitalopram (10–20)	• Begin with low dose and titrate by response
SNRIs	• Venlafaxine (25–300) • Duloxetine (20–80)	• Psychological and analgesic effects
Atypical antipsychotics	• Quetiapine (25–100)	• Preliminary reports
Tetracyclic antidepressant	• Mirtazepine (15–45)	• Antiemetic properties
Azaspirodecanediones	• Buspiron (10–60)	• Improves gastric receptive relaxation

For example, TCAs tend to be more constipating and have less anxiolytic properties, so an SNRI would be the preferred option in a patient with constipation or prominent anxiety. However, in many cases a combination of two drugs or more is necessary. Instead of increasing the dose of a single drug to the maximum, the use of a combination of two or more drugs from different classes and in lower doses (e.g., a TCA and an SNRI or SSRI) is recommended. This approach known as "augmentation therapy," helps minimize adverse effects, to which patients with functional GI disorders are prone [45].

Mirtazapine is a tetracyclic antidepressant used primarily in the treatment of depression. It has serotonergic as well as noradrenergic properties. It has antagonistic alpha-2 receptor and 5HT1, 5HT2, and 5HT3 properties as well as moderated peripheral alpha-1 adrenergic and alpha-1 anticholinergic properties. Its 5HT3 antagonistic action is probably responsible for its antiemetic properties. In addition to its antidepressant effects, it is also used at times as a hypnotic, antiemetic, as an appetite stimulant, and for the treatment of anxiety. In IBS, it can be used to augment the antidepressant and anxiolytic properties of other agents (such as a TCA or an SNRI) and for nausea and vomiting or low body weight, as is often seen in patients with a comorbid eating disorder. Data regarding its use in IBS are limited and more studies are needed to explore its exact place.

Quetiapine is an atypical antipsychotic approved for the treatment of schizophrenia, bipolar disorder, and as an add-on to treat depression. It has potential benefits in IBS by reducing anxiety, restoring normal sleep patterns, and potentially through a direct analgesic effect. A recent paper reported a retrospective analysis of its use in low doses (50–200 mg) in patients with severe FGIDs. Of the 21 treated patients, 10 discontinued the drug due to adverse effects or lack of efficacy, but of the 11 patients who stayed on the drug 6 reported improvement [47]. Although this is a small and uncontrolled study, it is encouraging considering that these were patients with extremely severe IBS who did not respond to any previous treatment modality. A larger, prospective, open-label study is currently underway.

Finally, Buspirone is a nonbenzodiazepine anxiolytic agent that is used in psychiatry to augment the effect of antidepressants. It also has a 5HT1 agonist effect, which may contribute to increasing gastric compliance/relaxation as has been shown to occur for functional dyspepsia. Therefore, it might be useful in patients with comorbid dyspeptic symptoms such as epigastric discomfort and early satiety.

There are two main barriers that clinicians face when trying to treat IBS patients with antidepressants. The first is the general reluctance of these patients to take "chemical" and "mind altering" agents. The second is patients' tendency to underestimate the psychological component of their symptoms. A thorough explanation regarding the mechanisms of pain (visceral hypersensitivity modulated by central mechanisms) and the drug's independent analgesic properties is enough in many cases. Some patients view the recommendation for a psychotropic drug as evidence that the doctor does not acknowledge their pain and thinks that they are "crazy." If we emphasize that we are recommending these drugs for their central analgesic effect, we can overcome much of this reticence to take them. This can be accomplished with a statement such as: "The same drug can be used for different reasons. For example, in the past aspirin was the leading drug for reducing fever and relieving pain, but currently it is the number one drug for the prevention of heart disease. Similarly, antidepressant drugs are effective in the treatment of depression at higher doses, but are also effective in lower dosages for pain relief". The patient should always make the final decision regarding the drug. This can be achieved by fostering a feeling of therapeutic partnership instead of an authoritative relationship where the patient has no say about the way he is treated. An example for such an approach would be: "In IBS there are many therapeutic options, with and without drugs. Each has its advantages and disadvantages. Do you want me to tell you about options that could help you with your symptoms?" By making the drug the patient's choice, we can augment adherence to treatment. Finally, in our experience, the adherence rate for drug therapy increases if the physician is available to address, in real time, early adverse effects, and other concerns that otherwise may lead the patient to discontinue therapy on their own.

Nonpharmacologic Therapy for IBS

Nonpharmacological treatments for IBS include stress reduction, and behavioral and psychological interventions.

Behavioral Interventions

Behavioral interventions are commonly used to treat IBS. They are safe and their benefit may go beyond symptomatic treatment and induce positive physiological changes. They are particularly suited to patients who do not want to take drugs. The effect of different modalities, including cognitive behavioral therapy (CBT), interpersonal (psychodynamic) therapy, hypnosis, stress reduction, and mindfulness meditation, has been evaluated for IBS. All help patients deal with issues such as maladaptive illness beliefs and behaviors, and the relationship between stress, life events, and symptomatology.

CBT can help patients recognize misperceptions and maladaptive thoughts regarding their symptoms and enhance their coping abilities. It can be administered as individual or group therapy [48–50]. In the largest randomized placebo-controlled study conducted to date, the investigators found that 12 weekly CBT sessions were significantly more beneficial than placebo for female patients with moderate-to-severe FGIDs [51].

Interpersonal (psychodynamic) therapy presumes that symptoms are associated with difficulties in interpersonal relationships. Its focus is on the identification of interpersonal situations that lead to symptom exacerbation. The treatment itself involves psychotherapy. The symptoms improve when the conflicts are resolved. Interpersonal dynamic psychotherapy has been shown to improve symptoms and to reduce disability and healthcare costs in IBS [52–54].

The aim of stress reduction (relaxation training) is to counteract the physiologic effects of stress. Reduction in skeletal muscle tension can decrease autonomic arousal and subjective tension/anxiety and may improve gut motility. Stress reduction and relaxation training includes modalities such as guided imagery, relaxation response, meditation, yoga, and biofeedback. Muscle relaxation alone or in combination with CBT and other techniques was shown to reduce IBS symptoms [55]. Mindfulness meditation is a form of relaxation involving an active nonjudgmental awareness of body sensations and emotions. Group mindfulness meditation resulted in improved IBS symptoms and health-related quality of life as well as reduced stress levels in women with IBS [56], effects that persisted at a three-month follow-up assessment.

Hypnosis is a form of guided imagery that uses muscle relaxation and gut-targeted suggestions to improve the gut function and reduce symptoms. Hypnosis involves nonspecific effects of relaxation, stress management, ego strengthening, and gut-directed suggestions of normal functioning and pleasant feeling. Data gathered from studies in different centers support the use of hypnosis as an effective, viable treatment option in IBS [57] that improves IBS symptoms and quality of life and reduces stress and anxiety. Moreover, the beneficial effects of hypnosis have been shown to persist at long-term follow-up [58–60].

The predictors of a favorable outcome in behavioral interventions include confidence in treatment success, perceived sense of control over symptoms, a good relationship with the therapist, and early response [61]. The choice of intervention depends on local expertise and availability as well as patient preference.

Summary and Conclusions

IBS is a common medical problem, which, although not life threatening, has a significant negative impact on patients' quality of life. Its range of severity ranges from mild intermittent symptoms to a disabling condition with a considerable loss of daily function. Pain in IBS is the result of peripheral afferent stimulation and CNS processing. A biopsychosocial perspective, taking into account the patient's psychological status, life experiences, beliefs, and concerns can help doctors provide optimal care. The primary goal of treatment is care rather than cure, and the various treatment options can be highly effective in reducing suffering and improving quality of life. The doctor–patient relationship is the foundation of successful treatment and should be supplemented by pharmacological or nonpharmacological treatments in accordance with the clinical situation and the patient's preference.

References

1. Longstreth GF, Thompson WG, Chey WD, Houghton LA, Mearin F, Spiller RC. Functional bowel disorders. In: Drossman SA, Corazziari E, Delvaux M, Spiller RC, Talley NJ, Thompson WG, et al., editors. Rome III the functional gastrointestinal disorders. McLean, VA: Degnon Associates; 2006. p. 487–555.
2. Sperber AD. The challenge of cross-cultural, multi-national research: potential benefits in the functional gastrointestinal disorders. Neurogastroenterol Motil. 2009;21:351–60.
3. Drossman DA, Camilleri M, Mayer EA, Whitehead WE. AGA technical review on irritable bowel syndrome. Gastroenterology. 2002;123:2108–31.
4. Andrews EB, Eaton SC, Hollis KA, Hopkins JS, Ameen V, Hamm LR, et al. Prevalence and demographics of irritable bowel syndrome: results from a large web-based survey. Aliment Pharmacol Ther. 2005;22:935–42.

5. Sperber AD, Friger M, Shvartzman P, Abu-Rabia M, Abu-Rabia R, Abu Rashed M, et al. Rates of functional bowel disorders among Israeli Bedouins in rural areas compared to those who moved to permanent towns. Clin Gastroenterol Hepatol. 2005;3:342–8.
6. Lu CL, Chang FY, Lang HC, Chen CY, Luo JC, Lee SD. Gender difference on the symptoms, health-seeking behaviour, social impact and sleep quality in irritable bowel syndrome: a Rome II-based survey in an apparent healthy adult Chinese population in Taiwan. Aliment Pharmacol Ther. 2005;21:1497–505.
7. Xiong LS, Chen MH, Chen HX, Xu AG, Wang WA, Hu PJ. A population-based epidemiologic study of irritable bowel syndrome in South China: stratified randomized study by cluster sampling. Aliment Pharmacol Ther. 2004;19:1217–24.
8. Makharia GK, Verma AK, Amarchand R, Goswami A, Singh P, Agnihotri A, et al. Prevalence of irritable bowel syndrome: a community based study from northern India. J Neurogastroenterol Motil. 2011;17:82–7.
9. Ghoshal UC, Abraham P, Bhatt C, Choudhuri G, Bhatia SJ, Shenoy KT, et al. Epidemiological and clinical profile of irritable bowel syndrome in India: report of the Indian Society of Gastroenterology Task Force. Indian J Gastroenterol. 2008;27:22–8.
10. Hungin AP, Chang L, Locke GR, Dennis EH, Barghout V. Irritable bowel syndrome in the United States: prevalence, symptom patterns and impact. Aliment Pharmacol Ther. 2005;21:1365–75.
11. Hungin AP, Whorwell PJ, Tack J, Mearin F. The prevalence, patterns and impact of irritable bowel syndrome: an international survey of 40,000 subjects. Aliment Pharmacol Ther. 2003;17:643–50.
12. Minocha A, Chad W, Do W, Johnson WD. Racial differences in epidemiology of irritable bowel syndrome alone, un-investigated dyspepsia alone, and "overlap syndrome" among African Americans compared to Caucasians: a population-based study. Dig Dis Sci. 2006;51:218–26.
13. Wilson S, Roberts L, Roalfe A, Bridge P, Singh S. Prevalence of irritable bowel syndrome: a community survey. Br J Gen Pract. 2004;54:495–502.
14. Maxion-Bergemann S, Thielecke F, Abel F, Bergemann R. Costs of irritable bowel syndrome in the UK and US. Pharmacoeconomics. 2006;24:21–37.
15. Sandler RS, Everhart JE, Donowitz M, adams E, Cronin K, Goodman C, et al. The burden of selected digestive diseases in the United States. Gastroenterology. 2002;122:1500–11.
16. Vanner SJ, Depew WT, Paterson WG, DaCosta LR, Groll AG, Simon JB, et al. Predictive value of the Rome criteria for diagnosing the irritable bowel syndrome. Am J Gastroenterol. 1999;94:2912–7.
17. Hammer J, Eslick GD, Howell SC, Altiparmak E, Talley NJ. Diagnostic yield of alarm features in irritable bowel syndrome and functional dyspepsia. Gut. 2004;53:666–72.
18. Longstreth GF, Thompson WG, Chey WD, Houghton LA, Mearin F, Spiller RC. Functional bowel disorders. Gastroenterology. 2006;130:1480–91.
19. Talley NJ, Zinsmeister AR, Melton 3rd LJ. Irritable bowel syndrome in a community: symptom subgroups, risk factors, and health care utilization. Am J Epidemiol. 1995;142:76–83.
20. Owens DM, Nelson DK, Talley NJ. The irritable bowel syndrome: long-term prognosis and the physician-patient interaction. Ann Intern Med. 1995;122:107–12.
21. Harvey RF, Mauad EC, Brown AM. Prognosis in the irritable bowel syndrome: a 5-year prospective study. Lancet. 1987;1(8539):963–5.
22. Svendsen JH, Munck LK, Andersen LR. Irritable bowel syndrome: prognosis and diagnostic safety. A 5-year follow up study. Scand J Gastroenterol. 1985;20:415–8.
23. Drossman DA. Functional abdominal pain syndrome. Clin Gastroenterol Hepatol. 2004;2:353–65.
24. Coderre TJ, Katz J, Vaccarino AL, Melzack R. Contribution of central neuroplasticity to pathological pain: review of clinical and experimental evidence. Pain. 1993;52:259–85.
25. Sperber AD, Drossman DA. Review article: the functional abdominal pain syndrome. Aliment Pharmacol Ther. 2011;33:514–24.
26. Yunus MB. Central sensitivity syndromes: a new paradigm and group nosology for fibromyalgia and overlapping conditions, and the related issue of disease versus illness. Semin Arthritis Rheum. 2008;37:339–52.
27. Elsenbruch S. Abdominal pain in irritable bowel syndrome: a review of putative psychological, neural and neuro-immune mechanisms. Brain Behav Immun. 2011;25:386–94.
28. Halvorson HA, Schlett CD, Riddle MS. Postinfectious irritable bowel syndrome–a meta-analysis. Am J Gastroenterol. 2006;101: 1894–9.
29. Dunlop SP, Jenkins D, Neal KR, Spiller RC. Relative importance of enterochromaffin cell hyperplasia, anxiety, and depression in postinfectious IBS. Gastroenterology. 2003;125:1651–9.
30. Guilarte M, Santos J, de Torres I, Alonso C, Vicario M, Ramos L, et al. Diarrhoea-predominant IBS patients show mast cell activation and hyperplasia in the jejunum. Gut. 2007;56:203–9.
31. Marshall JK, Thabane M, Garg AX, Clark WF, Salvadori M, Collins SM. Incidence and epidemiology of irritable bowel syndrome after a large waterborne outbreak of bacterial dysentery. Gastroenterology. 2006;131:445–50.
32. Sperber AD, Blank Morris C, Greemberg L, Bangdiwala SI, Goldstein D, Sheiner E, et al. Development of abdominal pain and IBS following gynecological surgery: a prospective, controlled study. Gastroenterology. 2008;134:75–84.
33. Mertz H, Morgan V, Tanner G, Pickens D, Price R, Shyr Y, et al. Regional cerebral activation in irritable bowel syndrome and control subjects with painful and nonpainful rectal distention. Gastroenterology. 2000;118:842–8.
34. Naliboff BD, Derbyshire SW, Munakata J, Berman S, Mandelkern M, Chang L, et al. Cerebral activation in patients with irritable bowel syndrome and control subjects during rectosigmoid stimulation. Psychosom Med. 2001;63:365–75.
35. Chang L, Berman S, Mayer EA, Suyenobu B, Derbyshire S, Naliboff B, et al. Brain responses to visceral and somatic stimuli in patients with irritable bowel syndrome with and without fibromyalgia. Am J Gastroenterol. 2003;98:1354–61.
36. Ringel Y, Drossman DA, Turkington TG, Bradshaw B, Hawk TC, Bangdiwala S, et al. Regional brain activation in response to rectal distension in patients with irritable bowel syndrome and the effect of a history of abuse. Dig Dis Sci. 2003;48:1774–81.
37. Drossman DA, Ringel Y, Vogt BA, Leserman J, Lin W, Smith JK, et al. Alterations of brain activity associated with resolution of emotional distress and pain in a case of severe irritable bowel syndrome. Gastroenterology. 2003;124:754–61.
38. Tanaka Y, Kanazawa M, Fukudo S, Drossman DA. Biopsychosocial model of irritable bowel syndrome. J Neurogastroenterol Motil. 2011;17:131–9.
39. Drossman DA. Psychosocial sound bites: exercises in the patient-doctor relationship. Am J Gastroenterol. 1997;92:1418–23.
40. Drossman DA. Diagnosing and treating patients with refractory functional gastrointestinal disorders. Ann Intern Med. 1995;123:688–97.
41. Wong RK, Tan JS, Drossman DA. Here's my phone number, don't call me: physician accessibility in the cell phone and e-mail era. Dig Dis Sci. 2010;55:662–7.
42. Sperber AD, Dekel R. Irritable bowel syndrome and co-morbid gastrointestinal and extra-gastrointestinal functional syndromes. J Neurogastroenterol Motil. 2010;16:113–9.
43. Cremonini F, Delgado-Aros S, Camilleri M. Efficacy of alosetron in irritable bowel syndrome: a meta-analysis of randomized controlled trials. Neurogastroenterol Motil. 2003;15:79–86.
44. Drossman DA, Chey WD, Johanson JF, Fass R, Scott C, Panas R, et al. Clinical trial: lubiprostone in patients with constipation-associated irritable bowel syndrome–results of two randomized,

placebo-controlled studies. Aliment Pharmacol Ther. 2009;29: 329–41.

45. Drossman DA. Beyond tricyclics: new ideas for treating patients with painful and refractory functional gastrointestinal symptoms. Am J Gastroenterol. 2009;104:2897–902.
46. Ford AC, Talley NJ, Schoenfeld PS, Quigley EMM, Moayyedi P. Efficacy of antidepressants and psychological therapies in irritable bowel syndrome: systematic review and meta-analysis. Gut. 2009;58:367–78.
47. Grover M, Dorn SD, Weinland SR, Dalton CB, Gaynes BN, Drossman DA. Atypical antipsychotic quetiapine in the management of severe refractory functional gastrointestinal disorders. Dig Dis Sci. 2009;54:1284–91.
48. Lackner JM, Jaccard J, Krasner SS, Katz LA, Gudleski GD, Holroyd K. Self-administered cognitive behavior therapy for moderate to severe irritable bowel syndrome: clinical efficacy, tolerability, feasibility. Clin Gastroenterol Hepatol. 2008;6:899–906.
49. Kennedy TM, Chalder T, McCrone P, Darnley S, Knapp M, Jones RH, et al. Cognitive behavioural therapy in addition to antispasmodic therapy for irritable bowel syndrome in primary care: randomised controlled trial. Health Technol Assess. 2006;10:iii–iv, ix–x, 1–67.
50. Blanchard EB, Lackner JM, Sanders K, Krasner S, Keefer L, Payne A, et al. A controlled evaluation of group cognitive therapy in the treatment of irritable bowel syndrome. Behav Res Ther. 2007;45:633–48.
51. Drossman DA, Toner BB, Whitehead WE, Diamant NE, Dalton CB, Duncan S, et al. Cognitive-behavioral therapy versus education and desipramine versus placebo for moderate to severe functional bowel disorders. Gastroenterology. 2003;125:19–31.
52. Hyphantis T, Guthrie E, Tomenson B, Creed F. Psychodynamic interpersonal therapy and improvement in interpersonal difficulties in people with severe irritable bowel syndrome. Pain. 2009;145: 196–203.
53. Creed F, Fernandes L, Guthrie E, Palmer S, Ratcliffe J, Read N, et al. The cost-effectiveness of psychotherapy and paroxetine for severe irritable bowel syndrome. Gastroenterology. 2003;124: 303–17.
54. Guthrie E, Creed F, Dawson D, Tomenson B. A controlled trial of psychological treatment for the irritable bowel syndrome. Gastroenterology. 1991;100:450–7.
55. Blanchard EB, Schwarz SP, Suls JM, Gerardi MA, Scharff L, Greene B, et al. Two controlled evaluations of multicomponent psychological treatment of irritable bowel syndrome. Behav Res Ther. 1992;30:175–89.
56. Gaylord SA, Palsson OS, Garland EL, Faurot KR, Coble RS, Mann JD, et al. Mindfulness training reduces the severity of irritable bowel syndrome in women: results of a randomized controlled trial. Am J Gastroenterol. 2011;106:1678–88.
57. Lindfors P, Unge P, Arvidsson P, Nyhlin H, Bjornsson E, Abrahamsson H, et al. Effects of gut-directed hypnotherapy on IBS in different clinical settings-results from two randomized, controlled trials. Am J Gastroenterol. 2012;107:276–85.
58. Gonsalkorale WM, Houghton LA, Whorwell PJ. Hypnotherapy in irritable bowel syndrome: a large-scale audit of a clinical service with examination of factors influencing responsiveness. Am J Gastroenterol. 2002;97:954–61.
59. Lindfors P, Unge P, Nyhlin H, Ljotsson B, Bjornsson ES, Abrahamsson H, et al. Long-term effects of hypnotherapy in patients with refractory irritable bowel syndrome. Scand J Gastroenterol. 2012;47:414–20.
60. Gonsalkorale WM, Miller V, Afzal A, Whorwell PJ. Long term benefits of hypnotherapy for irritable bowel syndrome. Gut. 2003; 52:1623–9.
61. Lackner JM, Gudleski GD, Keefer L, Krasner SS, Powell C, Katz LA. Rapid response to cognitive behavior therapy predicts treatment outcome in patients with irritable bowel syndrome. Clin Gastroenterol Hepatol. 2010;8:426–32.

7 Gastroparesis: Pathophysiology of Chronic Abdominal Pain and Current Treatment

Marcum Gillis and Kenneth L. Koch

Abbreviations

CNS	Central nervous system
cpm	Cycles per minute
EGG	Electrogastrogram
FDA	Food and Drug Administration
GERD	Gastroesophageal reflux disease
GES	Gastric electrical stimulation
H_2 blocker	Histamine$_2$ receptor antagonist
IBS	Irritable bowel syndrome
ICC	Interstitial cells of Cajal
LUQ	Left upper quadrant
NSAIDS	Non-steroidal anti-inflammatory drugs
PPI	Proton-pump inhibitor
PUD	Peptic ulcer disease
RUQ	Right upper quadrant

Definition of Gastroparesis

Gastroparesis is a neuromuscular disorder of the stomach characterized by delayed emptying of food from the stomach in the absence of mechanical obstruction. Gastric neuromuscular dysfunction results in an inability of the stomach to properly receive a food bolus (receptive relaxation of the fundus), inability of the corpus-antrum to triturate the food into appropriately sized particles and inability to empty the nutrient suspension through the pylorus into the duodenum for further digestion [1]. These dysfunctions of the neuromuscular activity of the stomach result in gastroparesis. The most common causes of gastroparesis are diabetes and post-gastric surgery procedures; however, the largest group of gastroparesis patients has no identifiable cause and is termed idiopathic gastroparesis.

Symptoms Associated with Gastroparesis

Patients with gastroparesis present with a wide variety of symptoms, including abdominal pain that ranges from mild to severe. Approximately 20 % of gastroparesis patients describe abdominal pain as their predominant symptom, although bloating, early satiety, epigastric fullness, nausea, and vomiting are more often the predominant symptoms [1]. Data from a large gastroparesis cohort indicates 91 % of patients with diabetic gastroparesis and 90 % of patients with idiopathic gastroparesis experience nausea. Vomiting was present in 72 and 55 % of these patient populations with gastroparesis, respectively [2]. Gastroparesis symptoms are usually absent or minimal when the patient is fasting but are triggered by ingestion of food or liquids, which stimulate the abnormal gastric neuromuscular activity.

Abdominal pain in patients with gastroparesis is usually experienced in the epigastric region but may also be located in the periumbilical region [3]. Unfortunately, the mechanisms underlying abdominal pain in gastroparesis are not well understood. Gastroparesis patients may have visceral hypersensitivity which predisposes them to experience pain compared to control patients [4]. Another hypothesis states that gastroparesis patients, particularly those with diabetes, have a lower threshold for pain as their stomach distends [4]; thus, these patients typically have no pain during fasting, but pain is brought on during and after meals. A recent study from the Gastroparesis Clinical Research Consortium found that 66 % of patients experience epigastric discomfort, but only 21 % of this patient population reported abdominal pain was their predominant symptom. In contrast, 44 % of these patients reported nausea and vomiting was their predominant symptoms [5].

M. Gillis, M.D. • K.L. Koch, M.D. (✉)
Section on Gastroenterology, North Carolina Baptist Medical Center, Medical Center Blvd., Winston-Salem, NC 27157, USA
e-mail: mgillis@wakehealth.edu; kkoch@wakehealth.edu

L. Kapural (ed.), *Chronic Abdominal Pain: An Evidence-Based, Comprehensive Guide to Clinical Management*,
DOI 10.1007/978-1-4939-1992-5_7, © Springer Science+Business Media New York 2015

Tests Available to Diagnose Gastroparesis

The current gold standard test to diagnose gastroparesis is a four-hour, solid-phase gastric emptying study [6, 7]. The patient ingests an Egg Beater® sandwich labeled with the radioisotope technetium-99 m. Following ingestion of this standard meal, a one-minute scan is performed at specific time intervals after ingestion of the meal (1 min; 30 min; 1, 2, 3, and 4 h) [6, 7]. Delayed gastric emptying is present if greater than 60 % of the meal is retained at 2 h or 10 % is retained at 4 h. This study must be completed with the patient of medications known to affect gastric motility, such as narcotics, anticholinergics, and pro-kinetic agents (e.g., metoclopramide).

Several other tests are used to diagnose gastroparesis and associated gastric neuromuscular dysfunction which leads to gastroparesis. The wireless motility capsule determines gastric emptying time by measuring intraluminal pH and pressure within the stomach. The capsule measures the drop in pH upon entering. When the capsule exits the stomach into the duodenum, a sharp rise in pH is measured that reflects the alkaline pancreatic bicarbonate secretions. A normal emptying time is 5 h after ingestion of a standard test meal and correlates with 90 % emptying of solidphase scintigraphy [8]. Breath tests can also be used to measure gastric emptying by monitoring levels of labeled carbon-13 [9]; however, they are not Food and Drug Administration (FDA)-approved and are not as accurate in patients with chronic hepatic and pulmonary conditions which prevent normal metabolism and excretion of the radio-labeled carbon [10].

Electrogastrography is a noninvasive method used to evaluate gastric myoelectrical activity in a fashion similar to an electrocardiogram recording. The slow wave or pacemaker potential of the stomach ranges from 2.5 to 3.7 cycles per minute (cpm). Rhythms lower than 2.5 cpm or higher than 3.7 cpm are bradygastrias and tachygastrias, respectively. There is a high prevalence of gastric dysrhythmias in patients with gastroparesis [11], but the presence of gastric dysrhythmias does not necessarily predict the rate of gastric emptying. For example, some patients with pyloric dysfunction or obstruction have gastroparesis and normal 3 cpm electrogastrogram (EGG) signals [12]. Gastric dysrhythmias correlate well with the nausea related to nausea of pregnancy [13], motion sickness [14], and functional dyspepsia [15]. Gastric dysrhythmias do not correlate with abdominal pain. The water load test is typically performed as part of the EGG test [16]. Patients drink water until they feel "comfortably full" within a 5-min period. The water load distends the stomach and stimulates gastric neuromuscular activity and may evoke gastric dysrhythmias. The volume of water ingested provides insight into the gastric capacity of the stomach. Patients with poor water load tests (<550 ml) have poor gastric capacity/distention.

Table 7.1 Causes and common symptoms of gastroparesis

Causes	Common symptoms
Diabetes mellitus	Nausea, early satiety, epigastric pain
Idiopathic	Epigastric pain, fullness, nausea, and vomiting
Postsurgical	Upper abdominal pain, nausea, vomiting
Ischemic	Post-prandial abdominal pain or discomfort, nausea, vomiting, fullness
Obstructive	RUQ pain, nausea, vomiting

Causes of Gastroparesis

There are three common categories of gastroparesis ((1) idiopathic, (2) diabetic, and (3) postsurgical) and several uncommon but reversible forms of gastroparesis (Table 7.1). Idiopathic gastroparesis is the most common group and includes patients without a clinical history of diabetes or collagen-vascular diseases, medications, or gastric surgeries to explain gastroparesis [17]. The pathophysiology of idiopathic gastroparesis is not known, but a correlation between the onset of symptoms and an acute viral illness which leads to gastric dysrhythmias and gastroparesis has been reported [1]. Damage to the neuromuscular apparatus and/or interstitial cells of Cajal (ICCs) during viral infections may lead to gastric dysrhythmias and gastroparesis [18]. Patients with idiopathic gastroparesis are more likely to experience abdominal pain compared with patients with diabetic gastroparesis [19].

The second most common cause of gastroparesis is from diabetes mellitus. The prevalence of gastroparesis is estimated between 30 and 50 % in patients with type 2 diabetes mellitus [20]. Type 1 diabetics are more likely to develop gastroparesis compared with type 2 diabetics (5.5 vs. 1.0 % over a 10-year period); however, both are likely to cause gastroparesis compared with control subjects (hazard ratio of 33 and 7.5, respectively) [21]. Symptoms of delayed gastric emptying typically develop after approximately 10 years of poorly controlled hyperglycemia and are most severe during periods of moderate to severe hyperglycemia [22].

Diabetes causes gastroparesis through effects on gastric enteric neurons, ICCs, and smooth muscle. Prolonged hyperglycemia affects the enteric nervous system and ICCs in the stomach leading to gastric dysrhythmias and poor gastric neuromuscular coordination. Gastric dysrhythmias result in abnormal intra-gastric distribution of food [23, 24] and decreased antral motility [25]. Additionally, animal models of diabetic gastroparesis show decreased contractility in gastric smooth muscle following prolonged periods of hyperglycemia [26]. Therefore, patients with diabetes may have decreased enteric nerve and ICC function and decreased smooth muscle function.

Postsurgical gastroparesis occurs after surgical operations on the stomach with or without vagotomy. After antrectomy and vagotomy or esophagectomy, the remaining

Table 7.2 Differential diagnosis of gastrointestinal sources of abdominal pain in patients with gastroparesis

Anatomical location	Diseases causing pain	Key abdominal aspects of pain
Esophagus	GERD	Substernal burning
Stomach	Gastroparesis	Pressure, bloating, fullness, nausea
	PUD	Epigastric burning
	Pylorospasm	Sharp, RUQ pain
	Chronic mesenteric ischemia	Increased after meals
Gallbladder	Cholecystitis	RUQ
	Choledocholithiasis	Fever
Pancreas	Acute pancreatitis	Sharp, epigastric and LUQ → back pain
	Chronic pancreatitis	
Colon	IBS	Pain in all areas of abdomen; Altered bowel habits

RUQ right upper quadrant, *LUQ* left upper quadrant, *PUD* Peptic ulcer disease, *IBS* Irritable bowel syndrome

stomach frequently cannot relax, contract, or empty food normally. These patients are at risk for mechanical obstruction strictures at the anastamosis. Patients who have vagotomies may develop gastric neuromuscular dysfunction leading to delayed gastric emptying [27]. Operations that result in resection of the corpus and/or antral wall, which contain the gastric pacemaker and mixing regions, may lead to gastric neuromuscular dysfunction and gastroparesis. Fundoplication for gastroesophageal reflux disease (GERD) is also a cause of postsurgical gastroparesis. These patients are at risk for rapid antral filling due to decreased fundus relaxation following meal ingestion [28], inadvertent vagal nerve injury, and poor antropyloroduodenal coordination [29].

Reversible, but much less common, causes of gastroparesis include chronic mesenteric ischemia and pyloric obstruction. Chronic mesenteric ischemia associated with gastroparesis typically causes mild, vague abdominal pain after meals (not the severe abdominal pain associated with small bowel infarction secondary to acute mesenteric ischemia). Fortunately, chronic mesenteric ischemic gastroparesis is rare but should be considered in patients with known peripheral vascular disease, abdominal aortic atherosclerosis, and coronary artery disease [30]. Ischemic gastroparesis results from poor blood flow through the celiac and superior mesenteric arteries and causes dysfunction of the ICCs, enteric nervous system, and smooth muscle leading to gastric dysrhythmias and poor gastric emptying.

Obstructive gastroparesis refers to mechanical obstruction to gastric emptying due to either a tumor or fibrosis in the pylorus or duodenum [12]. Fibrotic tissue may be present in the pylorus in patients with a history of peptic or duodenal ulcers or radiation therapy. Gastroparesis related to obstruction of the pylorus is associated with right upper quadrant (RUQ) discomfort and pain. Unlike other forms of gastroparesis with gastric dysrhythmias, these patients have normal or high-amplitude 3 cpm EGG signals and gastroparesis [12]. Gastric outlet obstruction is also caused by pylorospasm when the pyloric channel is intermittently obstructed due to spasm of the pyloric sphincter [31]. These patients have gastroparesis, normal or high-amplitude EGG signals, and epigastric or RUQ pain related to spasm of the pylorus.

Differential Diagnosis of Abdominal Pain in Patients with Gastroparesis

There are numerous diseases to consider when evaluating patients with upper abdominal pain and gastroparesis (Table 7.2). In this section, the causes of pain will be discussed by specific organs: (1) stomach, (2) pylorus, (3) non-gastric GI organs, and (4) non-GI causes. There are three major stomach diseases to consider when evaluating a patient with abdominal pain and gastroparesis. These include GERD, peptic ulcer disease (PUD), and gastroparesis itself.

GERD is extremely common in the general population with an incidence of 42 % [32]. The basic cause of GERD symptoms is exposure of the esophageal mucosa to gastric acid. The lower esophageal sphincter (LES) creates a pressure zone that prevents acid and gastric contents from refluxing into the esophagus during digestion. Barriers to prevent acid injury also include secretion of saliva and secondary esophageal peristalsis. The diagnosis of GERD requires a clinical history of heartburn symptoms. Empiric trials of acid suppression medications such as proton-pump inhibitors (PPI) or histamine$_2$ receptor antagonists (H_2 blockers) that eliminate heartburn symptoms confirm the diagnosis. Diagnostic tests for GERD include upper endoscopy and 24-h esophageal pH monitoring. Endoscopic evaluation reveals normal or inflamed esophageal mucosa, structures, or masses. During esophageal pH monitoring, a decrease in esophageal pH to less than four is considered a positive reflux event [33]. Approximately 25 % of GERD patients also have gastroparesis, so abdominal pain symptoms can overlap in these patients [34].

PUD is defined as a break in the mucosal barrier within the stomach and duodenum that evolves into erosions and ulcers. Gastric acid remains the primary cause of these erosions, but non-steroidal anti-inflammatory drugs (NSAIDs), aspirin, alcohol, and Helicobacter pylori all increase the risk for PUD. Patients with PUD have burning epigastric pain that does not typically radiate into the chest. Hematemesis or melena is more consistent with PUD compared with GERD. The diagnosis of PUD involves taking a thorough history coupled with endoscopic or radiographic evaluation of the stomach and duodenum.

Gastroparesis itself may have an abdominal pain component. In contrast to the burning pain associated with GERD and PUD, these patients may experience a vague discomfort or pain in the epigastric region that is highly variable in severity. Some patients classify their symptoms of nausea, vomiting, and bloating as "pain." Patients with gastroparesis also may have PUD and/or GERD. One study found that 25 % of patients with GERD had delayed gastric emptying, while 70 % of these patients also had gastric dysrhythmias [34]. These overlap syndromes of GERD and gastroparesis may be differentiated to some degree with empiric treatment with PPIs. Adequate suppression of acid eliminates the symptoms related to GERD/PUD but do not improve symptoms related to gastroparesis.

Spasm of the pyloric sphincter can also cause abdominal pain and is a cause of functional obstructive gastroparesis. Patients with pylorospasm and gastroparesis typically have normal 3 cpm EGG signals because the neuromuscular apparatus in the gastric body and antrum is intact. The pyloric spasm creates an obstruction to emptying of chyme from the stomach. Therefore, these patients have the symptoms of gastroparesis including nausea, bloating, and gastric distention. However, these patients often also report post-prandial pain in the RUQ due to pylorospasm. This symptom complex may also mimic chronic cholecystitis.

There are numerous *non-gastric causes* of abdominal pain in patients with gastroparesis; however, only a few of these causes will be discussed here. One of the common diseases seen in patients with gastroparesis is irritable bowel syndrome (IBS). Patients with IBS experience abdominal pain or discomfort associated with constipation, diarrhea, or both [35]. These symptoms are usually increased after a meal. IBS patients must, by definition, have altered bowel habits. Though patients with gastroparesis and IBS may have abdominal distention and bloating, patients affected by gastroparesis alone do not have altered bowel habits. Patients with IBS may experience pain in the epigastrium but are more likely to report lower abdominal discomfort compared to patients with gastroparesis. Furthermore, IBS patients typically have a decrease in abdominal pain following defecation, which is not the case with gastroparesis symptoms. Similar to those gastroparesis patients experiencing pain, IBS patients often have a higher sensitivity to visceral stimuli than control patients [36].

Biliary and pancreas diseases must be ruled out in gastroparesis patients with upper abdominal pain. Gallstone disease is one of the most prevalent illnesses in the western world involving almost 20 million Americans in 2004 [37]. The RUQ pain results from intermittent obstruction of the cystic or common bile ducts. Similar to patients with gastroparesis, patients with gallstone disease exhibit post-prandial nausea and epigastric pain. However, patients with gallbladder disease also report radiation of this pain to the RUQ, back, and scapulas which is not typical in patients with gastroparesis. If gallstone disease progresses to cholecystitis, choledocholithiasis, or cholangitis, then fevers and leukocytosis develop, and these features are not present in patients experiencing pain from gastroparesis. Pancreatitis presents with post-prandial epigastric pain that radiates to the back, nausea, and vomiting. Acute pancreatitis can be differentiated from gastroparesis given its acute onset and elevated serologic tests including lipase and amylase. Patients with chronic pancreatitis may also have crampy epigastric pain and bloating, symptoms of pancreatic exocrine dysfunction including weight loss and diarrhea, and radiologic evidence of pancreatic calcifications, all of which will differentiate pancreatitis from gastroparesis.

Finally, epigastric pain may be related to causes *outside the GI tract*. The most common causes of non-GI tract pain in patients with gastroparesis are the abdominal wall pain syndromes. Careful history reveals these pains increase with physical movement and *not* ingestion of meals or defecation. Typically, a very localized "trigger point" is identified on physical examination by performing Carnett's test [38]. During Carnett's test, the localized tender area is compressed with a single finger and the patient raises his head or both feet off the exam table. Contraction of the abdominal wall muscles during this maneuver results in an immediate increase of abdominal pain if the pain source is within the abdominal wall [38]. These focal abdominal wall pains most often occur at sites of abdominal incisions where scar tissue or underlying adhesions have developed following abdominal surgery [39]. These focal trigger points are treated with subcutaneous injections of local anesthetics and/or steroids.

Treatment

The treatment of patients with gastroparesis and abdominal pain requires a thorough understanding of specific characteristics of the pain: timing, radiation, and factors that exacerbate and relieve pain. The first step is to determine the true cause of the abdominal pain because the patient's pain/discomfort symptoms may not be related to gastroparesis. The pain characteristics should help to identify the cause of pain and

Table 7.3 Treatment modalities for patients with gastroparesis and abdominal pain

Disease causing pain	Treatment class	Comments
Gastroparesis with gastric dysrhythmias	Prokinetic agents	Metoclopromide, domperidone
	Antispasmodic	Dicyclomine
	Nutrition	Gastroparesis diet
	Tricyclic	Amitriptyline
	Antidepressants	Nortriptyline
Gastroparesis with normal 3 cpm gastric rhythm	Endoscopic	Botox injections (pylorus)
		Balloon dilation (pylorus)
IBS	Antispasmodic	Dicyclamine
	Fiber supplementation	Psyllium
	Laxative	Miralax
	Tricyclic antidepressants	Nortriptyline
		Amitriptyline
Abdominal wall pain syndrome	Local injections	Lidocaine, bupivicaine
		Steroid injection

guide treatment. If doubt remains after review of the clinical history and physical examination, then further testing should be performed. An abdominal wall syndrome should be considered and evaluated in all patients with gastroparesis and abdominal pain. Treatments for gastroparesis are outlined below and summarized in Table 7.3, and focus on reducing the discomfort and pain related to neuromuscular dysfunction of the stomach [1].

First, antiemetic therapies are used to reduce the nausea and vomiting of gastroparesis. 5-HT3 receptor antagonists such as ondansetron and granisetron are commonly used. Phenothiazines such as promethazine and benzodiazepines such as lorazepam are also commonly used for nausea and act within the central nervous system (CNS). These medications may treat the symptoms associated with delayed gastric emptying but not the underlying pathophysiology of gastroparesis.

Prokinetic agents may correct the underlying neuromuscular dysfunctions of gastroparesis [40]. There are three subclasses of medications within this group. First, the macrolide antibiotic erythromycin has prokinetic properties and increases the rate of gastric emptying in patients with gastroparesis. Unfortunately, patients frequently experience abdominal pain and nausea while taking erythromycin. The second subgroup of prokinetic medications are the substituted benzamides metoclopramide and domperidone, which is not approved by the FDA for treatment of gastroparesis. These medications are dopamine receptor antagonists and increase gastric emptying and improve gastric dysrhythmias. Metoclopramide can induce serious extrapyramidal side effects and tardive dyskinesia with long-term use. Domperidone is a peripheral dopamine receptor antagonist and thus does not have the same risk for extrapyramidal side effects. Domperidone can cause hyperprolactinemia and has increased risk for cardiac dysrhythmias which should be considered before prescribing this medication. The third prokinetic drug group, which is not available in the United States, includes the serotonin agonists cisapride and tegaserod. These medications are 5-HT4 receptor agonists and increase gastric motility. These drugs may cause cardiac dysrhythmias and ischemic bowel disease, however, and were removed from the market.

An important treatment for gastroparesis involves dietary counseling and the gastroparesis diet [1, 41]. Since most symptoms associated with gastroparesis such as nausea and abdominal pain increase after meals, gastric wall stretch or distention may play a role in the generation of these symptoms. The three-step gastroparesis diet decreases the overall volume of the ingested meal by increasing the number of meals to six, encouraging small volume meals, and changing the types of foods ingested—all to match the poor neuromuscular function of the stomach. During step one only small volumes of salty liquids such as sports drinks or bouillon are consumed. The goal of this step is to avoid dehydration while providing some caloric intake. Once the patients tolerate step one, they advance to step two. Step two involves soups, crackers, and small amounts of softer solids such as cheese and peanut butter. The "liquid nutrients" of soup provide calories but require little gastric contraction to mix and empty the meal. Once the patients are able to tolerate step two, they advance to step three. Step three includes noodles, pasta, potatoes, chicken, and fish; these solid foods are low in fat and require modest triturate to mill the food before emptying. Patients are counseled to avoid fried foods and red meat and high fiber vegetables which require long duration trituration.

Endoscopic and electrical stimulation therapies are available to treat gastroparesis and may reduce abdominal pain and nausea. A diagnostic upper endoscopy should be performed in the gastroparesis patient with prolonged abdominal pain, nausea, or vomiting to rule out PUD, gastritis, and gastric outlet obstruction. For patients with gastroparesis and a normal 3 cpm EGG rhythm (and no mechanical cause for gastroparesis), Botox injection in the pyloric sphincter during endoscopy may induce sphincter relaxation and improve gastric emptying rates and abdominal pain. Endoscopic balloon dilation of the pylorus is another option to decrease pyloric resistance in patients with gastroparesis and 3 cpm EGG signals. In 18 patients with gastroparesis and normal 3 cpm EGG rhythm treated endoscopically with either Botox injections or balloon dilation of the pylorus, 83 % of patients gained relief from their gastroparesis symptoms at their four-month follow-up evaluation [31]. Pyloroplasty improved gastric emptying and reduced symptom burden including abdominal pain, nausea, and vomiting in patients with gastroparesis [42].

Gastric electrical stimulation (GES) is a device treatment for patients with drug-refractory symptoms of gastroparesis. The electrical stimulation reduces nausea and vomiting but does not improve gastric emptying or gastric dysrhythmias [43]. In patients with diabetic gastroparesis, GES treatment decreased abdominal pain as well as nausea and vomiting [44]. In contrast to GES, gastric pacing devices stimulate the gastric electrical system and convert gastric dysrhythmias to the normal 3 cpm pattern and improve gastric emptying [45]; however, these studies involve small numbers of patients.

In conclusion, a minority of patients with gastroparesis report abdominal pain as their predominant symptom. Postprandial distention of the gastric wall and pylorospasm may cause abdominal pain in patients with gastroparesis. However, a relevant differential diagnosis of the pain must be considered and investigated to exclude non-gastric causes of pain like the abdominal wall syndrome or IBS. Once gastroparesis is identified as the cause of abdominal pain, treatments should be tailored according to the cause of the patient's gastroparesis and should involve a combination of drug and dietary therapies. In selected patients, endoscopic therapies such as Botox of the pylorus or GES for drug-refractory gastroparesis may reduce pain and symptoms of gastroparesis.

References

1. Koch KL. Gastric neuromuscular function and neuromuscular disorders. In: Feldman M, Friedman LS, Brandt LJ, editors. Sleisenger and Fordtran's gastrointestinal and liver disease: pathophysiology/diagnosis/management. Philadelphia: Elsevier; 2010. p. 789–815.
2. Cherian D, Parkman HP. Nausea and vomiting in diabetic and idiopathic gastroparesis. Neurogastroenterol Motil. 2012;24:217–e103.
3. Cherian D, Sachdeva P, Fisher RS, Parkman HP. Abdominal pain is a frequent symptom of gastroparesis. Clin Gastroenterol Hepatol. 2012;8:676–81.
4. Kumar A, Attaluri A, Hashmi S, Schulze KS, Rao SS. Visceral hypersensitivity and impaired accommodation in refractory diabetic gastroparesis. Neurogastroenterol Motil. 2008;20:635–42.
5. Hasler WL, Koch KL, et al. Importance of abdominal pain as a symptom in gastroparesis: relation to clinical factors, disease severity, quality of life, gastric retention, and medication use. Gastroenterology. 2010;138:S461.
6. Tougas G, Eaker EY, Abell TL, et al. Assessment of gastric emptying using a low fat meal: establishment of international control values. Am J Gastroenterol. 2000;95:1456–62.
7. Guo JP, Maurer AH, Fisher RS, Parkman HP. Extending gastric emptying scintigraphy from two to four hours detects more patients with gastroparesis. Dig Dis Sci. 2011;46:24–9.
8. Kuo B, McCallum RW, Koch KL, et al. Comparison of gastric emptying of a non-digestible capsule to a radio-labeled meal in healthy and gastroparetic patients. Aliment Pharmacol Ther. 2008;27: 186–96.
9. Viramontes BE, Kim DY, Camilleri M, et al. Validation of a stable isotope gastric emptying test for normal, accelerated, or delayed gastric emptying. Neurogastroenterol Motil. 2001;13:567–74.
10. Szarka LA, Camilleri M, Vella A, et al. A stable isotope breath test with a standard meal for abnormal gastric emptying of solids in the clinic and in research. Clin Gastroenterol Hepatol. 2008;6:635–43.
11. Chen JD, Lin Z, Pan J, McCallum RW. Abnormal gastric myoelectrical activity and delayed gastric emptying in patients with symptoms suggestive of gastroparesis. Dig Dis Sci. 1996;41:1538–45.
12. Brzana RJ, Koch KL, Bingaman S. Gastric myoelectrical activity in patients with gastric outlet obstruction and idiopathic gastroparesis. Am J Gastroenterol. 1998;93:1803–9.
13. Koch KL, Stern RM, Vasey M, et al. Gastric dysrhythmias and nausea of pregnancy. Dig Dis Sci. 1990;35:961–8.
14. Stern RM, Koch KL, Sterwart WR, Lindblad IM. Spectral analysis of tachygastria recorded during motion sickness. Gastroenterology. 1987;92:92–7.
15. Parkman HP, Miller MA, Trate D, et al. Electrogastrography and gastric emptying scintigraphy are complementary for assessment of dyspepsia. J Clin Gastroenterol. 1997;24:214–9.
16. Koch KL, Hong S-P, Xu L. Reproducibility of gastric myoelectrical activity and the water load test in patients with dysmotility-like dyspepsia symptoms and in control subjects. J Clin Gastroenterol. 2000;31:125–9.
17. Soykan I, Sivri B, Saroseik I, et al. Demography, clinical characteristics, psychological and abuse profiles, treatment, and long term follow-up of patients with gastroparesis. Dig Dis Sci. 1998;43: 2398–404.
18. Zarate N, Mearin F, Wang XY, et al. Severe idiopathic gastroparesis due to neuronal and interstitial cells of Cajal degeneration: pathological findings and management. Gut. 2003;52:966–70.
19. Parkman HP, Yates K, Hasler WL, Nguyen L, et al. Similarities and differences between diabetic and idiopathic gastroparesis. Clin Gastroenterol Hepatol. 2011;12:1056–64.
20. Intagliata N, Koch KL. Gastroparesis in type 2 diabetes mellitus: prevalence, etiology, diagnosis, and treatment. Curr Gastroenterol Rep. 2007;9:270–9.
21. Choung RS, Locke 3rd GR, Schleck CD, et al. Risk of gastroparesis in subjects with type 1 and 2 diabetes in the general population. Am J Gastroenterol. 2012;107:82–8.
22. Soykan I, Lin Z, Sarosiek I, McCallum RW. Gastric myoelectrical activity, gastric emptying, and correlations with symptoms and fasting blood glucose levels in diabetic patients. Am J Med Sci. 1999;317:226–31.
23. Troncon LE, Rosa Silva L, Oliveira RB, et al. Abnormal intragastric distribution of a liquid nutrient meal in patients with diabetes mellitus. Dig Dis Sci. 1998;43:1421–9.
24. Samson M, Roelofs JM, Akkermans LM, et al. Proximal gastric motor activity in response to a liquid meal in type 1 diabetes with autonomic neuropathy. Dig Dis Sci. 1998;43:491–6.

25. Koch KL. Diabetic gastropathy: gastric neuromuscular dysfunction in diabetes mellitus. A review of symptoms, pathophysiology, and treatment. Dig Dis Sci. 1999;44:1061–75.
26. Takahaski T, Kojima Y, Tsunoda Y, et al. Impaired intracellular signal transduction in gastric smooth muscle of diabetic BB/W rats. Am J Physiol. 1996;270:G411–7.
27. Hinder RA, Kelley KA. Human gastric pacesetter potential: site of origin, spread, and response to gastric transection and proximal vagotomy. Am J Surg. 1997;133:29–33.
28. Crowley DJ, Vernon P, Jones T, et al. Gastric emptying of solid meals after truncal vagotomy and pyloroplasty in human subjects. Gut. 1972;13:176–81.
29. Sun WM, Doran SM, Jones KL, et al. Long-term effects of pyloromyotomy on pyloric motility and gastric empting in humans. Am J Gastroenterol. 2000;95:92–100.
30. Liberski SM, Koch KL, Atnip RG, Stern RM. Ischemic gastroparesis: resolution after revascularization. Gastroenterology. 1990;99:252–7.
31. Kundu S, Koch KL. Effect of balloon distension or botulinum toxin A injection of the pylorus on symptoms and body weight in patients with gastroparesis and normal 3 cycle per minute gastric electrical activity. Am J Gastroenterol. 2011;84:535.
32. Locke GR, Talley NJ, Fett SL, et al. Prevalence and clinical spectrum of gastroesophageal reflux: a population-based study in Olmsted County, Minnesota. Gastroenterology. 1997;112:1448.
33. Kim GH. How to interpret ambulatory 24 hr esophageal pH monitoring. J Neurogastroenterol Motil. 2010;16:207–10.
34. Koch KL, Xu L, Noar M. Gastric myoelectrical and emptying activity in patients with gastroesophageal reflux disease (GERD) and dysmotility like functional dyspepsia (GERD+): effect of water load test. Am J Gastroenterol. 2001;96:526.
35. Longstreth GF, Thompson WG, Chey WD, Houghton LA, Mearin F, Spiller RC. Functional bowel disorders. In: Drossman DA, Corrazziari E, Delvaux M, et al., editors. Rome III: the functional gastrointestinal disorders. 3rd ed. McLean: Degnon Associates; 2006. p. 487–555.
36. Bouin M, Plourde V, Boivin M, et al. Rectal distention testing in patients with irritable bowel syndrome: sensitivity, specificity, and predictive values of pain sensory thresholds. Gastroenterology. 2002;122:1771–7.
37. Russo MW, Wei JT, Thiny MT, et al. Digestive and liver diseases statistics. Gastroenterology. 2004;126:1448–53.
38. Carnett JB. Intercostal neuralgia as a cause of abdominal pain and tenderness. Surg Gynecol Obstet. 1926;12:625–34.
39. Costanza CD, Longstreth GF, Liu AL. Chronic abdominal wall pain: clinical features, health care costs and long-term outcome. Clin Gastroenterol Hepatol. 2004;2:395–9.
40. Parkman HP, Hasler WL, Fisher RS. American Gastroenterological Association technical review on the diagnosis and treatment of gastroparesis. Gastroenterology. 2004;127:1592–622.
41. Parrish CR. Nutrition concerns for the patient with gastroparesis. Curr Gastroenterol Rep. 2007;9:295–302.
42. Hibbard ML, Dunst CM, Swanstrom LL. Laparoscopic and endoscopic pyloroplasty for gastroparesis results in sustained symptom improvement. J Gastrointest Surg. 2011;15:1513–9.
43. Abell T, McCallum R, Hocking M, et al. Gastric electrical stimulation for medically refractory gastroparesis. Gastroenterology. 2003;125:421–8.
44. McCallum RW, Snape W, Brody F, Wo J, Parkman HP, Nowak T. Gastric electrical stimulation with Enterra therapy improves symptoms from diabetic gastroparesis in a prospective study. Clin Gastroenterol Hepatol. 2010;8:947–54.
45. Lin Z, Sarosiek I, Forster J, Ross RA, Chen JD, McCallum RW. Two-channel gastric pacing in patients with diabetic gastroparesis. Neurogastroenterol Motil. 2011;23:912–e396.

Chronic Pain Due to Postsurgical Intra-abdominal Adhesions: Therapeutic Options

8

Jose De Andres, Luciano Perotti, Stefano Palmisani, Vicente Luis Villanueva Perez, Juan Marcos Asensio-Samper, and Gustavo Fabregat

Introduction

Peritoneal adhesions are bands of fibrous tissue that join abdominal organs to each other or to the abdominal wall [1]. Adhesion formation is a common complication after abdominal and pelvic surgery and is source of considerable morbidity [2].

Peritoneal adhesions can originate from any surgical procedures [1, 2]; literature reports that the incidence of peritoneal adhesions range from 67 to 93 % after abdominal surgery reaching an incidence of 97 % after open gynecological surgery [2].

Adhesions can be congenital or acquired, and the latter are divided in inflammatory or postsurgical [2]. Inflammatory adhesions derive from an intra-abdominal inflammatory process, whereas postsurgical adhesions result from the scar fusion of the tissue injured from surgical manipulation such as incision, cauterization, and suturing [2].

The morbidity of postsurgical adhesions increases with patient's age, number of surgical interventions, and complexity of surgical procedures [2], leading to complications such as small-bowel obstruction, female infertility, abdominal, and pelvic chronic pain [1, 2]. Operations that frequently lead to adhesions formation include colon and rectal surgery, gynecologic surgery, and nonelective appendectomy [2].

Tension, stretching, and traction of abdominal and pelvic organs, which stimulate peritoneal pain receptors, associated with limitation in organs mobility and distensibility are considered as the main causes of adhesion-related pain [1, 2]. Moreover, adhesions themselves are directly implicated in pain generation: a mapping study of reported pain performs in patients with pelvic adhesions have demonstrated that these structures can directly generate pain stimuli, and the presence of nervous fibers in human pelvic and abdominal adhesions has been recently demonstrated [1, 2]. In fact, peritoneal adhesions are formed by vascularized collagenous bands, adipose tissue, and a certain amount of nerve fibers (both myelinated and nonmyelinated). Although the role of these nervous structures is still not clear, they seems to be involved in the regulation of regional blood flow, neurogenic inflammation, and in the wound-healing process [1].

Pathogenesis on Pain from Adhesions

Peritoneal healing differs from skin healing. Peritoneum becomes mesothelilysed simultaneously regardless the size of the injury; new mesothelium develops from islands of mesothelial cells which proliferate into sheath of cells [2].

Injury or inflammation of peritoneum triggers a coagulative state that releases multiple chemical mediators and activates leukocytes and mesothelial cells, whereas macrophages change their function releasing inflammatory mediators (cyclooxygenase and lipoxygenase metabolites, plasminogen

J. De Andres, M.D., Ph.D., F.I.P.P., E.D.R.A. (✉)
Anesthesia Division, Department of Surgery, Valencia School of Medicine, Valencia, Spain

Anesthesia Critical Care and Pain Management Department, General University Hospital, Avda Tres Cruces s/n, Valencia, Spain 46014
e-mail: deandres_jos@gva.es

L. Perotti, M.D.
Department of Anesthesia and Intensive Care, Civil Hospital of Vigevano, Corso Milano, 19, Vigevano, Pavia, Italy 27029
e-mail: lucianoperotti@inwind.it

S. Palmisani, M.D.
Guy's and St Thomas' Hospital, London, UK

Pain Management and Neuromodulation Centre, Westminster Bridge Road, London, UK SE1 7EH
e-mail: palmisani@gmail.com

V.L.V. Perez • J.M. Asensio-Samper • G. Fabregat
Anesthesia Critical Care and Pain Management Department, General University Hospital, Avda Tres Cruces s/n, Valencia, Spain 46014
e-mail: v.villanuevaperez@gmail.com; jmasensiosamper@hotmail.com; gfabregat@gmail.com

L. Kapural (ed.), *Chronic Abdominal Pain: An Evidence-Based, Comprehensive Guide to Clinical Management*, DOI 10.1007/978-1-4939-1992-5_8, © Springer Science+Business Media New York 2015

activator, PAI, IL-1, IL-6, and TNF) and recruiting new mesothelial cells [2].

A key role in adhesions formation is played by the fibrin matrix, as shown by the significant correlation between low fibrinolytic activity and increased adhesion production in several animal models. That explains why peritoneal specific fibrinolytic activity has been often targeted in the attempt of preventing adhesions development [2].

Inadequate bloody supply and reduced tissue oxygenation, common situations during surgical injuries, can also inhibit the physiologic fibrinolysis and fibrinolytic activity, resulting in the persistence of the fibroproliferative structure which eventually leads to fibrovascular adhesions development [2].

Ischemia caused by peritoneal suture, pressure and excessive handling or drying of peritoneum, presence of foreign materials (i.e., starch powder from surgical gloves), and intraperitoneal bleeding are further factors considered to promote adhesion formation through a significant antifibrinolytic activity [2].

Visceral Sources

Sixteen million patients in the USA complain of abdominal pain every year; of these, two million continue to have persistent severe abdominal pain despite pharmacological interventions, minimally invasive techniques, and surgical interventions.

The principal function of visceral receptors and their associated afferent axons is to convey information from the viscera to the central nervous system; most of such information is rarely perceived and the principal conscious sensations arising from the viscera are discomfort and pain [3]. An altered sensation from the viscera, which characterizes functional alterations such as bowel disorders, interstitial cystitis, and ureteric colic, is usually considered as visceral hyperalgesia; hyperalgesia often involves both peripheral and central nervous system components and theoretically can be initiated and maintained entirely by peripheral or central mechanisms [3].

Visceral nociceptors located in the serosa, mucosa, and muscle of hollow viscus have no morphologic specialization and are associated with unmyelinated C-fiber and thinly myelinated Aδ fibers. Studies report that visceral nociceptors respond only to intense mechanical stimuli and to chemical stimuli, such as the products of inflammation; in the hollow viscus are also present a certain amount of "silent" nociceptors, which are switched to an active status only by mucosa's inflammation of the innervated organs [3].

Abdominal Wall

Chronic abdominal wall pain (CAWP) is defined as constant or intermittent pain of more than one month of duration which fulfills the following characteristics: much localized pain or fixed location of tenderness and superficial tenderness or point of tenderness with less than 2.5 cm of diameter or increased point tenderness with abdominal wall muscle tension [4].

Patients with CAWP often receive inaccurate diagnosis and are therefore treated as suffering from visceral pain [4]; it has been estimated that in about 10–30 % of the patients with chronic abdominal pain, the abdominal wall is the main pain generator [5].

Commonly, these patients underwent repeated clinical examinations and expensive invasive investigations, which results in a waste of economic health resources and significant delays in diagnosis and treatments [4, 5].

Sources of abdominal wall pain usually include: pain referred from abdominal or thoracic viscera by neural convergence in the spinal cord with somatic sites, T7–T12 radicular lesions or peritoneal/abdominal wall lesion leading to nerve injury are sources of abdominal wall pain [4]. However, surgical iatrogenic peripheral nerve injury is considered one of the most common causes of CAWP [6, 7].

Surgery may provoke nerve injury either when the surgical incision directly damages a cutaneous nerve or when a cutaneous nerve is indirectly trapped by scar tissue formation or directly by surgical sutures [7]. The entrapment of cutaneous abdominal nerve branches, known also as Abdominal Cutaneous Nerve Entrapment Syndrome (ACNES), can be due not only to the surgical trauma but also to the peculiar anatomy of the cutaneous nervous branches [4, 5, 7]. Peripheral nerve entrapment occurs at anatomic sites where the nerve changes direction to enter a fibrous or osseofibrous tunnel or where the nerve passes over a fibrous or muscular band and that entrapment can be at these sites because mechanically induced irritation is most likely to occur at these locations. The most common cause of abdominal wall pain is nerve entrapment at the lateral border of the rectus muscle. In the rectus channel, the nerve and its vessels are surrounded by fat and connective tissue that bind the nerve, artery, and vein into a discrete bundle capable of functioning as a unit independently from surrounding tissue. At a point located about three quarters of the way through the rectus muscle (from back to front), there is a fibrous ring that provides a smooth surface through which the bundle can slide. Positioned anterior to the ring, the rectus aponeurosis provides a hiatus for the exiting bundle [8].

Differential Diagnosis

The first step in evaluating a patient with suspect of abdominal wall pain is the exclusion of intra-abdominal pathology. Accurate medical history, exhaustive medical examination, endoscopic screening options, proper diagnostic imaging, and laboratory tests should aid in excluding other conditions causing abdominal pain, such as intra-abdominal malignancy; Spigelian hernia, that mimics entrapment of T10 nerve branch; referred pain from other pathology of nervous system, excluded by Carnett's test; ilioinguinal and/or iliohypogastric nerve entrapment, usually related with previous groin surgery [5].

Organic intra-abdominal pathology and intra-abdominal malignancy might be detected also with the use of Guaiac or immunochemical fecal occult blood tests (FOBTs), bearing in mind sensitivity, specificity, and positive predictive value of FOBT are difficult to estimate from screening programs [9]. Thus, all patients with chronic abdominal pain that result negative to intra-abdominal pathology and with negative fecal occult blood test should be investigated for ACNES [5].

It is important to remember that entrapment of ilioinguinal and/or iliohypogastric and genital branch of the genitofemoral nerve is a condition difficult to separate from true ACNES in the lower abdomen [5]. Other condition difficult to separate from nerve entrapment is the presence of myofascial trigger point in the abdominal wall because these patients can result positive also at Carnett's test; however, patients with myofascial pain generally complain multiple trigger points in the musculoskeletal system [4, 5].

Differential diagnosis of ACNES also include nervous lesion due to neoplasia, herpes zoster, traumatic radiculitis, rectus sheet hematoma, hernias, and painful rib [4].

Therapeutical Options (Fig. 8.1)

Role of Tests in Defining Visceral Pain

Carnett's Test

The Carnett's test is an easy way to determine if abdominal pain arises from abdominal wall. The test is considered positive when the palpation of the tender area results painful only when the patient tenses the abdominal wall by elevating the head and the shoulder or straight leg rising [4].

As suggested by Gallegos and Hobsley, the Carnett's test well fits in the algorithm for abdominal wall pain diagnosis [6]. When positive and the painful area is located near or in correspondence of a surgical scar, an injection of local anesthetic is usually performed. The diagnosis of abdominal wall pain is confirmed only if the patient experiences immediate pain relief after the injection, otherwise different sources of pain should be sought [4, 6].

Epidural Differential Block

Differential neuraxial blockade is a temporary diagnostic block that takes the advantage of the variable effect of local anesthetics on different nerve fibers to identify the etiology of pain. The differential block is performed by the administration of placebo or local anesthetic through an epidural catheter. Hypothetically, leads to selection of the pain origin as psychogenic, sympathetic, nociceptive (sensory based), or central [10]. Nevertheless, the ability of differential neuraxial blocks to diagnosis of various categories of pain generation is unproved. Reviewing the literature, only two reports were found with weak evidence that differential neuraxial blockade can predict treatment response. Both reports evaluated only a small cohort of patients [11, 12].

The rationale behind the test is based on the different sensitivity of nerve fibers to local anesthetics, which is ultimately related to the anatomical and functional differences among the fibers such as size and myelination degree.

An alternative of the classical differential epidural block evaluates the ratio between pain and pinprick skin sensitivity during the recovery phase from a complete (surgical) epidural anesthesia in the dermatomes corresponding to pain

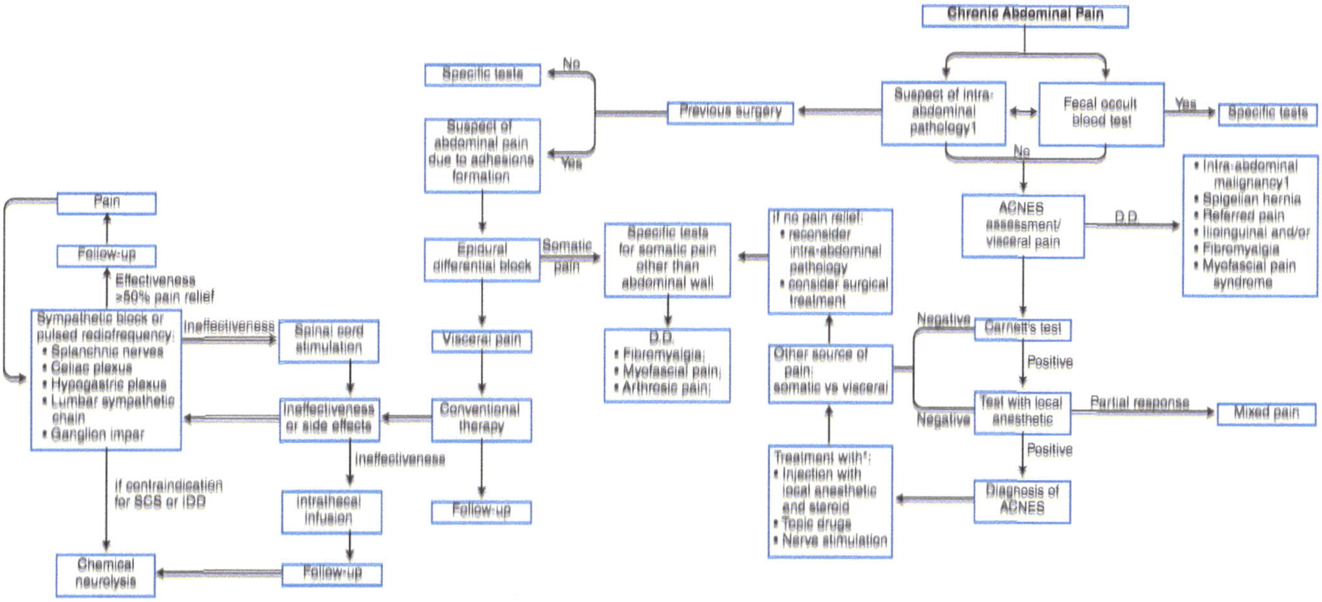

Fig. 8.1 Proposal for algorithm of therapeutic decision making in the treatment of chronic pain due to postsurgical intra-abdominal adhesions

irradiation [10]. A persistent pain relief when skin dermatomes anesthesia recedes indirectly defines the pain as visceral and points toward an intra-abdominal organ as the pain generator [10].

Limitation of this kind of test is due to the interaction between local anesthetic and nerve fibers and is a dynamic and unpredictable phenomenon that may be influenced by a multitude of factors. Also, overlap in the range of nerve fiber sizes makes it unlikely that any fiber type can be reliably isolated by this procedure. There is no guarantee that the surgical anesthesia achieved during the procedure blocks nervous transmission in all fibers and there is no evidence that sympathetic and visceral fibers are always slowest to return to normal function after the block [10]. For these reasons, the interpretation of differential epidural test sometimes could be very difficult. In addition, epidural differential block is a time-consuming technique and exposes patients to side effects and complications of a neuraxial blockade [13].

Rectus Sheath Block and Transversus Abdominis Plane Block

The distribution of sensory blockade is different with rectus sheath block, and Transversus abdominis plane (TAP) block. Rectus Sheath block is performed just below the costal margin at an angle of approximately 45° to the skin, in a plane between posterior rectus sheath and fascia transversalis. It provides sensory block for the whole midline of the abdomen. TAP block is performed laterally on the abdomen by placing local anesthetic in the plane between the internal oblique and transversus abdominis muscles. Bilaterally performed provides only reliable analgesia below the umbilicus.

Recent studies investigating abdominal wall pain in pediatric population proposed the use of Rectus Sheath Block and TAP Block in the management of CAWP [7, 14].

When the exact localization of the painful spot results to be difficult (i.e., pediatric population), Rectus Sheath Block represents an alternative to the infiltration of the point of maximum tenderness for both diagnosis and treatment of abdominal wall pain [7]. Due to the peculiar anatomical conformation of cutaneous nerve branches, however, Rectus Sheath Block might fail in one-third of the patients [7].

The TAP block allows to obtain a blockade of afferents originating from the anterior abdominal wall (skin, muscle, and parietal peritoneum), interrupting signals transmission travelling along the anterior rami of the lower six thoracic nerves and first lumbar nerve [15]. These nerves pass through a fascial plane between the internal oblique and transversus abdominis muscle called transversus abdominis fascial plane [9, 15]. A cadaveric and human study performed by McDonnell et al. shows that deposition of local anesthetic within the TAP produces a reproducible sensory block over the anterior abdominal wall from T7 to L1 dermatomes [11]. After initial description of, ultrasound-guided techniques of TAP, variation from the classic TAP block, the subcostal TAP block, has also been described; it is designed to provide more reliable coverage of the upper abdominal wall [16].

In 2009, Soliman and Narouze proposed the use of TAP block as a substitute of differential epidural block to distinguish visceral pain from abdominal wall (somatosensory) pain [12]. TAP block could replace differential epidural block because of a better side effects profile, especially when performed by an experienced physician with an ultrasound-guided technique [13]. Limitation of the use of TAP block in the differential diagnosis of abdominal pain is the difficulty to detect somatosensory pain which does not originate from the anterior abdominal wall.

Paravertebral Block

Paravertebral nerve block is a useful tool in the differential diagnosis of abdominal pain [17]. The spinal nerves in this space are devoid of a fascial sheath, making them exceptionally susceptible to local anesthetics.

Paravertebral analgesia is achieved placing local anesthetic alongside the vertebral column in the paravertebral space, which is crossed by intercostal and sympathetic nerves [17]. Paravertebral block also provide large unilateral somatic (mean of five dermatomes) and sympathetic block (mean of eight dermatomes), including the posterior ramus in multiple contiguous thoracic dermatomes. Apart from strict longitudinal spread, other forms of distribution have also been observed [18].

Relative contraindications for paravertebral block are coagulation disorders, anticoagulation, tumor in the paravertebral space, and empyema [16]. Complications are mainly due to the close relationship of the paravertebral space to the pleura and the neuraxial structures: an incorrect needle placement might result in pneumothorax or in epidural/intrathecal spread of the local anesthetic [17, 19].

Richardson et al. performed a review of 12 published studies with a total of 538 patients, underwent bilateral paravertebral block indicates paravertebral block as a procedure that can provide good intra and postoperative analgesia not only for thoracic surgery but also during abdominal and gynecological surgery and in chronic pain management [17]. Naja et al. reported the effectiveness of paravertebral block in the treatment of refractory myofascial pain syndrome at thoracic level [20].

Pain from Visceral Sources

Spinal Cord Stimulation

Chronic visceral pain was usually considered a "somatic" pain and therefore not suitable to be treat with treatments such as Spinal Cord Stimulation (SCS), which have been shown to be quite effective in treating chronic pain with the presence of a significant neuropathic component. Recent evidence, on the contrary, support the neuropathic origin of chronic visceral pain [21], which seems to be due to chronic sensitization of peripheral visceral nociceptors and wide dynamic range (WDR) neurons within the spinal cord [21].

Conventional pharmacologic therapy, sympathetic blocks, and radiofrequency application are important tool for the treatment of chronic visceral pain, but unfortunately they offer only a transient pain relief [22]. Since the first animal report of viscera-motor reflex suppression by the application of electrical stimulation to the spinal cord, the use of SCS for the treatment of chronic visceral pain significantly increased over time, in the attempt of providing a long-term therapeutic option [22].

In 2006, Tiede et al. [23] published a case report of two patients with refractory abdominal pain effectively treated with SCS; both patients had a history of several complicated abdominal surgeries with adhesions formation and subsequent surgical interventions for lysis of adhesions did not provide satisfactory pain relief.

Kapural et al. [22] in a survey performed in 2010 reported how the most common pathologies in patients with chronic abdominal pain treated with SCS were chronic pancreatitis, postsurgical intra-abdominal adhesions, and gastroparesis. Authors underlined how SCS is still rarely used despite its high therapeutic success rate; although the high cost of the device could be a matter of concern, there are scientific evidence to show how usually the cost of an SCS implant are overcome by the reduction in postimplant healthcare associated costs.

Kapural et al. also presented a case series of 35 patients suffering from chronic abdominal pain effectively treated with SCS, including 7 patients with chronic abdominal pain related to postsurgical adhesions [24]. A midline electrode placement at T5–T6 level was used in the majority of the patients, with exception of those patients suffering from lower abdominal quadrant pain who had electrodes placed at T11–T12 level. The reported percentage of patients with satisfactory (>50 %) pain relief was around 86 %, and the success rate of the trial phase (86 %) was higher than the usual 60–70 % typical of other fields of SCS application [18]. Most of the patients who failed the SCS trial also had poor response to sympathetic nerves block and, on the contrary, patients with good response to sympathetic block also showed good results during the SCS trial [24]. As with most of the invasive chronic pain interventions, patient's selection is crucial. In his works, Kapural selected the patients suitable for an SCS trial on the basis of the result of a differential epidural block and sympathetic block [22, 24, 25].

Sympathetic Blocks

A percutaneous block of the sympathetic chain is often performed as a therapeutic measure in patients suffering from chronic pain refractory to conventional drug therapy or to avoid intolerable drugs side effects [26]. However, it can be also used in the decision-making process to assess patients suitability for SCS therapy, as previously suggested [22, 24], or to identify potential responders to thermal or chemical sympathetic neurolysis [25].

Thoracic sympathetic block are selected infrequently for neural blockade [27]. Indications for sympathectomy of thoracic ganglia include: CRPS I and II, neuropathic pain in thorax, chest wall, thoracic viscera, upper abdominal viscera, herpes zoster, postherpetic neuralgia, phantom breast pain after mastectomy, ischemia due to arterial occlusion, drug-resistant Raynaud's disease, Burger's disease, and injuries of upper extremities [24, 27]. Complications of thoracic sympathetic blocks are nerve root injury, spinal cord damage, and pneumothorax [27].

The preganglionic axons from T5 to T9 coalesce to form the Greater Splanchnic Nerve at the level of T9–T10, course through the diaphragm and end in the Celiac plexus, which extends for several centimeters in front and laterally around the aorta [25]; the preganglionic axons from T10 to T11 form the Lesser Splanchnic Nerve while the Least Splanchnic Nerve arise from T12 [25].

The *Celiac Plexus block* and the *Splanchnic Nerves block* are usually performed in case of upper abdominal pain due to malignant or nonmalignant conditions involving the gastrointestinal tract from the distal third of the esophagus to the transverse colon, the liver, the biliary tract, the adrenals, and the mesentery [25, 26]. Side effects include hypotension and diarrhea, whereas complications include but are not limited to nerve injury, paralysis, pneumothorax, bowel injury, and bleeding [25, 26, 28].

Indications for blockade of the celiac plexus or splanchnic nerves include cancer of the abdominal viscera [28] to the splenic flexure, and chronic benign abdominal pain refractory to pharmacological treatment [25]. Side effects include hypotension and diarrhea. Complications include but are not limited to nerve injury, paralysis, pneumothorax, bowel injury, and bleeding. A review of 31 studies involving 1,599 patients who received 2,750 Neurolytic celiac plexus block (NCPB) found that 85–90 % achieved good to excellent pain relief after NCPB. The efficacy and safety of NCPB are also supported by a meta-analysis of 24 studies [29]. Only two were randomized controlled trials, however. Long-term benefit was achieved in 79–90 % with upper abdominal pain, most frequently from pancreatic cancer. Six percent to 8 % may require a second block to achieve pain control. Some suggest that the efficacy of NCPB has not been established given that pre- and post-NCPB pain assessment data are lacking in many studies [30]. There is substantial published work to validate a grade "B" recommendation based on the validity of available evidence for NCPB as a reasonable therapy for cancer pain [31, 32].

Previous studies estimated the positive predictive value, for pain relief, of a diagnostic block to be 85 % and the negative predictive value to be 58 %; this implies that a negative diagnostic block may discourage a physician to consider a procedure on Celiac Plexus that could provide a more lasting pain relief [28]. Carroll indicates NCPB provides persistent augmented analgesia when used as an adjunct to systemic opiates, when used as a part of a comprehensive analgesic plan, but does not reliably decrease opiate requirements. He also indicates that splanchnicectomy under fluoroscopic guidance is the optimal approach to perform this block [28]. Also, Day [25] in his review about sympathetic block indicates the Celiac Plexus block/Sympathetic Nerves block as the only block concluded grade of recommendation 1B, with moderate-quality evidence for abdominal pain management.

The lumbar sympathetic chain lies at the anterolateral border of the lumbar vertebral bodies, and a block of the *Lumbar Sympathetic Chain* is usually indicated for CRPS I and II, peripheral neuropathy pain, and for ischemia-related pain [25]. A common side effect is hypotension due to peripheral vasodilatation; complications include bleeding, nerve root injury, genitofemoral neuralgia, paralysis, neuraxial injection, and renal puncture [25].

The *Superior Hypogastric Plexus* is a retroperitoneal structure located slightly left off the midline from the level of lower L3 lumbar vertebral body to the upper S3 sacral vertebral body near the bifurcation of the common iliac vessels [25]. The plexus branches descend into the pelvis as the *Inferior Hypogastric Plexus receives* parasympathetic fibers from S2 to S4 sacral level and forms the pelvic, middle rectal, vesicle, prostatic, and uterovaginal plexus [25].

The *Superior Hypogastric Plexus block* is indicated for cancer and noncancer pain originating from the descending colon to the rectum and for the urogenital system in the pelvis [25, 26]. Complications include intravascular injection, discitis, neuraxial injection, urinary tract injury, and bladder/bowel incontinence [25].

Ganglion impar (also known as Walther's Ganglion) is the terminal ganglion of the sympathetic chain. Its anatomy is variable but is usually located caudal of the sacrococcygeal junction [25]. The *Ganglion Impar block* is usually indicated for vulvar pain, chronic perineal pain, and sacrococcygeal pain [25, 26].

Literature report few data about effectiveness of sympathetic blocks in the treatment of abdominal pain, and the available studies are of poor quality (only Celiac Plexus block/Splanchnic Nerves block shows a grade of evidence 1B) [25], due to the lack of well-designed RCTs with a sufficient number of enrolled patients. Pain physicians however continue to perform these procedures based on their everyday clinical experience with good results, reduced oral drug requirements, and safe side effects profile. In addition, sympathetic blocks play a fundamental role in the decisional algorithm for chronic abdominal pain treatments—as proposed by Kapural for the selection of potential SCS candidates [24].

Neurolytic Blocks

The use of neurolytic blockade is still playing an important role in controlling cancer pain in selected patients. No neurolytic agent or technique has been proven to be superior to another. Current evidence suggests that patients with pain of malignant origin may benefit from a variety of neurolytic techniques, mainly for visceral pain control [33]. Those techniques interrupt the sympathetic nervous system at the ganglion level, where afferent fibers from abdominal and pelvic organs converge, to treat chronic pain of various, but cancer-related etiologies [33].

A neurolytic block of the sympathetic plexus or chain may maximize the analgesic effects of conventional therapy, reducing the opioid daily dose and consequently minimizing the side effects. This approach is particularly useful in the treatment of cancer pain patients, where an aggressive therapeutic approach may be justified despite significant side effects [34].

There are several techniques on how to perform truncal neurolytic ablation, among which the Interpleural Phenol block, the Celiac Plexus block, Superior Hypogastric Plexus block, and the Ganglion Impar block, are used the most [34].

To perform an Interpleural block, local anesthetic is injected into the thoracic cage between the parietal and visceral pleura, producing an ipsilateral somatic block of multiple thoracic dermatomes together with the block of the sympathetic chains and the splanchnic nerves [35]. It provides relief from surgical and nonsurgical pain originating from the chest and upper abdomen in acute and chronic settings [35].

The Interpleural Phenol block provides an effective technique for the treatment of visceral pain originating from esophagus, liver, biliary tree, stomach, and pancreas [34, 35]. Although the evidence for interpleural block in chronic pain derives mostly from case reports and care series, the use of interpleural block is suggested in many painful disorders including esophageal cancer pain and chronic pain in patients with upper abdominal cancer and chronic benign and neoplastic pancreatic pain [36]. It can also represent a valid alternative to celiac plexus block in selected patients where the celiac block is too difficult or unsafe to perform [36]. Pneumothorax and phrenic nerve palsy resulting in respiratory failure are known complications of the technique [34]. To make a block neurolytic, increased concentration of phenol are used, starting with the initial recommended dose of 5–10 ml of 6 % phenol with a subsequent progression in concentration (up to 10 %) [34]. In patients with severe refractory pain (i.e., cancer patients) are also suggested the use of local anesthetics (i.e., Bupivacaine or Ropivacaine) with continuous or intermittent bolus infusion [34].

The celiac plexus receives parasympathetic fibers from the vagus nerve and autonomic fibers supplying liver, pancreas, gallbladder, stomach, spleen, kidneys, intestines, and adrenal glands [34].

Celiac Plexus block is effective in patients with pain from chronic pancreatitis or with visceral pain from cancer in the upper abdomen [34]. Among the known complications, hypotension, diarrhea, dysesthesia, interscapular back pain, reactive pleuritis, hiccups, hematuria, retroperitoneal hemorrhage, paraplegia or transient motor paralysis, and abdominal aortic dissection are worth to mention [34]. For this block either alcohol or phenol is the suggested drug. Alcohol block—with concentrations ranging from 50 to 100 %—produces severe pain before destroying the nerve fibers; hence, a local anesthetic injection (5–10 ml) is recommended before proceeding with alcohol block. On the other side phenol, in a concentration of 10 %, produces a painless injection; this is the reason why phenol, which seems to have the same neurolytic activity of alcohol, could be considered as the agent of choice [34].

Superior hypogastric plexus block can be utilized in patients with chronic pelvic pain [34]. Analgesia to the organs situated in the pelvis is possible because afferent fibers from these structures travel along the sympathetic nerves, trunks, ganglia, and rami [34]. The superior hypogastric plexus is situated in the retroperitoneum, bilaterally, extending from the lower third of the fifth lumbar vertebral body to the upper third of the first sacral vertebral body [34].

Several authors acknowledge to neurolytic sympathetic blocks a significant capacity to improve quality of life and pain control in patients with abdominal and pelvic malignant pain, and suggest to consider those interventions earlier in the management algorithm for visceral cancer pain [33].

Unfortunately, due to the severity of some side effects and complications which are not tolerable in the long run, the same considerations cannot be applied to other different categories of patients.

Radiofrequency

Radiofrequency is a useful tool in the treatment of chronic pain. Besides being a minimally invasive, outpatient treatment, it has some advantage over surgical resection, phenol or alcohol neurolysis because of its "target-selective" approach which cause fewer complications [37]. In fact, radiofrequency lesion is circumscribed around the needle's tip (about 5–6 mm of cross-sectional diameter) and the area to be lesioned is pretested with electric stimulation and impedance monitoring [38].

Interruption of the sympathetic chain has been used for a long time to treat intractable pain in the sacral-pelvic region, for the management of visceral pain and *CRPS*. The application of RF in this indication differs from its use for other targets such as sensory nerve tissue because no sensory threshold can be achieved in the sympathetic nerves [37]. The use of RF treatment in *Visceral pain* due to chronic pancreatitis, pancreatic cancer, liver cancer, or postabdominal surgery pain that is not or no longer responding to pharmacological treatment can be managed, according to available experience, more selective, and with fewer complications, using RF lesioning of the splanchnic nerves [38].

The available evidence varies in quality and level from one indication to another, mainly because of the ethic concerns of performing completely blinded and placebo-controlled trials in the chronic pain setting [37]. Success of the technique is related to the accuracy of the diagnosis, to the correct identification of the causal nerve structure, to the technical experience of the physician performing the technique and to the patient's expectations [37]. An accurate use of radiofrequency as part of multimodal and multidisciplinary approach could avoid or delay the use of more invasive and more expansive treatment options.

The recent introduction of "pulsed" radiofrequency (PRF), a nonablative radiofrequency modality, has now expanded the therapeutic RF applications in the management of chronic pain; with its nonablative modulating effect, PRF allows to avoid some serious side effects typical of conventional RF performed at visceral level, and its efficacy can now be safely tested even in chronic noncancer pain patients. In the WHO treatment ladder pulsed radiofrequency should be considered in the second step as first procedure for pain treatment [37].

The available documentation on the RF treatment in abdominal pain syndrome indicates that this option will only be considered when conservative causative and symptomatic treatment has been used to its full extent and fails to provide satisfactory pain relief. The success rate will depend on the accuracy of the diagnosis and the identification of the causal nerve structure, the experience of the physician using the technique, and the patient's expectations. The optimal environment for applying RF treatment is a multidisciplinary setting facilitating diagnosis, treatment, and guidance in terms of expectations and coping with the rest pain [37].

Pain from Abdominal Wall

Currently, there is a lack of evidence-based indications for the treatment of CAWP [5]. Conservative treatments are the first option and can be useful also for the differential diagnosis.

If the pain does not reduce patient's quality of life, a simple diagnostic block with local anesthetic may be enough to establish the pain source; most patients with not-disabling chronic pain are satisfied just to know the source of their pain [5].

In the treatment of mild–severe pain, a long-acting corticosteroid can be added to local anesthetic to provide a longer-lasting pain relief [5].

A case series of seven pediatric patients reports the effectiveness of Rectus Sheath Block for the treatment of refractory CAWP. All patients underwent to Rectus Sheath block with local anesthetic and corticosteroid and showed a significant initial improvement in pain and quality of life [7]. Three patients required only the RSB to enable them to be pain-free and return to normal schooling and physical activities. Two children received complete relief for more than 1 year.

Stretching of abdominal musculature, topic drugs, and nerve stimulation are other conservative treatments in case of mild–severe pain. Unfortunately, only 40 % of patients suffering from severe CAWP are satisfied by conservative therapy. Specific disorder, such as myofascial trigger points, may benefit from botulinum toxin injection or local phenol injection. In case of peripheral nerve entrapment (ACNES, entrapment of ilioinguinal, and iliohypogastric nerves), the surgical release of entrapped nerves might be considered [5].

In addition to the above-mentioned techniques, refractory abdominal wall pain could benefit from the therapeutic options also indicated for visceral pain. Central and peripheral neuromodulation (spinal cord stimulation, peripheral nerve stimulation, and peripheral field stimulation) has a role in the treatment of neuropathic pain deriving from nerve entrapment or radiculopathy. Pulsed radiofrequency may provide pain relief in thoracic radicular pain and Intrathecal Drug Delivery is an effective tool to be considered when other therapeutic options fail to provide adequate pain relief.

Miscellaneous

Intrathecal Drug Delivery

Differently from somatic structures, the viscera receive dual innervations from vagal and spinal primary afferent neurons; the cellular bodies of these neurons lie in the brain stem (vagal afferents) and segmentally in the dorsal root ganglia (spinal afferents) [3].

Normally sensations originating from viscera are not perceived and the principal conscious sensations that arise from viscera are discomfort and pain. Visceral pain can therefore derive from the perception of altered sensation from the viscera; one of the mechanisms that contribute to develop this altered sensation is represented by visceral hyperalgesia. The mechanisms of visceral hyperalgesia are not completely understood but probably consist in both peripheral and central nervous system components [3].

Intrathecal Drug Delivery (IDD) is thought to work by interfering with pain signals before they reach the brain. However, there has been only limited research into the precise mode of action of the various agents reported to provide analgesia following intrathecal delivery, as most use has been empiric [39].

IDD find a wide range of applications in the treatment of chronic abdominal pain due to the possibility to cover different source of pain with the employment of different classes of drugs. Several drugs are usually administered through an IDD, but only for opioids the mechanism of action is clearly known [39]. Opioids modulate pain signals transmission binding presynaptic and postsynaptic mu-receptors in the substantia gelatinosa in the dorsal horns and thus inhibiting C-fibers transmission. Local anesthetics impair the pain signal transmission blocking the action potential progression; Ziconotide blocks signal transmission blocking the N-type calcium channel; and adjuvant drugs direct inhibit the neurotransmission and the release of C-fibers transmitters [39].

Combinations of different drug classes, such as opioids/local anesthetics, opioids/clonidine, and opioids/local anesthetics/clonidine, are currently being used in clinical practice. The efficacy reports appear favorable, but are based largely on case studies and retrospective analysis. No information is available on the long-term compatibility of these combinations. Several new agents appear to have the interest of both practitioners and researchers for the future treatment of intractable pain disorders [40].

Further research is needed to determine the best clinical application for many of the compounds currently used in clinical practice. Data are limited on the compatibility of drug combinations, dose ranges, and long-term safety for many agents currently identified as clinically relevant. Information is also incomplete regarding the long-term effects of drug combinations on intrathecal catheters and infusion devices. Opioids and local anesthetics can be used to modulate visceral pain taking advantage of their action on primary spinal afferent but could also be useful in the treatment of intractable abdominal pain of somatic origin. Opioids, local anesthetics, and ziconotide are also indicated for the treatment of neuropathic pain [39] such as CAWP originating from nerve entrapment (ACNES, ilioinguinal, and/or iliohypogastric nerve entrapment).

Despite those theoretical bases supporting the use of IDD for the treatment of chronic abdominal pain, its use appears very limited in reality. At our knowledge, no reports of IDD use in chronic abdominal pain related to adhesion have been published so far. However, Guttman et al. [41] reported satisfactory pain relief in a patient suffering from refractory chronic abdominal pain due to Median Arcuate Ligament (MAL) syndrome. In this report, all the therapeutic options, including oral opioids, sympathetic block, SCS and surgery, failed to provide adequate pain relief. After a successful trial with intrathecal morphine, an IDD system was implanted with catheter's tip at T11 vertebral body level; 10 days after the implant, the patient reported a complete pain relief and returned to a normal quality of life.

Despite it is obviously impossible to draw any conclusion from the experience of a single case reported, it suggests that IDD may represent an important therapeutic option when all other conventional therapies had failed to control visceral pain.

Chronic Abdominal Pain After Gastric Bypass

The number of gastric bypass surgeries performed in the USA every year is about 200,000 [42]. Chronic abdominal pain is one of the most frequent complications of this type of surgery, and it is the first cause of admission in emergency room in this patient population for the first 3 years after surgery [42]. Clinical presentation of chronic abdominal pain in patients who underwent gastric bypass varies and still represents a diagnostic and therapeutic challenge.

Causes of abdominal pain after gastric bypass include behavioral and dietary disorders (overeating and nutrient deficiencies), functional disorders (constipations, motility disorders, dumping syndrome, and irritable bowel syndrome), biliary disorders (cholelithiasys, cholecystitis, and sphincter of Oddi dysfunction), pouch or remnant stomach disorders (ulcer disease and GERD hiatus hernia), and small-intestine disorders (abdominal wall hernias, internal hernia, intussusception, and adhesions). The diagnostic approach should consider the broad range of causes of abdominal pain after gastric bypass and the diagnostic algorithms must be flexible considering the clinical history and physical examination. In the absence of a clear diagnosis, physicians may opt for surgical exploration [42].

Preventing Strategies

Intra-abdominal adhesions are usually iatrogenic, occurring in 95 % of patients who underwent previous abdominal or pelvic surgery, even though adhesions have been reported also in patients without a history of previous laparotomy [43, 44]. Pelvic surgery is associated with both de novo adhesions formation and adhesions reformation rate around 80 % [45], but pelvic inflammation or endometriosis are other frequent causes.

The treatment of adhesions-related complications represents a significant burden for the healthcare system; they can make a further surgical access more complex, prolong operative times, and hospital stay, require readmissions and repeated surgical interventions for bowels obstruction, and be a significant cause of chronic abdominal pain, dyspareunia, and female infertility [43, 46]. Certain surgical operations are associated with higher risk of adhesions-related complications such as ileoanal pouch procedures and non-elective appendicectomy. In order to reduce related complications, adhesions prevention seems to be the most viable strategy [43].

Adhesions are highly cellular, vascularized, and dynamic structures and should not be considered as simple inactive scar tissue [45]. Because acute inflammation, peritoneal injury, fibrin deposition, and subsequent fibrinolysis are key events in the genesis of adhesions, factors that determine cell proliferation, migration, differentiation, angiogenesis, apoptosis, and host defense have been considered as potential targets to control and limit the process of adhesions formation [43].

The potential role of pharmacological agents in reducing or preventing adhesions formation has been examined in several published studies. Drugs to mediate the inflammatory response has been tried: steroid and nonsteroidal anti-inflammatory drugs have been used in the past with equivocal results because of systemic side effects such as bleeding (NSAIDs) or impaired wound healing (steroids). Inhibiting fibroblasts proliferation with agents, such as Mitomycin C, have been also tried but side effects limited its use. The use of therapeutic anticoagulants to reduce fibrin deposition did not show any statistically significant effect on adhesions reduction and has been also associated with an increased risk of postoperative bleeding.

Recently, free radical scavengers, such as methylene blue, inhibitors of proinflammatory cytokines, and antihistamine agents, have been reported to yield good results in preclinical studies but still without reaching the clinical daily practice [43].

Experimental studies have shown a reduction in adhesions formation with the use of broad-spectrum chemokine's blockade; also selective chemokine inhibition seems to have an effective role in adhesion prevention in animal models and clinical trials are awaited [45].

Topical products have been also tested to reduce adhesions formation, and they can be divided mainly in two groups. The first group includes liquids that are instilled in the abdominal cavity to reduce the contact surface; liquids similar to peritoneal dialysis solution have shown a certain degree of effectiveness in adhesions prevention [45]. The second group consists of inert physical materials (gels or films) directly applied in the abdominal cavity to mechanically separate and to prevent contact between the damaged serosal surfaces [44, 45]. Among these, Chitosan, a natural polysaccharide, has shown promising results in animal studies and also has hemostatic and antimicrobial properties [45]. The use of synthetic barriers is however limited to site-specific adhesions prevention; a high risk of generalized adhesions formation should suggest fluid and/or pharmacological agents as the main adhesions prevention strategy [44].

A recent Cochrane review investigated the efficacy of *intra-abdominal prophylactic agents* for adhesions prevention in *nongynecologic abdominal surgery* and found hyaluronic acid/carboxymethyl cellulose membrane (HA/CMC) has been the only agent with some effectiveness [43]. The use of HA/CMC membrane may be considered in the prophylaxis of intraperitoneal adhesions as it seems to reduce the incidence, extent, and severity of adhesions, even if does not reduce the incidence of abdominal obstruction.

Another Cochrane review analyzed the use of *synthetic barriers* to reduce the incidence of postoperative adhesions in *pelvic surgery* and reported *Gore-Tex* and *Oxidized Regenerated Cellulose* as effective in reducing the incidence of postoperative adhesions without eliminating adhesions formation for any patients [44].

Several medications have been tested over the years; steroids, antihistamine agents, dextran solution, and hyaluronic acid are only few examples of the drugs tested [46].

In women underwent pelvic surgery, the use of steroids SprayGel, 4 % Icodextrin (ADEPT), Noxytiolone, Heparin, and Promethazine does not prevent or reduce adhesions following surgery, according to a Cochrane review [46]. The use of hyaluronic acid agents may decrease adhesion formation and prevent the deterioration of preexisting adhesions.

Unfortunately, the limited number of studies available does not permit an univocal and certain data interpretation, but there is insufficient evidence for the use of the following agents: steroids, icodextrin 4 %, SprayGel, and dextran in improving adhesions following surgery. In addition, the use of Dextran is related to several and serious side effects such as pleural effusion, anaphylaxis, labial edema, and abnormal liver enzyme.

Source(s) of financial support: none
Authorship statement: none
Conflict interest statement: none

References

1. Sulaiman H, Gabella G, Davis C, Mutsaers SE, Boulos P, Laurent GJ, et al. Presence and distribution of sensory nerve fibers in human peritoneal adhesions. Ann Surg. 2001;234:256–61.
2. Liakakos T, Thomakos N, Fine PM, Dervenis C, Young RL. Peritoneal adhesions: etiology, pathophysiology, and clinical significance. Dig Surg. 2001;18:260–73.
3. Gebhart GF. Pathobiology of visceral pain: molecular mechanisms and therapeutic implications IV. Visceral afferent contributions to the pathobiology of visceral pain. Am J Physiol Gastrointest Liver Physiol. 2000;278(6):G834–8.
4. Srinivasan R, Greenbaum DS. Chronic abdominal wall pain: a frequently overlooked problem. Am J Gastroenterol. 2002;97: 824–30.
5. Lindsetmo R, Stulberg J. Chronic abdominal wall pain—a diagnostic challenge for the surgeon. Am J Surg. 2009;198:129–34.
6. Gallegos NC, Hobsley M. Abdominal wall pain: an alternative diagnosis. Br J Surg. 1990;77:1167–70.
7. Skinner AV, Lauder GR. Rectus sheath block: successful use in the chronic pain management of pediatric abdominal wall pain. Paediatr Anaesth. 2007;17:1203–11.
8. Applegate WV, Buckwalter NR. Microanatomy of the structures contributing to abdominal cutaneous nerve entrapment syndrome. J Am Board Fam Pract. 1997;10(5):329–32.
9. Guittet L, Bouvier V, Mariotte N, et al. Comparison of a guaiac based and an immunochemical faecal occult blood test in screening for colorectal cancer in a general average risk population. Gut. 2007;56:210–4.
10. Hogan Q, Abram S. Neural blockade for diagnosis and prognosis: a review. Anesthesiology. 1997;86:216–41.
11. Conwell DL, Vargo JJ, Zuccaro G, et al. Role of differential neuroaxial blockade in the evaluation and management of pain in chronic pancreatitis. Am J Gastroenterol. 2001;96:431–6.
12. Bradley EL, Reynhout JA, Peer GL. Thorascopic splanchnicectomy for a small duct chronic pancreatitis: case selection by differential epidural anesthesia. J Gastrointest Surg. 1998;2:88–94.
13. Soliman LM, Narouze S. Ultrasound-guided transversus abdominus plan block for the management of abdominal pain: an alternative to differential epidural block. Thech Reg Anesth Pain Manag. 2009;13:117–20.
14. Simpson DM, Tyrrel J, De Ruiter J, Campbell FA. Use of ultrasound-guided treansversus abdominis plane blocks in a pediatric patient with chronic abdominal wall pain. Paediatr Anaesth. 2011;21:88–90.
15. McDonnell JG, O'Donnell BD, Farrell T, Gough N, Tuite D, Power C, Laffey JG. Transversus abdominis plane block: a cadaveric and radiologic evaluation. Reg Anesth Pain Med. 2007;32: 399–404.
16. Hebbard P. Subcostal transversus abdominis plane block under ultrasound guidance. Anesth Analg. 2008;106(2):674–5.
17. Richardson J, Lonnqvist PA, Naja Z. Bilateral thoracic paravertebral block: potential and practice. Br J Anaesth. 2011;106(2):164–71.
18. Richardson J, Sabanathan S, Jones J, Shah RD, Cheema S, Mearns AJ. A prospective randomized comparison of preoperative and continuous balanced epidural or paravertebral bupivacaine on post-thoracotomy pain, pulmonary function and stress responses. Br J Anaesth. 1999;83:387–92.
19. Thavaneswaran P, Rudkin G, Cooter RD, Moyes DG, Perera CL, Maddern GJ. Brief reports: paravertebral block for anesthesia: a systematic review. Anesth Analg. 2010;110(6):1740–4.
20. Naja ZM, Al-Tannir MA, Zeidan A, El-Rajab M, Ziade F, Baraka A. Nerve stimulator-guided repetitive paravertebral block for thoracic myofascial pain syndrome. Pain Pract. 2007;7:348–51.
21. Krames ES, Foreman R. Spinal cord stimulation modulates visceral nociception and hyperalgesia via the postsynaptic dorsal column pathways: a literature review and hypothesis. Neuromodulation. 2007;10:224–37.
22. Kapural L, Deer T, Yakovlev A, Bensitel T, Hayek S, Pyles S, et al. Technical aspects of spinal cord stimulation for managing chronic visceral abdominal pain. Pain Med. 2010;11:685–91.
23. Tiede JM, Ghazi SM, Lamer TJ, Obray JB. The use of spinal cord stimulation in refractory abdominal visceral pain: case reports and literature review. Pain Pract. 2006;6:197–202.
24. Kapural L, Nagem H, Tlucker H, Sessler DI. Spinal cord stimulation for chronic vesceral abdominal pain. Pain Med. 2010;11:347–55.
25. Day M. Sympathetic blocks: the evidence. Pain Pract. 2008;8(2): 98–109.
26. Reisfield GM, Wilson GR. Blocks of the sympathetic axis for visceral pain, 2nd edition #97. J Palliat Med. 2011;14(4):510–1.
27. Plancarte-Sanchez R, Guajardo-Rosas J, Guillen-Nunez R. Sympathetic block: thoracic and lumbar. Tech Reg Anesth Pain Manag. 2005;9:91–6.

28. Carroll I. Celiac plexus block for visceral pain. Curr Pain Headache Rep. 2006;10:20–5.
29. Eisenberg E, Carr DB, Chalmers TC. Neurolytic celiac plexus block for treatment of cancer pain: a meta-analysis. Anesth Analg. 1995;80:290–5.
30. Staats PS, Yearwood T, Charapata SG, et al. Intrathecal ziconotide in the treatment of refractory pain in patients with cancer or AIDS: a randomized controlled trial. JAMA. 2004;291:63–70.
31. Wong GY, Schroeder DR, Carns PE, et al. Effect of neurolytic celiac plexus block on pain relief, quality of life, and survival in patients with unresectable pancreatic cancer. JAMA. 2004;291:1092–9.
32. Sharfman WH, Walsh TD. Has the analgesic efficacy of neurolytic celiac plexus block been demonstrated in pancreatic cancer pain? Pain. 1990;41:267–371.
33. de Oliveira R, dos Reis MP, Prado WA. The effect of early or late neurolytic sympathetic plexus block on the management of abdominal or pelvic cancer pain. Pain. 2004;110:400–8.
34. de Leon-Casasola OA. Critical evaluation of chemical neurolysis of the sympathetic axis for cancer pain. Cancer Control. 2000;7:142–8.
35. Dravid RM, Paul RE. Interpleural block—part 1. Anaesthesia. 2007;62:1039–49.
36. Dravid RM, Paul RE. Interpleural block—part 2. Anaesthesia. 2007;62:1143–53.
37. Van Zundert J, Raj P, Erdine S, van Kleef M. Application of radiofrequency treatment in practical pain management: state of the Art. Pain Pract. 2002;2:269–78.
38. Raj PP, Sahinler B, Lowe M. Radiofrequency lesioning of splanchnic nerves. Pain Pract. 2002;2:241–7.
39. Erdine S, De Andres J. Drug delivery system. Pain Pract. 2006;6:51–7.
40. Bennett G, Serafini M, Burchiel K, Buchser E, Classen A, Deer T, Du Pen S, Ferrante FM, Hassenbusch SJ, Lou L, Maeyaert J, Penn R, Portenoy RK, Rauck R, Willis KD, Yaksh T. Evidence-based review of the literature on intrathecal delivery of pain medication. J Pain Symptom Manage. 2000;20(2):S12–36.
41. Guttman OT, Rosenblatt MA, Mims T. Median arcuate ligament syndrome—a novel treatment using an intrathecal morphine pump to relieve intractable visceral pain. Pain Pract. 2008;8:133–7.
42. Greenstein AJ, O'Rourke RW. Abdominal pain after gastric bypass: suspect and solutions. Am J Surg. 2011;201:819–27.
43. Kumar S, Wong PF, Leaper DJ. Intra-peritoneal prophylactic agents for preventing adhesions and adhesive intenstinal obstruction after non-gynaecological abdominal surgery. Cochrane Database Syst Rev. 2009;1, CD 005080. doi:10.1002/14651858.pub2.
44. Ahmad G, Duffy JMN, Vail A, Vanderkerchove P, Watson A, Wiseman D. Barrier agents for adhesion prevention after gynaecological surgery. Cochrane Database Syst Rev. 2008;2, CD000475. doi:10.1002/14651858.CD00475.pub2.
45. Lauder CIW, Garcea G, Strickland A, Maddern GJ. Abdominal adhesion prevention: still a sticky subject? Dig Surg. 2010;27:237–58.
46. Metawally M, Watson A, Lilford R, Vanderkerchove P. Fluid and pharmacological agents for adhesion prevention after gynaecological surgery. Cochrane Database Syst Rev. 2006;2, CD001298. doi:10.1002/14651858. CD001298.pub3.

Chronic Pancreatitis With or Without Acute Exacerbations: Novel Options for Pain Control

9

Martine Puylaert

Introduction

Chronic pancreatitis (CP) is a benign inflammatory disease of the pancreas. The mechanism of pain is incompletely understood although the knowledge of this disease starts with the study of Sarles et al. in 1965 [5]. Irreversible morphological changes take place with progressive loss of the exocrine and later endocrine function of the gland due to fibrosis [1].

Pancreatitis: inflammation of the pancreas is subdivided clinically in its chronic and acute form.

This subdivision between acute and chronic pancreatitis into completely different entities must be revised. It is rather a continuum of disease where patients may show an evolution from acute to chronic pancreatitis. Both stages were described as being at separate ends of the same spectrum [2].

Acute pancreatitis, mainly characterized by acute pain due to inflammation and tissue necrosis is a transient event. During the recovery period, the gland is more vulnerable to alcohol, metabolic and oxidative stress, and cytokines. The latter are up regulated during the acute inflammation. Chronic pancreatitis is the ongoing process which consists of permanent and irreversible damage [3]. Estimates of annual incidence of chronic pancreatitis provide ranges from 5 to 12 cases per 100,000 persons. Furthermore depending on the population being studied, wide variations in incidence and prevalence figures are found [4].

Chronic pancreatitis is considered a common disease state, but it has different causes, 51 % is attributed to alcohol abuse. There are also autoimmune, hereditary (like in cystic fibrosis), metabolic (hypercalcemia, hyperlipidemia), tropical, and idiopathic forms of pancreatitis. Idiopathic chronic pancreatitis reflects that no cause can be identified. Chronic renal failure and hypercalcemia are described as risk factors.

According to the etiology, the different forms of pancreatitis have a difference in age of onset, sexual differentiation, and life expectancy. Hereditary pancreatitis starts between 10 and 14 years, whereas alcoholic pancreatitis starts around 40 years of age [5].

Predisposing Factors

Major predisposing risk factors for chronic pancreatitis may be categorized as either (1) toxic-metabolic, (2) idiopathic, (3) genetic, (4) autoimmune, (5) recurrent and severe acute pancreatitis, or (6) obstructive (TIGAR-O system). After classification, staging of pancreatic function, injury, and fibrosis becomes the next major concern.

Etiology

The different possible etiologies for chronic pancreatitis are listed in Table 9.1

Increased daily alcohol intake has been linked to a higher risk for chronic pancreatitis. There is, however, no known threshold value below which the disease does not occur [6, 7].

Although it is difficult to determine with certainty the involvement of alcohol intake in the pathogenesis of pancreatitis, in almost all patients at least 5 years (and sometimes 10 years) of excessive intake preceded the development of chronic pancreatitis [4]. A strong association of simultaneous alcohol intake and smoking has been demonstrated to increase the risk for chronic pancreatitis.

For *alcoholic pancreatitis* the age of onset is between 40 and 50 [5]. There is mortality within 10 years after the diagnosis of 30 %. After a period of acute inflammation, stellate cells in the pancreas get activated due to cytokines as a product of the inflammation but also by ethanol and its metabolites. Secondary this induces the increased fibrosis of the pancreas [7–9].

M. Puylaert, M.D., F.I.P.P. (✉)
Ziekenhuis Oost-Limburg, Anesthesiology and Multidisciplinary Pain Center, Bessemersstraat 478, Lanaken 3620, Belgium
e-mail: martine.puylaert@skynet.be

L. Kapural (ed.), *Chronic Abdominal Pain: An Evidence-Based, Comprehensive Guide to Clinical Management*, DOI 10.1007/978-1-4939-1992-5_9, © Springer Science+Business Media New York 2015

Table 9.1 Possible etiologies of chronic pancreatitis

Alcoholic pancreatitis
Hereditary pancreatitis
Autoimmune pancreatitis
Metabolic pancreatitis (hypercalcemia, hyperlipidemia)
Tropical pancreatitis
Idiopathic

One could presume that death is caused by multiple organ failure, sepsis, surgical complications or late complications of diabetes mellitus. But, the most prominent cause of death is the patients' lifestyle and alcohol related accidents. There is also an increased risk of lung cancer, esophageal cancer, and pancreatic cancer. These patients also seem to have an increased risk for cardiovascular disease.

Tropical pancreatitis, as the name suggests, is predominantly found in tropic regions such as Southwest India, Africa, Southeast Asia, and Brazil. Initially it was judged that tropical pancreatitis was restricted to areas within 30° latitude from the equator. The mean age of onset is 24 years. In endemic areas the prevalence may be as high as 1 in 500 persons. The pathophysiology is unclear, genetic mutations, environmental triggers, viral and parasitic infections have been suggested.

Clinical manifestations of tropical pancreatitis are: abdominal pain, severe malnutrition, and exocrine or endocrine insufficiency. Endocrine insufficiency seems to be directly related to diabetes. Steatorrhea is rare because of the very low-fat diet. In more than 90 % of the cases pancreatic calculi are present [4].

In families with *hereditary pancreatitis*, mutations in PRSS1 [protease, serine, 1 (trypsin 1) belong to a family of genes called serine peptidases] may cause chronic pancreatitis. Other mutations are considered cofactors to the development of chronic pancreatitis by increasing the susceptibility, or as modifier genes that increase the pace or severity of the disease. Several studies suggested that less severe CFTR (cystic fibrosis transmembrane conductance regulator) gene mutations and SPINK1 (Pancreatic secretory trypsin inhibitor (PSTI) also known as serine protease inhibitor Kazal Type 1) mutations may be associated with idiopathic chronic pancreatitis.

There are three major genetic factors that may play a role in chronic pancreatitis.

PRSS1

In normal conditions, trypsinogen is converted to the active trypsin. Three versions of trypsinogen can be identified: cationic, anionic, and mesotrypsinogen with respectively the involvement of PRSS1 gene, PRSS2 gene, and PRSS3 gene.

In 1996 Whitcomb et al. [10] isolated the first responsible mutation in the cationic trypsinogen gene (*PRSS1*). In the mutated families, there is an enhanced intra-pancreatic trypsinogen auto activation with secondary initiation of chronic pancreatitis. On the other hand, mutations of the chromosome PRSS2 have a disease protective effect for chronic pancreatitis.

SPINK 1

Serine protease inhibitor Kazal Type 1 or pancreatic trypsin inhibitor is an important inhibitor of the intra-pancreatic conversion of trypsinogen to trypsin. Mutation in the gene reduces the inhibition of auto activation with sequential activation of the zymogenes and auto digestion.

CFTR

Cystic fibrosis transmembrane conductance regulator (CFTR) regulates ductal bicarbonate secretion in the pancreas. Mutations of the CFTR gene are associated with cystic fibrosis, an autosomal recessive disease, with pulmonary and pancreatic dysfunction.

Autoimmune pancreatitis refers to a distinct chronic inflammatory and sclerosing disease of the pancreas. It is accompanied by dense infiltration of the pancreas, and sometimes other organs with lymphocytes and plasma cells that express IgG4 on the surface H. pylori infection can play a potential role in autoimmune pancreatitis. Because of the frequent presence of extra pancreatic manifestations such as biliary strictures, hilar lymphadenopathy, sclerosing sialadenitis, retroperitoneal fibrosis, and tubulointerstitial nephritis, it is assumed that autoimmune pancreatitis may be one manifestation of what has been called IgG4-related sclerosing disease or IgG4-related systemic disease.

The disease occurs most often after the age of 50 years and touches twice as much men than women. Clinically it presents as painless obstructive jaundice due to obstruction of the intra-pancreatic bile duct. It responds rapidly to glucocorticoid therapy. Most reports on autoimmune pancreatitis come from Japan and Asia. The overall prevalence is estimated to be 0.82.

Obstructive Chronic Pancreatitis

Obstruction of the main pancreatic ducts may be caused by different factors such as tumors, scars, ductal stones, duodenal wall cysts, or stenosis of the papilla of Vater or the minor papilla. Obstructive chronic pancreatitis is, however, a distinct entity produced by a single dominant narrowing or stricture of the main pancreatic duct.

Clinical Presentation

Pain

Pain and more specifically abdominal pain is the most predominant symptom that is responsible for the decreased quality of life, a reduced appetite and consequently reduced food intake and malnutrition leading to dramatic weight loss. Chronic severe pain is often responsible for the progressive social isolation of the patients. The addictive behavior and the difficulty to control chronic pancreatic pain may lead to addiction for narcotic analgesics.

There are no firm pain patterns. Patients report mostly epigastric pain that may radiate into the back. Pain often increases after ingestion of high fat food. It is described as boring, deep, and penetrating. It is often associated with nausea and vomiting. Bending forward and assuming the knee-chest position on one side or clasping the knees to the chest may alleviate the pain. No clear evolution pattern of the pain can be found.

Steatorrhea

When pancreatic lipase secretion is reduced to less than 10 % of the maximum output *steatorrhea* will occur. This is a feature of far-advanced chronic pancreatitis in which most of the acinar cells have been injured or destroyed. Maldigestion of fat, protein, and carbohydrates occur, but the maldigestion of fat occurs earlier and is more severe than protein or carbohydrate maldigestion. The median time to development of exocrine insufficiency has been reported as low as 5.6 years, but most studies report 13.1 years in patients with alcoholic chronic pancreatitis; 16.9 years in patients with late-onset idiopathic chronic pancreatitis, and 26.3 years in patients with early-onset idiopathic chronic pancreatitis. Significant weight loss due to maldigestion is uncommon. This is most commonly seen during painful flare ups, when pain, nausea, and vomiting prevent accurate food intake. In patients with chronic pancreatitis and steatorrhea deficiencies in fat soluble vitamins and specifically vitamin D may be observed.

Diabetes Mellitus

Endocrine insufficiency is also a consequence of long-standing chronic pancreatitis and results in diabetes mellitus in approximately 80 % of the patients with chronic pancreatitis. This diabetes is classified as type 3.

Less Common Symptoms

Jaundice
Skin nodules
Painful joints
Abdominal distension
Shortness of breath
Pleural effusions and ascites

Diagnostic Process

Physical Examination

Physical examination does not give much additional information that allows fine tuning the diagnosis of chronic pancreatitis. Aside from the abdominal tenderness a palpable pseudocyst may occasionally be found and jaundice may be seen in presence of coexisting alcoholic liver disease or bile duct compression within the head of the pancreas.

Diagnostic Tests

Laboratory

Serum Test

In contrast with acute pancreatic disease, where serum lipases and amylase are elevated, these tests stay normal in chronic pancreatitis, and thus have no diagnostic value.

Complete blood count, electrolytes, and liver function tests are normal. Elevated serum bilirubin and alkaline phosphates can be indicative for compression of the intrapancreatic part of the bile duct or pancreatic cancer.

In cases of autoimmune chronic pancreatitis, an elevated ESR, IgG4, rheumatoid factor, ANA, and anti-smooth muscle antibody titer can be detected.

Deficiencies of maldigestion of fat and proteins or vitamins like vitamin A, B12, and D can only be seen if 90 % of the glandular function is lost [11].

Pancreatic Functional Testing

Exocrine Function

The pancreas secretes daily 1.5 L of fluid rich in pancreatic enzymes for the digestion of fats, starch, and proteins. Secretin and cholecystokinin (CCK) play a key role in the regulation by a hormonal and neuronal feedback mechanism. Testing the functional activity of the pancreas can be done directly or indirectly. In advanced chronic pancreatitis these tests are unnecessary as imaging tests reveal structural changes. On the contrary, these tests can be helpful to diagnose the disease in an early stage. They can also be used as a guidance for adapting enzyme therapy.

For direct testing, the pancreas is stimulated by administration of a meal (Lundh test) or hormonal secretion stimulating products (CCK or secretin). Secretin stimulates the duct cells while CCK stimulates the acinar cells. Pancreatic fluid is collected by means of double lumen gastrointestinal tubes. Duodenal fluids are collected over 90 min. The fluids are analyzed to quantify enzymes (tryptase, amylase, lipase) and bicarbonate. The value of the bicarbonate and enzymes is a parameter to quantify the functional mass of pancreatic tissue. This test can reveal early stage chronic pancreatitis before the development of steatorrhea [12].

Endoscopic secretin test is now the reference test. Comparison of Pancreatic Functional Testing (PFT) with histological changes showed 67 % sensitivity and 90 % specificity of the secretin CCK test for chronic pancreatitis [13].

Indirect tests measuring the consequence of pancreatic insufficiency are more widely available. These tests are less sensitive and less specific in the diagnosis of chronic pancreatitis.

Serum trypsinogen: a low level trypsinogen has a high specificity for chronic pancreatitis. In case of normal level but with a clinical presentation of chronic pancreatitis the test should be repeated [14].

Fecal Tests

Fecal fat is tested in a stool sample. Steatorrhea is suggestive for a loss of more than 90 % of the normal pancreatic exocrine enzyme secretory output. Fat malabsorption may also occur in cases of disease of the small intestinal mucosa.

Fecal Chymotrypsin, Fecal Elastase 1

These tests show a poor sensitivity in early chronic pancreatitis and false positive testing in gastro intestinal disease.

Endocrine Function

Serum glucose HBA1 determination can be used to assess the endocrine function. This is often sooner affected than the exocrine function.

Genetic Analysis

Five pancreatitis susceptibility genes are established: cystic fibrosis transmembrane conductance regular gene (CFTR), pancreas secretory trypsin inhibitor gene (SPINK-1), chymotrypsinogen Cgene (CTRC), calcium sensing receptor gene (CASR), and cationic trypsinogen gene (PRSS), linked to hereditary pancreatitis.

Routine full genetic analyses are not recommended since they are not necessarily for the diagnosis of chronic pancreatitis, they are expensive and generally do not alter management. On indication CFTR and SPINK are mostly performed.

Imaging Studies

Plain Film of the Abdomen

Plain X-rays of the abdomen can show diffuse calcifications this is pathognomonic in chronic pancreatitis but, it occurs late. Calcification primarily represents intraductal calculi, either in the main pancreatic duct or in the smaller pancreatic ductal radicles. Clinical relevance is very low.

Ultrasound

Trans-abdominal ultrasound is a highly specific, inexpensive, and noninvasive screening test. In patients with thin bodies, trans-abdominal ultrasound can show the anatomy of the pancreas, parenchymal changes (atrophic and fibrosis), and ductal features suggestive of chronic pancreatitis (sensitivity, 60–70 %; specificity, 80–90 %).

Ultrasound also helps in ruling out other causes of epigastric pain, such as gallstones and aneurysmata. Complications of chronic pancreatitis, such as arterial pseudoaneurysms, left-sided portal hypertension (i.e., splenic venous thrombosis), and pleural effusions are readily detected with ultrasound. The pancreas is not always visualized if there is gas or in obese patients. Differential diagnosis between inflammatory processes and carcinomas are difficult [15].

Endoscopic Ultrasonography

Endoscopic ultrasonography (EUS) is more sensitive in showing changes of the hyper echoic foci, hyperechoic strands, lobularity, hyperechoic duct, irregular duct, visible side-branches, ductal dilation, calcification, and cysts. The diagnosis of chronic pancreatitis can be made at an earlier stage of the disease, if more than two criteria for pancreatitis are in place [16].

Findings of pancreatic function tests, which can be considered standard for detecting early changes of chronic pancreatitis, have been compared with those of endoscopic ultrasound. Overall, endoscopic ultrasound and pancreatic function tests agreed in approximately 75 % of cases [17].

The feasibility of performing both endoscopic ultrasound and endoscopic pancreatic function tests during the same endoscopic session as a simultaneous assessment of pancreatic structure and function was demonstrated [17].

Computed Tomography with Contrast

Computed tomography (CT) with contrast gives adequate information about pancreatic volume, calcifications, duct dilation when performed using thin slices through the pancreas, it is a reliable test for the diagnosis of advanced chronic pancreatitis.

CT has a sensitivity rate for advanced chronic pancreatitis of 74–90 % and a specificity of 84–100 %.

CT also allows the detection of complications, including pseudocysts, splenic artery pseudoaneurysm, and biliary duct involvement, pancreatic cancer or inflammatory masses and surrounding anatomical involvement [18].

Currently, CT is regarded as the imaging modality of choice for the initial evaluation of suggested chronic pancreatitis.

Magnetic Resonance Imaging, Magnetic Resonance Cholangiopancreatography

Magnetic resonance imaging (MRI) has no radiation risk. It can demonstrate calcifications, atrophy, ductal abnormalities, and fluid filled cysts in T2 weighted images and may offer improved differentiation of neoplastic and inflammatory masses.

Contrast-enhanced MRI weighted images may offer improved differentiation of neoplastic and inflammatory masses.

Magnetic resonance cholangiopancreatography (MRCP) allows a noninvasive alternative to Endoscopic retrograde cholangiopancreatography (ERCP) for imaging the pancreatic duct.

When no abnormalities can be shown in physiologic conditions, and there is a clinical presentation indicative for chronic pancreatitis, secretin-enhanced MRCP might improve the detection of diseased pancreatic ducts. It also provides additional functional information regarding pancreatic exocrine function. As experience grows, MRI imaging, particularly MRCP, may be increasingly used for assessment and screening for chronic pancreatitis.

Endoscopic Retrograde Cholangiopancreatography

Endoscopic retrograde cholangiopancreatography (ERCP) plays a role in gallstone pancreatitis and complicated acute and chronic pancreatitis. It is a highly sensitive radiographic test for chronic pancreatitis (sensitivity, 71–93 %; specificity, 89–100 %). As with most diagnostic tests, studies comparing ERCP with histology, the true gold standard, are lacking.

ERCP is not only a diagnostic tool but can also be used for therapeutic purposes. Pancreatic duct leaks or strictures can be stented as a bridge to surgery, common bile duct stones can be removed, pseudopancreatic cysts can be treated by stents, papillotomy for drainage or cystogastro of duodenostomy can be performed.

Pancreatic cancer diagnosis is possible but the accuracy of this technique is lower than with endoscopic ultrasonography.

ERCP carries a 5–10 % risk of inducing acute pancreatitis. Other less common risks include bleeding, infection, and perforation. In recent years, the role of ERCP in the diagnosis of pancreatic disease has decreased because safer and less invasive techniques have been developed [19].

PET Scan

Patients with chronic pancreatitis are at risk of developing pancreatic cancer. Fluoro deoxy glycose positron emission tomography (FDG-PET) has a potential role as a diagnostic tool for detecting pancreatic cancer in long-standing chronic pancreatitis.

However in the specific case of autoimmune-related pancreatitis, an intense uptake is observed, which disappears after steroid therapy. This fact should be kept in mind because it may lead to falsely making a diagnosis of a neoplastic process [20–22].

Pain Evaluation/Testing

Differential epidural anesthesia (DEA) is a test used for initial evaluation of the neural mechanism of the pain problem. In patients with visceral pain, DEA is used as a diagnostic modality to identify which patients should get celiac, splanchnic, or hypogastric blocks. It is not as precise as sometimes claimed [23, 24].

Pathophysiology of Pain Induced by Chronic Pancreatitis

The study of the pain mechanisms has been complicated by the difficulties in producing animal models that mimic chronic pancreatitis [25].

Pain in pancreatitis may be caused by different mechanisms. Over the years the theories shifted from mechanical to neurobiological pathogenesis. Pain can be divided into: [26].

1. Nociceptive pain
2. Neuropathic pain
3. Neurogenic inflammation

Nociceptive pain occurs after the activation of primary afferent neurons that respond to chemical or mechanical stimuli. The pain is proportional to the degree of stimulation. Chronic pancreatitis involves inflammatory infiltration of sensory nerves. In human and animal models with chronic pancreatitis, perineural infiltrates are found with a high percentage of eosinophils in which the degree of infiltrative disorder correlates with the severity of the pain [27]. In the presence of inflammation, ischemia, increased pressure and release of, for instance, bradykinins, prostaglandins, and substance P, nociceptors are activated, generating action potentials, and nociceptive pain thus develops [27].

One theory argues that increased pressure in the pancreatic duct leads to pain due to obstruction. Obstruction of the pancreatic duct can cause an "overpressure" proximally. This explanation of the pain is the basis for endoscopic and surgical drainage procedures.

Subsequently, several studies into overpressure in the pancreatic duct were carried out (preoperatively and during endoscopic retrograde cholangio pancreatography with manometry of the pancreatic duct), which showed inconsistent results. There are three studies in which the pressure in the pancreatic parenchyma was determined before surgery or partial pancreatic resection. Although higher pressures were found in the patients' parenchyma and the pressures were lower

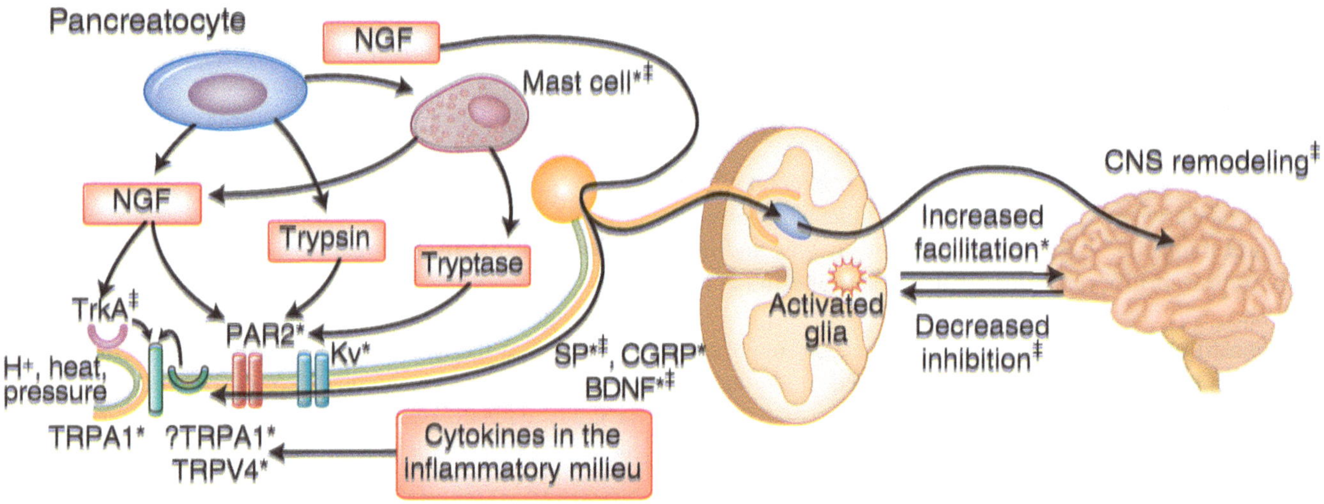

Fig. 9.1 Peripheral mechanisms of pain in chronic pancreatitis. A conceptual paradigm for the pathogenesis of pain in pancreatitis. Biological factors such as NGF that are produced in chronic pancreatitis can sensitize the nociceptor neuron by upregulating several key molecules, such as the receptor TRPV1 and neurotransmitters such as SP and CGRP, as well as by downregulating potassium channels. NGF is produced by pancreatic cells as well as mast cells. Mast cells also produce tryptase that along with trypsin can activate the PAR2 receptor, which is also expressed by nociceptors. In addition to TRPV1, several other receptors, such as TRPA1 and TRPV4 are capable of inducing noxious thermal, chemical, or mechanical stimuli. The inflammatory milieu in chronic pancreatitis also contains many different kinds of cytokines and other inflammatory mediators that act on the neurons and further sensitize and/or activate them. Superscripts denote whether these factors have been implicated in the pathogenesis of pain in animal (*asterisk*) or human studies (*double dagger*). *BDNF* brain-derived neurotrophic factor, *CGRP* calcitonin gene-related peptide, K_v voltage activated potassium channels, *NGF* nerve growth factor, *PAR2* protease activated receptor 2, SP_I substance P, *TrkA* trypomyosin-related kinase A receptor, *TRPV2* vanilloid receptor

after the procedure, there was no consistent correlation with the pain [28].

Other factors that may cause nociceptive pain in chronic pancreatitis include: obstruction of the duodenum or common bile duct (ductus choledochus), infiltration of the retroperitoneum, pseudocyst formation with compression of the surrounding organs, obstruction of the ductus pancreaticus due to fibrosis/stones/protein plugs, pancreatic ischemia due to atherosclerosis, gastric or duodenal ulcers, and meteorism due to malabsorption [29].

Neuropathic pain involves a change of the sensory nerves or the central nervous system itself. This change or damage is caused by (but is not dependent on for perpetuation) nociceptive activation. It has been shown that changes occur in the neurons innervating the pancreas that are located in the spinal ganglia (dorsal root ganglia) [29]. Patients with chronic pancreatitis appear to show generalized hyperalgesia, possibly based on deep sensitization [29]. Neurogenic inflammation is another proposed mechanism for pain. Cell death and tissue inflammation cause changes in the pH and the release of ions and inflammatory products such as cytokines and ATP. These inflammatory substances have direct as well as indirect effects on the nerve fibers and their ganglia once neuropathic pain develops. Neurogenic inflammation itself induces the production and increased release of neuropeptides, which then reinforces the inflammatory reaction in the tissues [30].

Since 2005, however, a rodent model for chronic pancreatitis with face and predictive validity was published. Studies with this model allowed to identify mechanisms of pancreatic pain. Its signals are transmitted via primary afferent nociceptors and induced by inflammation, morphological changes in peripheral nerves, and damage to the tissue. A cascade of events is initiated that includes central and peripheral sensitization and upregulation of various molecules. The group of Pasca di Magliano [25] formulated a paradigm with regard to the pain mechanisms of chronic pancreatitis (Fig. 9.1).

Treatment Options

The treatment of chronic pancreatitis should ideally be disease oriented, however, because pain is often the first sign that stimulates the patient to search medical help, and the disease has already reached a stage where it is irreversible, the management will be predominantly palliative. We focus here on the pain management of chronic pancreatitis.

Lifestyle Adjustments

Because chronic pancreatitis is in the majority of the cases due to alcohol abuse, the first treatment step is complete abstinence of alcohol, even in those forms of pancreatitis that are not linked with alcohol consumption. Smoking cessation is highly recommended. These changes in lifestyle reduce pain and improve life expectancy [31–33].

Pharmacological Pain Management

Pharmacological management of pain induced by chronic pancreatitis follows the guidelines of the three step WHO pain ladder for the management of cancer pain [34]. It must be stressed that, particularly the patients who suffer pain from alcoholic chronic pancreatitis, there is a propensity toward addiction. Moreover these patients have frequently liver and renal insufficiency, factors that must always be considered when establishing the treatment schedule.

Peripheral Analgesics

For the management of mild to moderate pain, paracetamol (acetaminophen) is the medication of first choice. It has good analgesic and antipyretic properties and few side effects, especially no gastrointestinal side effects in the recommended dosage.

Nonsteroidal Anti-Inflammatory Drugs

The inflammatory component of pain may justify the use of nonsteroidal anti-inflammatory drugs (NSAIDs) that exert an inhibitory activity on COX-1 and COX-2 to varying degrees. Attention should be paid to the potential side effects that vary from dyspepsia and skin disorders to gastric ulcerations and renal toxicity.

Overexpression of COX-2 in chronic pancreatitis has been shown [35]. The use of selective COX-2 inhibitors may be considered, however, the contribution of COX-1 inhibitors should not be underestimated in the treatment of pronociceptive factors such as prostaglandins as part of the treatment of chronic pancreatitis. Moreover, long-term use of selective COX-2 inhibitors presumably increases the risk of cardiac disease, which makes these drugs less suitable for the treatment of chronic pancreatitis. There are even case reports that suggest COX-2 inhibitors induce flares of acute pancreatitis.

Opioids

The second and third step in the WHO pain ladder consists in the use of opioids of varying strength. In-line with the WHO recommendations a treatment around the clock with long acting preparations is recommended. To obtain stable plasma levels and hence pain control, but also to limit the risk of addiction. Fast acting opioid preparations may be used for the management of breakthrough pain. Common side effects such as nausea and vomiting usually disappear when the treatment is continued. At the start of treatment those side effects can be managed with low doses of a centrally acting antiemetic or haloperidol. Constipation, a common side effect of opioid treatment is preferentially managed with laxatives started together with the opioid therapy.

Opioids act by binding to one of the opioid receptor (mu, kappa, and delta). These receptors are found on the neuronal cell membranes, but also in other organs, such as the mu receptors in the gut and the mu, kappa, and delta receptors in the sphincter of Oddi. This is a muscular valve in the duodenal wall that controls the release of bile and pancreatic juice, which is influenced by the hormone CCK. There is a tonic rest pressure as well as phasic antegrade contractions in this sphincter. Opioids result in an increase of the contraction frequency, amplitude, and rest pressure. As this effect can only partly be counteracted by naloxone (an opioid-antagonist), it is likely that the effect of morphine on the sphincter of Oddi is mediated by several opioid receptors. The degree to which various morphinomimetics influence the pressure in the sphincter of Oddi has been studied. The results vary, partly also because different manometric techniques were used. From the different studies, it can be concluded that all opioids cause an increase of the sphincter pressure. However, there are no studies that justify the conclusion that increased pressure of the sphincter of Oddi has an effect on the development or deterioration of acute or chronic pancreatitis [36–39].

Co-analgesics

The co-analgesics, act predominantly on neuropathic pain but are also used for the treatment of chronic pancreatitis. Tricyclic antidepressants, SSRI and SNRI act on the peripheral of a central nerve stimulation. Calcium channel blockers like pregabalin, gabapentin are strongly recommended in chronic pancreatitis. Olesen's group demonstrated in a randomized controlled trial that pregabalin reduces pain in chronic pancreatitis [40–43].

Additionally the tricyclic antidepressants act on the depressive symptoms that are frequently seen in chronic pain patients.

Ketamine

Ketamine has been used in many chronic pain states and also in cancer pain [44]. S-ketamine infusion in chronic pancreatitis pain patients reduces hyperalgesia immediately after infusion. It might have a role in those patients who have exhausted the full range of medical and surgical options [44, 45].

Non-analgesics

Pancreatic Enzyme Supplements

The rationale behind the use of pancreatic enzymes is found in the fact that they degrade the CCK releasing factor, thus lowering the CCK levels. Through this mechanism pain is reduced. It is important to note that only enteric coated formulations, that allow liberation of the enzymes in the duodenum, have a positive effect on pain. The best results of pancreatic enzyme supplementation are noted in small duct disease or minimal change chronic pancreatitis [28].

Octreotide

Octreotide is an inhibitor of the exocrine secretion of the pancreas. It has an anti-inflammatory action, reduces the pressure in the sphincter of Oddi and inhibits neural stimulation. Small studies have suggested a dose dependent effect with slightly better results in the highest dosing (200 μg) group. This difference was not clinically significant [28].

Antioxidants

The observation that patients with chronic pancreatitis show low plasma levels of antioxidants. It is hypothesized that free radicals play a role in pancreatic injury. Administration of antioxidants seemed to be promising in animal models. These findings could however not be confirmed in human studies [46, 47].

Non-pharmacological

The use of non-pharmacological treatment options is based on the anatomical origin of pain.

Gastroenterological

Endoscopy

Endoscopic therapy is considered as first choice in uncomplicated chronic pancreatitis.

If the pancreatitis is induced by intraductal *stones*, extracorporal shock wave lithotripsy is recommended, combined with endoscopic extraction of the stones. This was confirmed in a systematic review on 1,149 patients with success in 89 % [48]. Best results are obtained if stones are located in the head of the pancreas and in case of solitary stones. A morbidity is described of 6 % [48].

If the pancreatitis is caused by *strictures* of the main duct, stenting with endoscopic drainage is the therapy of choice according to the guidelines of the European Society of Gastrointestinal Endoscopy (ESGE) [49]. Strictures may be single or multiple. Depending on the size of dilatation of the main pancreatic duct behind the stricture, they are divided into dominant or non-dominant strictures (size >6 mm) and stenting can be performed by plastic or self-expandable metallic stents.

The technical success rate of stenting is high with an immediate pain relief in more than 65 % of the cases. During follow up of 14–58 months, this pain relief is ongoing in 32–68 % of the patients.

Pancreatic sfincterotomy should be performed prior to any stricture treatment [50–52].

Pancreas Pseudocysts are collections of pancreatic fluid surrounded by fibrous granulation tissue. These cysts can be drained by tubing transmural to the digestive wall or through the papil on EUS guidance. This is indicated if the cysts are enlarging or infected or in case of intracystic hemorrhage.

In case of clinical signs like pain, gastric outlet obstruction, weight loss, jaundice or early satiety, vascular compression of pancreaticopleural fistula drainage is indicated. Endoscopic drainage is the first choice therapy. It has a low cost with similar effects as surgery.

After 6–8 weeks, re-evaluation is performed and further treatment should be discussed multidisciplinary with endoscopists, surgeons, internists, radiologists, and pain specialists [53].

Biliary Strictures

Up to 20 % of the chronic pancreatitis cases are related to biliary strictures. Treatment is necessary in case of elevated alkaline phosphatase and serum bilirubin for more than a month, biliary cirrhosis, stones or progression of biliary strictures.

Biliary obstruction complicates the course of chronic pancreatitis in 3–23 % of patients [54].

Different cholangiographic types of chronic pancreatitis-related biliary strictures have been described, the type being suggestive of the etiology of biliary obstruction (fibrosis, compression by a pseudocyst or cancer); therefore cancer should always be excluded by a brush cytology or PET scan [48].

Surgery

Surgery is often performed when medical and endoscopic treatment did not provide satisfactory pain relief, or to exclude neoplasm and complications in the surrounding organs.

Pain can be treated by means of various techniques involving drainage (Puestow procedure) or resection (pancreaticoduodenectomy, total pancreatectomy with autotransplantation of the islets of Langerhans) or a combination of both (Frey procedure). Drainage procedures intend to reduce the pain by decompression of the pancreatic duct. The theory behind pain relief due to resection is that inflammatory activity causes pain as a result of qualitative and quantitative changes of the nerve fibers.

Endoscopy Versus Surgery

The current guidelines for the management of pain due to chronic pancreatitis, recommend a step up approach, starting with lifestyle changes, pharmacological treatment including opioid analgesics and when these treatments provide insufficient effect, the endoscopic interventions are performed. Surgery is considered the last resort [26].

The choice between surgical or endoscopic treatment is difficult. Comparative studies do not provide exclusion on this point. In a 5-year follow up study, pain relapsed in 15 % in endoscopic treatment and 34 % in surgical treatment [49].

Recent studies have shown that preoperative opioid use predisposes to failure of long-term pain relief after surgical or endoscopic interventions. Peripheral and central sensitization is a possible explanation for this phenomenon [55, 56].

Although the European Society for Gastro-Enterology (ESGE) recommends endoscopic therapy as the first-line therapy for painful uncomplicated chronic pancreatitis, there is growing evidence that early surgical interventions have a better outcome concerning hospital stay, subsequent interventions, and relapse free interval with similar complication rates. Early surgical interventions give better pain relief but also better conservation of exocrine and endocrine function. Two recent randomized studies show that these surgical procedures lead to better results compared with endoscopic treatment [57, 58]. The percentage of patients who are free of pain after 5 years is 40 %.

The reluctance of gastroenterologist for surgery might be due to the high morbidity and mortality associated with pancreatic surgery in the setting of chronic pancreatitis. In contrast, morbidity and mortality rates for endoscopic therapy for chronic pancreatitis are rather low [48, 53, 58, 59].

Ablative and Neuromodulation Techniques

Nerve blocks, ablative procedures, and neuromodulating techniques aim at interrupting or alternating the pain conduction. The neuro-anatomy is therefore important.

Relevant Neuro-Anatomy

The sympathetic innervation of the abdominal organs starts from the anterolateral horn in the spinal cord. Preganglionar fibers of Th5–Th12 leave the spinal column after merging with the ventral ramus. Together with these communicating rami they course in the direction of the sympathetic chain. The fibers do not form synapses in the sympathetic chain, but run through it. The formation of synapses occurs more peripheral to the level of the ganglia: celiac ganglion, aorticorenal ganglion, superior mesenteric ganglion.

Preganglionar nerves confluence into three splanchnic nerves (greater, lesser, lowest) that course along the paravertebral border (Table 9.2).

Just below the level of the crus of the diaphragm, the splanchnic nerves confluence with the vagal preganglionar parasympathetic fibers, sensory fibers of the phrenic nerve, and postganglionar sympathetic fibers to the celiac plexus that are draped around the abdominal aorta, especially at the anterior side. Figure 9.2 provides an image of the innervation of abdominal organs.

Table 9.2 Splanchnic nerves and preganglionar fiber level

Splanchnic nerve division	Preganglionar fiber level
Greater splanchnic nerve	Th5–Th9
Lesser splanchnic nerve	Th10–Th11
Lowest splanchnic nerve	Th11–Th12

The splanchnic nerves are localized in a narrow pyramid of which the medial edge is formed by the lateral border of the vertebra, the lateral edge by the medial pleura and the crus of the diaphragm forms the basis of the triangle. The anterior side is formed by the posterior wall of the mediastinum and the posterior wall by the attachment of the parietal pleura on the lateral wall of the vertebrae [26].

Nerve Blocks

Celiac Plexus Block

Nerve blocks are widely used to reduce pain due to pancreatic cancer and pancreatitis. In patients with cancer, lysis (alcoholization or phenolization) of the celiac plexus has a positive recommendation [60]. Injection of a neurolytic agent around the nerve has several risks, such as uncontrolled flow of the neurolytic and uncontrolled lesion size. Using a neurolytic agent around the celiac plexus may cause paraplegia and retroperitoneal fibrosis. Therefore the use of neurolytic agents is restricted to the management of patients with cancer. Injection of local anesthetics, steroids, and antibiotics was recommended instead, based on the idea of decreasing neuronal inflammation [61]. Even for a chronic disease state as pancreatitis, the reported experiences do not present long-term effects [62]. The effect is not prolonged in time if repeated blocks are performed.

Techniques

Most of the Celiac Plexus Block (CPB) is performed percutaneous under fluoroscopic guidance. Transaortic and retrocrural techniques are described with both paravertebral and transdiscal approaches.

CT Guided

CT allows excellent visualization of the anatomic structures that lie in close proximity to the target site during neurolytic CPB. Performing the procedure this way might avoid complications such as hematuria, intravascular injection, and pneumothorax [63].

Ultrasound Guided

The introduction of endoscopic ultrasound led to the ultrasound guided celiac plexus block (with local anesthetic and corticosteroid) or celiac plexus neurolysis [64]. A prospective study on the efficacy and safety of endoscopic ultrasound guided celiac plexus block showed good pain relief in 55 % of the patients at 4 and 8 weeks follow-up.

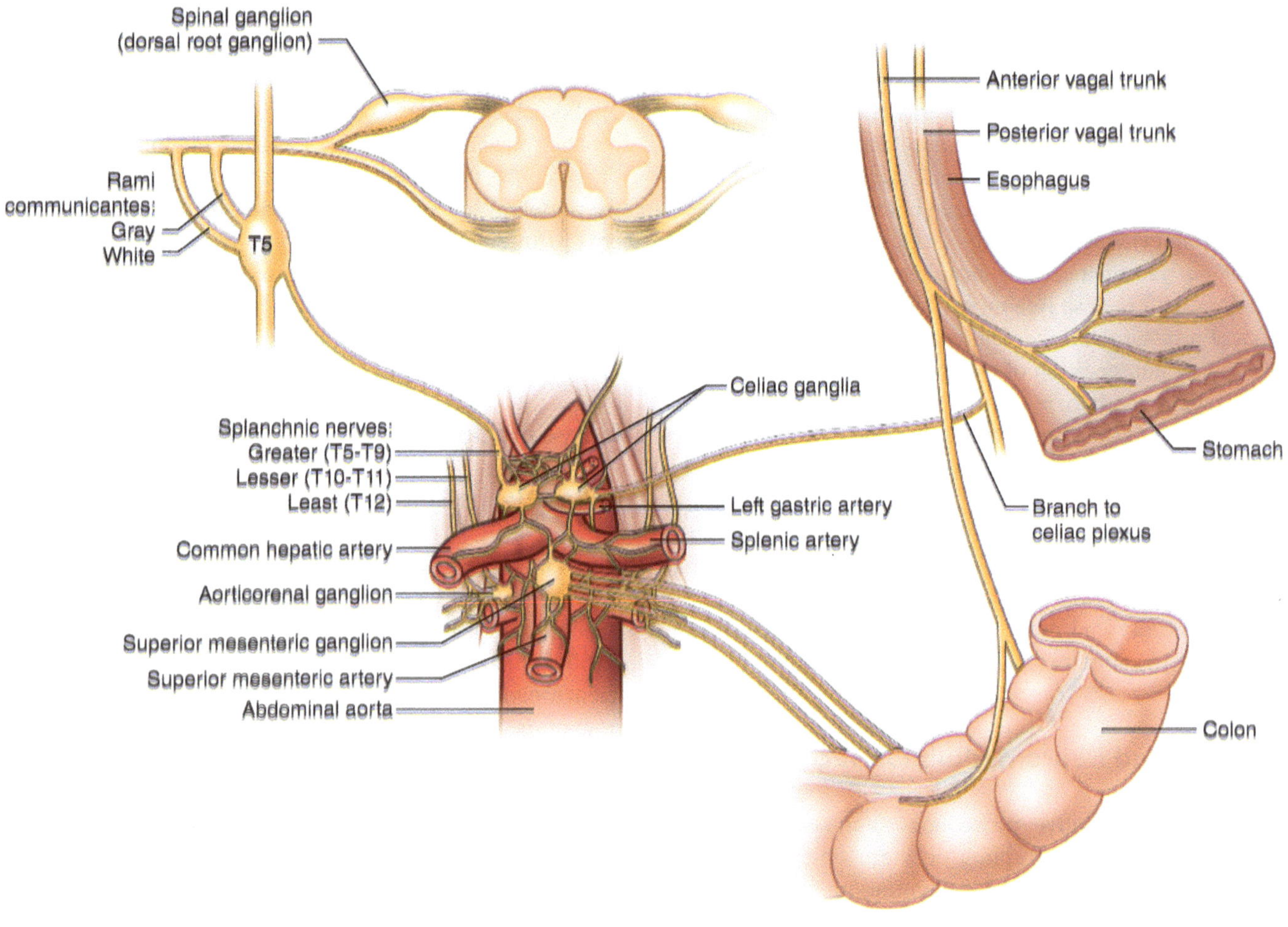

Fig. 9.2 Innervation of the abdominal organs. Modified from Rogier Trompert, Medical Art

At 12 and 24 weeks follow-up, 26 and 10 % of patients respectively had ongoing pain relief [65].

Complications of endoscopic ultrasound guided celiac plexus block and neurolysis were studied in a large series of prospectively collected information. The overall complication rate was 1.8 %; only one major complication occurred in the neurolysis group [66]. Recently case reports on spinal cord infarction after the endoscopic procedure with major complications, such as paraplegia, were published [67, 68].

Splanchnic Nerve Block

The specific anatomy in which the splanchnic nerves are located in a narrow compartment allows a targeted denervation. Radiofrequency (RF) thermolesioning, in which denervation only takes place at the tip of the electrode, seems more suitable than injection of neurolytics for this indication.

The use of RF nervus splanchnicus treatment has been described in two patient series [69, 70]. Raj [69] reports on 107 patients who underwent RF treatment of the splanchnic nerves as a treatment of upper abdominal pain. The involvement of the splanchnic nerves was confirmed by means of a diagnostic block with a local anesthetic. Seventy-three patients were followed prospectively. Thirty-eight patients only received a block with a local anesthetic and 31 received RF treatment. In both groups a pain relief of >50 % was found in 40 % of the patients.

Garcea [70] describes ten patients who underwent RF splanchnic nerve denervation as a treatment of chronic pancreatitis with a mean follow-up of 18 months (12–24 months). A significant pain reduction was observed, accompanied by a clear decrease in the need for opiates and acute hospitalization. Moreover, the parameters of the quality of life improved as well.

A recently published retrospective review of patients with chronic pancreatitis treated with RF thermolesioning of the splanchnic nerve showed a significant pain reduction for a mean period of 45 weeks in 11/18 patients. The analgesic use was significantly reduced in 4 patients and 4 other patients stopped completely analgesic intake [71]. In the patients that had a good pain relief, the procedure could be repeated in case of relapse and the effect was comparable. Presumably the pain returns due to nerve regeneration.

Complications of RF thermolesioning of the splanchnic nerve cannot be derived from the small series published up till now. As with the neurolytic blocks, postprocedural neuritis is possible. Hypotension and diarrhea may occur shortly after the intervention, but can be treated easily.

The use of pulsed radiofrequency of the splanchnic nerve was described in two cases by Brennan et al. [72]. The effect of this treatment needs further study in large patient groups.

Thoracoscopic Splanchnicectomy

The nociceptive fibers from the pancreas are incorporated in the splanchnic nerves. By transecting these nociceptive fibers, a long-term pain relief was expected.

Recent advances in laparoscopic techniques formed new developments in the field of thoracoscopy. The first report on successful thoracoscopic splanchnicectomy for pancreatic pain was published in 1993. This procedure is performed under general anesthesia with double lumen intubation and one lung ventilation in prone position. With this thoracoscopic operation technique, dissection of the parietal pleura was performed from Th5 till Th10 to identify and transect the nociceptive splanchnic nerve fibers at the thoracic level. This technique was promising because of sufficient pain relief with maintenance of the pancreatic function. In some cases the transection was uncertain and this resulted in a fast recurrence of pain.

Early pain relief was significant in several studies but after 6 months, there was recurrence in 25 % of the patients. Pain recurred in 50 % of the patients followed for a long term.

Bilateral thoracoscopic splanchnicectomy appears to work best in patients who have had no prior operative or endoscopic interventions. Prior opioid abuse results in reduced efficacy.

Other nerve structures may take over nociception. The vagal nerve was presumed to be a partner for this transmission, but, additional vagotomy does not give any additional effect. Due to sensitization, inactive nociceptors can become activated due to certain stimuli and play an active role in chronic visceral pain [73–77].

Spinal Cord Stimulation

The use of Spinal Cord Stimulation (SCS) to treat visceral pain was initially described in several case reports [78–83]. A recent publication of a retrospective review of 35 patients who received a trial with SCS reported that 30 % experienced ≥50 % pain relief at the end of the trial [84]. In the 28 patients who received a permanent implant, 1 was lost to follow-up and 5 had the lead and generator removed for various reasons. Nineteen of the 22 patients were followed for more than 1 year. Over the complete evaluation period pain scores and opioid use remained low, suggesting that SCS for chronic abdominal pain of various causes may provide consistent long-term improvements. A national survey on SCS for chronic abdominal pain that followed this retrospective study included 76 case reports and its results were consistent in technical aspects of SCS implantation, as well as the opioid use and pain score improvements [84, 85]. Both studies described SCS leads positioned with their tips mostly at the level of Th5 vertebral body. Pain relief exceeded 50 % in most of the patients and long-term opioid use decreased by more than 2/3 [84, 85]. Another interesting fact from both studies is the presence of the large treated population of the patients with severe chronic pancreatitis [84, 85]. There were 26 out of 35 patients in a retrospective, and 26 out of 70 patients in survey study who had diagnosis of chronic pancreatitis. Analyzed effects of SCS in this subgroup of the patients helped to conclude that the improvements in opioid use and pain scores were similar to those of patients with other sources of their chronic visceral abdominal pain.

In a retrospective analysis 30 patients with severe pain due to chronic pancreatitis underwent a trial with spinal cord stimulation during 7–14 days. Twenty-four patients reported 80 % pain relief at the end of the trial period. After definite implantation 1 patient was lost to follow-up and in 3 patients the system had to be removed because of infection. At 1-year follow-up VAS score pain and opioid consumption were significantly reduced [86].

The main complications of SCS are migration and breakage of the electrode. In addition infection which includes anything from cellulites to epidural abscess is possible.

Splanchnic Nerve Stimulation

A case report of a young patient with painful chronic pancreatitis of more than 5 years duration and refractory to all conservative treatments illustrates a considerable pain reduction, and diminished opioid consumption for at least 18 months was obtained with neuromodulation of the splanchnic nerves with two permanently implanted octopolar leads at the Th11/Th12 area connected to an implantable pulse generator [87]. This experimental treatment may by further investigated.

Conclusion

Pain due to chronic pancreatitis is severe and reduces the quality of life tremendously. At this moment, treatment is partially dependent on the medical specialists that look to the problem.

Better understanding of the potential mechanisms of pain can elucidate the effect or ineffectiveness of current therapy and result in the development of better treatment options. Resection or decompression of ductal stenosis will have no impact on the hypersensitivity of the gland. Ineffectiveness of neuro-ablative procedures can be an anatomical issue but also due to regeneration of nerve fibers with secondary extra input to the cord. Another important issue is the central sensitization which is a strong indication for spinal cord stimulation.

Comparative studies between surgery and endoscopic treatment show evidence in favor of early surgery.

When establishing a pharmacological pain treatment, non-analgesic drugs should also be considered. Antiepileptic drugs type Ca channel blocking agents seem to be effective. Other potential targets in the treatment are NGF inhibitors, TRPV1 antagonist. Concerning the central sensitization, the NMDA receptor blocking agents might play an important role but at this moment, they are impracticable because of the narrow therapeutic window and administration difficulties.

Studies on nerve blocks mainly focus on the CPB, which should be reserved for the treatment of cancer patients. Radiofrequency treatment of the splanchnic nerves seems promising but further RCT is needed to confirm the effect. These blocks could reduce pain to an extent that surgery can be delayed or even prevented.

References

1. Sarles H, Sarles JC, Camatte R, Muratore R, Gaini M, Guien C, Pastor J, Le Roy F. Observations on 205 confirmed cases of acute pancreatitis, recurring pancreatitis, and chronic pancreatitis. Gut. 1965;6:545–59.
2. Whitcomb DC. Mechanisms of disease: advances in understanding the mechanisms leading to chronic pancreatitis. Nat Clin Pract Gastroenterol Hepatol. 2004;1:46–52.
3. Apte MV, Pirola RC, Wilson JS. Mechanisms of alcoholic pancreatitis. J Gastroenterol Hepatol. 2010;25:1816–26.
4. Forsmark CE. Chronic pancreatitis. In M Feldman et al. (eds.) Sleisenger and Fordtran's Gastrointestinal and Liver Disease, 9th ed., 2010;1:985–1015. Philadelphia: Saunders.
5. Yadav D, Lowenfels AB. The epidemiology of pancreatitis and pancreatic cancer. Gastroenterology. 2013;144:1252–61.
6. Bourliere M, Barthet M, Berthezene P, Durbec JP, Sarles H. Is tobacco a risk factor for chronic pancreatitis and alcoholic cirrhosis? Gut. 1991;32:1392–5.
7. Witt H, Apte MV, Keim V, Wilson JS. Chronic pancreatitis: challenges and advances in pathogenesis, genetics, diagnosis, and therapy. Gastroenterology. 2007;132:1557–73.
8. Apte MV, Pirola RC, Wilson JS. Molecular mechanisms of alcoholic pancreatitis. Dig Dis. 2005;23:232–40.
9. Apte MV, Wilson JS. Mechanisms of pancreatic fibrosis. Dig Dis. 2004;22:273–9.
10. Whitcomb DC, Gorry MC, Preston RA, Furey W, Sossenheimer MJ, Ulrich CD, Martin SP, Gates Jr LK, Amann ST, Toskes PP, Liddle R, McGrath K, Uomo G, Post JC, Ehrlich GD. Hereditary pancreatitis is caused by a mutation in the cationic trypsinogen gene. Nat Genet. 1996;14:141–5.
11. DiMagno EP, Go VL, Summerskill WH. Relations between pancreatic enzyme ouputs and malabsorption in severe pancreatic insufficiency. N Engl J Med. 1973;288:813–5.
12. Chowdhury RS, Forsmark CE. Review article: pancreatic function testing. Aliment Pharmacol Ther. 2003;17:733–50.
13. Hayakawa T, Kondo T, Shibata T, Noda A, Suzuki T, Nakano S. Relationship between pancreatic exocrine function and histological changes in chronic pancreatitis. Am J Gastroenterol. 1992;87:1170–4.
14. Steinberg WM, Anderson KK. Serum trypsinogen in diagnosis of chronic pancreatitis. Dig Dis Sci. 1984;29:988–93.
15. Khan AN, Sheen AJ. Chronic pancreatitis imaging. In: Karani J, Krinsky G, Coombs BD, Schmiedl UP, Krasny RM, editors. Medscape. Vol.; 2011.
16. Sahai AV, Zimmerman M, Aabakken L, Tarnasky PR, Cunningham JT, van Velse A, Hawes RH, Hoffman BJ. Prospective assessment of the ability of endoscopic ultrasound to diagnose, exclude, or establish the severity of chronic pancreatitis found by endoscopic retrograde cholangiopancreatography. Gastrointest Endosc. 1998;48:18–25.
17. Stevens T, Dumot JA, Zuccaro Jr G, Vargo JJ, Parsi MA, Lopez R, Kirchner HL, Purich E, Conwell DL. Evaluation of duct-cell and acinar-cell function and endosonographic abnormalities in patients with suspected chronic pancreatitis. Clin Gastroenterol Hepatol. 2009;7:114–9.
18. Kim HC, Yang DM, Kim HJ, Lee DH, Ko YT, Lim JW. Computed tomography appearances of various complications associated with pancreatic pseudocysts. Acta Radiol. 2008;49:727–34.
19. Ulrich II C, Martin S. ERCP and chronic pancreatitis. http://pancreasfoundation.org/2010/04/ercp-and-pancreatic-disease/ May 2013. Accessed May 2013.
20. Rasmussen I, Sorensen J, Langstrom B, Haglund U. Is positron emission tomography using 18F-fluorodeoxyglucose and 11C-acetate valuable in diagnosing indeterminate pancreatic masses? Scand J Surg. 2004;93:191–7.
21. Nakamoto Y, Saga T, Ishimori T, Higashi T, Mamede M, Okazaki K, Imamura M, Sakahara H, Konishi J. FDG-PET of autoimmune-related pancreatitis: preliminary results. Eur J Nucl Med. 2000;27:1835–8.
22. van Kouwen MC, Jansen JB, van Goor H, de Castro S, Oyen WJ, Drenth JP. FDG-PET is able to detect pancreatic carcinoma in chronic pancreatitis. Eur J Nucl Med Mol Imaging. 2005;32:399–404.
23. Rizk MK, Tolba R, Kapural L, Mitchell J, Lopez R, Mahboobi R, Vrooman B, Mekhail N. Differential epidural block predicts the success of visceral block in patients with chronic visceral abdominal pain. Pain Pract. 2012;12:595–601.
24. Bielefeldt K, Gebhart G. Visceral pain. In: Benzon HT, Rathmell J, Wu C, Turk D, Argoff C, editors. Raj's practical management of pain. 4th ed. Philadelphia: Mosby/Elsevier; 2008.
25. Pasca di Magliano M, Forsmark C, Freedman S, Hebrok M, Pasricha PJ, Saluja A, Stanger BZ, Holt J, Serrano J, James SP, Rustgi AK. Advances in acute and chronic pancreatitis: from development to inflammation and repair. Gastroenterology. 2013;144:e1–4.
26. Puylaert M, Kapural L, Van Zundert J, Peek D, Lataster A, Mekhail N, van Kleef M, Keulemans YC. Pain in chronic pancreatitis. Pain Pract. 2011;11(5):492–505.
27. Anaparthy R, Pasricha PJ. Pain and chronic pancreatitis: is it the plumbing or the wiring? Curr Gastroenterol Rep. 2008;10:101–6.
28. Lieb 2nd JG, Forsmark CE. Review article: pain and chronic pancreatitis. Aliment Pharmacol Ther. 2009;29:706–19.
29. DiMagno MJ, DiMagno EP. Chronic pancreatitis. Curr Opin Gastroenterol. 2012;28:523–31.
30. Pasricha PJ. Unraveling the mystery of pain in chronic pancreatitis. Nat Rev Gastroenterol Hepatol. 2012;9:140–51.
31. Lankisch PG, Lohr-Happe A, Otto J, Creutzfeldt W. Natural course in chronic pancreatitis. Pain, exocrine and endocrine pancreatic insufficiency and prognosis of the disease. Digestion. 1993;54:148–55.

32. Miyake H, Harada H, Kunichika K, Ochi K, Kimura I. Clinical course and prognosis of chronic pancreatitis. Pancreas. 1987;2:378–85.
33. Layer P, Yamamoto H, Kalthoff L, Clain JE, Bakken LJ, DiMagno EP. The different courses of early- and late-onset idiopathic and alcoholic chronic pancreatitis. Gastroenterology. 1994;107:1481–7.
34. WHO. Cancer pain relief: with a guide to opioid availability. 2nd ed. Geneva: WHO; 1996.
35. Schlosser W, Schlosser S, Ramadani M, Gansauge F, Gansauge S, Beger HG. Cyclooxygenase-2 is overexpressed in chronic pancreatitis. Pancreas. 2002;25:26–30.
36. Radnay PA, Brodman E, Mankikar D, Duncalf D. The effect of equi-analgesic doses of fentanyl, morphine, meperidine and pentazocine on common bile duct pressure. Anaesthesist. 1980;29: 26–9.
37. Helm JF, Venu RP, Geenen JE, Hogan WJ, Dodds WJ, Toouli J, Arndorfer RC. Effects of morphine on the human sphincter of oddi. Gut. 1988;29:1402–7.
38. van Voorthuizen T, Helmers JH, Tjoeng MM. Otten MH [meperidine (pethidine) outdated as analgesic in acute pancreatitis]. Ned Tijdschr Geneeskd. 2000;144:656–8.
39. Gachago C, Draganov PV. Pain management in chronic pancreatitis. World J Gastroenterol. 2008;14:3137–48.
40. Olesen SS, Bouwense SA, Wilder-Smith OH, van Goor H, Drewes AM. Pregabalin reduces pain in patients with chronic pancreatitis in a randomized, controlled trial. Gastroenterology. 2011;141:536–43.
41. Olesen SS, Graversen C, Olesen AE, Frokjaer JB, Wilder-Smith O, van Goor H, Valeriani M, Drewes AM. Randomised clinical trial: pregabalin attenuates experimental visceral pain through subcortical mechanisms in patients with painful chronic pancreatitis. Aliment Pharmacol Ther. 2011;34:878–87.
42. Bouwense SA, Olesen SS, Drewes AM, Poley JW, van Goor H, Wilder-Smith OH. Effects of pregabalin on central sensitization in patients with chronic pancreatitis in a randomized, controlled trial. PLoS One. 2012;7:e42096.
43. Olesen SS, Graversen C, Bouwense SA, van Goor H, Wilder-Smith OH, Drewes AM. Quantitative sensory testing predicts pregabalin efficacy in painful chronic pancreatitis. PLoS One. 2013;8:e57963.
44. Hocking G, Cousins MJ. Ketamine in chronic pain management: an evidence-based review. Anesth Analg. 2003;97:1730–9.
45. Bouwense SA, Buscher HC, van Goor H, Wilder-Smith OH. S-ketamine modulates hyperalgesia in patients with chronic pancreatitis pain. Reg Anesth Pain Med. 2011;36:303–7.
46. Kirk GR, White JS, McKie L, Stevenson M, Young I, Clements WD, Rowlands BJ. Combined antioxidant therapy reduces pain and improves quality of life in chronic pancreatitis. J Gastrointest Surg. 2006;10:499–503.
47. Uden S, Bilton D, Nathan L, Hunt LP, Main C, Braganza JM. Antioxidant therapy for recurrent pancreatitis: placebo-controlled trial. Aliment Pharmacol Ther. 1990;4:357–71.
48. Nguyen-Tang T, Dumonceau JM. Endoscopic treatment in chronic pancreatitis, timing, duration and type of intervention. Best Pract Res Clin Gastroenterol. 2010;24:281–98.
49. Dumonceau JM, Delhaye M, Tringali A, Dominguez-Munoz JE, Poley JW, Arvanitaki M, Costamagna G, Costea F, Deviere J, Eisendrath P, Lakhtakia S, Reddy N, Fockens P, Ponchon T, Bruno M. Endoscopic treatment of chronic pancreatitis: European society of gastrointestinal endoscopy (ESGE) clinical guideline. Endoscopy. 2012;44:784–800.
50. Cremer M, Deviere J, Delhaye M, Baize M, Vandermeeren A. Stenting in severe chronic pancreatitis: results of medium-term follow-up in seventy-six patients. Endoscopy. 1991;23:171–6.
51. Ponchon T, Bory RM, Hedelius F, Roubein LD, Paliard P, Napoleon B, Chavaillon A. Endoscopic stenting for pain relief in chronic pancreatitis: results of a standardized protocol. Gastrointest Endosc. 1995;42:452–6.
52. Smits ME, Badiga SM, Rauws EA, Tytgat GN, Huibregtse K. Long-term results of pancreatic stents in chronic pancreatitis. Gastrointest Endosc. 1995;42:461–7.
53. Ahmed Ali U, Pahlplatz JM, Nealon WH, van Goor H, Gooszen HG, Boermeester MA. Endoscopic or surgical intervention for painful obstructive chronic pancreatitis. Cochrane Database Syst Rev. 2012;1, CD007884.
54. Abdallah AA, Krige JE, Bornman PC. Biliary tract obstruction in chronic pancreatitis. HPB (oxford). 2007;9:421–8.
55. Ahmed Ali U, Nieuwenhuijs VB, van Eijck CH, Gooszen HG, van Dam RM, Busch OR, Dijkgraaf MG, Mauritz FA, Jens S, Mast J, van Goor H, Boermeester MA, Dutch Pancreatitis Study G. Clinical outcome in relation to timing of surgery in chronic pancreatitis: a nomogram to predict pain relief. Arch Surg. 2012;147:925–32.
56. Negi S, Singh A, Chaudhary A. Pain relief after Frey's procedure for chronic pancreatitis. Br J Surg. 2010;97:1087–95.
57. Dite P, Ruzicka M, Zboril V, Novotny I. A prospective, randomized trial comparing endoscopic and surgical therapy for chronic pancreatitis. Endoscopy. 2003;35:553–8.
58. Cahen DL, Gouma DJ, Nio Y, Rauws EA, Boermeester MA, Busch OR, Stoker J, Lameris JS, Dijkgraaf MG, Huibregtse K, Bruno MJ. Endoscopic versus surgical drainage of the pancreatic duct in chronic pancreatitis. N Engl J Med. 2007;356:676–84.
59. Neal CP, Dennison AR, Garcea G. Surgical therapy in chronic pancreatitis. Minerva Gastroenterol Dietol. 2012;58:377–400.
60. Vissers KC, Besse K, Wagemans M, Zuurmond W, Giezeman MJ, Lataster A, Mekhail N, Burton AW, van Kleef M, Huygen F. 23. Pain in patients with cancer. Pain Pract. 2011;11(5):453–75.
61. Pap A, Nauss L, DiMagno E. Is percutaneous celiac plexus block (PCPB) associated with pain relief in chronic pancreatitis? a comparison among analgesic, alcohol and steroid PCPB (abstr). Pancreas. 1990;725.
62. Kaufman M, Singh G, Das S, Concha-Parra R, Erber J, Micames C, Gress F. Efficacy of endoscopic ultrasound-guided celiac plexus block and celiac plexus neurolysis for managing abdominal pain associated with chronic pancreatitis and pancreatic cancer. J Clin Gastroenterol. 2010;44:127–34.
63. Rathmell JP, Gallant JM, Brown DL. Computed tomography and the anatomy of celiac plexus block. Reg Anesth Pain Med. 2000; 25:411–6.
64. Wiersema MJ, Wiersema LM. Endosonography-guided celiac plexus neurolysis. Gastrointest Endosc. 1996;44:656–62.
65. Gress F, Schmitt C, Sherman S, Ciaccia D, Ikenberry S, Lehman G. Endoscopic ultrasound-guided celiac plexus block for managing abdominal pain associated with chronic pancreatitis: a prospective single center experience. Am J Gastroenterol. 2001;96:409–16.
66. O'Toole TM, Schmulewitz N. Complication rates of EUS-guided celiac plexus blockade and neurolysis: results of a large case series. Endoscopy. 2009;41:593–7.
67. Mittal MK, Rabinstein AA, Wijdicks EF. Pearls & oy-sters: acute spinal cord infarction following endoscopic ultrasound-guided celiac plexus neurolysis. Neurology. 2012;78:e57–9.
68. Fujii L, Clain JE, Morris JM, Levy MJ. Anterior spinal cord infarction with permanent paralysis following endoscopic ultrasound celiac plexus neurolysis. Endoscopy. 2012;44(2 UCTN):E265–6.
69. Prithvi Raj P, Sahinder B, Lowe M. Radiofrequency lesioning of splanchnic nerves. Pain Pract. 2002;2:241–7.
70. Garcea G, Thomasset S, Berry DP, Tordoff S. Percutaneous splanchnic nerve radiofrequency ablation for chronic abdominal pain. ANZ J Surg. 2005;75:640–4.
71. Verhaegh BP, van Kleef M, Geurts JW, Puylaert M, van Zundert J, Kessels AG, Masclee AA, Keulemans YC. Percutaneous radiofrequency ablation of the splanchnic nerves in patients with chronic pancreatitis: results of single and repeated procedures in 11 patients. Pain Pract. 2013;13(8):621–6.

72. Brennan L, Fitzgerald J, McCrory C. The use of pulsed radiofrequency treatment for chronic benign pancreatitis pain. Pain Pract. 2009;9:135–40.
73. Buscher HC, Jansen JB, van Dongen R, Bleichrodt RP, van Goor H. Long-term results of bilateral thoracoscopic splanchnicectomy in patients with chronic pancreatitis. Br J Surg. 2002;89:158–62.
74. Howard TJ, Swofford JB, Wagner DL, Sherman S, Lehman GA. Quality of life after bilateral thoracoscopic splanchnicectomy: long-term evaluation in patients with chronic pancreatitis. J Gastrointest Surg. 2002;6:845–52; discussion 853–844.
75. Stone HH, Chauvin EJ. Pancreatic denervation for pain relief in chronic alcohol associated pancreatitis. Br J Surg. 1990;77:303–5.
76. Malec-Milewska MB, Tarnowski W, Ciesielski AE, Michalik E, Guc MR, Jastrzebski JA. Prospective evaluation of pain control and quality of life in patients with chronic pancreatitis following bilateral thoracoscopic splanchnicectomy. Surg Endosc. 2013; 27(10):3639–45.
77. Bouwense SA, Buscher HC, van Goor H, Wilder-Smith OH. Has central sensitization become independent of nociceptive input in chronic pancreatitis patients who fail thoracoscopic splanchnicectomy? Reg Anesth Pain Med. 2011;36:531–6.
78. Tiede JM, Ghazi SM, Lamer TJ, Obray JB. The use of spinal cord stimulation in refractory abdominal visceral pain: case reports and literature review. Pain Pract. 2006;6:197–202.
79. Kapural L, Rakic M. Spinal cord stimulation for chronic visceral pain secondary to chronic non-alcoholic pancreatitis. J Clin Gastroenterol. 2008;42:750–1.
80. Ceballos A, Cabezudo L, Bovaira M, Fenollosa P, Moro B. Spinal cord stimulation: a possible therapeutic alternative for chronic mesenteric ischaemia. Pain. 2000;87:99–101.
81. Krames ES, Mousad D. Spinal cord stimulation reverses pain and diarrheal episodes of irritable bowel syndrome: a case report. Neuromodulation. 2004;7:82–8.
82. Khan I, Raza S, Khan E. Application of spinal cord stimulation for the treatment of abdominal visceral pain syndromes: case reports. Neuromodulation. 2005;8:14–27.
83. Kapural L. Neuromodulation for chronic visceral pelvic pain. In: Neuromodulation society, ed. *North American Neuromodulation society: 9th annual meeting*, Vol. Whashington DC; 2005.
84. Kapural L, Nagem H, Tlucek H, Sessler DI. Spinal cord stimulation for chronic visceral abdominal pain. Pain Med. 2010;11: 347–55.
85. Kapural L, Deer T, Yakovlev A, Bensitel T, Hayek S, Pyles S, Khan Y, Kapural A, Cooper D, Stearns L, Zovkic P. Technical aspects of spinal cord stimulation for managing chronic visceral abdominal pain: the results from the national survey. Pain Med. 2010;11: 685–91.
86. Kapural L, Cywinski JB, Sparks DA. Spinal cord stimulation for visceral pain from chronic pancreatitis. Neuromodulation. 2011;14:423–6; discussion 426–7.
87. Goroszeniuk T, Khan R. Permanent percutaneous splanchnic nerve neuromodulation for management of pain due to chronic pancreatitis: a case report. Neuromodulation. 2011;14:253–7; discussion 257.

Chronic Abdominal Pain of Gynecologic Causes: Diagnosis and Treatment

10

Miya P. Yamamoto, Jorge F. Carillo, and Fred M. Howard

Introduction

Chronic pelvic pain (CPP) is a common disorder in women. One study from the United Kingdom found a prevalence of 3.8 % in women aged 15–73; higher than the prevalence of migraine (2.1 %) and similar to that of asthma (3.7 %) and back pain (4.1 %) [1]. Similarly a study in Seveso, Italy, found that 4 % of all women had moderate to severe, noncyclic pelvic pain [2]. A US study suggested a higher prevalence of 16 %, with a mean average pain score of 5, and 4 % with pain severe enough to cause them to miss work [3]. This study estimated that 9.2 million US women suffer from chronic pelvic pain.

Chronic pelvic pain leads to many medical interventions. It is the indication for 17 % of all hysterectomies in the United States [4] and more than 40 % of gynecologic diagnostic laparoscopies [5]. Overall, it is estimated that direct and indirect costs of chronic pelvic pain in the United States are over $2 billion per year [3]. At an individual level, chronic pelvic pain frequently leads to years of disability and suffering, with loss of employment, marital discord, and divorce.

CPP as defined by the American College of Obstetrics and Gynecology is nonmenstrual pain of 6 or more months' duration that localizes to the anatomic pelvis, anterior abdominal wall below the umbilicus, or the lumbosacral back and causes functional disability or requires medical or surgical treatment [6]. This definition excludes vulvar pain and cyclical pain of dysmenorrhea. However, it is important to recognize that women with chronic pelvic pain often have vulvar pain or dysmenorrhea as part of their symptom complex.

There are a myriad number of gynecologic disorders that are associated with chronic pelvic pain [7]. As stated above, chronic abdomino-pelvic pain patients should always be evaluated for non-gynecologic sources of pain which are commonly found in this population. This chapter will focus on the more commonly diagnosed gynecologic diseases associated with pelvic pain: endometriosis, leiomyomata or uterine fibroids, adenomyosis, pelvic congestion syndrome, ovarian remnant syndrome, ovarian retention syndrome, and pelvic inflammatory disease (PID). Adhesive disease can also occur in the pelvis but postsurgical adhesions will be addressed in chapter XX. Because many women with CPP also have vulvar or vaginal pain, a few of the more important disorders associated with vulvar or vaginal pain—provoked vestibulodynia (previously vulvar vestibulitis) and pudendal neuralgia—will also be briefly reviewed.

History and Physical Examination

More than 80 % of women with chronic pelvic pain have had pain for longer than 1 year when they seek medical care, and about one third have had pain for longer than 5 years [8]. Making accurate diagnoses in women with CPP can be perplexing. There are numerous possible gynecologic and non-gynecologic diagnoses (only gynecologic will be covered in this chapter). Additionally, it is likely that chronic pain itself may need to be considered a diagnosis [9].

The history and physical examination are powerful diagnostic and therapeutic tools in chronic pelvic pain. As diagnostic tools, a thorough history and examination may lead to an accurate diagnosis. This process minimizes the need for

M.P. Yamamoto, M.D.
Department of Obstetrics & Gynecology, Minimally Invasive Gynecologic Surgeon, Kaiser Permanente, 27400 Hesperian Blvd., Hayward, CA 94545, USA
e-mail: miya.yamamoto@gmail.com

J.F. Carillo, M.D.
Department of Obstetrics and Gynecology, University of Rochester Medical Center, 601 Elmwood Avenue, Box 668, Rochester, NY 14642, USA
e-mail: Jorge_Carrillo@urmc.rochester.edu

F.M. Howard, M.S., M.D. (✉)
Department of Obstetrics and Gynecology, Strong Memorial Hospital, University of Rochester Medical Center, 601 Elmwood Avenue, Box 668, Rochester, NY 14642, USA
e-mail: fred_howard@urmc.rochester.edu

L. Kapural (ed.), *Chronic Abdominal Pain: An Evidence-Based, Comprehensive Guide to Clinical Management*,
DOI 10.1007/978-1-4939-1992-5_10, © Springer Science+Business Media New York 2015

expensive laboratory testing and imaging or risky operative interventions. It is important that the clinician remember that even a "routine" pelvic examination is emotionally stressful for many patients with chronic pelvic pain as the exam can be painful and often the patient has undergone numerous undesired exams. Establishing rapport and trust with a patient by a compassionately taken history and a sensitively performed examination ensures that the patient feels the physician is caring and competent. Often just by giving the patient the opportunity to express her frustration with the course of her symptoms and validating her suffering allows the patient to leave the physician's office feeling better.

The history of the patient's pain must be thoroughly explored and all pertinent imaging, pathology and operative records obtained, along with the review of systems, with particular attention to the gastrointestinal, reproductive, urologic, and musculoskeletal systems. Because of the complexity of the history in most patients, intake questionnaires are extremely helpful in obtaining details of the history (see, for example, www.pelvicpain.org). However, they should not replace allowing the patient to tell her story.

Establishing the location of the patient's pain may be an important key to accurate diagnosis. Women with somatic nociceptive pain usually describe pain that is well localized to the area of disease. Such clear pinpointing is not always the case, however. For example, sometimes with levator ani pain the symptoms are described as deep, aching, heavy pain along with sharp, shooting pain and the patient cannot accurately localize the pain to the pelvic floor muscles. With visceral nociceptive pain there is almost always poor localization and usually a description of deep, dull, and cramping pain. Furthermore, because the cervix, uterus, and adnexae have the same metameric innervation as the bladder, distal ureter, lower ileum, colon, and rectosigmoid, it is often difficult to determine if visceral abdomino-pelvic pain is of gynecologic, urologic, or intestinal origin. For these reasons it is important to have the patient complete a pain map.

Women with CPP are more likely to have dysmenorrhea and dyspareunia. For example, dysmenorrhea is present in more than 80 % of women with chronic pelvic pain (in contrast to about 50 % in the general population), and dyspareunia is present in at least 40 % (in comparison to 10–15 % in the general population) [8]. The presence of either of these symptoms often is assumed to indicate a gynecological diagnosis, but dyspareunia and increased pain perimensturally are as common with IBS, IC/PBS, and pelvic floor muscle pain as with gynecological disorders.

Exploring the nature of the onset of pain may aid in diagnosis. For example, an immediately antecedent trauma, such as a fall, surgery, or motor vehicle accident, suggests a musculoskeletal cause. Pain that started with a pregnancy or immediately postpartum may suggest peripartum pelvic pain syndrome. Pain that started at or soon after menarche as dysmenorrhea, progressed to premenstrual pain, and then became constant suggests endometriosis or adenomyosis. If pain started soon after a physical or sexual assault, it may have significant musculoskeletal or psychological components.

Elucidation of any temporal pattern to the pain may be helpful. Cyclicity related to menses suggests gynecologic pain, but is not pathognomonic of gynecologic disease. The same pattern may occur with pain of intestinal, urologic, or musculoskeletal origin also. For example, symptoms of IBS or IC/PBS frequently increase premenstrually. A history of pain or increased pain with coitus is frequently present, as just discussed, and may be due to a variety of disorders, including psychological disease, marital problems, endometriosis, vulvodynia, IC/PBS, and IBS. If intercourse is painful, it is important to find out if pain is with entry at the outermost part of the vagina or if it is with deeper penetration high in the vagina or pelvis, or both. Diseases associated with chronic pelvic pain are not generally associated with entry dyspareunia, except as provoked vulvodynia or vaginismus.

The quality or nature of pain should be sought. For example, neuropathic pain is often described as burning or sharp and piercing, with an electric shock-like quality. Muscular pain may be aching in quality, with sharp, lancing pain with changes in position. Similar qualities of aching with occasional intermittent sharp and radiating pains may also be described with visceral pain. Pain with endometriosis is usually described as cramping [10].

Finding out about any prior treatments for chronic pelvic pain and the response to those treatments is a crucial part of the history. It may be important to know about any prior surgery, not just surgical treatment for pain, because postoperative surgical pain may be a risk factor for or proximate cause of chronic pelvic pain.

Ideally a thorough psychosocial history should be obtained on every patient with chronic pelvic pain. An extensive evaluation by a psychologist or similarly educated professional cannot always be done—nor is it always necessary. However, a basic psychosocial history is always important, especially asking about catastrophizing, anxiety, and depression. Depression, in particular, is one of several predictors of pain severity in women with chronic pelvic pain, and it is also a significant indicator of responsiveness to treatment. Asking about abuse, which may be difficult, is nevertheless another important part of the psychosocial history [11, 12]. Although there is a significant association between physical and sexual abuse and the development of chronic pelvic pain, the presence of moderate or severe depression or history of abuse was not found to be associated with a decreased response to treatment [13].

The physical examination should seek to find the exact anatomic locations of any areas of tenderness and, as much as possible, correlate these with areas of pain. This type of "pain mapping" examination requires a systematic and methodical attempt to duplicate the patient's pain by palpation, positioning, or bodily movement. At any tender areas or painful positions, the patient should be asked whether the pain produced is the same as her chronic pain. The examination should evaluate the musculoskeletal, gastrointestinal, urinary, and neurological systems, not just the reproductive tract. It facilitates the examination to divide it into standing, sitting, supine, and lithotomy components. As the pelvic examination is particularly relevant for gynecologic disorders, it is the only component of the examination that will be reviewed in detail in this chapter.

Vulvar examination starts with careful visualization of the vulva, perineal body, and anus looking for abnormal pigmentation or erythema, masses, and any skin or mucosal lesions. Palpation should be done with a moistened cotton-tipped swab to evaluate the vulva and the vulvar vestibule for tenderness. This is particularly useful in patients with localized provoked vestibulodynia (vulvar vestibulitis), who have exquisite tenderness in localized areas at the minor vestibular glands just external to the hymen, with normal sensation in adjacent vulvar areas.

A single-digit examination, using only one hand, is how the pelvic examination should be initiated. The introital bulbo-carvenosus and transverse perineal muscles, then the levator ani muscles, should be palpated for tone, spasm, and tenderness. In patients with pelvic floor pain this palpation may cause pain consistent with at least part of the patient's clinical pain symptoms. Pelvic floor pain may also result from trigger points of one or more of the muscles of the pelvis. The piriformis, coccygeal, and internal obturator muscles should be thoroughly evaluated using single-digit examination. The piriformis muscles can be difficult to evaluate transvaginally, however. Rectal examination may allow an easier evaluation than vaginal examination. Transvaginally or transrectally the examining finger is pressed posterolaterally just superior to the ischial spine. In the lithotomy position, if the patient is asked to abduct the thigh against resistance as the piriform muscle is palpated, the muscle may be more easily palpated, and there is exquisite tenderness of the muscle if there is spasm or tension myalgia involving the piriform muscle (piriformis syndrome).

The anterior vaginal, urethral, and trigonal areas should be palpated to elicit any areas of tenderness, induration, discharge, or thickening suggestive of chronic urethritis, chronic urethral syndrome, urethral diverticulum, vaginal wall cyst, trigonitis, or interstitial cystitis. With deeper palpation the cervix, paracervical areas, and vaginal fornices should be palpated with the single digit for tenderness or trigger points suggestive of problems such as repeated cervical trauma (usually from intercourse), pelvic infection, endometriosis, ureteral pain, or trigger points.

The uterus usually can be adequately evaluated for tenderness by direct palpation with a single digit. Significant uterine tenderness may be consistent with diseases such as adenomyosis, pelvic congestion syndrome, pelvic infection sequelae, endometriosis, or premenstrual syndrome. A uterus that is immobile and fixed in position, especially a retroflexed one, may suggest endometriosis or adhesions. The coccyx can be palpated with the single digit, and an attempt should be made to move it 30° or less. This may be easier to evaluate during the rectovaginal examination. Normally the coccyx moves 30° without eliciting pain, but in patients with coccygodynia this movement elicits pain. The ureteral and the adnexal areas should be palpated next, still using a single digit without the use of the abdominal hand. All of the preceding evaluations are "monomanual-monodigital"—that is, only one finger of one hand is used. No abdominal palpation with the other hand is involved.

The traditional visual, speculum, and bimanual examinations are still needed for a thorough evaluation, but they should usually follow the single-digit examination. A cotton-tipped swab can be used to evaluate the cervical os and the paracervical and cervical tissues for tenderness. In posthysterectomy patients the full vaginal cuff should be similarly palpated for tenderness with a cotton-tipped swab. Any pain or tenderness elicited with the bimanual examination is less specific, because it involves stimulation of all layers of the abdominal wall, the parietal peritoneum, and the palpated organ or organs. Including the rectovaginal examination is important in most women with chronic pelvic pain, looking particularly for nodularity and tenderness.

Endometriosis

Endometriosis is often blamed as the source for CPP but several studies on CPP patients found that the most frequent disorders were in fact non-gynecologic [8, 13] and one found that only 15 % of patients had confirmed endometriosis [13]. Endometriosis is the presence of ectopic endometrial glands and stroma, that is, endometrium located outside of the endometrial cavity. Endometriosis may be found in many locations in the body, but most often is found in the pelvis. Endometriosis remains an enigmatic disorder in that the etiology, the natural history, and the precise mechanisms by which it may cause pain are not completely understood.

Endometriosis has a well-recognized direct association with dysmenorrhea, dyspareunia, and chronic pelvic pain, but it may also be a risk factor for the development of nonreproductive tract CPP. For example, women with dysmenorrhea or endometriosis have increased episodes and increased severity of pain related to urinary calculi than women without dysmenorrhea or endometriosis [14]. Similar results have been experimentally demonstrated for vaginal pain, as well [15]. Also, women with endometriosis have been shown to have an increased incidence of interstitial cystitis/bladder pain syndrome [10, 16]. Such viscero–visceral interactions may have a significant role in chronic pelvic pain in women, reflecting central sensitization, explaining in part why some women with a past history of endometriosis have persistent pelvic pain after their endometriosis is gone [17].

Neither the etiology of endometriosis nor the etiology of pain associated with endometriosis is completely understood. The etiology of endometriosis is complex, but it seems certain that both genetic and environmental factors contribute to the disease phenotype. There are several general theories regarding the etiology of endometriosis but none is sufficient to explain the protean manifestations of endometriosis, or the predilection of some women, but not others, to develop symptomatic endometriosis. For example, the most widely spread theory, retrograde menstruation occurs in most women but only 5–10 % develop endometriosis [18].

Endometriosis is a disease of women of reproductive age, so most women with endometriosis-associated pain are 20–45 years of age. However it has been reported in girls as young as 10 years and it may be a more common cause of pain in teenagers than is generally recognized. It may also occur in postmenopausal women, particularly if they are on estrogen replacement.

Classically the woman with endometriosis presents with one or more of the following triad: an adnexal mass (endometrioma), infertility, or pelvic pain [19]. Pelvic pain most often starts as dysmenorrhea and about 90 % of women with endometriosis-associated pelvic pain have dysmenorrhea as a component of their pain symptoms. Perimenstrual exacerbation of pain can also occur in women with IC/PBS and IBS, and menstrual suppression may reduce these pain flares. Also, these three common diagnoses frequently occur together. One study found only 18 % of women with endometriosis had only endometriosis, while 32 % also had IC/PBS and 31 % had IBS [10]. Dyspareunia with deep penetration is also a frequent component of endometriosis-associated pain, occurring in about 40 % of cases. Intestinal involvement occurs in about 15 % of women with endometriosis and may be associated with gastrointestinal symptoms of tenesmus, dyschezia, constipation, diarrhea, low back pain, and, rarely, hematochezia or symptoms of bowel obstruction. Urinary tract involvement occurs in about 10 % of women with endometriosis and may be associated with urinary frequency, pressure, dysuria, or hematuria.

In many women with endometriosis-associated pelvic pain the physical examination is completely normal. In others there is tenderness and other findings only during menses. For this reason it is sometimes helpful to do the examination during the first day or two of menstrual flow in women with suspected endometriosis. Some women with endometriosis have persistent areas of tenderness in the pelvis, whether or not they are menstruating. Classic physical findings include a fixed retroverted uterus with tenderness posteriorly, tender nodularity of the uterosacral ligaments, and cul-de-sac on rectovaginal examination. However, these are not commonly found. Asymmetrically enlarged, tender ovaries that are fixed to the broad ligaments or pelvic sidewalls may also occasionally be found.

The symptoms and signs that lead to a clinical diagnosis of endometriosis are reliable 65–80 % of the time, but an accurate diagnosis can only be made by surgical excision with histologic confirmation [20, 21]. Accurate diagnosis requires microscopic visualization of both endometrial glands and stroma [22, 23]. As endometriosis may have a wide variety of gross appearances, it is essential that the surgeon be familiar with the variety of potential appearances of endometriosis for accurate diagnosis [22, 23]. A negative laparoscopy should lead one to consider other diagnoses such as IC/PBS or IBS, which can mimic endometriosis.

Treatment of endometriosis is also complex and many factors must be considered in planning treatment. It is important to educate the patient about endometriosis and the treatment options and actively involve her in decision-making. The patient's age, reproductive plans, duration of infertility, and attitude toward surgery or toward hormonal medications may be vital components of the patient's needs or concerns.

There are many medical treatments available for endometriosis-associated pelvic pain. Only those used commonly with good evidence of efficacy will be reviewed here.

Medical treatment with gonadotropin releasing hormone (GnRH) agonists, progestins, danazol or combined oral contraceptives effectively relieves endometriosis-associated pelvic pain. The number needed to treat for medical treatments is 2–2.5 [24].

GnRH agonists shut down LH and FSH production and release, leading to a dramatic decline in estradiol levels, induction of amenorrhea, and improvement of pain levels. When patients have a recurrence of pain within 1 year after treatment with GnRH analogs, re-treatment appears to be reasonably effective, with about two-thirds of patients showing a significant reduction of pain levels during re-treatment [25]. Loss of bone density with GnRH analogs is a serious concern. Clinical trials with GnRH agonists show that add-back therapy with conjugated equine estrogen and/or norethindrone acetate significantly decreases bone loss [26].

Danazol, a 17-ethinyl-testosterone derivative, has efficacy similar to that of GnRH agonists, but is not as frequently used due to possible androgenic side effects, including significant weight gain, mood changes, and masculinizing symptoms [27].

Medroxyprogesterone acetate (MPA) has been a recommended treatment for many years. Although a high dose of 100 mg/day was used in the only placebo-controlled trial of MPA, lower doses are used generally in clinical practice [28]. A randomized study of a 104 mg dose of a subcutaneous formulation of MPA compared to depot-leuprolide showed similar efficacies for both treatments [29].

Oral contraceptive (OCP) treatment of endometriosis is a long-standing approach, using either cyclical or continuous dosing. Efficacy appears to be similar or somewhat less than the other hormonal treatments [30]. There is some evidence that women not responsive to cyclical administration of OCPs may respond to continuous administration [31].

Some more recent approaches to medical treatment of endometriosis-associated pelvic pain are levonorgestrel-releasing intrauterine device [32], aromatase inhibitors [33, 34], selective progesterone modulators [35], and other progestins [36]. Although data from randomized clinical trials are needed, current evidence suggests that these therapies are effective alternatives in many patients.

Surgical treatment can be done at the time of laparoscopic diagnosis in symptomatic women. Organ-preserving, laparoscopic surgical treatment has been shown to significantly improve pain in women with stage II, III, and IV endometriosis, with a number needed to treat of 2–2.5 [37, 38]. Surgery for advanced-stage endometriosis can be challenging, tedious, frustrating, and prone to complications, so it is likely that surgical outcomes are surgeon dependent. Postoperative medical treatment may improve pain relief for the duration of the medical treatment, but does not improve long-term outcomes [28, 39, 40].

Presacral neurectomy (resection of the superior hypogastric plexus) and uterosacral neurectomy (uterine nerve resection or transection of the uterosacral ligament) have been recommended for relief of CPP associated with endometriosis. Data from clinical trials show that presacral neurectomy somewhat improves pain relief obtained with surgical treatment of endometriosis [41, 42]. However, presacral neurectomy may lead to intractable constipation and urinary urgency in up to 5 % of patients [43]. Data from clinical trials show that uterosacral neurectomy does not improve pain relief when included in surgical treatment of endometriosis and should not be performed [44].

If fertility is not desired, then hysterectomy, with or without bilateral salpingo-oophorectomy, is often recommended for endometriosis-associated pelvic pain. There is no consensus as to the advisability of removal of both ovaries if one or both are not directly involved by endometriosis. In one study evaluating this dilemma, recurrence of pain when one or both ovaries are preserved has been reported to occur in 62 % of cases compared to 10 % when both ovaries are removed, giving a relative risk for pain recurrence of 6.1 (confidence interval 2.5–14.6) [45]. Reoperation for pain was also more likely with ovarian preservation, with 31 % requiring reoperation compared to 4 % if both ovaries were removed at the time of hysterectomy for endometriosis. Although uncommon, endometriosis has been reported to recur after hysterectomy and bilateral salpingo-oophorectomy, with and without estrogen replacement therapy [46–48]. The potential risks of cardiovascular and bone density risks in premenopausal women without estrogen replacement likely outweigh the rare risk of recurrence.

Complications of medical treatment include side effects of weight gain, edema, hot flushes, headaches, nausea, acne, hirsutism, hot flushes, abnormal uterine bleeding, decreased breast size, decreased libido, vaginal dryness, weakness, decreased bone density, and thromboembolic disease. Surgical complications vary with the severity of disease, but injury to pelvic viscera is a potential risk in women with endometriosis. Endometriosis untreated is rarely life threatening, although there are cases of ureteral and bowel obstruction due to endometriosis, as well as invasion of the urinary and gastrointestinal tracts.

It is important to recognize that the finding of endometriosis in women with CPP does not ensure that medical or surgical treatment of endometriosis will result in effective relief of pain. To the contrary, although treatment of endometriosis in women with pelvic pain is clearly indicated based on randomized, placebo-controlled, double-blind clinical trials, pain relief of 6 or more months duration due to treatment can be expected in only about 40–70 % of women with endometriosis-associated CPP (number needed to treat of 2.0–2.5). Also, recurrence rates of CPP are high after medical and surgical therapy. Women with endometriosis need to be evaluated for all sources of chronic abdomino-pelvic pain including non-gynecologic to ensure appropriate treatment for their CPP.

Pelvic Congestion Syndrome

Pelvic Congestion Syndrome (PCS) is defined by the triad of pelvic pain, pelvic varicosities, and abnormal venous function. Abnormal venous function is identified by a pelvic venogram. Pain is usually worse premenstrually rather than having pain levels which peak with menses. Symptoms often develop after childbirth as pregnancy increases the capacity of pelvic veins by 60 % [49]. Both of these are likely related to hormonal changes affecting the pelvic vasculature associated with pregnancy and the menstrual cycle. In addition, pain is usually dull and aching, with exacerbations related to movement and position changes such as sitting to standing, or left to right side while supine, and may improve after lying down. Backaches which worsen with standing can also occur. Deep dyspareunia and post-coital pain are very common. On abdominal exam, palpation at the ovarian point can reproduce the pelvic pain in 80 % of women with PCS. This point lies at the junction of the upper and middle thirds of a line drawn from the anterior superior iliac spine to the pubic symphysis. Adnexal tenderness may also be found on bimanual exam.

Diagnosis is confirmed by a pelvic venogram showing venous stasis, dilation, delayed emptying and a plexus formation of the ovarian or uterine vessels. This can be performed by transcervical injection of the myometrium at the fundus or by transcutaneous retrograde injection of the ovarian veins by interventional radiology with serial imaging. It cannot be diagnosed on ultrasound or CAT scan or by simply visualizing enlarged pelvic vessels on laparoscopy.

Treatment can be either menstrual suppression with progestin hormonal therapy, with GnRH agonists, with embolization of the dilated ovarian vessels by interventional radiology, or for refractory cases in women done with childbearing, hysterectomy and bilateral salpingo-oophorectomy. Symptom improvement after embolization may take several months, but long-term success in a small 3-year follow up study has been reported at 76 % [50].

Pelvic Inflammatory Disease and Adhesions

Up to 36 % of women with acute PID may develop chronic abdomino-pelvic pain [51]. This association has been reported on extensively in the literature and is widely accepted. Recurrent PID appears to exponentially increase the risk of subsequent CPP in a follow up of patients for 84 months [52]. PID is more commonly found among teenagers and women under age 25; this may be related to increased susceptibility of the endocervical glandular cells to ascending infection in addition to behavioral differences related to age and maturity. PID can result in adnexal adhesions, tubo-ovarian abcesses, hepatic adhesions, and hydrosalpinges.

The mechanism by which 1/3 of women with PID develop CPP is not well understood but may be related to inflammatory mediated changes and possibly adhesion formation. PID can also result from pelvic tuberculosis in women from high-risk countries or those with high-risk status such as HIV patients. PID can be associated with appendicitis due to appendiceal-adnexal adhesions and similarly with colonic diverticulitis or inflammatory bowel disease. The connection between adhesions and CPP is also not well understood as findings of intra-abdominal and pelvic adhesions are often incidental in many patients without pain. Adhesions are easily blamed for pain if present in patients with CPP, but not clearly demonstrated as the etiology. Two randomized trials of adhesiolysis for CPP failed to show significant improvement after both a laparoscopic and laparotomy approach [53, 54]. Additionally, there are no randomized controlled trials which demonstrate effective means by which to prevent adhesions from reforming. Adhesions and chronic abdominal pain will be addressed in more detail in Chapter XX.

Adenomyosis

Adenomyosis is a benign condition in which endometrial glands and stroma are found in the myometrium (invading past the endometrial cavity). The true incidence is unknown, since the final diagnosis is done with a pathological specimen after hysterectomy. It is thought to be most prevalent among perimenopausal and multiparous women between 40 and 50 years of age.

The classic presentation is a triad of abnormal uterine bleeding (50 %), prolonged dysmenorrhea (30 %), and an enlarged "globular," tender uterus [55]. Menorrhagia (heavy menstrual bleeding) can cause dysmenorrhea from stimulation and edema of endometrial tissue within the myometrium [56]. These symptoms start perimenstrually, are cyclic and often last the entire reproductive age if untreated. They can be associated with chronic pelvic pain. The symptoms can overlap with uterine fibroids and endometriosis. However, endometriosis does not classically present with heavy menses or irregular menses, and fibroids are easily distinguished on ultrasound.

As described previously the final diagnosis is made with histological examination after hysterectomy, which would not be an option for someone desiring future fertility. Imaging is another modality that is used. Clinically, dysmenorrhea, a tender, boggy uterus on bimanual exam, the absence of fibroid uterus on ultrasound and an ultrasound report of heterogenous texture of the myometrium with an enlarged uterus without discrete masses will help make the diagnosis. While MRI has a sensitivity of 88–93 % and specificity 66–91 % compared to transvaginal ultrasound, it rarely uses first line as the final diagnosis is made by histology, not by a suspicious MRI, and treatment does not vary by MRI findings.

The definitive treatment for this adenomyosis is hysterectomy. When hysterectomy is contraindicated *or undesired by the patient*, a number of conservative treatments can be considered [55]. Hormonal therapy with oral progestins, levonorgestrel IUD, gonadotropin releasing hormones agonists, aromatase inhibitors, and danazol may control symptoms. Oral contraceptives have not generally been effective, but may be tried as a simple first time treatment. Uterine artery embolization (UAE) has been used for treatment of adenomyosis but is less well studied and accepted than for uterine fibroids. A review article of the six small uncontrolled studies of an UAE for adenomyosis, with a total of 208 patients, found only a 65 % improvement at 40 months of follow up [57]. Larger studies are needed before UAE will be accepted widely as a viable treatment option for adenomyosis. In patients with focal disease or with adenomyomas, surgical excision can be done but long-term recurrence risks are unknown.

Uterine Leiomyomas

Uterine leiomyomas or fibroids are the most common benign gynecologic tumors. They originate from the myometrial cells. These tumors are hormonal dependent, associated mainly with estrogen, growth hormone, and progesterone, estrogen being the main growth stimulator and progesterone appearing to inhibit growth. There are different types of fibroids: subserosal, intramural and submucosal. Subserosal myomas originate at the external portion of the myometrium and often protrude out into the abdominal cavity and may cause pressure symptoms or be asymptomatic. Intramural myomas are located completely within the muscular wall of the uterus and often cause menorrhagia and dysmenorrhea. Submucous myomas are close to the endometrium and impinge on the endometrial cavity and are associated with irregular heavy bleeding (menometrorrhagia).

The estimated incidence of fibroids ranges from 33 to 77 %, depending on the method of diagnosis (i.e., clinical, ultrasound, pathology) [58]. Fibroids are most often identified during reproductive age, with most women being in their 30–40s. Symptoms vary depending on the number, size, and location. Abnormal uterine bleeding is the most frequent symptom associated with fibroids, usually presenting as cyclical, heavy menses. The mechanism of abnormal uterine bleeding is unknown but may be caused by dysregulation of angiogenic factors [59].

Pelvic pressure and pain are symptoms associated with the size of the uterus and the location of the fibroids. A fibroid pressing on the bladder or ureter may cause symptoms of urinary frequency, incontinence urinary retention, or hydronephrosis secondary to ureteral obstruction. A fibroid pressing on the rectum may cause symptoms of constipation or low back pain.

The diagnosis is made by physical examination and imaging studies. A pelvic exam will show a pelvic mass that can be identified as a large irregular uterus. Transvaginal and transabdominal ultrasounds have a high sensitivity (90–100 %) and specificity (87–98 %), with a positive predictive value of 81–93 % and a negative predictive value of 98–100 % [60]. Ultrasound is the preferred and most cost effective way of diagnosing a fibroid uterus. MRI can be used to assess the location, size, and depth of fibroids for surgical planning of a myomectomy but it is not used for routine screening. Treatment options for leiomyomas are expectant, medical, interventional radiologic, or surgical.

Expectant management or observation is reasonable in patients with fibroids that are asymptomatic.

There are a number of options for medical management. Combined oral contraceptive pills and oral or injectable progestins are useful for menstrual abnormalities. Levonogestrel intrauterine device also is useful for menstrual abnormalities and has the benefit of reduced hormonal side effects in the patients.

Gonadotropin releasing hormone (GnRH) agonists lead to amenorrhea in most patients and provide a 35–65 % reduction in fibroid volume over 3 months. GnRH agonists are most often used as presurgical therapy for 3–6 months to treat anemia and facilitate less complicated surgery. Long-term use has bone density reduction risk which can be prevented with add-back hormone therapy such as an oral norethindrone acetate.

Mifepristone, a progesterone modulator, has been shown to reduce the volume of fibroids and to induce amenorrhea without the concern for bone density loss. Known side effects include risk of endometrial hyperplasia and transient elevation of transaminase levels. Further studies are needed to evaluate the safety and best use of this class of treatment [61].

Uterine artery embolization (UAE) is performed primarily by interventional radiologists. The approach is done via transcutaneous femoral artery approach to identify and embolize major blood supply to fibroids and is performed primarily by interventional radiologists. It should be used with caution when the patient desires to retain her ability to conceive because age-related amenorrhea can occur and abnormal placentation is possible [62]. Several long-term studies show a higher rate of clinical failures and reoperation rates [61]. Higher success rates may be possible in well selected patients who are closer to menopause.

MRI-guided focused ultrasound was FDA approved in 2004 but there are no long-term studies over 24 months. A high intensity, directed ultrasound approach.

Surgical treatment is by myomectomy or hysterectomy. Myomectomy is surgical excision of the fibroids and is an option for patients who wish to retain their fertility or their uterus. However, it is not definitive treatment as studies have shown that the increased risk of recurrence is associated with the number of fibroids present [61]. Women should be appropriately counseled about their recurrence risk and subsequent need for another procedure or hysterectomy in the future. Hysterectomy is the definitive surgical treatment for a symptomatic fibroid uterus.

Provoked Vulvodynia or Vestibulodynia

This disorder (previously called vulvar vestibulitis) is associated with dyspareunia (pain with intercourse). Pain is present on light touch of the vulvar vestibule in the absence of other findings [63]. Vulvovaginal infections such as candida vaginitis and dermatoses such as lichen sclerosis should be ruled out and any visible lesions biopsied to rule out malignancy. The etiology is unknown. Patients with this disorder also are commonly found to have disorders associated with chronic pain such as depression, IBS, and pelvic floor tension myalgia [64]. Dyspareunia related to this disease can severely affect a patient's quality of life.

Due to the unclear etiology, treatments vary widely and include oral tricyclic antidepressants, gabapentin, steroid creams or injections, topical anesthetic ointments, physical therapy, avoidance of surface irritants, biofeedback, and cognitive-behavioral therapy. In patients with hypoestrogenic vaginal atrophy, correction with topical estrogen may be helpful. Such cases may be postmenopausal women or, more rarely, women on hormones such as depo-medroxyprogesterone acetate. A recent randomized controlled trial of oral desipramine with and without lidocaine ointment failed to show a reduction in pain compared to placebo [65]. This approach is commonly used to treat both central and peripheral neuropathology. The authors theorized that cream massage of the control ointment, weekly telephone surveys, and medication instructions by an RN and the natural history of the disease process may all have contributed to the improvement among the control group.

For treatment of cases resistant to these therapies, surgical treatment with vulvar vestibulectomy is often performed. It involves excision of the vestibule and possibly vaginal advancement. It does result in disfigurement of the introitus; so many patients are hesitant to use this as a first line treatment. A randomized trial by Bergeron et al. compared vestibulectomy, group cognitive-behavioral therapy and surface electromyographic biofeedback. They found that at 6 months, while all groups had decreased pain, vestibulectomy patients had significantly lower pain levels and were significantly more improved than the other two [66]. They cautioned that there were a larger number of patients who refused vestibulectomy and dropped out before treatment. Their subsequent follow up study at 2.5 years (51 of the original 78) found that these results persisted [67]. They found that higher pretreatment pain intensity predicted poorer 2.5 year outcomes for all groups and that erotophobia predicted a poorer outcome for vestibulectomy [67].

Pelvic Floor Tension Myalgia

Musculoskeletal disorders were found in 23 % patients with CPP in a specialty practice [13]. These include disorders such as sacroiliac joint dysfunction, coccygodynia, and low back pain as well, the more common diagnoses of pelvic floor tension myalgia (PFTM) and abdominal wall myofascial pain (which is covered in Chapter XX). A myofascial trigger point is a hyperirritable spot within a tense band of muscle or fascia which is painful on compression and usually causes referred pain and other sensory disturbances [68, 69]. Patients with CPP most often have these points in their abdominal wall (i.e., rectus abdomins and external and internal obliques) and in their pelvic floor muscles (i.e., levator ani and obturator internus). However, they can also have trigger points in the lumbar and gluteus muscles. Trigger points of the abdomen and pelvic floor are found with a single digit palpation applying pressure along the muscle belly and fascial insertions. These myofascial pain syndromes can occur as a sole cause of CPP or can occur in conjunction with other disorders such as IC/PBS, provoked vestibulodynia, or endometriosis [70]. They can develop after an acute traumatic event, repetitive microtrauma or as a result of chronic muscle tension and muscle shortening from pelvic girdle dysfunction, short leg syndrome or as a response to visceral pain and inflammation from the above disorders (IC/PBS, endometriosis, etc.).

Successful long-term treatment can occur with weekly physical therapy to recruit inadequately used muscles of the pelvic girdle and relax the contracted muscles. Other treatment options are trigger point injections with local anesthetics, dry needling, muscle relaxants, moist heating pads, transcutaneous electrical stimulation (TENS) units, yoga or stretching exercises, and massage, and specifically for PFTM, vaginal diazepam 5–10 mg twice daily. The therapist treating PFTM should be specially trained because exercises such as Kegels will worsen their symptoms and the patient will stop therapy. Trigger point injections are less frequently used for PFTM as the pain associated the injections is severe. Adequate treatment by trigger point injections usually requires visits every 2–6 weeks, which may not be feasible for many patients. Successful injections often result in decreased pain for several weeks. There are no prospective randomized trials to promote one treatment over another.

Pudendal Neuralgia

Pudendal neuralgia (PN) currently is best diagnosed using the "Nantes Criteria" [71, 72]: (1) pain in the anatomical distribution of the pudendal nerve; (2) pain more severe when sitting; (3) pain that does not awaken from sleep; (4) no objective sensory loss on clinical examination; and (5) pain that is relieved by diagnostic pudendal nerve blocks. The diagnosis is likely if all five criteria are met. The differential diagnosis includes recurrent vaginitis, non-provoked vulvodynia, provoked vestibulodynia, pelvic congestion syndrome, and pelvic floor tension myalgia.

There sometimes is a history of long bicycle rides, episodes of prolonged sitting, or pelvic trauma immediately preceding the onset of symptoms. Also, reconstructive pelvic surgery can damage the nerve from compression, scar tissue formation, or direct impingement. Frequently no reason for the onset of symptoms can be elicited.

PN can affect any part of the perineum and vagina including the labia, vestibule, mons pubis, urethra, clitoris, anus and rectum, following the distribution of the nerve. Symptoms can include hyperesthesia, allodynia, paresthesias, burning or stabbing pain, sensations of incomplete voiding or inability to void normally, overactive bladder, dyspareunia, along with bowel dysfunction, sensation of a mass in the rectum, and pain with defecation; sensory symptoms should follow the course of the nerve.

The pudendal nerve arises from the sacral plexus (S2–S4) and then splits up into the anorectal, perineal, and clitoral branches. There are several locations in the course of the nerve where damage or entrapment occurs, but the most common are between the sacrospinous and sacrotuberous ligaments around the ischial spine (80 %), and in Alcock's canal (20 %). Image-guided pudendal nerve blocks at these locations are both diagnostic and therapeutic. If the pain is relieved after injections (local anesthetic with or without steroids) then most likely the source of the pain has been identified. If the pain is not relieved, then the source may not be the pudendal nerve. A series of nerve blocks can be completed which may control the symptoms. Medical management with a tricyclic antidepressant or gabapentin is often helpful. Other treatment options include pelvic floor physical therapy, pregabalin, oral muscle relaxants, and local muscle relaxants (vaginal diazepam and rectal belladonna and opium suppositories) [72]. Surgical decompression of the nerve has been reported to have a number needed to treat of 1.7, compared to medical treatment only, in a non-blinded randomized trial [73].

Ovarian Remnant Syndrome and Ovarian Retention Syndrome (Residual Ovary Syndrome)

Ovarian remnant syndrome is the presence of painful, histologically documented ovarian tissue in a patient who has undergone a previous bilateral oophorectomy. Often the ovarian tissue is found adherent to bowel or to the pelvic sidewall. It may rarely occur in a patient who has undergone a unilateral oophorectomy, with painful, persistent ipsilateral ovarian tissue. It may be a more common cause of chronic pelvic pain than is generally recognized [74]. A normal FSH hormone level in a patient of hormone replacement often can make the diagnosis. In some cases ovarian stimulation with GnRH agonist documented by increase of estradiol levels before and after stimulation will confirm the diagnosis. If the location of the remnant ovarian tissue is unclear despite imaging, stimulation of ovarian tissue to enlarge it with clomiphene citrate may aid in finding the remnant tissue both on imaging and on diagnostic laparoscopy. Although laparoscopy may have a role in diagnosis and treatment in some cases these remnants are often embedded in dense scar tissue and the surgery is difficult laparoscopically unless extensive adhesiolysis is performed [75].

Another uncommon cause of ovarian cysts and chronic pelvic pain is residual ovary syndrome or ovarian retention syndrome. Ovarian retention syndrome is the presence of persistent pelvic pain, dyspareunia, or a pelvic mass after conservation of one or both ovaries at the time of hysterectomy.

Pain due to ovarian retention syndrome is usually relieved by complete ovarian suppression with GnRH-a treatment [76]. Definitive treatment is surgical extirpation and can sometimes be performed laparoscopically [77]. In young women in whom preservation of ovarian function is desired, adhesiolysis and ovarian cystectomy may be tried, but appears to be less likely to relieve pain.

References

1. Zondervan KT, Yudkin PL, Vessey MP, Dawes MG, Barlow DH, Kennedy SH. Prevalence and incidence of chronic pelvic pain in primary care: evidence from a national general practice database. Br J Obstet Gynaecol. 1999;106(11):1149–55.
2. Lippman SA, Warner M, Samuels S, Olive D, Vercellini P, Eskenazi B. Uterine fibroids and gynecologic pain symptoms in a population-based study. Fertil Steril. 2003;80(6):1488–94.
3. Mathias SD, Kuppermann M, Liberman RF, Lipschutz RC, Steege JF. Chronic pelvic pain: prevalence, health-related quality of life, and economic correlates. Obstet Gynecol. 1996;87(3):321–7.
4. Hillis SD, Marchbanks PA, Peterson HB. The effectiveness of hysterectomy for chronic pelvic pain. Obstet Gynecol. 1995;86(6):941–5.
5. Howard FM. The role of laparoscopy in chronic pelvic pain: promise and pitfalls. Obstet Gynecol Surv. 1993;48(6):357–87.
6. ACOG Committee on Practice Bulletins—Gynecology. ACOG Practice Bulletin No. 51. Chronic pelvic pain. Obstet Gynecol. 2004;103(3):589–605.
7. Howard FM. Chronic pelvic pain. Obstet Gynecol. 2003;101(3):594–611.
8. Zondervan KT, Yudkin PL, Vessey MP, et al. Chronic pelvic pain in the community—symptoms, investigations, and diagnoses. Am J Obstet Gynecol. 2001;184(6):1149–55.
9. Tracey I, Bushnell MC. How neuroimaging studies have challenged us to rethink: is chronic pain a disease? J Pain. 2009;10(11):1113–20.
10. Droz J, Howard FM. Use of the short-form McGill Pain Questionnaire as a diagnostic tool in women with chronic pelvic pain. J Minim Invasive Gynecol. 2011;18(2):211–7.
11. Elzevier HW, Voorham-van der Zalm PJ, Pelger RC. How reliable is a self-administered questionnaire in detecting sexual abuse: a retrospective study in patients with pelvic-floor complaints and a review of literature. J Sex Med. 2007;4(4 Pt 1):956–63.
12. Wijma B, Schei B, Swahnberg K, et al. Emotional, physical, and sexual abuse in patients visiting gynaecology clinics: a Nordic cross-sectional study. Lancet. 2003;361(9375):2107–13.
13. Lamvu G, Williams R, Zolnoun D, et al. Long-term outcomes after surgical and nonsurgical management of chronic pelvic pain: one year after evaluation in a pelvic pain specialty clinic. Am J Obstet Gynecol. 2006;195(2):591–8; discussion 598–600.
14. Giamberardino MA, De Laurentis S, Affaitati G, Lerza R, Lapenna D, Vecchiet L. Modulation of pain and hyperalgesia from the urinary tract by algogenic conditions of the reproductive organs in women. Neurosci Lett. 2001;304(1–2):61–4.
15. Berkley KJ, Cason A, Jacobs H, Bradshaw H, Wood E. Vaginal hyperalgesia in a rat model of endometriosis. Neurosci Lett. 2001;306(3):185–8.
16. Chung MK, Chung RR, Gordon D, Jennings C. The evil twins of chronic pelvic pain syndrome: endometriosis and interstitial cystitis. JSLS. 2002;6(4):311–4.
17. Stratton P, Berkley KJ. Chronic pelvic pain and endometriosis: translational evidence of the relationship and implications. Hum Reprod Update. 2011;17(3):327–46.
18. Guo SW. Epigenetics of endometriosis. Mol Hum Reprod. 2009;15(10):587–607.
19. Adamson GD. Diagnosis and clinical presentation of endometriosis. Am J Obstet Gynecol. 1990;162(2):568–9.

20. Sutton C, Hill D. Laser laparoscopy in the treatment of endometriosis. A 5-year study. Br J Obstet Gynaecol. 1990;97(2):181–5.
21. Walter AJ, Hentz JG, Magtibay PM, Cornella JL, Magrina JF. Endometriosis: correlation between histologic and visual findings at laparoscopy. Am J Obstet Gynecol. 2001;184(7):1407–11; discussion 1411–3.
22. Martin DC, Hubert GD, Vander Zwaag R, el-Zeky FA. Laparoscopic appearances of peritoneal endometriosis. Fertil Steril. 1989;51(1):63–7.
23. Jansen RP, Russell P. Nonpigmented endometriosis: clinical, laparoscopic, and pathologic definition. Am J Obstet Gynecol. 1986;155(6):1154–9.
24. Ling FW. Randomized controlled trial of depot leuprolide in patients with chronic pelvic pain and clinically suspected endometriosis. Pelvic Pain Study Group. Obstet Gynecol. 1999;93(1):51–8.
25. Hornstein MD, Yuzpe AA, Burry K, et al. Retreatment with nafarelin for recurrent endometriosis symptoms: efficacy, safety, and bone mineral density. Fertil Steril. 1997;67(6):1013–8.
26. Hornstein MD, Surrey ES, Weisberg GW, Casino LA. Leuprolide acetate depot and hormonal add-back in endometriosis: a 12-month study. Lupron Add-Back Study Group. Obstet Gynecol. 1998;91(1):16–24.
27. Farquhar C, Sutton C. The evidence for the management of endometriosis. Curr Opin Obstet Gynecol. 1998;10(4):321–32.
28. Telimaa S, Puolakka J, Ronnberg L, Kauppila A. Placebo-controlled comparison of danazol and high-dose medroxyprogesterone acetate in the treatment of endometriosis. Gynecol Endocrinol. 1987;1(1):13–23.
29. Schlaff WD, Carson SA, Luciano A, Ross D, Bergqvist A. Subcutaneous injection of depot medroxyprogesterone acetate compared with leuprolide acetate in the treatment of endometriosis-associated pain. Fertil Steril. 2006;85(2):314–25.
30. Vercellini P, Trespidi L, Colombo A, Vendola N, Marchini M, Crosignani PG. A gonadotropin-releasing hormone agonist versus a low-dose oral contraceptive for pelvic pain associated with endometriosis. Fertil Steril. 1993;60(1):75–9.
31. Vercellini P, Frontino G, De Giorgi O, Pietropaolo G, Pasin R, Crosignani PG. Continuous use of an oral contraceptive for endometriosis-associated recurrent dysmenorrhea that does not respond to a cyclic pill regimen. Fertil Steril. 2003;80(3):560–3.
32. Petta CA, Ferriani RA, Abrao MS, et al. Randomized clinical trial of a levonorgestrel-releasing intrauterine system and a depot GnRH analogue for the treatment of chronic pelvic pain in women with endometriosis. Hum Reprod. 2005;20(7):1993–8.
33. Mousa NA, Bedaiwy MA, Casper RF. Aromatase inhibitors in the treatment of severe endometriosis. Obstet Gynecol. 2007;109(6):1421–3.
34. Nawathe A, Patwardhan S, Yates D, Harrison GR, Khan KS. Systematic review of the effects of aromatase inhibitors on pain associated with endometriosis. BJOG. 2008;115(7):818–22.
35. Chwalisz K, Perez MC, Demanno D, Winkel C, Schubert G, Elger W. Selective progesterone receptor modulator development and use in the treatment of leiomyomata and endometriosis. Endocr Rev. 2005;26(3):423–38.
36. Vercellini P, Pietropaolo G, De Giorgi O, Pasin R, Chiodini A, Crosignani PG. Treatment of symptomatic rectovaginal endometriosis with an estrogen-progestogen combination versus low-dose norethindrone acetate. Fertil Steril. 2005;84(5):1375–87.
37. Sutton CJ, Ewen SP, Whitelaw N, Haines P. Prospective, randomized, double-blind, controlled trial of laser laparoscopy in the treatment of pelvic pain associated with minimal, mild, and moderate endometriosis. Fertil Steril. 1994;62(4):696–700.
38. Abbott J, Hawe J, Hunter D, Holmes M, Finn P, Garry R. Laparoscopic excision of endometriosis: a randomized, placebo-controlled trial. Fertil Steril. 2004;82(4):878–84.
39. Parazzini F, Fedele L, Busacca M, et al. Postsurgical medical treatment of advanced endometriosis: results of a randomized clinical trial. Am J Obstet Gynecol. 1994;171(5):1205–7.
40. Hornstein MD, Hemmings R, Yuzpe AA, Heinrichs WL. Use of nafarelin versus placebo after reductive laparoscopic surgery for endometriosis. Fertil Steril. 1997;68(5):860–4.
41. Candiani GB, Fedele L, Vercellini P, Bianchi S, Di Nola G. Presacral neurectomy for the treatment of pelvic pain associated with endometriosis: a controlled study. Am J Obstet Gynecol. 1992;167(1):100–3.
42. Zullo F, Palomba S, Zupi E, et al. Effectiveness of presacral neurectomy in women with severe dysmenorrhea caused by endometriosis who were treated with laparoscopic conservative surgery: a 1-year prospective randomized double-blind controlled trial. Am J Obstet Gynecol. 2003;189(1):5–10.
43. Zullo F, Palomba S, Zupi E, et al. Long-term effectiveness of presacral neurectomy for the treatment of severe dysmenorrhea due to endometriosis. J Am Assoc Gynecol Laparosc. 2004;11(1):23–8.
44. Vercellini P, Aimi G, Busacca M, Apolone G, Uglietti A, Crosignani PG. Laparoscopic uterosacral ligament resection for dysmenorrhea associated with endometriosis: results of a randomized, controlled trial. Fertil Steril. 2003;80(2):310–9.
45. Namnoum AB, Hickman TN, Goodman SB, Gehlbach DL, Rock JA. Incidence of symptom recurrence after hysterectomy for endometriosis. Fertil Steril. 1995;64(5):898–902.
46. Metzger DA, Lessey BA, Soper JT, McCarty Jr KS, Haney AF. Hormone-resistant endometriosis following total abdominal hysterectomy and bilateral salpingo-oophorectomy: correlation with histology and steroid receptor content. Obstet Gynecol. 1991;78(5 Pt 2):946–50.
47. Redwine DB. Endometriosis persisting after castration: clinical characteristics and results of surgical management. Obstet Gynecol. 1994;83(3):405–13.
48. Dmowski WP, Radwanska E, Rana N. Recurrent endometriosis following hysterectomy and oophorectomy: the role of residual ovarian fragments. Int J Gynaecol Obstet. 1988;26(1):93–103.
49. Perry CP. Current concepts of pelvic congestion and chronic pelvic pain. JSLS. 2001;5(2):105–10.
50. Creton D, Hennequin L, Kohler F, Allaert FA. Embolisation of symptomatic pelvic veins in women presenting with non-saphenous varicose veins of pelvic origin—three-year follow-up. Eur J Vasc Endovasc Surg. 2007;34(1):112–7.
51. Haggerty CL, Peipert JF, Weitzen S, et al. Predictors of chronic pelvic pain in an urban population of women with symptoms and signs of pelvic inflammatory disease. Sex Transm Dis. 2005;32(5):293–9.
52. Trent M, Bass D, Ness RB, Haggerty C. Recurrent PID, subsequent STI, and reproductive health outcomes: findings from the PID evaluation and clinical health (PEACH) study. Sex Transm Dis. 2011;38(9):879–81.
53. Swank DJ, Swank-Bordewijk SC, Hop WC, et al. Laparoscopic adhesiolysis in patients with chronic abdominal pain: a blinded randomised controlled multi-centre trial. Lancet. 2003;361(9365):1247–51.
54. Peters AA, Trimbos-Kemper GC, Admiraal C, Trimbos JB, Hermans J. A randomized clinical trial on the benefit of adhesiolysis in patients with intraperitoneal adhesions and chronic pelvic pain. Br J Obstet Gynaecol. 1992;99(1):59–62.
55. Matalliotakis IM, Katsikis IK, Panidis DK. Adenomyosis: what is the impact on fertility? Curr Opin Obstet Gynecol. 2005;17(3):261–4.
56. Dietrich JE. An update on adenomyosis in the adolescent. Curr Opin Obstet Gynecol. 2010;22(5):388–92.
57. Popovic M, Puchner S, Berzaczy D, Lammer J, Bucek RA. Uterine artery embolization for the treatment of adenomyosis: a review. J Vasc Interv Radiol. 2011;22(7):901–9; quiz 909.

58. Cramer SF, Patel A. The frequency of uterine leiomyomas. Am J Clin Pathol. 1990;94(4):435–8.
59. Chavez NF, Stewart EA. Medical treatment of uterine fibroids. Clin Obstet Gynecol. 2001;44(2):372–84.
60. Salman T, Davis C. Uterine fibroids, management and effect on fertility. Curr Opin Obstet Gynecol. 2010;22(4):295–303.
61. American College of Obstetricians and Gynecologists. ACOG practice bulletin. Alternatives to hysterectomy in the management of leiomyomas. Obstet Gynecol. 2008;112(2 Pt 1):387–400.
62. Goodwin SC, Spies JB. Uterine fibroid embolization. N Engl J Med. 2009;361(7):690–7.
63. Foster DC. Vulvar disease. Obstet Gynecol. 2002;100(1):145–63.
64. Edwards L. New concepts in vulvodynia. Am J Obstet Gynecol. 2003;189(3 Suppl):S24–30.
65. Foster DC, Kotok MB, Huang LS, et al. Oral desipramine and topical lidocaine for vulvodynia: a randomized controlled trial. Obstet Gynecol. 2010;116(3):583–93.
66. Bergeron S, Binik YM, Khalife S, et al. A randomized comparison of group cognitive—behavioral therapy, surface electromyographic biofeedback, and vestibulectomy in the treatment of dyspareunia resulting from vulvar vestibulitis. Pain. 2001;91(3):297–306.
67. Bergeron S, Khalife S, Glazer HI, Binik YM. Surgical and behavioral treatments for vestibulodynia: two-and-one-half year follow-up and predictors of outcome. Obstet Gynecol. 2008;111(1):159–66.
68. Lavelle ED, Lavelle W, Smith HS. Myofascial trigger points. Anesthesiol Clin. 2007;25(4):841–51. vii–iii.
69. Lavelle ED, Lavelle W, Smith HS. Myofascial trigger points. Med Clin North Am. 2007;91(2):229–39.
70. Hoffman D. Understanding multisymptom presentations in chronic pelvic pain: the inter-relationships between the viscera and myofascial pelvic floor dysfunction. Curr Pain Headache Rep. 2011;15(5):343–6.
71. Labat JJ, Riant T, Robert R, Amarenco G, Lefaucheur JP, Rigaud J. Diagnostic criteria for pudendal neuralgia by pudendal nerve entrapment (Nantes criteria). Neurourol Urodyn. 2008;27(4):306–10.
72. Hibner M, Desai N, Robertson LJ, Nour M. Pudendal neuralgia. J Minim Invasive Gynecol. 2010;17(2):148–53.
73. Robert R, Labat JJ, Bensignor M, et al. Decompression and transposition of the pudendal nerve in pudendal neuralgia: a randomized controlled trial and long-term evaluation. Eur Urol. 2005;47(3):403–8.
74. Steege JF. Ovarian remnant syndrome. Obstet Gynecol. 1987;70(1):64–7.
75. Nezhat C, Kearney S, Malik S, Nezhat F. Laparoscopic management of ovarian remnant. Fertil Steril. 2005;83(4):973–8.
76. Siddall-Allum J, Rae T, Rogers V, Witherow R, Flanagan A, Beard RW. Chronic pelvic pain caused by residual ovaries and ovarian remnants. Br J Obstet Gynaecol. 1994;101(11):979–85.
77. El-Minawi AM, Howard FM. Operative laparoscopic treatment of ovarian retention syndrome. J Am Assoc Gynecol Laparosc. 1999;6(3):297–302.

Pediatric Chronic Abdominal Pain: Etiology, Diagnosis, and Treatment

11

Dawn A. Sparks, Monica P. Garin-Laflam, and Joseph P. Cravero

Background

Children with chronic abdominal pain continually mystify healthcare professionals worldwide. There has been a wealth of data produced over nearly seven decades of chronic pain research. Yet, children, as well as some adolescent chronic abdominal pains remain complicated to diagnose or treat. Abdominal pain accounts for over 2.5 million visits to primary care providers, pediatricians, and other office-based physicians per year. This makes abdominal pain one of the most common chief complaints to various medical practices.

The cost, in medical expenses, and the missed days of school secondary to abdominal pain remain excessive. Recurrent abdominal pain (RAP) described in John Paley's study in 1958 is defined as continuous pain of more than 2 weeks duration [1]. To be defined as having chronic abdominal pain the child must meet a criterion of at least three pain episodes over 3 months that interfere with function [2]. Most children, who present with chronic abdominal pain, do not have a clear organic cause to their pain. In fact, their pain is frequently functional in nature.

D.A. Sparks, D.O. (✉)
Pacific Pain Free Professionals,
Seattle, WA98121
e-mail: sparkycare@gmail.com

M.P. Garin-Laflam, M.D.
Pediatric Gastroenterology, Virginia Tech-Carilion School of Medicine, Carilion Clinic Children's Hospital,
102 Highland Avenue, Suite 305, Roanoke, VA 24013, USA
e-mail: mgarinlaflam@icloud.com

J.P. Cravero, M.D.
Anesthesiology, Perioperative and Pain Medicine,
Boston Children's Hospital, D300 Longwood Ave,
Boston, MA 02115, USA
e-mail: Joseph.Cravero@childrens.harvard.edu

Incidence

Chronic abdominal pain occurs in about 9–15 % of all children. One study suggested that approximately 13% of middle-school students and 17 % of high-school students experience weekly abdominal pain [3]. Girls have a bimodal distribution in which chronic abdominal pain is most common between the ages of 5–6 and 10–11. On the other hand, boys tend to develop chronic abdominal pain most commonly between the ages of 5 and 6 [4]. Up to 46 % of school-age children experience recurrent abdominal pain during their childhood.

Etiology

Etiology of chronic abdominal pain in the pediatric population may be difficult to elucidate. This chapter will also use the term "functional" with reference to abdominal pain which lacks a clear organic etiology. About 20 % [5] of school-age children have abdominal pain. . It is important to provide parents and patients with data on the frequency and generally benign nature of abdominal pain in the age group in order to reassure them and allow appropriate treatment.

Pathophysiology

Functional abdominal pain syndrome (FAPS) is pain that persists for greater than 6 months without evidence of physiologic disease, shows no relationship to physiologic events, and interferes with daily functioning. This condition is poorly understood but certainly seems to involve altered nociceptive pathways. This hyperalgesic state can be caused, or enhanced by multiple factors. These factors include: cognitive or psychological issues, and the chronic pain itself. The psychological factors can vary between individuals and may be directly related to coping mechanisms.

L. Kapural (ed.), *Chronic Abdominal Pain: An Evidence-Based, Comprehensive Guide to Clinical Management*,
DOI 10.1007/978-1-4939-1992-5_11, © Springer Science+Business Media New York 2015

Underlying depression may be expressed through abdominal symptoms. In stressful situations, patients may have potential secondary gains achieved through the expression of persistent pain. This internal system may result in external perception of pain even after any stimulus has dissipated (Box 11.1).

Chronic abdominal pain can furthermore be divided into three types: visceral pain, somatosensory pain, and referred pain [6]. Visceral pain is most likely described as dull pain which is poorly localized, and can originate in the midline. The unmyelinated nerve endings that are the source of the pain have a low threshold for firing in the face of stretching, ischemia, or intra-abdominal inflammatory causes. Different structures with similar embryonic origin cause pain in various areas of the abdomen. For example, lower esophageal and gastric pain originates from the foregut structures and the pain is perceived in the epigastrium. Visceral pain also occurs when noxious stimuli affect a viscus, such as the stomach or intestines. Tension, stretching, and ischemia stimulate visceral pain fibers. Tissue congestion and inflammation tend to sensitize nerve endings, and lower the threshold for stimuli [5]. Visceral pain-afferent fibers transmit a nonspecific pain pattern along both sympathetic and parasympathetic pathways. Parietal pain results from ischemia, inflammation, or stretching of the parietal peritoneum. This pain occurs through the myelinated afferent fibers that correspond to the specific dorsal root ganglia on the same side [5]. This type of pain is usually sharp and intense and likely to be intensified by movement. The corresponding dermatomal levels are consistent with the origin of the pain.

Box 11.1. Key Terms

**Recurrent abdominal pain (RAP)* is defined as pain for more than 2 weeks duration and a child must meet a certain criterion which includes: at least three pain episodes over at least 3 months in which the pain interferes with function.

**Functional abdominal pain (FAP)* is defined as pain with poor relation to gut function and decreased activities of daily living.

**Functional abdominal pain syndrome (FAPS)* is defined as pain that persists for greater than 6 months without evidence of physiologic disease, and shows no relationship to physiologic events and also interferes with daily functioning.

**Functional gastrointestinal disorders (FGIDs)* is defined by the ROME III criteria to include: functional dyspepsia, functional abdominal pain, abdominal migraine, and irritable bowel syndrome.

Referred pain is due to the segmental distribution of spinal nerves, and is an example of a convergence of various fibers. In chronic abdominal pain, the origin is mainly visceral. This type of pain can be referred to remote areas of the body if visceral impulses enter the spinal cord at the same level as afferent nerves from another area [6]. It results from shared central pathways of the afferent neurons leading to different sites. A classic example of the phenomena is a patient with pneumonia who presents with abdominal pain because the T9 dermatome distribution is shared by the lung and the abdomen [6].

Diagnosis

The criteria used to define childhood abdominal pain has expanded since Apley and Naish's seminal paper in 1958. The Rome II criteria developed in 1999, an effort organized by the Rome Coordinating Committee, defined childhood functional gastrointestinal disorders (FGIDs) to include: functional dyspepsia, functional abdominal pain, abdominal migraine, and irritable bowel syndrome. Subsequently in 2006, Rome III criteria were established, expanding abdominal pain-related functional gastrointestinal disorders to include not only functional abdominal pain, but also functional abdominal pain syndrome. The criteria's purpose was to decrease cost for the patients, as well as minimize suffering by encouraging a defined-criteria diagnosis of the disorder instead of obtaining one by exclusion [7]. The term "Recurrent Abdominal Pain" is a vague description of symptoms, rather than objective diagnosis. Furthermore, the term "functional" can be misinterpreted as pejorative because it seems to underscore the lack of any organic etiology (See Table 11.1). By redefining FGIDs in childhood, the authors of these definitions hope to promote the concept that a positive

Table 11.1 FAP symptoms

Symptoms	Diagnosis
Umbilical pain	Organic problem the further, the pain originates from umbilicus
Early satiety, nausea, sour breath, belching	Peptic origin
Crampy pain and/or bloating and/or intestinal gas related to meals, as well as pain with dairy	Lactose intolerance, giardiasis
Cough, wheezing, laryngitis, pain supine	Gastroesophageal reflux
Irregular bowel movements, encopresis, mass in the left lower abdominal quadrant and abdominal distension	Constipation
Blood in stool	Inflammatory bowel disease
Bulimia behavior with or without weight loss	Gastroesophageal reflux from an eating disorder
Reference [2]	

Box 11.2. ROME III [70]

ROME III classification of childhood functional abdominal pain disorders

H1. Vomiting and aerophagia
  H1a. Adolescent rumination syndrome
  H1b. Cyclic vomiting syndrome
  H1c. Aerophagia
H2. Abdominal pain-related FGIDs
  H2a. Functional dyspepsia
  H2b. Irritable bowel syndrome
  H2c. Abdominal migraine
  H2d. Childhood functional abdominal pain
H3. Constipation and incontinence
  H3a. Functional constipation
  H3b. Nonretentive fecal incontinence

diagnosis can be reached in these cases. It has been also suggested that such reclassification allows for increased frequency of labeling children with FGIDs [8] (Box 11.2).

Keeping in mind not all abdominal pain is gastrointestinal in origin; establishing differential diagnosis of abdominal pain begins with good history and physical exam. The first step should identify any "red flags" that could suggest organic disease. Many clinicians, in the setting of chronic abdominal pain, will consider a baseline workup to include: CBC with differential, sedimentation rate, urine analysis, pancreatic enzymes, and fecal screening for ova and parasites.

The utility of imaging would depend on patients' symptomatology. Abdominal ultrasound is a noninvasive test and does not expose the child to unnecessary radiation. The frequency of identification of abnormalities is as low as 1 % in children without "red flag" symptoms [5]. Still, many centers rely on ultrasound diagnosis to rule out possible organic cause of abdominal pain. Ultrasonographic examination of the abdomen is noninvasive and inexpensive test which is relatively painless for the pediatric patient. It can be utilized to detect irregularities of the kidneys, gallbladder, liver, pancreas, appendix, intestines, ovaries, and uterus. When symptoms and signs are present; such as jaundice, back pain, flank pain, vomiting, and/or abnormal physical exam findings, the probability of identifying organic abdominal abnormalities is increased to 11 %. In addition, imaging may have additional benefits as it can provide reassurance for both patients and families that some catastrophic event is not likely. In each case the potential for cost containment should be balanced against the risk that an incidental finding may cause more undue concern.

An example of a clinical entity where X-ray examination may be helpful is that of constipation. The contribution of constipation to chronic abdominal pain is often underappreciated. Abdominal pain caused by constipation is frequently left-sided, or suprapubic. A low-residue diet is the top cause of functional constipation, especially when it is greater than 3 months in duration. In some children whose habitus may prevent optimal abdominal evaluation, it may be reasonable to consider abdominal X-ray, if clinical history for constipation is highly suggestive.

Once the diagnosis of functional gastrointestinal disorder is made, it can be further categorized by the new Rome III criteria [9]. Into Functional Dyspepsia, Irritable Bowel Syndrome, Abdominal Migraine, Chronic Functional Abdominal Pain, and Chronic Functional Abdominal Pain Syndrome.

Functional Dyspepsia

To fulfill criteria for the functional dyspepsia symptoms listed below must occur at least once per week for at least 2 months. Those include:

1. Persistent/recurrent pain or discomfort centered in upper abdomen (above the umbilicus).
2. Pain not relieved by defecation, or associated with changes in stool frequency, or stool form.
3. No evidence of inflammatory, anatomic, metabolic, or neoplastic process to explain such symptoms.

Upper endoscopy is no longer mandatory to establish diagnosis of functional dyspepsia. A recent publication describes the long-term outcomes in pediatric patients who underwent endoscopy and continue with dyspeptic symptoms and those with presence or absence of reflux esophagitis symptoms persisted for similar periods of time [10]. This prospective cohort study also demonstrated a strong association between pediatric dyspepsia and anxiety. About half of these patients studied had a lifetime history of one or more anxiety disorders [10].

Irritable Bowel Syndrome

Irritable Bowel Syndrome (IBS) is defined as an abdominal discomfort or pain associated with two or more of the following occurring at least 25 % of the time:

1. Improvement with defecation
2. Onset associated with change in frequency of stool
3. Onset associated with change in form (appearance) of stool

Patients should not have any evidence of inflammatory, anatomic, metabolic, or neoplastic process to explain symptoms and such symptoms should be present once a week for at least 2 months.

Serious symptoms, like rectal bleeding, involuntary weight loss, growth retardation, and unexplained fevers necessitate further evaluation before consideration of functional gastrointestinal disease FGIDs such as IBS. Physical exam and history may assist in such determination. There can also be several other disease states that can mimic symptoms of IBS including lactose intolerance, sucrase-isomaltase deficiency, celiac disease, small bowel bacterial overgrowth, microscopic colitis, and bile acid malabsorption.

Post-infectious IBS is believed to be due to mild inflammation with subsequent visceral hypersensitivity that continues after infectious process has abated [11, 12]. It is therefore believed that childhood conditions associated with intestinal inflammation including cow's milk allergy, and even recovery from pyloric stenosis may contribute to increased probability of developing childhood FGIDs including FAP and IBS [13, 14].

Abdominal Migraine

Abdominal migraine (AM) is defined as

1. Pain severe enough to interfere with normal daily activities.
2. Pain dull, or colicky in nature.
3. Periumbilical, or poorly localized pain.
4. Any two of anorexia, nausea, vomiting, headache, photophobia, or pallor.
5. Attacks lasting for at least 1 h.
6. Complete resolution of symptoms between attacks.

Children with migraine headaches are twice as likely to develop abdominal migraines, and children with abdominal migraines were twice as likely to have migraine headaches compared to general pediatric population [15].

In order to establish diagnosis of AM patients must have at least two or more of the following conditions over the preceding 12 months and are characterized by:

1. Paroxysmal episodes of intense, acute periumbilical pain that lasts for 1 h or more.
2. Intervening periods of usual health lasting weeks to months.
3. Pain that interferes with normal activities.
4. Pain associated with two of the following:
   - 4.1 Anorexia
   - 4.2 Nausea
   - 4.3 Vomiting
   - 4.4 Headache
   - 4.5 Photophobia
   - 4.6 Pallor
   - 4.7 In addition, there should be no evidence of an inflammatory, anatomic, metabolic, or neoplastic process to explain above symptoms.

Abdominal migraines are present in 1–4 % of children, and more commonly in girls [16, 17]. Age of presentation is between 7 and 12 years of age. In one study investigators found that, in spite of the fact the patient may have met the diagnosis based on above criteria, in at least 4 % of patients presenting with chronic abdominal pain, none of the patients received the diagnosis of AM [18]. It is particularly important to recognize these cardinal features as well as the prevalence given for the potential abortive and preventative migraine-specific therapies [18].

Childhood Functional Abdominal Pain

Childhood functional abdominal pain is suspected if episodic or continuous pain is present once or twice a week for at least 2 months and there are insufficient criteria for other FGIDs, there should be no evidence of inflammatory, anatomic, metabolic, or neoplastic processes to explain such symptoms.

There is some degree of controversy concerning classification when children with FAP/IBS demonstrate low-grade inflammation. In 2008, Shulman published a prospective study of 65 children with FAP/IBS and compared to 39 age- and gender-matched controls. Their results suggested that proximal GI and colonic permeability were increased in affected patients. Furthermore, the same study proposed that the frequency low-grade inflammation in FAP/IBS patients was increased as measured rising fecal calprotectin levels [19]. This is in contrast to the previously published Norwegian study in which 90 % of pediatric patients with FGIDs had normal levels of fecal calprotectin.

Childhood Functional Abdominal Pain Syndrome

This new diagnostic category enables the inclusion of children with recurrent abdominal pain that does not impair the ability to maintain regular activities under the traditional diagnosis of functional abdominal pain. This syndrome includes both the impairment of some daily routine as well as extraintestinal symptoms.

Symptoms are present at least once a week for duration of at least 2 months and must satisfy the above listed criteria for childhood functional abdominal pain. In addition patients must have one or more of the following at least 25 % of the time:

1. Some loss of daily functioning.
2. Additional somatic symptoms including headache, limb pain, or difficulty sleeping.

Prospective studies with incorporation of Rome III criteria will make possible to gauge whether this distinction equates with significant changes in treatment strategies, and long-term prognosis [8].

Treatment

There is no clear, effective, evidence-based algorithm for treatment of abdominal pain. Currently, a multimodal, non-invasive treatment strategy is recommended. In a small percentage of cases, patients may benefit from an interventional or surgical intervention.

The goal of any therapy for functional abdominal pain is to reduce stress and alleviate tension for both the child and the parents. At the same time the therapy should promote normal patterns of physical activity, social interaction, and school attendance. The therapy must involve the patient, parents, pediatrician, gastroenterologist, psychologist, social workers, and teachers. It should be noted and reinforced that the pain the patient is experiencing is *real* pain; however, treatment goals should be aimed at minimizing disruption of daily activities due to abdominal complaints rather than treating a specific source of the pain. The overall focus should be on managing pain rather than eliminating pain as the target goal.

Families need to be cognizant that up to one-third of children with FGIDs may continue with symptoms in 5 years [20]. They need to feel comfortable validating the child's abdominal complaints, without encouraging or reinforcing symptoms. For some children, discounting their symptoms leads to exacerbation of physical pain. An alliance with the child's school nurse can impact ability to encourage school attendance. Specifically, familiarity with cognitive-behavioral therapy and guided imagery techniques, can allow for implementation of coping strategies for children with chronic abdominal pain.

In addition, arranging for a consultation with a child psychologist is a vital part to the treatment of functional abdominal pain. Psychological support is a valuable tool for the entire team in order to assist the child to cope with chronic pain. The goal of treatment is not the total elimination of symptoms, but rather the acquisition of strategies for coping with the pain and getting on with their life [21]. For children who miss school because of their symptoms, going back to school is a prime objective [22]. It is crucial to convey to the parents and the child that the pain is not due to organic causes. The family should be informed that stress may trigger symptoms. Taking time to explain to children (especially teens) about visceral hypersensitivity will help them to understand the physiology behind their pain. Patients should be given the details about the stretching of the bowel wall in relation to the child's low-pain threshold. At the same time it should be reinforced that this does not mean that abdominal pain is not a real problem (or imply that the pain is "just in your head") [21]. Often the explanation itself usually leads to a marked improvement of symptoms [23].

Cognitive-behavioral strategies have an established role in the treatment of children with anxiety and depression disorders. This seems appropriate given that cognitive-behavioral therapy is the treatment best supported by presently available evidence and is considered "probably efficacious" according to the widely established empirically supported treatment criteria [22, 24]. Due to the previously discussed observation of coexistence of anxiety and depressive disorders, such non-pharmacologic approaches are recommended as first-line therapy [8, 25, 26]. Two RCTs [27, 28] evaluated the efficacy of a cognitive-behavioral program and a cognitive-behavioral family intervention for the treatment of nonspecific abdominal pain. In the first study, results showed that both the experimental and the control groups had decreased levels of pain. However, the treated group improved more quickly, the effects generalized to the school setting, and a larger proportion of subjects were completely pain-free by 3 months of follow-up. In the second study, the children and mothers who were taught coping skills had a higher rate of complete elimination of pain, lower levels of relapse at 6 and 12 months' follow-up, and lower levels of interference with their activities as a result of pain, and parents reported a higher level of satisfaction with the treatment. After controlling for pretreatment levels of pain, children's active self-coping and mothers' care giving strategies were significant independent predictors of pain behavior after treatment.

Multiple studies also indicate a supportive role for hypnotherapy [29–31]. Studies including hypnotherapy tend to show patients having decreased anxiety and improved activities of daily living. Dietary changes including increasing daily fiber are often recommended. Studies including adult experience are inconclusive. As it is inexpensive, it may be reasonable to consider. If added fiber therapy is chosen, the target goal for fiber (in grams per day) is the child's age + 5—up to 30 g/day. Parents are advised that the introduction of fiber needs to be gradual, over several weeks.

If there are specific food triggers for pain that are identified, selective avoidance is indicated. Many families believe that food antigens are contributing to their child's abdominal pain, as they perceive the pain is triggered by eating. Current evidence suggests that food allergies represent less than 5 % of children with FGIDs [32]. As such extensive allergy testing is unlikely to help with diagnosis or treatment of function chronic abdominal pain.

The use of probiotics has been shown to be beneficial in children with IBS [33–35]. Conversely studies have failed to find an association between pediatric functional abdominal pain and lactose intolerance [36]. Elimination of lactose from the diet is unlikely to improve chronic functional pain symptoms.

Pharmacological therapy for treatment of FGIDs has included the use of low-dose tricyclic antidepressants (TCAs);

however, their utility is uncertain. A Cochrane review examined improvement of functional abdominal pain in children and adolescents and found that improvement did not differ substantially between those who did or did not receive medical therapy. The review was based on two small RCTs [37, 38]. There has since been a retrospective study of 98 patients in which there seemed to be an improvement in abdominal complaints with use of low-dose TCAs. On the other hand, Saps [38] published a study that included a placebo-controlled arm and did not demonstrate a difference between the treatment and control group.

Support for the possible utility of TCA's stems from several investigations have reported higher levels of life stress in children with chronic abdominal pain compared with children without abdominal pain. Two studies compared pediatric abdominal pain patients with healthy school-age children and found significantly higher levels of life-event stress in patients with pain [39, 40] A separate diary study found that patients with recurrent episodes of abdominal pain reported significantly more daily stressors than school children without abdominal pain; moreover, the relation between daily stressors and somatic complaints was significantly stronger for patients with abdominal pain than for healthy school children [41].

Serotoninergic agents were studied via a small randomized-controlled trial of pizotifen in children with abdominal migraine. While the authors were able to demonstrate improvement in treatment versus placebo group, study size was limited to 14 patients, and this medication is not currently approved in the United States [42]. Another study evaluated the selective serotonin reuptake inhibitor Citalopram. Authors conducted a small ($N=25$) 12-week open-label trial. The medication was well tolerated, but larger randomized controlled trials need to be considered [43].

In rare cases diagnostic laparoscopy may be indicated. Sringel et al. reported a case series of 13 children with chronic severe episodes of abdominal pain who were subjected to diagnostic laparoscopy. Laparoscopic findings identified the cause of abdominal pain in 12 of 13 patients. Laparoscopic appendectomy was performed in all patients. Abdominal pain resolved in ten patients. These authors concluded that diagnostic laparoscopy is a beneficial procedure in the management of some children with chronic recurrent abdominal pain resistant to other treatments [44].

When the child receives a diagnosis of a functional abdominal pain disorder, the rationale behind this diagnosis must be made evident. If a child continues to have abdominal pain, and organic causes have been excluded the physician should discuss cognitive-behavioral therapy with the patient and their family. Next, all treatment options must be discussed and appropriate time allowed for questions and answers. The family needs to understand that the symptoms support the diagnosis for the criteria pointing to the functional condition. All supporting data from the physical exam, laboratory results, and studies should be made readily available to reassure the patient and family. Any additional medical information that relates to the diagnosis should be provided to the family.

Perioperative Abdominal Pain in Children

Postoperative abdominal pain is variable in intensity and presentation depending on the nature of the operation, the exact location of the procedure, the techniques used for surgery, and the underlying health/psychological status of the child. Open abdominal surgeries and urological procedures present considerable challenges in terms of pain management. There is an abundance of descriptive studies that address pain control after abdominal surgery. A multitude of techniques including opioid infusions, patient-controlled analgesia, nurse-controlled analgesia, epidural infusions, and regional nerve blocks have been reported. Unfortunately there are very few well-designed clinical trials of these techniques. The procedures involved vary widely from primarily esophageal procedures such as fundoplications, to renal procedures (pyeloplasties, etc.), to primary gastrointestinal procedures. While the current trend to perform many of these surgeries with laparoscopic techniques has served to decrease the total burden of perioperative pain, attention to abdominal discomfort is still required.

Systemic Agents

Unlike abdominal pain stemming from medical or psychological causes, in most cases the origin of abdominal pain in the postoperative patient is obvious. Open surgical procedures of the abdomen should be considered in terms of the size and location of the abdominal wall incision and the muscle/tissue interruption that has occurred. Pain after these procedures is significant and requires the use of potent systemic analgesics. Multiple techniques for managing pain after open abdominal procedures are possible. There are many delivery methods for systemic opiates including intermittent boluses, patient-controlled analgesia, nurse-controlled analgesia, and continuous opiate infusions. Ben-Meir et al. reported on postoperative pain management of post-pyeloplasty patients with systemic opiates and non-opiate oral agents vs. epidural analgesia [45]. Both strategies offered effective pain control. No difference was found in the amount of rescue pain medicine required or overall pain assessment between the two methodologies. Multiple studies have documented that opiates can effectively be delivered by patient-controlled analgesia (PCA) in children 5 years of age and older [46, 47]. The use of bolus dose alone vs. bolus dose with background infusion does not appear to greatly affect the degree of pain control after abdominal procedures. Furthermore, Monitto et al. reported on the safety and efficacy of parent/nurse-controlled analgesia with opioids [48]. In their observational report, postoperative analgesia was as effective as that provided by PCA in older patients but respiratory depression was noted more commonly—in 1.7 % of subjects.

The addition of non-steroidal anti-inflammatory drugs (NSAIDs) to opiate therapy for analgesia after abdominal surgery has been shown to significantly reduce the overall pain experienced and the amount of opiate consumed during the postoperative period by 20–30 % [49].

Neuraxial Analgesia

Regional caudal/epidural blocks can be performed with "single shot" neuraxial administration of local anesthetics, continuous infusions of epidural anesthetics, and patient-controlled epidural analgesia (PCEA). In addition, ultrasound-guided nerve blocks such as transverses abdominus plane blocks, ilioinguinal–iliohypogastric nerve blocks, and rectus sheath blocks can also be employed. As a general rule, analgesia can be best offered by blocking the spinal nerves in the dermatomes that are most directly involved with the sensation in that region. Historically, this has been undertaken with epidural anesthesia. An example of effective use of epidural analgesia includes the Nissen Fundoplication procedure. In this case, epidural pain control aimed at the level of the incision has been shown to decrease perioperative complications, improve respiratory performance postoperatively, and shorten hospital stays [50, 51]. Epidural analgesia has been effectively utilized in all age groups including neonates [52].

Epidural infusions consisting of local anesthesia plus opioid have been reported as highly effective in providing pain control after abdominal surgery. Common opiate additives include morphine, hydromorphone, and fentanyl. Side effects of extradural opiate infusions such as nausea, vomiting, and pruritus, are related to drug dose and tend to be more common with hydrophilic opiates [53]. Clonidine has also been shown to be an effective additive for abdominal analgesia in postoperative patients. It has been associated with dose-dependent sedation and hypotension; however, without the opiate-related side effects [54].

Patient-controlled epidural anesthesia (PCEA) has been shown to be effective in pediatric aged patients [55]. Somri et al. studied 128 postsurgical patients (predominantly abdominal procedures) and found 90.1 % achieved satisfactory analgesia for up to 103 h with no episodes of desaturation. There were no other clinical signs of toxicity. Children as young as 5 years old were found to have the cognitive ability to appropriately use the PCEA. The authors warn that total hourly dose should be carefully checked to prevent local anesthetic overdose.

Abdominal Wall Blocks

Abdominal wall nerve blocks are effective in ameliorating the discomfort that stems from the tissue disruption and muscle dissection associated with superficial procedures involving the abdominal wall. These blocks are not capable of eliminating the visceral pain that accompanies peritoneal traction or distention of abdominal viscera. Caudal or epidural anesthesia are associated with lower levels of catecholamine response after major abdominal or inguinal procedures [56]. On the other hand, there are many circumstances under which epidural anesthesia, is contraindicated or impossible due to anatomical considerations or coagulation issues. Furthermore, abdominal wall analgesia can be preferable in situations such as ambulatory surgery where delayed ambulate on or urinary retention could be particularly problematic.

Rectus Sheath Block

This block can be used to effectively manage pain from a number of surgeries that involve the midline of the abdominal wall. This block is growing in popularity for procedures such as epigastric hernia repairs, umbilical hernia repairs, and laparoscopic surgery [57, 58]. The growth of the popularity of this block is almost certainly due to the availability of ultrasound as an adjunct for placing the block. This ultrasound can be used with one of two techniques—an out of plane approach that approaches from the top of the posterior rectus sheath under real-time visualization [57] or as an in-plane approach with the probe as close as possible to the lateral edge of the rectus muscle and advanced slowly between the internal oblique and the transversus muscles [59, 60]. Both blocks allow block of the terminal branches of the 9th, 10th, and 11th intercostal nerves. This block improves operative anesthesia and provides effective postoperative pain control.

Transverse Abdominal Plane (TAP) Block

This block has become widely popular for pain control after abdominal surgery. The block is intended to deposit local anesthesia between the transverse abdominal and internal oblique abdominal muscles with a single injection. In so doing, this maneuver should block the segmental nerves T9, 10, 11, 12, and L1. Once again this block has been made infinitely safer and simpler to perform with the advent of ultrasound guidance. The technique is slightly different for adults and children. As described by Hebbard in 2007, with the ultrasound probe in the midaxillary line the muscles of the abdominal wall are identified and local anesthetic is deposited between the internal oblique muscle and the transversus

Table 11.2 Cognitive therapeutic methods for the treatment of chronic FAP

Element	Examples	Goal
Psychoeducation	Education about the causes of abdominal pain	Promoting patient cooperation and supporting self-responsibility
	Teaching of coping strategies	
Relaxation	Progressive muscle relaxation	Reduction of tension due to pain and creation of a relaxed state
	Autogenic training	
Cognitive Techniques	Distraction techniques	Learn to deal with the pain with a positive attitude

References [21, 68]

abdominus muscle [61]. More recently Suresh and colleagues described the use of ultrasound for TAP blocks in infants and children using a slightly different technique [62]. These authors recommend identifying the rectus sheath and rectus abdominis muscle immediately lateral to the umbilicus. The probe is then slid laterally until the lateral border of the latissimus dorsi can be visualized. At this point the origin of the transversus abdominis can be easily identified and local anesthetic deposited. This allows the operator to deposit local closer to the origin of the thoracolumbar roots and allows greater spread of local anesthetic. This approach is described in next chapter by the authors.

Unfortunately there are no prospective studies evaluating the relative effectiveness of the TAP block in children after abdominal surgery although there are studies in adults that report variable effectiveness [61]. Anecdotal reports support its use and efficacy although lack of evidence does not allow firm recommendations about its use at this time.

Laparoscopic Surgery

Laparoscopic surgery is associated with less overall pain than open procedures that are aimed at the same surgical result. The duration of postoperative pain is decreased [63], but the amount of pain experienced on the first day postoperatively is often similar to open procedures [64, 65]. There is almost no comparative literature on the options for postoperative pain control in this setting but, in general, a multimodal pain management is recommended. Strategies that include local anesthesia infiltration of port sites, intravenous or oral opioids, non-steroidal anti-inflammatory drugs (NSAIDs), and acetaminophen are acceptable. Epidural infusions are not required after these types of surgeries.

Summary

Abdominal pain in a child or adolescent typically affects the entire family. The physician and other caregivers need to be trusted by the patient and family to provide a supportive environment. The patient must understand and receive the same assurances from the physician and family members that his/her pain issues are important. Family participation may also be needed to collect information to help accurately diagnose the pain. The well-informed family plays a significant role in the entire process through diagnosis that will assist the patient [66].

The physician must rule out significant pathology first and foremost in the child with abdominal pain. Red flags include (but are not limited to) weight loss, deceleration of linear growth velocity, significant vomiting, chronic severe diarrhea, evidence of gastrointestinal blood loss, persistent right upper or right lower quadrant pain, unexplained fever, family history of inflammatory bowel disease, or abnormal or unexplained physical findings [67].

Children with recurrent abdominal pain may be predisposed to certain psychosocial pathology. While no evidence-based studies have implicated parents in the causality of RAP, these parents on average tend to have a higher likelihood of psychological issues. However, families of children with chronic abdominal pain do not seem to differ from families of children without abdominal pain or families of patients with acute illness in broad areas of family functioning [67].

Chronic abdominal pain remains a diagnostic dilemma and treatment is still evolving. A multimodal approach that includes cognitive-behavioral therapy may be useful in improving pain and disability outcome in the short term [67]. Continued research and varying treatment strategies will allow for quicker diagnosis and enhanced treatment plans (Table 11.2).

References

1. Apley J. The child with abdominal pains. 2nd ed. Oxford: Blackwell Scientific; 1975.
2. Lake A. Chronic abdominal pain in childhood: diagnosis and management. Am Fam Physician. 1999;59:1823–30.
3. Chacko MR. Chronic abdominal pain in children and adolescents: beyond the basics. UpToDate [article]. 2011. http://www.uptodate.com/contents/patient-information-chronic-abdominal-pain-in-children-and-adolescents-beyond-the-basics.

4. Hyam JS, Burke G, Davis PM, Rzepski B, Andrulonis PA. Abdominal pain and irritable bowel syndrome in adolescents: a community-based study. J Pediatr. 1996;129:220–6.
5. Leung AKC, Sigalet DL. Acute abdominal pain in children. Am Fam Physician. 2003;67(11):2321–7.
6. Warfield CA, Bajwa ZH. Principles and practices of pain medicine. 2004;34:350–2.
7. Pediatrics. 2005; 115(3);e370–81. doi:10.1542/peds2004-2523.
8. Baber KF, Anderson J, Puzanovova M, Walker LS. Rome II versus Rome III classification of functional gastrointestinal disorders in pediatric chronic abdominal pain. J Pediatr Gastroenterol Nutr. 2008;47:299–302.
9. Walker LS, Baker GJ, Smith CA. A typology of pain coping strategies in pediatric patients with chronic abdominal pain. Pain. 2008;137:266–75.
10. Rippel SW, Acra S, Correa H, Vaezi M, Di Lorenzo C, Walker LS. Pediatric patients with dyspepsia have chronic symptoms, anxiety, and lower quality of life as adolescents and adults. Gastroenterology. 2012;142:754–61.
11. Dupont AW. Post-infectious irritable bowel syndrome. Curr Gastroenterol Rep. 2007;9:378–84.
12. Spiller RC, Jenkins D, Thornley JP, Hebden JM, Wright T, Skinner M. Increased rectal mucosal enteroendocrine cells, T lymphocytes, and increased gut permeability following acute Campylobacter enteritis and in post-dysenteric irritable bowel syndrome. Gut. 2000;47:804–11.
13. Saps M, Lu P, Bonilla S. Cow's milk allergy is a risk factor for the development of FGIDs in children. J Pediatr Gastroenterol Nutr. 2011;52:166–9.
14. Saps M, Bonilla S. Early life events: infants with pyloric stenosis have a higher risk of developing chronic abdominal pain in childhood. J Pediatr. 2011;159:551–4. e1.
15. http://www.uptodate.com/contents/evaluation-of-the-child-and-adolescent-with-chronic-abdominal-pain?source=search_result&search=abdominal+pain&selectedTitle=4%7E150.
16. Mortimer MJ, Kay J, Jaron A. Clinical epidemiology of childhood abdominal migraine in an urban general practice. Dev Med Child Neurol. 1993;35:243–8.
17. Abu-Arafeh I, Russell G. Prevalence and clinical features of abdominal migraine compared with those of migraine headache. Arch Dis Child. 1995;72:413–7.
18. Carson L, Lewis D, Tsou M, McGuire E, Surran B, Miller C. Abdominal migraine: an under-diagnosed cause of recurrent abdominal pain in children. Headache. 2011;11(51):707–12.
19. Shulman R, Eakin M, Czyewski D, Jarrett M, Ou C. Increased gastrointestinal permeability and Gut inflammation in children with functional abdominal pain and irritable bowel syndrome. J Pediatr. 2008;153:646–50.
20. Youssef NN, Murphy TG, Schuckalo S, Intile C, Rosh J. School, nurse knowledge and perceptions of recurrent abdominal pain: opportunity for therapeutic alliance? Clin Pediatr. 2007; 46:340–4.
21. Bufler P, Gross M, Uhlig HH. Recurrent abdominal pain in childhood. Dtsch Arztebl Int. 2011;108(17):295–304. doi:10.3238/arztebl.2011.0295.
22. Di Lorenzo C, Colletti RB, Lehmann HP. Chronic abdominal pain in children: a technical report of the American Academy of Pediatrics and the North American society for pediatric gastroenterology, hepatology and nutrition. J Pediatr Gastroenterol Nutr. 2005;40:249–61.
23. Berger MY, Gieteling MJ, Benninga MA. Chronic abdominal pain in children. BMJ. 2007;334:997–1002.
24. Janicke DM, Finney JW. Empirically supported treatments in pediatric psychology: recurrent abdominal pain. J Pediatr Psychol. 1999;24:115–27.
25. Campo JV, Bridge J, Ehmann M, Altman S, Lucas A, Birmaher B. Recurrent abdominal pain, anxiety, and depression in primary care. Pediatrics. 2004;113:817–24.
26. Ramchandani PG, Fazel M, Stein A, Wiles N, Hotopf M. The impact of recurrent abdominal pain: predictors of outcome in a large population cohort. Acta Paediatr. 2007;96:697–701.
27. Sanders MR, Rebgetz M, Morrison M. Cognitive-behavioral treatment of recurrent nonspecific abdominal pain in children: an analysis of generalization, maintenance, and side effects. J Consult Clin Psychol. 1989;57:294–300.
28. Sanders MR, Shepherd RW, Cleghorn G, Woolford H. The treatment of recurrent abdominal pain in children: a controlled comparison of cognitive-behavioral family intervention and standard pediatric care. J Consult Clin Psychol. 1994;62:306–14.
29. Anbar RD. Self-hypnosis for the treatment of functional abdominal pain in childhood. Clin Pediatr. 2001;40:447–51.
30. Vlieger AM, Menko Frankenhuis C, Wolfkamp SCS, Tromp E, Benninga MA. Hypnotherapy for children with functional abdominal pain or irritable bowel syndrome: a randomized controlled trial. Gastroenterology. 2007;133:1430–6.
31. Vlieger AM, Rutten JMTM, Govers AMAP, Frankenhuis C, Benninga MA. Long-term follow-up of gut-directed hypnotherapy vs standard care in children with functional abdominal pain or irritable bowel syndrome. Am J Gastroenterol. 2012;107:627–31.
32. Kokkonen J, Haapalahti M, Tikkanen S, Karttunen R, Savilahti E. Gastrointestinal complaints and diagnosis in children: a population-based study. Acta Paediatr. 2007;93:880–6.
33. Horvath A, Dziechciarz P, Szajewska H. Meta-analysis: Lactobacillus rhamnosus GG for abdominal pain-related functional gastrointestinal disorders in childhood. Aliment Pharmacol Ther. 2011;33:1302–10.
34. Gawronska Awrońska A, Dziechciarz P, Horvath A, Szajewska H. A randomized double-blind placebo-controlled trial of Lactobacillus GG for abdominal pain disorders in children. Aliment Pharmacol Ther. 2006;25:177–84.
35. Bausserman M, Michail S. The use of lactobacillus GG in irritable bowel syndrome in children: a double-blind randomized control trial. J Pediatr. 2005;147:197–201.
36. Weydert JA, Ball TM, Davis MF. Systematic review of treatments for recurrent abdominal pain. Pediatrics. 2003;111:e1–11.
37. Bahar RJ, Collins BS, Steinmetz B, Ament ME. Double-blind placebo-controlled trial of amitriptyline for the treatment of irritable bowel syndrome in adolescents. J Pediatr. 2008;152:685–9.
38. Saps M, Youssef N, Miranda A, Nurko S, Hyman P, Cocjin J. Multicenter, randomized, placebo-controlled trial of amitriptyline in children with functional gastrointestinal disorders. Gastroenterology. 2009;137:1261–9.
39. Hodges K, Kline JJ, Barbero G, Flanery R. Life events occurring in families of children with recurrent abdominal pain. J Psychosom Res. 1984;28:185–8.
40. Robinson JO, Alverez JH, Dodge JA. Life events and family history in children with recurrent abdominal pain. J Psychosom Res. 1990;34:171–81.
41. Kokkonen J, Ruuska T, Karttunen TJ, Niinimaki A. Mucosal pathology of the foregut associated with food allergy and recurrent abdominal pains in children. Acta Paediatr. 2001;90:16–21.
42. Symon DN, Russell G. Double blind placebo controlled trial of pizotifen syrup in the treatment of abdominal migraine. Arch Dis Child. 1995;72:48–50.
43. Teitelbaum JE, Arora R. Long-term efficacy of low-dose tricyclic antidepressants for children with functional gastrointestinal disorders. J Pediatr Gastroenterol Nutr. 2011;53:260–4.
44. Sringel G, Berezin SH, Bostwick HE, Halata MS. Laparoscopy in the management of children with chronic recurrent abdominal pain. JSLS. 1999;3:215–9.

45. Ben-Meir D, Livne PM, Katz J, Gelman O, Efrat R. Continuous epidural versus nonepidural analgesia for post-pyeloplasty pain in children. J Urol. 2009;182:1841–4.
46. Yildiz K, Tercan E, Dogru K, Ozkan U, Boyaci A. Comparison of patient-controlled analgesia with and without a background infusion after appendicectomy in children. Paediatr Anaesth. 2003;13: 427–31.
47. Peters JW, Bandell Hoekstra IE, Huijer Abu-Saad H, Bouwmeester J, Meursing AE, Tibboel D. Patient controlled analgesia in children and adolescents: a randomized controlled trial. Paediatr Anaesth. 1999;9:235–41.
48. Monitto CL, Greenberg RS, Kost-Byerly S, Wetzel R, Billett C, Lebet RM, Yaster M. The safety and efficacy of parent-/nurse-controlled analgesia in patients less than six years of age. Anesth Analg. 2000;91:573–9.
49. Morton NS, O'Brien K. Analgesic efficacy of paracetamol and diclofenac in children receiving PCA morphine. Br J Anaesth. 1999;82:715–7.
50. McNeely JK, Farber NE, Rusy LM, Hoffman GM. Epidural analgesia improves outcome following pediatric fundoplication. A retrospective analysis. Reg Anesth. 1997;22:16–23.
51. Wilson GA, Brown JL, Crabbe DG, Hinton W, McHugh PJ, Stringer MD. Is epidural analgesia associated with an improved outcome following open Nissen fundoplication? Paediatr Anaesth. 2001;11:65–70.
52. Bosenberg AT. Epidural analgesia for major neonatal surgery. Paediatr Anaesth. 1998;8:479–83.
53. Moriarty A. Postoperative extradural infusions in children: preliminary data from a comparison of bupivacaine/diamorphine with plain ropivacaine. Paediatr Anaesth. 1999;9:423–7.
54. Klamt JG, Garcia LV, Stocche RM, Meinberg AC. Epidural infusion of clonidine or clonidine plus ropivacaine for postoperative analgesia in children undergoing major abdominal surgery. J Clin Anesth. 2003;15:510–4.
55. Birmingham PK, Wheeler M, Suresh S, Dsida RM, Rae BR, Obrecht J, Andreoni VA, Hall SC, Cote CJ. Patient-controlled epidural analgesia in children: can they do it? Anesth Analg. 2003; 96:686–91.
56. Somri M, Gaitini LA, Vaida SJ, Yanovski B, Sabo E, Levy N, Greenberg A, Liscinsky S, Zinder O. Effect of ilioinguinal nerve block on the catecholamine plasma levels in orchidopexy: comparison with caudal epidural block. Paediatr Anaesth. 2002;12:791–7.
57. Willschke H, Bosenberg A, Marhofer P, Johnston S, Kettner SC, Wanzel O, Kapral S. Ultrasonography-guided rectus sheath block in paediatric anaesthesia—a new approach to an old technique. Br J Anaesth. 2006;97:244–9.
58. Azemati S, Khosravi MB. An assessment of the value of rectus sheath block for postlaparoscopic pain in gynecologic surgery. J Minim Invasive Gynecol. 2005;12:12–5.
59. De Jose MB, Gotzens V, Mabrok M. Ultrasound-guided umbilical nerve block in children: a brief description of a new approach. Paediatr Anaesth. 2007;17:44–50.
60. Willschke H, Kettner S. Pediatric regional anesthesia: abdominal wall blocks. Paediatr Anaesth. 2012;22:88–92.
61. Hebbard P, Fujiwara Y, Shibata Y, Royse C. Ultrasound-guided transversus abdominis plane (TAP) block. Anaesth Intensive Care. 2007;35:616–7.
62. Suresh S, Chan VW. Ultrasound guided transversus abdominis plane block in infants, children and adolescents: a simple procedural guidance for their performance. Paediatr Anaesth. 2009;19:296–9.
63. Charlton S, Cyna AM, Middleton P, Griffiths JD. Perioperative transversus abdominis plane (TAP) blocks for analgesia after abdominal surgery. Cochrane Database Syst Rev. 2010;12, CD007705.
64. Howard R, Carter B, Curry J, Morton N, Rivett K, Rose M, Tyreel J, Walker S, Williams G. Postoperative pain. Paediatr Anaesth. 2008;18(1):36–63.
65. Sekaran P, MacKinlay GA, Lam J. Comparative evaluation of laparoscopic versus open nephrectomy in children. Scott Med J. 2006;51:15–7.
66. Subcommittee on Chronic Abdominal Pain. Chronic Abdominal Pain in Children. Pediatrics. 2005;115(3):e370–81.
67. Rasquin A, Di Lorenzo C, Forbes D, Guiraldes E, Hyams JS, Staiano A, Walker LS. Childhood functional gastrointestinal disorders: child/adolescent. Gastroenterology. 2006;130:1527–37.
68. Ruob M. Wirksamkeit, Wirkfaktoren, psychologischer Schmerztherapie. Eine Übersicht Verhaltenstherapie. 1998;8:14–25.

12 Interventional Treatment of Chronic Abdominal Pain in Children

Ravi Shah and Santhanam Suresh

Introduction

Chronic abdominal pain is a frequently encountered problem in the pediatric population. Children with recurrent abdominal pain are at increased risk for developing chronic pain in adulthood; this may be due to mechanisms associated with heightened central sensitization [1, 2]. Pediatric chronic abdominal pain syndromes frequently involve a somatosensory component. Peripheral nerve blockade has become an important therapeutic measure for the management of such abdominal pain states. Serial peripheral nerve blocks and continuous peripheral nerve blocks (CPNBs) have been reported effective in facilitating physical therapy [3] and controlling pediatric chronic pain in both inpatient and outpatient settings [4]. Peripheral nerve blocks may be utilized as adjunctive therapy in children with functional abdominal pain or to treat chronic abdominal wall pain. These procedures are traditionally performed using anatomic landmarks; however, the evolution of ultrasound-guided techniques has led to safer practices with improved success rates [5]. This chapter focuses on indications, procedure description, and potential complications of peripheral nerve blocks for the treatment of pediatric chronic abdominal pain.

Regional blocks of the trunk may be used to provide analgesia in children with chronic abdominal and groin pain. The utilization of ultrasound guidance has made these techniques more popular and effective [6, 7]. Somatosensory components of such pain states are effectively blocked via the transversus abdominis plane, the ilioinguinal/iliohypogastric nerves, or the rectus sheath. Choice of technique is primarily based on the anatomical distribution of pain.

R. Shah, M.D. (✉) • S. Suresh, M.D., F.A.A.P.
Department of Pediatric Anesthesiology, Ann & Robert H. Lurie Children's Hospital of Chicago, 225 East Chicago Avenue, Box 19, Chicago, IL 60611, USA

Feinberg School of Medicine, Northwestern University, 420 East Superior Street, Chicago, IL 60611, USA
e-mail: rdshah@luriechildrens.org; ssuresh@luriechildrens.org

Transversus Abdominis Plane Block

Anatomy and Indications

The transversus abdominis plane (TAP) block has been described for the treatment of pediatric patients with refractory abdominal neuropathic pain [8]. Postsurgical anterior cutaneous nerve entrapment syndrome (ACNES) has also been successfully managed with serial TAP blocks [9]. US guidance allows visualization of the virtual space between the internal oblique and the transversus abdominis muscles where the thoracolumbar nerve roots (T8–L1) lie. An indwelling catheter may be left in the space for continuous analgesia. We have also performed serial TAP blocks in children with chronic abdominal wall pain with favorable results.

Three muscle layers lie lateral to rectus abdominis muscles: the external oblique, internal oblique, and transversus abdominis (Fig. 12.1). The thoracolumbar nerve roots (T8–L1) traverse the space between the internal oblique and transversus abdominis muscle. These nerves provide sensory innervation to the muscles and skin of the anterior abdominal wall.

Technique

Various techniques have been described to use an in-plane approach with ultrasound guidance that allows needle advancement and placement of local anesthetic into the TAP [10, 11]. A high-frequency linear probe is placed lateral to the umbilicus and moved laterally to demarcate the three muscle layers of the abdominal wall. The needle is advanced using an in-plane technique to the space between the internal oblique and the transversus abdominis. Injection, with incremental aspiration, will create an elliptical pocket of local anesthetic into the space where the nerves traverse.

Complications

TAP blocks are easily performed in the outpatient setting. Potential complications include intravascular injection and peritoneal and/or bowel puncture. These complications are minimized with the use of ultrasound guidance.

L. Kapural (ed.), *Chronic Abdominal Pain: An Evidence-Based, Comprehensive Guide to Clinical Management*,
DOI 10.1007/978-1-4939-1992-5_12, © Springer Science+Business Media New York 2015

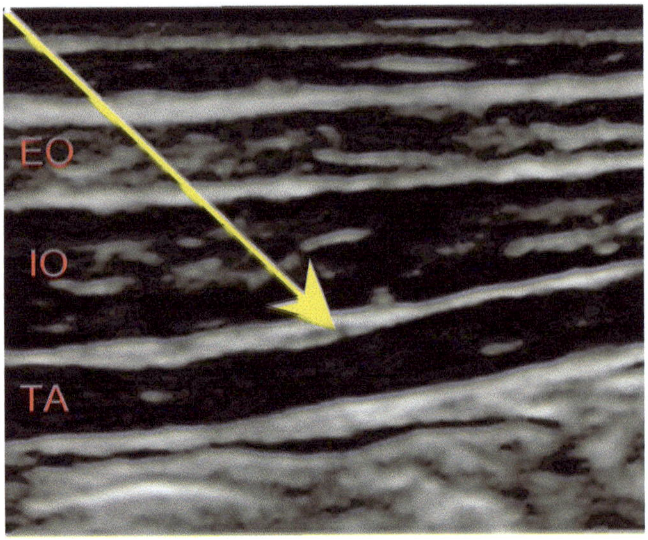

Fig. 12.1 *Arrow* indicating potential space where local anesthetic is deposited (*EO* external oblique, *IO* internal oblique, *TA* transversus abdominis muscle)

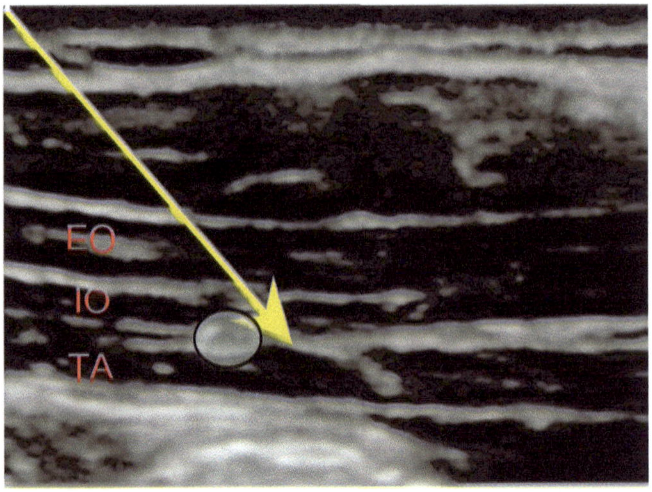

Fig. 12.2 *Arrow* indicating space where local anesthetic injected for ilioinguinal/iliohypogastric nerve block (*EO* external oblique, *IO* internal oblique, *TA* transversus abdominis muscle)

Ilioinguinal/Iliohypogastric (IL/IH) Nerve Block

Anatomy and Indications

The ilioinguinal nerve, which provides sensation to the groin area, can be blocked for therapeutic purposes in patients suffering from chronic pain after previous surgical procedures in the inguinal region. Ilioinguinal neuralgia following hernia repair is an underreported cause of abdominal pain in older children and adolescents [12] and is likely secondary to major nerve dissection during surgery. We have demonstrated the efficacy of performing serial ilioinguinal nerve blocks in adolescents with persistent groin pain following inguinal hernia repair [13]. For cases of refractory pain, we have placed a continuous infusion catheter in this space with good results.

The ilioinguinal/iliohypogastric (IL/IH) nerves originate from T12 and L1 of the thoracolumbar plexus. The nerves traverse the internal oblique aponeurosis just medial to the anterior superior iliac spine (ASIS). IL/IH nerve blocks provide analgesia to the inguinal region and anterior scrotum [14]. Successful placement of these blocks results in equivocal pain relief as provided by caudal blocks for inguinal procedures [15, 16].

Technique

A linear ultrasound probe is placed at the ASIS, in line with the umbilicus. The three abdominal muscle layers are identified (internal oblique, external oblique, and transversus abdominis), although at this level the external oblique muscle layer may be aponeurotic (Fig. 12.2). The inguinal nerve may appear as an ovular structure between the internal oblique and transverse abdominal muscles. The needle is inserted in plane from a lateral to medial approach with incremental needle repositioning. The volume of local anesthetic solution required to block conduction of both nerves has been reported as significantly less with the use of ultrasound when compared to landmark-based techniques [17, 18].

Complications

Bowel puncture and intravascular injection are rare but potentially severe complications. Isolated case reports exist of pelvic hematoma and femoral nerve palsy with performance of II/IH nerve blocks.

Rectus Sheath Block

Anatomy and Indications

Serial rectus sheath blocks have be described as an effective means of providing analgesia for children with chronic abdominal wall pain [19]. This technique is particularly useful for the treatment of periumbilical pain states.

The rectus abdominis muscle lies on the anterior abdominal wall and is separated in the midline by the linea alba. The thoracolumbar nerves (T7–T11) traverse the potential space between the rectus abdominis muscle and posterior rectus sheath. The rectus sheath block is commonly used to achieve periumbilical analgesia for surgical procedures including Single Incision Laparoscopic Surgery (SILS) and umbilical hernia repair [20].

Technique

A linear probe is placed transversely immediately lateral to the umbilicus (Fig. 12.3). The rectus abdominis muscle is visualized as the first major layer beyond the subcutaneous tissue (Fig. 12.4). The posterior sheath lies immediately below the rectus abdominis and sits above the peritoneum. The probe

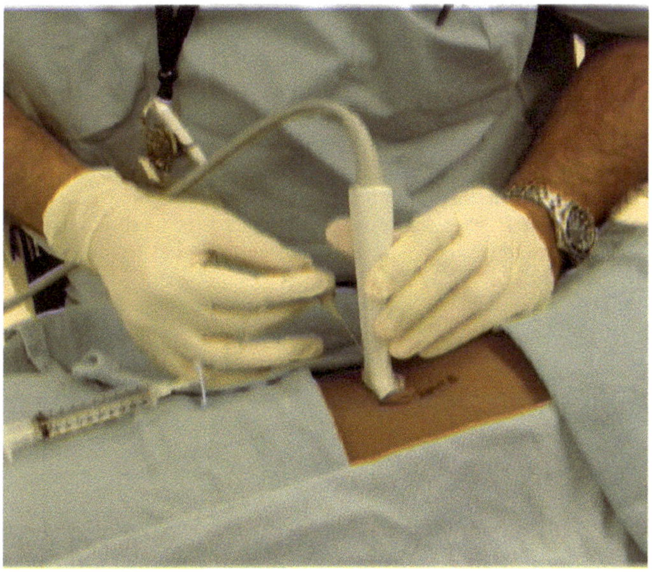

Fig. 12.3 Rectus sheath block performed in a child using an "in-plane" approach

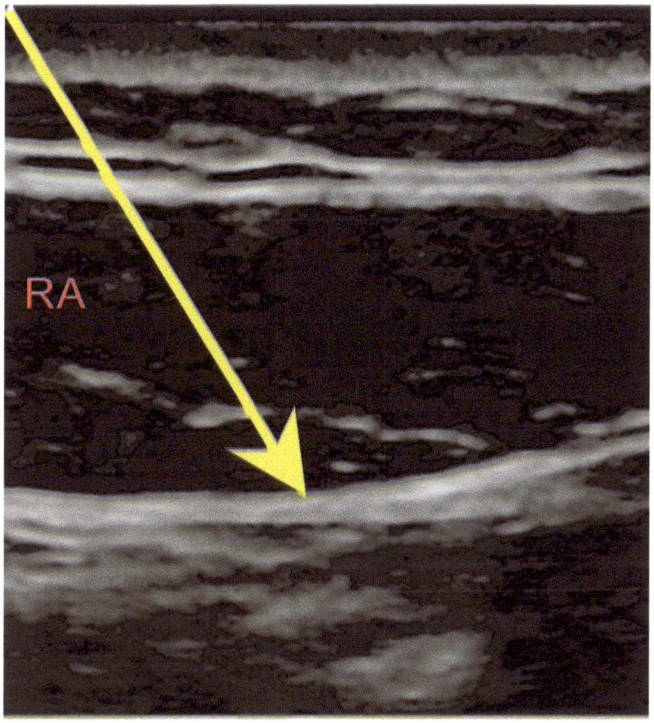

Fig. 12.4 Sonographic image with *arrow* indicating location for local anesthetic placement for rectus sheath block (*RA* rectus abdominis muscle)

is maintained immediately lateral to the umbilicus. A needle is placed in-plane from the lateral aspect of the probe and local anesthetic is deposited in the potential space between the rectus abdominis muscle and its posterior sheath. Approximately 0.1 ml/kg of local anesthetic is used to provide analgesia [21].

Complications

Bowel puncture is a potential complication as the needle is in close proximity to the peritoneum and bowel. Intravascular injection may occur with inadequate negative aspiration, as the inferior epigastric artery is also near needle trajectory.

Conclusion

Peripheral nerve blocks can serve as useful adjuncts to managing chronic abdominal pain conditions in children, especially in cases that are refractory to noninvasive treatments. In this chapter, we have summarized the current knowledge and reported practices of peripheral nerve blocks in managing pediatric chronic abdominal pain. The majority of literature on this topic consists of case reports and retrospective studies. Further evidence about the benefits or potential risks of performing such procedures in children and adolescents, including more prospective, randomized controlled trials, are needed to better guide therapy.

References

1. Walker LS, et al. Functional abdominal pain in childhood and adolescence increases risk for chronic pain in adulthood. Pain. 2010;150(3):568–72.
2. Walker LS, et al. Functional abdominal pain patient subtypes in childhood predict functional gastrointestinal disorders with chronic pain and psychiatric comorbidities in adolescence and adulthood. Pain. 2012;153(9):1798–806.
3. Carayannopoulos AG, et al. Use of regional blockade to facilitate inpatient rehabilitation of recalcitrant complex regional pain syndrome. PM R. 2009;1(2):194–8.
4. Dadure C, et al. Continuous peripheral nerve blocks at home for treatment of recurrent complex regional pain syndrome I in children. Anesthesiology. 2005;102(2):387–91.
5. Ponde V, Desai AP, Shah D. Comparison of success rate of ultrasound-guided sciatic and femoral nerve block and neurostimulation in children with arthrogryposis multiplex congenita: a randomized clinical trial. Paediatr Anaesth. 2013;23(1):74–8.
6. Bhalla T, et al. Ultrasound-guided trunk and core blocks in infants and children. J Anesth. 2013;27(1):109–23.
7. Gurnaney HG, et al. Prospective randomized observer-blinded study comparing the analgesic efficacy of ultrasound-guided rectus sheath block and local anaesthetic infiltration for umbilical hernia repair. Br J Anaesth. 2011;107(5):790–5.
8. Pak T, et al. Transverse abdominis plane block: a new approach to the management of secondary hyperalgesia following major abdominal surgery. Paediatr Anaesth. 2009;19(1):54–6.
9. Simpson DM, et al. Use of ultrasound-guided subcostal transversus abdominis plane blocks in a pediatric patient with chronic abdominal wall pain. Paediatr Anaesth. 2011;21(1):88–90.
10. Fredrickson M, Seal P, Houghton J. Early experience with the transversus abdominis plane block in children. Paediatr Anaesth. 2008;18(9):891–2.
11. Suresh S, Chan VW. Ultrasound guided transversus abdominis plane block in infants, children and adolescents: a simple procedural guidance for their performance. Paediatr Anaesth. 2009;19(4):296–9.
12. Kehlet H. Chronic pain after groin hernia repair. Br J Surg. 2008; 95(2):135–6.

13. Suresh S, et al. Ultrasound-guided serial ilioinguinal nerve blocks for management of chronic groin pain secondary to ilioinguinal neuralgia in adolescents. Paediatr Anaesth. 2008;18(8):775–8.
14. Jagannathan N, et al. Unilateral groin surgery in children: will the addition of an ultrasound-guided ilioinguinal nerve block enhance the duration of analgesia of a single-shot caudal block? Paediatr Anaesth. 2009;19(9):892–8.
15. Hannallah RS, et al. Comparison of caudal and ilioinguinal/iliohypogastric nerve blocks for control of post-orchiopexy pain in pediatric ambulatory surgery. Anesthesiology. 1987;66(6):832–4.
16. Markham SJ, Tomlinson J, Hain WR. Ilioinguinal nerve block in children. A comparison with caudal block for intra and postoperative analgesia. Anaesthesia. 1986;41(11):1098–103.
17. Willschke H, et al. Ultrasonography for ilioinguinal/iliohypogastric nerve blocks in children. Br J Anaesth. 2005;95(2):226–30.
18. Smith T, Moratin P, Wulf H. Smaller children have greater bupivacaine plasma concentrations after ilioinguinal block. Br J Anaesth. 1996;76(3):452–5.
19. Skinner AV, Lauder GR. Rectus sheath block: successful use in the chronic pain management of pediatric abdominal wall pain. Paediatr Anaesth. 2007;17(12):1203–11.
20. Willschke H, et al. Ultrasonography-guided rectus sheath block in paediatric anaesthesia–a new approach to an old technique. Br J Anaesth. 2006;97(2):244–9.
21. Ferguson S, Thomas V, Lewis I. The rectus sheath block in paediatric anaesthesia: new indications for an old technique? Paediatr Anaesth. 1996;6(6):463–6.

Cancer-Related Abdominal Pain

13

Thomas Chai, J. Gabriel Tsang, and Brian M. Bruel

Introduction

Among the most frequently occurring cancers worldwide is gastrointestinal cancer, with variable rates depending on region. For instance, the highest prevalence of colorectal cancer is found in countries such as the USA, Japan, and parts of Europe and Australia, whereas liver cancer and stomach cancer are relatively more prevalent in Asia and Africa [1]. It appears that environmental, lifestyle, and inherited factors contribute to tumor etiology.

According to the National Cancer Institute Surveillance Epidemiology and End Results (NCI SEER) incidence data, it is estimated that approximately 1.66 million Americans will be diagnosed with cancer in the year 2013 [2]. Among those, over 315,000 Americans are diagnosed with cancer involving the digestive system (oropharyngeal, esophageal, stomach, pancreatic, hepatobiliary, intestinal, colorectal, and anal cancer). Per SEER November 2012 submission data, the complete prevalence of Americans with digestive-system cancer is approximately 1.55 million.

Pain is common in cancer patients—it is reported by at least 50 % of cancer patients, and there is a positive correlation between pain and disease stage [3–5]. Cancer-related abdominal pain is a consequence both of alteration or disruption in somatic, neuropathic, and visceral structure and function by the malignant process, and of the associated treatment—whether it is chemotherapy, radiation, surgery, or other modalities—for the underlying malignancy. The intent of this chapter is to describe (1) the pertinent neuroanatomy and physiology of abdominal somatic, visceral, and neuropathic pain, (2) the characteristics of cancer-related abdominal pain, and (3) select cancer-related abdominal pain states both in active cancer patients and in cancer survivors.

Neuroanatomical and Physiological Considerations in Cancer-Related Abdominal Pain

Pain transmission from the abdomen to the central nervous system occurs through both the somatic sensory and visceral afferent pathways [6]. Somatic pain is characterized as sharp and well localized, whereas visceral pain is described more so as dull, vague, and poorly localized. Somatic and visceral pains are carried by thinly myelinated A-delta fibers and unmyelinated C-fibers. Both the abdominal somatic sensory and visceral afferent cell bodies are located in dorsal root ganglia, which project to the spinal cord dorsal horn, synapsing there with second-order neurons. The second-order dorsal horn cells then send ascending projections via, primarily, the spinothalamic tract (in the anterolateral quadrant of the cord) to the thalamus, which then in turn sends projections to the somatosensory cortex and limbic system. Other ascending pathways related to visceral nociception have been identified as well—the spinoreticular tract, dorsal columns, trigeminoparabrachioamygdaloid tract, and spinohypothalamic tract [7].

Somatic Sensory Innervation of the Abdominal Wall

The lower thoracic ventral spinal rami (from T6 to T11) and the subcostal nerves (T12 ventral rami) provide somatic sensory innervation to the anterolateral abdominal wall, including to the skin, subcutaneous tissue, fascia, muscle, and the parietal peritoneum that lines its internal surface. The T6–T11 ventral rami travel in a neurovascular plane between the transversus abdominis muscle and the internal

T. Chai M.D., J.Gabriel Tsang M.B.B.S. (✉)
B.M. Bruel M.D.
Department of Pain Medicine, University of Texas MD Anderson Cancer Center, 1515 Holcombe Blvd., Unit 409, Houston, TX 77030, USA
e-mail: jgtsang@mdanderson.org; TChai@mdanderson.org; bbruel@mdanderson.org

L. Kapural (ed.), *Chronic Abdominal Pain: An Evidence-Based, Comprehensive Guide to Clinical Management*, DOI 10.1007/978-1-4939-1992-5_13, © Springer Science+Business Media New York 2015

oblique muscle, whereas the T12 subcostal nerve enters the abdomen behind the lateral arcuate ligament, crosses the quadratus lumborum muscle, and enters the neurovascular plane by piercing the transversus abdominis muscle.

Abdominal Visceral Innervation

Receptors (such as end-organ-like Pacinian corpuscles) that respond to various stimuli, such as distention (stretch), spasm, ischemia, hypoxia, and inflammation, are found in the serosal, muscular, and mucosal layers of hollow visceral structures, as well as in the mesentery [8–10]. Silent nociceptors are also present, appearing only after *repeated* visceral nociceptive stimulation or irritation/inflammation. The capsule of solid organs contains visceral afferents; however, the parenchymata of solid organs appear not to be innervated by visceral nociceptors.

Visceral-nociceptive input is carried by peripheral fibers from pseudounipolar neurons, whose cell bodies are located in dorsal root ganglia. These first-order neurons have central projections that travel to the spinal cord, primarily to the marginal zone (lamina I) and substantia gelatinosa (lamina II), among other Rexed laminae (to also include laminae X surrounding the central canal) of the dorsal horn.

Visceral-nociceptive afferent fibers share the same pathways as do the autonomic nervous system fibers (both sympathetic and parasympathetic divisions) to reach the central nervous system. For example, the thoracic splanchnic nerves—which include the greater (from T5 to T9), lesser (from T10 to T11), and least (from T12) thoracic splanchnic nerves (with their cell bodies originating in the thoracic intermediolateral cell column)—are paired nerves that carry preganglionic sympathetic efferent fibers to synapse at prevertebral ganglia, specifically at the celiac plexus, which then in turn sends postganglionic sympathetic fibers to abdominal viscera—to the distal esophagus, stomach, liver, pancreas, biliary tract, gallbladder, kidney, adrenal glands, spleen, omentum, small intestines, and colon (to the splenic flexure distally). The thoracic splanchnic nerves and celiac plexus, however, *also* relay visceral *afferent* fibers from these abdominal viscera, back to the central nervous system, synapsing at the spinal dorsal horn.

The celiac plexus is the largest of the three prevertebral sympathetic plexuses (the other two, for completeness, being the cardiac plexus and the hypogastric plexuses). It is a retroperitoneal structure composed of 1–5 ganglia, and it is found anterolateral to the aorta (in close proximity to the origin of the celiac artery) at the level of the T12 or L1 vertebra (Fig. 13.1). Both the celiac plexus and thoracic splanchnic nerves serve as targets in interventional pain medicine for visceral analgesia in intra-abdominal cancer pain states.

Distal to the splenic flexure of the large intestine, the visceral-nociceptive fibers of the remaining colon (and of pelvic viscera) travel in part through the superior hypogastric plexus and lumbar splanchnic nerves to reach the spinal cord. The superior hypogastric plexus is a flattened band of intercommunicating fibers continuous with the intermesenteric plexus. It is found retroperitoneal and caudad to the inferior mesenteric artery near the abdominal aortic bifurcation, at the level of the L5/S1 vertebral junction. Since it carries not only sympathetic efferent fibers but also visceral afferent fibers from these abdominopelvic structures, the superior hypogastric plexus serves as a target in interventional pain medicine for visceral analgesia of the distal colon and pelvic viscera.

The pelvic splanchnic nerves (nervi erigentes) arise from S2 to S4 ventral rami and contribute parasympathetic fibers to the inferior hypogastric plexus (also called the pelvic plexus), where they continue through to ganglia in close proximity to

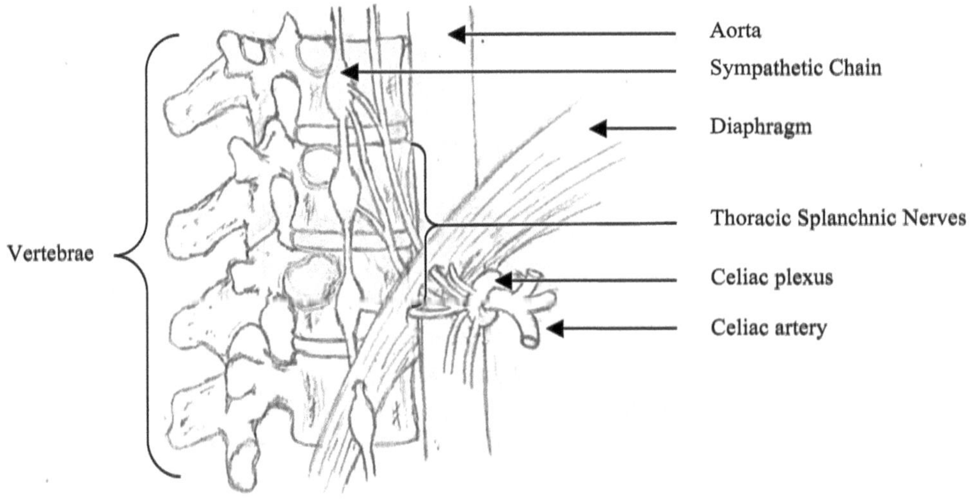

Fig. 13.1 Celiac plexus parasagittal anatomy

the target pelvic viscera. This parasympathetic autonomic route also serves as another afferent visceral pathway.

The vagus nerve also carries visceral afferent information to the central nervous system (with associated cell bodies residing in the nodose ganglia); however, this information is thought to be for autonomic (parasympathetic) regulation, primarily. It is unclear at this time the degree to which, if any, the vagus afferent pathway is related to visceral nociception. There is evidence to suggest that vagal nerve stimulation may attenuate both somatic and visceral pain, possibly through descending inhibitory influences on the spinal dorsal horn neuron responses [11, 12].

The reader should also be aware of the existence of the enteric nervous system, which is the intrinsic nervous system of the gastrointestinal tract that functions autonomously to regulate autonomic gut function.

Sensitization of Visceral Afferents

Visceral afferent nociceptive pathways, as is the case with somatic afferents, may undergo sensitization after disease or inflammation. A general state of neuronal hyperexcitability results in decreased thresholds for firing, increased number of suprathreshold responses, and an increase in spontaneous electrical activity [13].

Visceral Pain Due to Abdominal Malignancy

Tumor burden from primary or metastatic disease may result in visceral pain through a number of mechanisms. For instance, visceral nociceptors respond to noxious mechanical stretch. This can occur in the setting of tumor growth in solid organ parenchymata, causing capsular distention, or in hollow organs, causing narrowing or obstruction with proximal tissue stretch. Ischemia from tumor invasion or compression of blood supply leads to an inflammatory response, characterized by the release of inflammatory mediators such as prostaglandins, bradykinin, and cytokines. This can in turn sensitize the visceral afferent system, amplifying nociception both peripherally and centrally. Neuropathic-type pain is due to compression or dysfunction of neural structures from direct or indirect tumor involvement. Visceral pain can be sequelae of cancer treatment as well, since there is often significant alteration in both structure and function of involved treatment areas. The following sections describe select cancer-related abdominal pain states both from tumor and in survivorship.

Select Abdominal Cancer Pain States

Gastric Pain

For the year 2013, the estimated new cases of stomach cancer in the USA is over 21,000, per SEER [2]. The majority (90 %) of gastric cancer type is adenocarcinoma. Risk factors for stomach cancer include the presence of *Helicobacter pylori* infection, chronic gastritis, and lifestyle factors (such as smoking or consuming smoked and pickled foods) [14].

Gastric cancer patients complain of burning and dull pain along the epigastrium or left upper abdomen. Tumor burden produces pain from distention, ischemia, compression, erosion, inflammation, and obstruction of the stomach and surrounding structures. Gastric and gastroesophageal (GE) junction cancer pain is managed primarily with a combination of nonopioid analgesics, opioid analgesics, and adjuvants, per World Health Organization (WHO) Analgesic Ladder guidelines [15]. Abdominal pain from gastric and GE junction cancer can also be managed by interventional pain techniques [3, 16, 17] such as thoracic splanchnic neurolysis or celiac plexus neurolysis, with absolute alcohol or 10 % phenol (in 20 % glycerin) as the neurolytic agent.

Treatment for primary gastric cancer depends on the TNM (Tumor, Node, and Metastasis) staging [18]. Radical surgery (subtotal, or total gastrectomy with omentobursectomy and lymph node dissection) is the standard therapy for local disease, although neoadjuvant or adjuvant chemotherapy and/or radiation therapy may also be part of the treatment plan.

Hepatobiliary Pain

The estimated year 2013 incidence for hepatobiliary cancer in the USA is over 30,000 [2]. Hepatocellular carcinoma accounts for 90 % of all cases of primary liver cancer. Hepatitis viral infection (both hepatitis B and C) as well as hepatic cirrhosis from heavy alcohol consumption often precedes the diagnosis of liver cancer. Other associated risk factors include tobacco use and consumption of food contaminated with aflatoxins (metabolites produced from *Aspergillus* fungal strains) [1].

Gallbladder (and biliary tree) cancer, on the other hand, is less common, with year 2013 estimates for US incidence at approximately 10,000, nearly all adenocarcinomas [2]. Specific risk factors for both gallbladder and bile duct cancer have been identified as increased body mass index and cholelithiasis.

Visceral nociceptors found along the liver capsule, vasculature, and biliary system are activated by tumor growth through hepatic capsular distention, hepatobiliary duct obstruction, and vascular (portal vein or hepatic vein) obstruction. Patients describe the pain as located about the right side of the abdomen. It may be continuous or colicky. Pain may refer to the back or to the right shoulder, when diaphragmatic irritation occurs (this is referred pain via convergence of visceral and somatic afferents in the spinal cord dorsal horn).

Hepatobiliary cancer pain is managed primarily with a combination of nonopioid analgesics, opioid analgesics, and adjuvants, per WHO Analgesic Ladder guidelines. Patients with abdominal pain from hepatobiliary cancers may be candidates for thoracic splanchnic neurolysis or celiac plexus neurolysis with absolute alcohol or 10 % phenol (in 20 % glycerin) [19].

In patients with localized liver cancer, surgical resection is the recommended treatment. Depending on TNM staging, radiotherapy and/or chemotherapy systemically or via hepatic arterial infusion (HAI) may also be offered. Chemoembolization employs both HAI and distal hepatic arterial vessel occlusion (to increase regional drug distribution and dwell time in the tumor) by a number of materials, such as gelatin sponge, collagen, polyvinyl alcohol, starch microspheres, etc. and can be associated with abdominal pain during infusion. Local gallbladder cancer and bile duct cancer (cholangiocarcinoma) may be amenable to surgical resection; otherwise, unresectable disease may be treated with adjuvant chemotherapy and radiotherapy.

Pancreatic Pain

In 2013, over 45,000 people in the USA are diagnosed with pancreatic cancer. It is the fifth most common cause of cancer-related deaths in the USA [2]. Over 90 % of pancreatic cancers are ductal adenocarcinomas. Pancreatic cancer is associated with chronic pancreatitis, diabetes, smoking, obesity, and heavy alcohol use.

Patients with pancreatic cancer often complain of a dull epigastric pain radiating to the back. Visceral structures coaffected include the deep posterior abdominal wall, connective tissue, ducts, and vasculature. Pain is also attributed to associated bulky lymphadenopathy about the celiac axis.

Pancreatic cancer pain is managed primarily with a combination of nonopioid analgesics, opioids, and adjuvants, according to WHO Analgesic Ladder recommendations. Patients with abdominal pain from pancreatic cancer or pancreatitis often are good candidates for thoracic splanchnic neurolysis or celiac plexus neurolysis with absolute alcohol or 10 % phenol (in 20 % glycerin) [20, 21].

Surgical resection for pancreatic tumors is recommended for localized disease, although this accounts for less than one-fifth of patients at presentation [18]. Pancreatic cancer is generally considered resistant to standard chemotherapy and radiation therapy. Patients may be enrolled, however, in clinical trials.

Intestinal Pain

Cancer of the small intestines is relatively uncommon and accounts for less than 5 % of gastrointestinal malignancies. Colorectal cancer, however, is one of the *most* common cancers—in the year 2013 the number of those newly diagnosed in the USA with colon cancer is approximately 140,000, whereas the number for newly diagnosed rectal cancer is over 40,000 [2, 14]. Risk factors for colorectal cancer have been identified as follows: genetic predisposition, physical inactivity, high body mass index, high intake of alcohol and red meat, and low consumption of fruits and vegetables [1].

Patients with colorectal cancers present with pain from abdominal cramping and distention, and other symptoms such as bleeding. Treatment for abdominal pain from colorectal carcinoma includes standard therapy through WHO Analgesic Ladder guidelines. In addition, specific interventional pain techniques for the management of colorectal pain consist of the following: celiac plexus neurolysis for pain arising from small bowel cancer or large bowel cancer proximal to the colonic splenic flexure; superior hypogastric neurolysis for pain arising from colon cancer distal to the splenic flexure; and ganglion impar block for distal rectal pain [3, 22–24].

Surgical resection of primary tumor and adjuvant chemotherapy is the general treatment plan for local colorectal carcinoma [18]. For advanced disease treatment, algorithms become more complex in order to address widespread metastatic disease.

Peritoneal-Related Pain

Peritoneal carcinomatosis occurs when abdominal cancer spreads to the peritoneum. Peritoneal carcinomatosis can therefore cause diffuse abdominal pain through inflammation, adhesions between tumor and surrounding tissue with resultant stretching and stricture, and ascites with abdominal distention. Intraperitoneal chemotherapy as treatment for abdominopelvic malignancy produces a painful chemical serositis experienced by up to half of those treated [3].

Select Abdominal Pain States in Cancer Survivorship

Cancer survivors may continue to experience persistent pain beyond the course of the disease and its related treatment. Up to one-half of cancer survivors experience ongoing pain after cancer treatment [4, 17]. Iatrogenic visceral pain from mechanical, ischemic, inflammatory, and neuropathic insult to tissue and nerves in the treatment field is seen after surgery, chemotherapy, radiation therapy, and other modalities. This damage may lead to chronic somatic-nociceptive, visceral-nociceptive, and neuropathic pain, as well as sympathetically maintained pain.

There exist a number of chronic abdominal pain conditions experienced by cancer survivors who have undergone treatment for malignant disease, examples of which are described in the following text.

Gastrointestinal Graft-Versus-Host Disease

Hematopoietic stem cell transplantation (HSCT) is a procedure used to treat certain blood cancers. The procedure involves depleting or eradicating the recipient bone marrow with chemotherapy and/or radiation therapy and then repopulating the bone marrow by transplanting autologous or allogeneic hematopoietic stem cells. HSCT is associated with significant morbidity to include infection, mucositis, and graft-versus-host disease (GVHD), to name a few. GVHD is explained in basic terms as the donor bone marrow immune cells identifying the host tissue as foreign and subsequently mounting an immune attack against the host tissue. This condition or one indistinguishable to it may actually be seen in both autologous and allogeneic transplantation [25]. GVHD may become a chronic (defined as 100 days posttransplant) inflammatory condition, commonly affecting skin, liver, and the gastrointestinal tract [26]. GVHD has been known to cause chronic abdominal pain states in cancer survivors due to frequent, recurrent loose stools, nausea/vomiting, and abdominal cramping with distention. Both acute and chronic GVHD are generally managed with steroids and other immunosuppressive agents [27, 28].

Radiation Enteritis

The intent of radiation therapy for malignancy is either to cure primary or metastatic disease or to palliate symptoms associated with disease. This treatment modality employs high-energy radiation, which can be applied through external sources (external beam radiation therapy) or through radioactive material placed in the body (brachytherapy) or bloodstream (systemic). Radiation targets the cellular DNA of rapidly dividing malignant cells, leading to impairment in cell division and cell death.

Intestinal mucosal cells also undergo rapid turnover and therefore are at risk for damage by radiation. Radiation therapy side effects can occur during the treatment course (early) or months to years after treatment (late) [29–32]. Examples of radiotherapy side effects involving the abdomen include radiation dermatitis, gastroenteritis, bacterial overgrowth with resultant malabsorption, ulceration, perforation, fibrosis and stricture leading possibly to obstruction, and fistula formation. Patients may experience abdominal complaints, such as painful cramping, loose stools, nausea/vomiting, and anorexia, both during and after radiation therapy. These symptoms are managed conservatively through proper skin care, dietary modifications, antimotility and antiemetic agents, and analgesics. In some cases, surgical intervention for the affected bowel is indicated.

Chronic Abdominal Pain After Cancer Surgery

Iatrogenic tissue and nerve injury and other complications arise at times from surgical tumor resection. Abdominal surgery for malignancy may entail tumor resection or debulking, often sacrificing or altering adjacent healthy tissue and nerves. Acute postoperative pain usually resolves over days to weeks; however, painful neuroma formation, scarring, fibrosis, and stricture formation along the involved abdomen/abdominal wall may ensue. Development of postsurgical intra-abdominal adhesions could contribute to chronic abdominal pain from mesenteric stretching, gut narrowing or obstruction.

Conclusion

Cancer-related abdominal pain is sequelae of both the underlying malignant process and the cancer treatment modality. Abdominal somatic, visceral, and nerve structures affected can produce a nociceptive, neuropathic, or mixed pain state.

References

1. Kelsen D. Principles and practice of gastrointestinal oncology. 2nd ed. Philadelphia: Lippincott, Williams & Wilkins; 2008. p. xv–749. p. 16 of plates.
2. Howlader N, Krapcho M, Garshell J, Neyman N, Altekruse SF, Kosary CL, Yu M, Ruhl J, Tatalovich Z, Cho H, Mariotto A, Lewis DR, Chen HS, Feuer EJ, Cronin KA. SEER Cancer Statistics Review, 1975–2010. Based on November 2012 SEER data submission. 2013. http://seer.cancer.gov/csr/1975_2010/. Accessed 10 May 2013.
3. Fishman S, et al. Bonica's management of pain. 4th ed. Baltimore: Lippincott, Williams & Wilkins; 2010. p. xxxii–1661.
4. Bruera E, Portenoy RK. *Cancer pain: assessment and management*. 2nd ed. New York: Cambridge University Press; 2010. p. xi–643. p. 4 of plates.
5. Raphael J, et al. Cancer pain: part 1: Pathophysiology; oncological, pharmacological, and psychological treatments: a perspective from the British Pain Society endorsed by the UK Association of

Palliative Medicine and the Royal College of General Practitioners. Pain Med. 2010;11(5):742–64.
6. Bailey A, Bernstein C. Pain in women: a clinical guide. New York: Springer; 2013. p. 320.
7. Knowles CH, Aziz Q. Basic and clinical aspects of gastrointestinal pain. Pain. 2009;141(3):191–209.
8. Cervero F. Sensory innervation of the viscera: peripheral basis of visceral pain. Physiol Rev. 1994;74(1):95–138.
9. Grundy D. Neuroanatomy of visceral nociception: vagal and splanchnic afferent. Gut. 2002;51 Suppl 1:i2–5.
10. Sengupta JN. Visceral pain: the neurophysiological mechanism. Handb Exp Pharmacol. 2009;194:31–74.
11. Zhang X, et al. Vagus nerve stimulation modulates visceral pain-related affective memory. Behav Brain Res. 2013;236(1):8–15.
12. Zurowski D, et al. Effects of vagus nerve stimulation in visceral pain model. Folia Med Cracov. 2012;52(1–2):57–69.
13. Cervero F. Visceral pain: mechanisms of peripheral and central sensitization. Ann Med. 1995;27(2):235–9.
14. Rodriguez MA, Walters RS, Burke TW. 60 years of survival outcomes at the University of Texas M.D. Anderson Cancer Center. New York: Springer; 2012. p. xiii–346.
15. Forbes K. Pain in patients with cancer: the World Health Organization analgesic ladder and beyond. Clin Oncol (R Coll Radiol). 2011;23(6):379–80.
16. Miguel R. Interventional treatment of cancer pain: the fourth step in the World Health Organization analgesic ladder? Cancer Control. 2000;7(2):149–56.
17. Raphael J, et al. Cancer pain: part 2: physical, interventional and complimentary therapies; management in the community; acute, treatment-related and complex cancer pain: a perspective from the British Pain Society endorsed by the UK Association of Palliative Medicine and the Royal College of General Practitioners. Pain Med. 2010;11(6):872–96.
18. N.C.I. Cancer Treatment, Health Professional Version. 2013. http://cancer.gov. Accessed 10 May 2013.
19. Akinci D, Akhan O. Celiac ganglia block. Eur J Radiol. 2005;55(3):355–61.
20. Zhong W, et al. Celiac plexus block for treatment of pain associated with pancreatic cancer: a meta-analysis. Pain Pract. 2014; 14(1):43–51.
21. Bahn BM, Erdek MA. Celiac plexus block and neurolysis for pancreatic cancer. Curr Pain Headache Rep. 2013;17(2):310.
22. Plancarte R, et al. Neurolytic superior hypogastric plexus block for chronic pelvic pain associated with cancer. Reg Anesth. 1997;22(6):562–8.
23. Waldman SD. Atlas of interventional pain management. 3rd ed. Philadelphia: Saunders Elsevier; 2009. p. xvii–702.
24. Benzon HT. Essentials of pain medicine and regional anesthesia. 2nd ed. Philadelphia: Elsevier-Churchill Livingstone; 2005. p. xv–734.
25. Snover DC. Graft-versus-host disease of the gastrointestinal tract. Am J Surg Pathol. 1990;14 Suppl 1:101–8.
26. Ferrara JL, et al. Graft-versus-host disease. Lancet. 2009;373(9674): 1550–61.
27. Awan F, Hamadani M. Gastrointestinal chronic graft-versus-host disease: management options. J Oncol Pharm Pract. 2007; 13(1):49–51.
28. Holmberg L, et al. Gastrointestinal graft-versus-host disease in recipients of autologous hematopoietic stem cells: incidence, risk factors, and outcome. Biol Blood Marrow Transplant. 2006; 12(2):226–34.
29. Theis VS, et al. Chronic radiation enteritis. Clin Oncol (R Coll Radiol). 2010;22(1):70–83.
30. Sinko D, et al. Symptoms, diagnosis and treatment of radiation-induced enteritis. Orv Hetil. 2010;151(36):1450–4.
31. Ruiz-Tovar J, et al. Late gastrointestinal complications after pelvic radiotherapy: radiation enteritis. Clin Transl Oncol. 2009;11(8): 539–43.
32. Bismar MM, Sinicrope FA. Radiation enteritis. Curr Gastroenterol Rep. 2002;4(5):361–5.

Opioids in Abdominal Pain

14

Genaro J. Gutierrez, Justin Rodante, and Mehul J. Desai

Introduction

The use of opioids in abdominal pain is a commonplace practice, yet there is very little support for it in the literature. Often the treatment of "last-resort," opioids are typically considered in patients when other agents were unsuccessful. Alternatively, opioids are often initiated in the emergency room or inpatient hospitalization setting with the expectation of continuity upon discharge. In many respects, the treatment of abdominal pain with opioids presents an enormous burden on the healthcare system with these patients often experiencing poor follow-up and long-term care.

Epidemiology

Abdominal pain is one of the most common reasons patients seek medical care. According to National Center of Health Statistics, patients with abdominal pain account for more emergency room visits than any other condition with the exception of chest pain [1]. Other national statistical surveys identify abdominal pain as the most prevalent complaint in the outpatient setting, occurring in up to 75 % of adolescents and 50 % of adults [2]. Abdominal discomfort is typically the presenting symptom of the most abdominal diseases including gastric conditions, pancreatitis, appendicitis, cholestatic disease, and diverticulitis, as well as functional disorders such as irritable bowel syndrome and dyspepsia. Although at times challenging, appropriate diagnosis and management of abdominal pain has considerable impact on patient care.

Characteristics of Abdominal Pain

Treatment is often primarily determined by appropriately characterizing the symptoms of abdominal pain. Most commonly abdominal pain of <1 month duration is generally considered acute, whereas persistent abdominal pain of >3–6 months is deemed chronic. An additional means to define chronicity refers to discomfort that lasts beyond the resolution of the initial tissue injury. Although the management of both acute versus chronic abdominal pain might involve opioids, data support the utilization of opioid therapy in acute abdominal pain much more so than in chronic noncancer abdominal pain. Chronic abdominal pain for which no definite structural or organic etiology is identified is typically categorized as functional abdominal pain syndrome (FAPS). Although the mechanism of FAPS is not fully understood, considerable research suggests a brain–gut connection triggered by ongoing inflammatory mediators within the gastrointestinal (GI) tract resulting in visceral and central sensitization [3].

Abdominal pain may be visceral, parietal, or referred. Visceral afferent nerve fibers diverge over many more spinal segments than their parietal counterparts and therefore produce pain that is typically poorly localized, deep, and achy in nature. Visceral pain is induced by ischemia, inflammation, or increased tension upon an abdominal organ. Somatic or parietal pain, on the other hand, occurs when nociceptive signals are triggered at the peritoneal lining, mesenteric root, or abdominal wall. Pain that occurs at a site different than the source of the pain is known as referred pain and frequently

G.J. Gutierrez, M.D.
Anesthesiology and Critical Care Medicine, The George Washington University Hospital, 900 23rd Street NW, Washington, DC 20037, USA
e-mail: ggutier1@bidmc.harvard.edu; jerrygutz@gmail.com

J. Rodante, M.P.H.
Department of Spine and Pain Medicine, The George Washington University Hospital, 2131 K Street NW Suite 600, Washington, DC 20037, USA
e-mail: jrpamph@yahoo.com

M.J. Desai, M.D., M.P.H. (✉)
Pain Medicine & Research, International Spine, Pain & Performance Center, 8401 Colesville Road #50, Silver Spring, MD 20910, USA
e-mail: mehuljdesaimd@gmail.com

L. Kapural (ed.), *Chronic Abdominal Pain: An Evidence-Based, Comprehensive Guide to Clinical Management*,
DOI 10.1007/978-1-4939-1992-5_14,

follows a dermatomal or myotomal pattern [4–6]. Appropriate categorization of abdominal pain should allow for streamlined and rapid treatment whether medical or in some cases surgical.

Opioids in Acute Abdominal Pain

Historically, opioids were used predominantly for cancer pain, certain acute pain conditions such as severe musculoskeletal injuries, migraine headaches, and spinal trauma, and in the postoperative period. Until relatively recently, analgesia for patients with acute abdominal pain, however, was often withheld for fear of masking the symptoms and compromising diagnostic accuracy. It was not until the 1980s that studies were conducted to test this strategy. A Cochrane review of six RCT's comparing the use of opioid analgesics to placebo in patients with nontraumatic abdominal pain found no difference between opioid and control groups in changes in the physical examination, diagnostic or treatment errors, or morbidity. The study did show an improvement in pain intensity among those patients who received opioids [7]. A second systematic review of 9 trials (6 involving adult patients and 3 in pediatrics) published in *JAMA* in 2006 did reveal an increase in altered findings on physical examination in the opioid group; however, these changes did not result in an increase in diagnostic errors, although there was a trend toward more unnecessary operations in those who received opioids [8].

A review of five RCTs with a total of 227 subjects evaluated the effect of buprenorphine, pethidine, and pentazocine specifically on patients with acute pancreatitis. Although somewhat limited in nature, the review concluded that opioids might be an appropriate analgesic option, limiting the need for additional analgesic therapies compared to other nonopioid analgesic therapies. The review also found no difference among complications or serious adverse events between patients who received opioids and those who did not. Currently, it is generally accepted that the use of opioid analgesics for patients with acute nontraumatic abdominal pain does not increase the risk of diagnostic error and is often effective at lowering pain intensity levels [9, 10].

Opioids in Chronic Abdominal Pain

Over the past two-three decades, the use of opioid analgesics for chronic noncancer pain has significantly risen. Although the rate of pain visits remained largely the same, the rate of opioid prescribing nearly doubled in a decade, from 11.3 % in 2000 to 19.6 % in 2010 [11, 12]. From 1997 to 2008 data show that the adjusted prevalence of abdominal pain visits for which an opioid was prescribed also doubled from 5.9 to 12.2 %, despite an overall decrease in prevalence of abdominal pain during that time period [13]. This increase in opioid prescribing has also resulted in some unintended, but not unexpected, consequences. As the use of opioids for chronic noncancer pain steadily increased, the rates of abuse, misuse, and diversion have also risen. According to the CDC, visits to emergency departments for nonmedical use of opioids tripled from 2004 to 2008 and methadone-related fatalities increased sevenfold in 2009 [14]. Mortality rates from prescription drug overdose now surpass those of illicit drugs such as cocaine and heroin, accounting for up to 75 % of prescription drug-related deaths [15]. Such findings have prompted concern among the medical and legal communities resulting in calls for more expansive oversight of opioid analgesics.

Despite the increasingly widespread use of opioids, few studies have established their role in the treatment of chronic, nonmalignant pain. To help fill this gap, the American Pain Society in partnership with the American Academy of Pain Medicine conducted a systematic review of the current literature and published the Clinical Guidelines for Chronic Opioid Therapy in Patients with Noncancer Pain in 2009 [16].

These guidelines suggest a detailed history and physical as well as risk assessment and stratification prior to initiating chronic opioid therapy (COT). Several different screening questionnaires or tools are available to help provide risk stratification.

Among four prospective studies evaluating the diagnostic accuracy of risk assessment screens, two suggested that higher scores on the Screener and Opioid Assessment for Patients with Pain (SOAPP) Version 1 and the Revised SOAPP (SOAPP-R) may hold an increased correlation with aberrant behavior [17]. Factors associated with opioid abuse/misuse include a personal or family history of alcohol or substance abuse, younger age, history of sexual abuse, and the presence of comorbid psychiatric conditions. The decision to initiate COT should be made on an individual basis, when the cumulative benefits are likely to outweigh the cumulative risks. The patient should also sign an informed consent, often contained within an opioid agreement, outlining the risks and benefits of opioid therapy. Once opioid therapy is initiated, the need for and response to COT should be reevaluated at each visit. COT patients should be monitored regularly for evidence of both analgesic and functional benefit, appearance of any adverse effects, and demonstration of red flag behaviors. Annual and random performance of urine drug testing as well as the utilization of state prescription drug monitoring programs might also be helpful in identifying evidence of misuse or diversion [16–20].

Opioid Receptors

Opioids exert their effects on the body by binding to receptors distributed throughout the body, particularly within the central nervous system and peripheral tissues.

Four opioid receptor types have been identified thus far: mu, delta, and kappa, and ORL- I. Mu receptors are believed to mediate nociceptive pain at the supraspinal level and are the receptor type with the highest affinity for morphine. Kappa receptors are responsible for analgesia at the spinal level, whereas delta receptors mediate nociceptive and inflammatory pain transmission. The ORL-I receptor, newly discovered and least understood opioid receptor type, is believed to be involved in central modulation of pain. These receptors directly regulate calcium, sodium, and potassium ion channels resulting in generation of action potentials along the affected neurons. The three subtypes of the mu receptors form the basis by which group the different types of opioids are categorized [21]. These receptors types have also been identified within the enteric system not only resulting in both the analgesic effects of opioids but also highlighting issues with side effects such as constipation [21].

Opioid Options

In clinical practice, the use of opioids must be balanced between their analgesic effects and their potential side effects and risk of abuse and/or misuse. A variety of analgesic opioid agents continue to emerge into the market. Many come in both short-acting and long-acting forms. Analgesic responses to a drug may vary considerably due to a variety of patient factors including age and weight differences, prior opioid exposure and tolerance, and the differences in bioavailability of various opioid formulations.

Recent attention to pharmacogenomics, particularly of the genetic composition of an individual's enzymatic systems (e.g., cytochrome P450 pathway) reveals that drug metabolic rates vary widely from one patient to another. Keeping in mind such variation can play into what type of drug is selected for a patient. As opioid use has risen and broadened in scope over the past decade, much research has investigated their role in the treatment of chronic noncancer pain (CNCP). While no data suggests one type as more efficacious than the other, several studies do demonstrate that both short-acting opioids (SAO) and long-acting opioids (LAO) can be used to alleviate pain [22] (Table 14.1).

Opioids are often classified as short-acting versus long-acting based on their duration of action. Formulated to provide a more rapid rise and then decrease in blood plasma levels, SAOs are considered appropriate for acute or breakthrough pain of a transient nature. The LAOs which typically last 8–72 h provide more stable plasma level concentrations over a longer period thereby avoiding the peak—trough effect of SAOs. Because the LAOs have a longer and more gradual onset of action (typically >1–2 h), they are considered to have less reward-associated reinforcement and therefore less risk of dependence. LAO agents can, however, pose an increased risk of drug accumulation and need to be titrated, monitored, and rotated carefully. Many of the oral SAO agents have been modified into longer-acting versions (see opioid table). Unlike several of its nonopioid analgesic counterparts (e.g., NSAIDs), no evidence has shown any end-organ failure from long-term use of SAOs or LAOs.

Table 14.1 Commonly used opioids

Medication	Routes	SAO vs. LAO	Onset of action	Duration of action	PCA-availability
Fentanyl	TD, TM, IV, SC,	Both	12–24 h TD,	72 h TD,	Yes,
(Duragesic, Actiq, Fentora)			5–15 min TM	Highly variable TM	50 mcg/ml
Hydromorphone	PO	Both	15–30 min for SAO	4–6 h	Yes,
(Dilaudid, Exalgo)	IV/SC		1–3 h for LAO	24 h	1 mg/ml
Methadone	PO	LAO	30–90 min	8–12 h	No
(Dolophine)					
Morphine	PO, IV, SC	Both	30–60 min SAO	3–6 h MSIR, Roxanol	Yes,
(MSIR, Roxanol, MS Contin, Kadian, Avinza)			30–90 min LAO	8–12 h MS Contin, 12–24 h, Kadian and Avinza	1 mg/ml or 5 mg/ml
Oxycodone	PO	Both	10–15 min SAO	4–6 h	No
(Roxicodone as SAO, oxycontin as LAO)			60 min LAO	8–12 h	
Oxymorphone	PO	Both			
(Opana, Opana ER)					
Tapentadol	PO	Both	15–60 min	4–6 h	No
(Nucynta as SAO, Nucynta ER as LAO)				8–12 h Nucynta ER	

TD Transdermal, *TM* Transmucosal, *IV* Intravenous, *SC* Subcutaneous, *SAO* Short Acting Opioids, *LAO* Long Acting Opioids

Although no opioid has been shown to be superior to the others, each compound has its own specific characteristics.

Morphine is the prototypical *u*- opioid agonist against which all other opioids are compared. It serves at the standard currency by which equianalgesic conversions of other opioid medications are performed. As a hydrophilic compound, it has a slower onset of action than many other opioids. Although its activity is mediated primarily by morphine's parent molecule, morphine's effects can be perpetuated by its two metabolites: morphine 3-glucuronide (M3G) and morphine 6-glucoronide (M6G) [31]. M3G does not possess any opioid agonist activity and is thought to be chiefly responsible for morphine's side effects, which include generalized hyperalgesia, seizures, myoclonus, and histamine release causing pruritus. Although morphine is metabolized hepatically, both its metabolites rely on the kidneys for excretion and therefore its use must be monitored closely in patients with underlying CKD.

Oxycodone is a semisynthetic compound of morphine that is typically more potent and is associated with fewer side effects. It is metabolized by the hepatic CYP2D6 pathway into oxymorphone, the agent believed to produce its analgesic properties. Frequent abuse and misuse of this drug has resulted in a recent reformulation of the extended-release form known as oxycontin. In the estimated 10 % of patients with genetically low CYP2D6 levels, increasing doses of oxycodone or oxymorphone are not likely to produce adequate analgesia.

Hydromorphone is 5–7× more potent than morphine, but has been shown to cause less pruritus, sedation, histamine release, nausea, and emesis than morphine. Similar to morphine, hydromorphone is extensively metabolized in the liver. Its primary metabolite hydromorphone-3-glucuronide (H3G) is excreted renally and can cause neurotoxic effects (excitation syndrome: hyperalgesia, myoclonus, and epilepsy) upon accumulation.

Fentanyl, originally formulated for anesthetic purposes, has an inherently faster onset of action and is approximately 100× more potent than morphine. Its greater degree of potency compared to other opioids allow for very small amounts to be administered (micrograms versus milligrams). While primarily prescribed as a long-acting topical formulation (Duragesic) in the outpatient setting, it is also administered parentally, epidurally, and intrathecally. As a popular 72 h transdermal patch formulation, fentanyl's effect does not require GI absorption or hepatic activation thus theoretically resulting in less GI side effects than other opioids. Because the long-lasting relief of the transdermal patches work by releasing fentanyl into body fats, many factors such as skin temperature, fat content, and proper adherence of the patch profoundly affect absorption rates.

Methadone, often best known for its role in drug addiction, has seen renewed attention in management of chronic pain due to its relatively cheap cost, high bioavailability, multiple receptor site activation, and lack of neurotoxic side effects. Though it does exert some activity at NMDA receptors which recently has been implicated in many central, chronic pain states, methadone's unpredictable half-life (ranging from 12 to 190 h), and high interindividual pharmacokinetic variability render appropriate dosing difficult. Because the p450 pathway metabolizes it, methadone interacts with a variety of other medications. Gastric pH levels also significantly affect methadone's degree of absorption. Methadone has also been associated with cardiac toxicity, specifically an increase of the QT interval. Therefore, a baseline EKG should be considered before starting methadone.

Despite the variety of available opioid agents, guidelines continue to recommend that opioid therapy be tailored to each patient. Analgesic responses to a drug can vary considerably due to a variety of patient factors including age and weight differences, prior opioid exposure and tolerance, and the differences in bioavailability of various opioid formulations. Recent attention to pharmacogenomics, particularly of the genetic composition of an individual's enzymatic systems (e.g., cytochrome P450 pathway) reveals that drug metabolic rates vary widely from one patient to another. Keeping in mind such variation can certainly be helpful when selecting an appropriate analgesic agent for a patient. Although is difficult to recommend one specific molecule over others based on the data, many practitioners anecdotally report the use of fentanyl and hydromorphone in an effort to minimize cognitive and constitutional side effects.

Special Issues

Chronic opioid use is associated with a constellation of disruptive effects among the GI tract collectively known as opioid-induced bowel dysfunction (OIBD). Constipation is the most common opioid side effect, occurring in an estimated 15–90 % of patients receiving long-term opioids [23, 24]. Because a high concentration of opioid receptors reside in the gastric antrum and duodenum, opioid-induced constipation most likely occurs via a decrease in intestinal motility, and to a lesser degree, via reducing intestinal secretion [24]. Previously, the role of the mu receptor in the constipating role of opioids was identified, recent studies however have also implicated the delta receptor as playing a role [25]. Currently, there is insufficient data to suggest that one opioid agent is more likely to cause constipation than another, and unfortunately this side effect is the least likely to resolve after continued opioid use.

Other primary adverse GI affects due to opioid use include xerostomia (75 % prevalence), nausea (7–28 %), and emesis [12]. Nausea is believed to occur via stimulation of the chemoreceptor zone. Opioid agonists are shown to increase biliary duct pressure and sphincter of Oddi tone in a dose-dependent manner; however, clinical differences as a result of such have not been demonstrated. Opioid use is associated with dysfunction of nearly every part of the GI tract to some degree including the lower esophagus, the esophageal and pyloric sphincters, the stomach, small and large intestines, and even the rectum. Patients who receive long-term opioid therapy for abdominal pain (typically > 30 mg of morphine-equivalent per day) might develop narcotic bowel syndrome (NBS), characterized by chronic or frequently recurring abdominal pain that worsens with continued or escalating dosages of narcotics. Studies suggest that chronic narcotic use causes a dysregulation in the inhibitory and excitatory neural pathways resulting in visceral hyperalgesia, a state that subsequently presents its own treatment challenges [5].

The Sphincter of Oddi (SO) is a smooth muscle structure that regulates bile and pancreatic secretion flow, prevents reflux of duodenal contents into the pancreatic and biliary system, and also diverts hepatic bile into gall bladder. The Sphincter of Oddi interacts with neural and hormonal signals to reflexively contract.

Sphincter of Oddi Dysfunction (SOD) refers to two motility disorders affecting this structure: biliary dyskinesia and papillary stenosis. Of interest to pain physicians is Sphincter of Oddi dyskinesia or "spasms" caused by morphine and other opioids. The effect of opioids on the SO has long been described and in current practice, morphine is often used during biliary studies (HIDA and MRCP) to improve ductal distention.

The exact mechanism of SOD due to opioids is yet to be described; however, several hypotheses exist. It is suggested that the mechanism is not mediated via the parasympathetic system, as atropine does not appear to reverse the contractions. Naloxone has inconsistently shown to reverse opioid-induced SOD; however, this may suggest that mu or delta stimulation play a role in this reflex. Of interest, naloxone was successful in decreasing the amplitude of the phasic contractions, but showed no change in the SO basal pressure [26]. Meperidine's ability to induce SO spasms has also promoted the opioid receptor theory [27].

Several studies have better defined the effects of opioids on the SO and provided a better understanding of this reflex. In early studies using indirect methods, investigators found that bile flow was impeded with all narcotic agents and morphine seemed to invoke the largest degree of bile duct stasis [28]. Most modern studies have employed SO manometry during ERCP to measure SO basal pressure as well as phasic contraction frequency, amplitude, and duration of contractions. At least one study with 40 subjects showed a statistically significant increase in all parameters with morphine at high doses (10 mg IM) [29]. Another study with 19 patients with normal pancreatic and biliary function showed that lower dose morphine caused increased contraction frequency, whereas at higher doses it caused increased basal pressure and contraction amplitude [26]. The findings that show morphine to be the biggest culprit of SO spasm may suggest that alternative narcotics may be a better choice for treatment of abdominal pain due to pancreatic or hepatobiliary processes. This statement however may not be completely accurate because although morphine may in fact compromise bile duct emptying; how this affects pancreatic duct emptying or the course of acute pancreatitis is not documented [28]. There is also no comparison of the outcomes in patients with acute pancreatitis treated with morphine or other narcotics. Whether SO function itself remains the same during acute pancreatitis as at baseline also remains unknown.

Although all opioids are thought to cause SOD, at varying degrees, thus far it is suggested that morphine may have the strongest effect on SOD. There is no single-best opioid medication that eliminates the risk of SOD in patients with pancreatic–biliary-mediated pain; however, there have been some proponents of buprenorphine as the preferred agent [30]. Others have also suggested that meperidine may be a better choice in treating patients with pain due to pancreatitis or SOD. However, due to meperidine's side-effect profile as well as the extended half-life of normeperidine, morphine may offer more benefit than meperidine by offering more pain relief with a lower seizure risk [27, 28].

Opioid Rotation

For many patients treated with opioids, one of the frequently encountered caveats to treatment is a patient's nonresponse to the currently prescribed opioid. In the face of nonresponse to treatment, a frequent strategy that may be undertaken is switching to a different agent or a different route of administration, otherwise known as opioid rotation. Theoretically, opioid rotation takes advantage of individual differences in the presence and expression pattern of opioid receptor subtypes, for which each opioid has a different preference [32]. In rotating, or changing to a different opioid, one must calculate a dose of a different opioid based on equianalgesic equivalencies. Many patients will require additional considerations for adjustments made based on differences in potency, pain diagnosis/characteristics, and individual differences [33]. Opioid rotation is also frequently employed in patients whom experience significant side effects. This is no exception in treating chronic abdominal pain in which many patients may experience new or increased levels of nausea, vomiting, or constipation. This may result in worse or increased pain that may cloud or complicate the picture in their treatment.

Rotation Strategies

Based on the conclusions of an expert panel [34], selecting a new opioid requires tailoring to an individual's need based on: demographic factors, disease, comorbidities, concomitant pharmacology, drug sensitivities, and patient's previous drug experience.

Generally, switching between opioids tends to be done on a trial-and-error process. However, some issues to consider include: choosing a long-acting formulation may be chosen over a short-acting formulation because of convenience in dosing and better medication adherence, avoiding morphine in renal failure patients due to metabolite accumulation, transdermal fentanyl improve administration to patients with poor oral tolerances [35].

Knowledge of equianalgesic conversions between opioids is essential for safely rotating between opioids. Although these conversions have been established as a guideline for opioid switching or supplementation, one must know that conversion tables have many shortcomings and limitations, mainly because they were constructed based on studies that included very different populations and pain syndromes.

Evidence of Benefits in Rotation

According to the APS/AAPM 2009 Recommendations for Treating of Noncancer Pain with Long-term Opioid Therapy [17] clinicians should consider opioid rotation when patient on COT experience intolerable adverse effects. This however, is a weak recommendation based on low-quality evidence. There is at least one study that showed achievement of improved pain control in 59 % of patients when switched to a new long-acting opioid [36]. In the same study population, a switch from a short-acting to a long-acting formulation of the same opioid resulted in improved pain control in 73 % of the patients.

Pasternak et al. have shown that a single individual can have a markedly different response to different types of opioids [37]. This is postulated to occur due to a number of variants in cloned mu opioid receptors MOR-1 that he describes. These variants differ in their functional activation and localization within cells and regions of the brain and many of these variants are truncated and do not conform to the G-protein-coupled receptors that traditionally describes the mu family.

Opioid-Induced Bowel Dysfunction

There is much rationale for the use of opioid rotation in the treatment of chronic abdominal pain. Namely, due to the abundance of side effects on the gastrointestinal tract that can exacerbate the patient's current symptoms as well as bring new ones. These adverse effects of opioids on the GI tract collectively are referred to as Opioid-Induced Bowel Dysfunction (OBD).

In terms of OBD physiopathology, opioids produce their effect through central and peripheral mechanisms [38]. Evidence of centrally mediated effects has been described with intraventricular cerebral administration of morphine in experimental animals which shows reduced intestinal transit [39, 40]. In terms of peripherally mediated effects, morphine and other opioids inhibit the release of acetylcholine, thereby decreasing muscarinic receptor stimulation while increasing intestinal tone and reducing peristalsis [41]. Opioids affect all levels of the GI tract via stimulation of mu receptors [23]. The stimulation of mu receptors in the GI tract produces decreased intestinal motility and an antisecretory effect, the latter being considered the most important in OBD genesis [42]. Additionally, when stimulating gastrointestinal mu receptors, morphine releases serotonin from neurons in the myenteric plexus [23]. The activation of 5-HT receptors releases norepinephrine, which in turn activates sympathetic a2 receptors which also consequently inhibiting enterocyte secretion [23]. There are few recommendations regarding specific opiates with favorable GI profiles, a review of the literature would suggest that morphine, particularly orally administered morphine, has the highest incidence of unwanted GI effects. Tramadol is one opiate that has been shown to have less impact on the GI tract than morphine. However, the potency of tramadol is that of a weak analgesic, thus in some cases it would not meet the analgesic requirement [43]. Rotating to methadone has been shown to reduce laxative consumption [44]. This is probably because methadone has less affinity for peripheral receptors than morphine. Fewer occurrences of constipation and less laxative consumption have been observed with the use of transdermal fentanyl in comparison to oral morphine [45]. Kappa receptor agonists, such as nalbuphine and butorphanol, have been used to take advantage of the analgesic properties of the kappa receptors found on visceral afferents and the lower likelihood of adverse effects [46]. One small study reported kappa agonist-induced pain relief in patients with chronic pancreatitis who were previously refractory to mu-opioid agonists. This study however only studied its effect after only one administration. There have been at least three studies that examined the use and effects of fedotozine in chronic abdominal pain over a period of 6 weeks. These studies showed little overall benefit over placebo [46].

Conclusion

The use of opioid therapy in the setting of abdominal pain remains controversial in the patient with chronic pain. Although there is support for the use of opioids in the acute setting, this support has not translated to more chronic conditions. There has been a dramatic increase in the understanding of the pathophysiology of abdominal pain; however, the elucidation of the conversion of the acute to chronic process and the perpetuation of pain remains poor. Therefore, opioids become the treatment of choice in most cases where other options have not provided benefit and remain a treatment of symptoms rather than underlying pathophysiology.

Multiple issues that are salient in the management of chronic abdominal pain with opioids include choice of molecules, management of opioid-related side effects including constipation and visceral hyperalgesia.

In our experience, empirically derived we tend to favor molecules with "cleaner" metabolic profile and those that theoretically minimize opioid side effects such as cognitive impairment and constipation.

References

1. US Department of Health and Human Services, NCHS. Data from the National Hospital Ambulatory Medical Care Survey: 1999–2008. ER department visits for chest pain and abdominal pain: United States. http://www.cdc.gov/nchs/data/databriefs/db43.htm. Accessed 6 Sept 2013.
2. Peery AF, Dellon ES, Lund J, Crockett SD, McGowan CE, Bulsiewicz WJ, Gangarosa LM, Thiny MT, Stizenberg K, Morgan DR, Ringel Y, Kim HP, Dibonaventura MD, Carroll CF, Allen JK, Cook SF, Sandler RS, Kappelman MD, Shaheen NJ. Burden of gastrointestinal disease in the United States: 2012 update. Gastroenterology. 2012;143(5):1179–87.
3. International Association of Pain. Global year against visceral pain October 2012–October 2013. Epidemiology of abdominal pain. http://www.iasp-pain.org/AM/Template.cfm?Section=Fact_Sheets5&Template=/CM/ContentDisplay.cfm&ContentID=16193. Accessed 25 Sept 2013.
4. Jones AD, Ramakrishnan K. Analgesics in the initial management of acute abdominal pain. Internet J Emerg Med. 2005;2(2). Doi: 10.5580/22ac.
5. Drossman DA. Severe and refractory chronic abdominal pain: treatment strategies. Clin Gastroenterol Hepatol. 2008;6(9):978–82.
6. Smith HS, et al. Current therapy in pain. Philadelphia: Saunders Elsevier; 2009.
7. Manterola C, Astudillo P, et al. Analgesia in patients with acute abdominal pain. Cochrane Database Syst Rev. 2011;19(1), CD005660.
8. Ranji SR, Goldman LE, et al. Do opiates affect the clinical evaluation of patients with acute abdominal pain? JAMA. 2006;296(14):1764–74.
9. Australian Government National Health and Research Council. National Institute of Clinical Studies Emergency Care Community of Practice. Evidence in Practice Series 2008. Pain medication for acute abdominal pain, a summary of best available evidence and information on clinical practice. http://www.nhmrc.gov.au/_files_nhmrc/file/nics/material_resources/pain_medication_
10. aute_abdominal_pain.pdf. Accessed 24 Sept 2013.
11. Matthew D, Hsien-Yen C, Yuping Y, Viswanathan S, Shah N, Stafford R, et al. Ambulatory diagnosis and treatment of nonmalignant pain in the United States, 2000–2010. Med Care. 2013;51(10):870. doi:10.1097/MLR.0b013e3182a95d86.
12. Gerson LB. Gastroenterologists and the US epidemic. Gastroenterology. 2012;142(5):1250–3.
13. Dorn SD, Meek PD, Shah NS. Increasing frequency of opioid prescriptions for chronic abdominal pain in the US outpatient clinics. Clin Gastroenterol Hepatol. 2011;9(12):1078–85.
14. Centers for Disease Control and Prevention. Emergency department visits involving nonmedical use of selected prescription drugs - United States, 2004–2008. MMWR Morb Mortal Wkly Rep. 2010;59(23):705–9.
15. Berland D, Rodgers P. Rational use of opioids for management of chronic nonterminal pain. Am Fam Physician. 2012;86(3):252–8.
16. Chou FR, Fine PG, et al. Clinical guidelines for the use of chronic opioid therapy in chronic noncancer pain. J Pain. 2009;10(2):113–30.
17. Chou R, Fanciullo GJ, Fine PG, et al. Opioids for chronic noncancer pain: prediction and identification of a aberrant drug-related behaviors: a review of the evidence for an American Pain Society and American Academy of Pain Medicine clinical practice guideline. J Pain. 2009;10(2):131–46.
18. Ramsin B, Trescot AM, Datta SM, Benyamin MD, et al. Opioid complications and side effects. Pain Physician. 2008;11(2 Suppl):S105–20.
19. Benzon HT, Raja SN, Molloy RE, Liu SS, Fishman SM. Essentials of pain medicine and regional anesthesia. 2nd ed. Philadelphia: Elsevier; 2005.
20. Epstein H, Hansen C, Thorson D. A protocol for addressing acute pain and prescribing opioids. Minn Med. 2014;97(4):47–51.
21. Koneru A, Satyanarayana S, Rizwan S. Endogenous opioids: their physiological role and receptors. Global J Pharmacol. 2009;3(3):149–53.
22. Argoff CE, Silvershein DI. Comparison of long- and short-acting opioids for the treatment of chronic noncancer pain: tailoring therapy to meet patient needs. Mayo Clin Proc. 2009;84(7):602–12.
23. Pappagallo M. Incidence, prevalence, and management of opioid bowel dysfunction. Am J Surg. 2001;182(5A Supplement):11S–8.
24. Panchal SJ, Muller-Schwefe PM, Wurzelmann JI. Opioid-induced bowel dysfunction: prevalence, pathophysiology, and burden. Int J Clin Pract. 2007;61(7):1181–7.
25. Pasternak GW. Opioids and their receptors, are we there yet? Neuropharmacology. 2014;76 Pt B:198–203.
26. Helm JF, et al. Effects of morphine on the human sphincter of Oddi. Gut. 1988;29(10):1402–7.
27. Hubbard GP, Wolfe KR. Meperidine misuse in a patient with sphincter of Oddi dysfunction. Ann pharmacother. 2003;37(4):534–7.
28. Thompson DR. Narcotic analgesic effects on the sphincter of Oddi: a review of the data and therapeutic implications in treating pancreatitis. Am J Gastroenterol. 2001;96(4):1266–72.
29. Wu SD, et al. Effect of morphine and M-cholinoceptor blocking drugs on human sphincter of Oddi during choledochofiberscopy manometry. Hepatobiliary Pancreat Dis Int. 2003;2(1):121–5.
30. Davis MP. Twelve reasons for considering buprenorphine as a frontline analgesic in the management of pain. J Support Oncol. 2012;10(6):209–19.
31. Osborne R, Joel S, Trew D, Slevin M. Morphine and metabolite behavior after afferent routes of morphine administration: demonstration of the importance of the active metabolite morphine-6-glucuronide. Clin Pharmacol Ther. 1990;47:12–9.
32. Fine PG, Portenoy RK. Strategies for opioid rotation: decision support in chronic pain management. MedScape Education. 2010. http://cme.medscape.com/viewarticle/717832. Accessed 30 Mar 2010.
33. Fine PG, Portenoy RK. For the ad hoc expert panel on evidence review and guidelines for opioid rotation. Establishing "best practices" for opioid rotation: conclusions of an expert panel. J Pain Symptom Manage. 2009;38:418–25.
34. Fine PG, Portenoy RK. Establishing "best practices" for opioid rotation: conclusions of an expert panel. J Pain Symptom Manage. 2009;38(3):418–25.

35. Fine PG, Portenoy RK. A clinical guide to opioid analgesia. New York: Vendome Group, LLC; 2007.
36. Thomsen AB, Becker N, Eriksen J. Opioid rotation in chronic non-malignant pain patients: a retrospective study. Acta Anaesthesiol Scand. 1999;43(9):918–23.
37. Pasternak GW. Molecular biology of opioid analgesia. J Pain Symptom Manage. 2005;29(5 suppl):S2–9.
38. Shooke JE, Pelton JT, Hurby VJ, et al. Peptide opioid antagonist separates peripheral and central antitransit effects. J Pharmacol Exp Ther. 1987;243:492–500.
39. Manara L, Bianchetti A. The central and peripheral influences of opioids on the gastrointestinal propulsion. Annu Rev Pharmacol Toxicol. 1985;25:249–73.
40. Margolin S. Decreased gastrointestinal propulsive activity after intracranial morphine. Fed Proc. 1954;13:383–4.
41. Tamayo AC, Diaz-Zuluaga PA. Management of opioid-induced bowel dysfunction in cancer patients. Support Care Cancer. 2004;12:613–8.
42. Mancini I, Bruera E. Constipation in advanced cancer patients. Support Care Cancer. 1998;6:356–64.
43. Wilder-Smith C, Betiga A. The analgesic tramadol has minimal effect on motor gastrointestinal function. Br J Clin Pharmacol. 1997;43:71–5.
44. Daenick P, Bruera E. Reduction in constipation and laxative requirements following opioid rotation to methadone: a report of four cases. J Pain Symptom Manage. 1999;18:303–9.
45. Radbruch L, Sabatowski R, Loick G, et al. Constipation and the use of laxatives: a comparison between transdermal fentanyl and oral morphine. Palliat Med. 2000;14:111–9.
46. Benzon H, Rathmell J, et al. Raj's practical pain management. 4th ed. Philadelphia: Mosby Elsevier; 2008.

15 Sympathetic Blocks for Chronic Abdominal Pain

Robert Bolash and Bruce Vrooman

Introduction

The abdominal and pelvic sympathetic nervous system is comprised of a complex and richly innervated network of sympathetic fibers that are amenable to local anesthetic blockade at discrete sites as part of a multimodal analgesic strategy for malignant or non-malignant abdominal or pelvic pain conditions. Patients suffering from abdominal pain that is entirely, or in part, mediated by transmission through the sympathetic nervous system may decrease both pain scores and reliance on analgesic medications, while improving function and quality of life [1–4]. Through understanding the relevant anatomy, indications, and interventional approaches, the pain physician can offer a supplemental analgesic strategy to patients suffering from a wide variety of abdominal and pelvic pain conditions.

Anatomy of the Sympathetic Nervous System

The sympathetic trunk is a paired network that lies anterolateral to the vertebral column beginning in the cervical region and extending to the coccyx, at which point bilateral trunks converge to form the ganglion impar (ganglion of Walther). Interrupting transmission of painful stimulus is possible at multiple sites along this axis, and several approaches to performing diagnostic and therapeutic blocks of the splanchnic nerves, celiac plexus, superior and inferior hypogastric plexi, and ganglion impar are presented with an emphasis on practical application by the interventional pain physician (Table 15.1).

R. Bolash, M.D. (✉) • B. Vrooman, M.D.
Pain Management Department, Cleveland Clinic, 9500 Euclid Ave/C25, Cleveland, OH 44195, USA
e-mail: BOLASHR@ccf.org; vroomab@ccf.org; BruceVrooman@gmail.com

Splanchnic Nerve Block

Anatomy and Related Structures

The greater, lesser, and least splanchnic nerves arise from preganglionic fibers which course through the crura of the diaphragm before synapsing in the celiac ganglion. Cadaveric studies have revealed that the greater splanchnic nerves originate between T5 and T10, while the lesser splanchnic nerves arises from T9 to T10, and the least splanchnic nerves from T11 to T12. Interventional approaches target these nerves at T11 and T12 when the nerves are in close proximity to the vertebral body (Fig. 15.1). The splanchnic nerve block offers the ability to both avoid the celiac artery and selectively interrupt transmission before the splanchnic nerves reach the celiac plexus [5].

At the level of the lower thoracic vertebral bodies, the splanchnic nerves have a predictable course bordered laterally by the pleura, ventrally by the posterior mediastinum, medially by the thoracic vertebral bodies, and dorsally by the pleural attachments to the vertebra. Additionally, this location is amenable to radiofrequency neurotomy for those patients who experience temporary, but non-sustained relief with local anesthetic blocks [6–8].

Indications

The splanchnic nerve block can be performed for a wide variety of malignant and non-malignant pain conditions arising from the stomach, pancreas, small bowel, proximal large bowel, kidneys, and proximal ureters.

Interventional Approach

Since the splanchnic nerve block was initially described, the technique has evolved with respect to the amount of

L. Kapural (ed.), *Chronic Abdominal Pain: An Evidence-Based, Comprehensive Guide to Clinical Management*,
DOI 10.1007/978-1-4939-1992-5_15, © Springer Science+Business Media New York 2015

Table 15.1 The sympathetic nervous system richly innervates the abdominal and pelvic viscera with frequent overlap in target organs

Sympathetic nerve supply	Target organs
Splanchnic nerves	Stomach, duodenum, jejunum, ileum, pancreatic head, proximal colon, kidneys, proximal ureters
Celiac plexus	Stomach, duodenum, jejunum, ileum, pancreas, proximal colon, spleen, gall bladder, liver
Superior hypogastric plexus	Distal colon, bladder, uterus, vagina, cervix, prostate, testes, urethra
Inferior hypogastic plexus	Distal colon, urinary bladder, penis, vagina, rectum, anus, perineum
Ganglion impar	Coccyx, cervix, distal rectum, perineum, prostate, sigmoid colon, distal urethra, vulva, scrotum, distal vagina

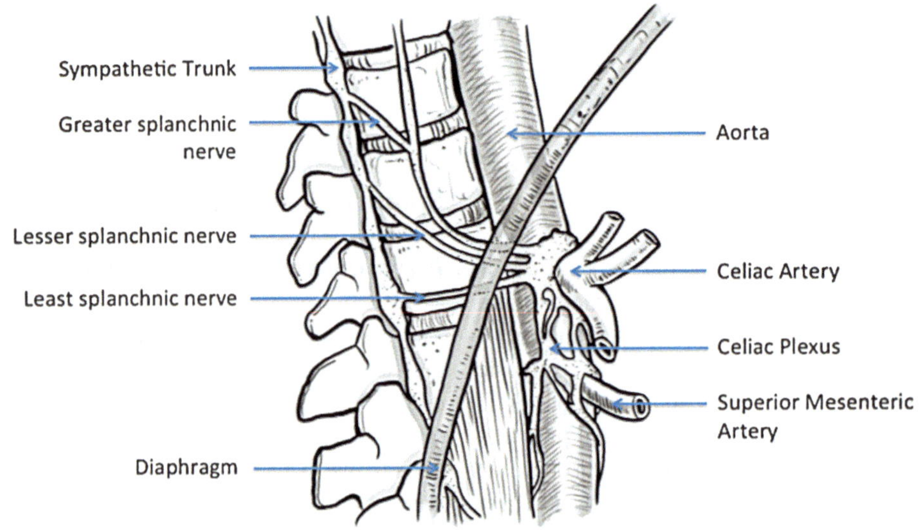

Fig. 15.1 Both the splanchnic and celiac nerve blocks can be performed via a dorsal approach. The sympathetic chain courses in a cranial-caudad direction and transmits the greater, lesser, and least splanchnic nerves which can be blocked at the level of the T11 and T12 vertebral bodies. The celiac plexus lies anterior to the abdominal aorta and is comprised of a complex meshwork of nerves surrounding the celiac and superior mesenteric arteries

fluoroscopic obliquity used, the needle chosen, the level of the block, and the type and volume of injectate.

A dorsal approach performed under fluoroscopy can be accomplished by positioning the patient prone and identifying the T11 and T12 vertebral bodies using radiographic imaging. After squaring the vertebral end plates and introducing a 10- to 15-degree ipsilateral oblique fluoroscopic angle, the junction of the ribs and the vertebral body will become visible. This junction marks the skin entry point for both the T11 and T12 levels (Fig. 15.2). Following aseptic preparation of the injection site, a local anesthetic skin wheal is created and a 22 or 25 gauge spinal needle is advanced coaxially hugging the lateral aspect of the mid-portion of the T11 or T12 vertebral body. The needle trajectory should course along the mid-portion of the vertebral body to avoid inadvertent entry into the disc or illiciting paresthesia along the nerve root. A cross-table lateral fluoroscopic image will reveal the needle advancing ventrally to lie at the middle third of the T11 vertebral body (Fig. 15.3). Two milliliters of radio-opaque contrast demonstrates spread posterior to the aorta and anterior to the foramen (Figs. 15.4 and 15.5). A diagnostic (and potentially therapeutic) unilateral or bilateral block of the greater, lesser, and least splanchnic nerves can be performed with 2 mL of injectate, such as 0.5 % bupivacaine, via a single injection. This volume is thought to be sufficient to block all three splanchnic nerves, though the technique can be modified by introducing a second needle at the T12 level and employing a smaller volume of injectate. Because of the proximity to the pleura, the final needle position should not extend beyond the anterior third of the vertebral body.

Choice of Injectate

Boas quantified the volume of the area bounded by the crura and the vertebral body to total 10 mL on each side and this represents a reasonable maximum injectate volume [1]. Local anesthetics, neurolytic agents, and steroids have all been used alone or in combination when performing the splanchnic nerve block. No randomized controlled trials have demonstrated the superiority of a single agent or combination agents over another [9].

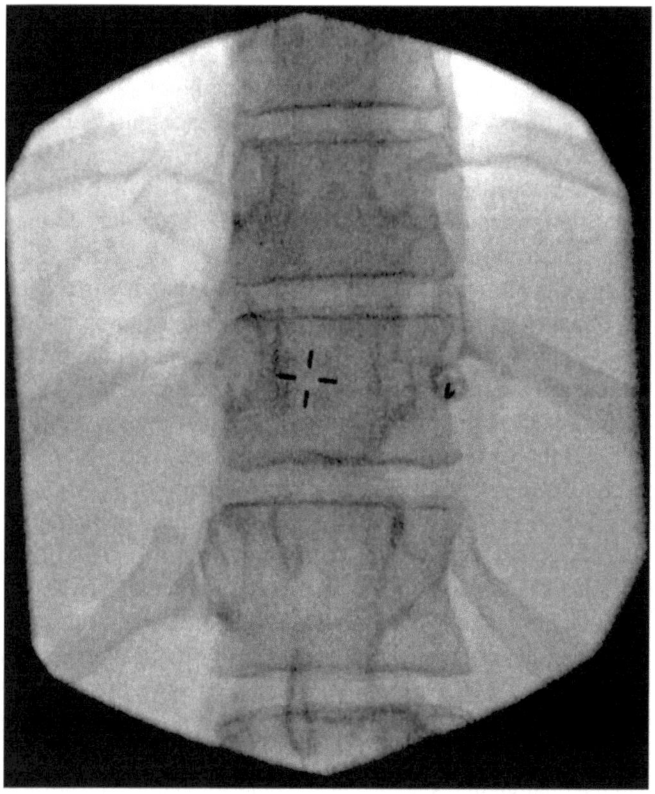

Fig. 15.2 Splanchnic nerve block. This anterior–posterior radiograph shows the needle advancing ventrally along the midpoint of the vertebral body to avoid contact with the spinal nerves and preventing penetration of the intervertebral disc

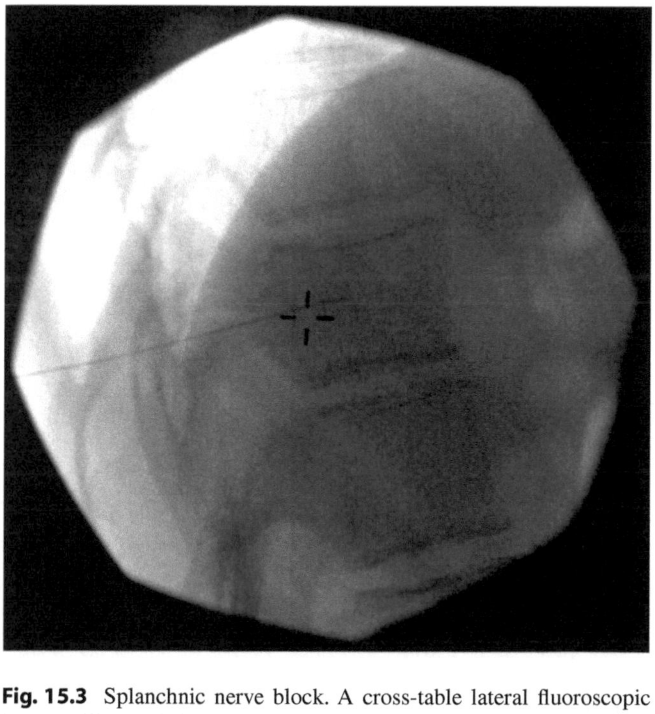

Fig. 15.3 Splanchnic nerve block. A cross-table lateral fluoroscopic image will reveal the needle advancing ventrally to lie at the middle third of the T11 vertebral body

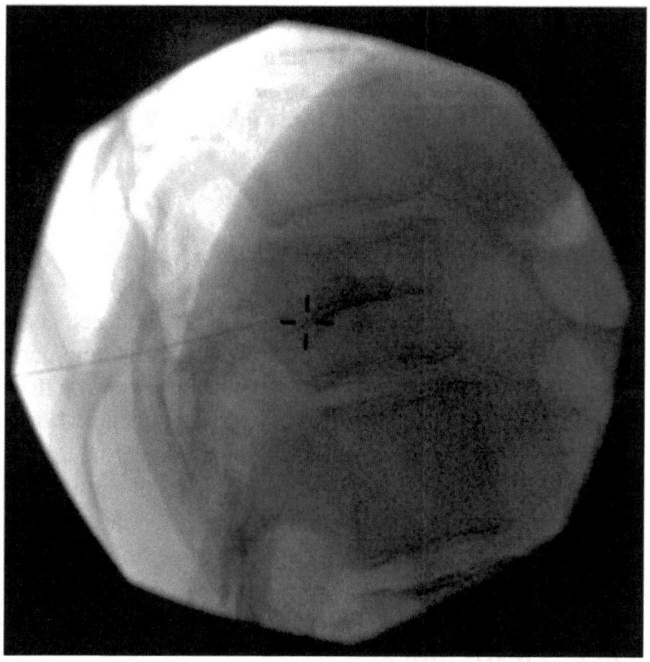

Fig. 15.4 Splanchnic nerve block. Contrast is seen hugging the vertebral body along the course of the splanchnic nerves. A small volume injectate targets the greater and lesser splanchnic nerves at the level of the T11 vertebral body. Performing the injection at the T12 level will target the least splanchnic nerve

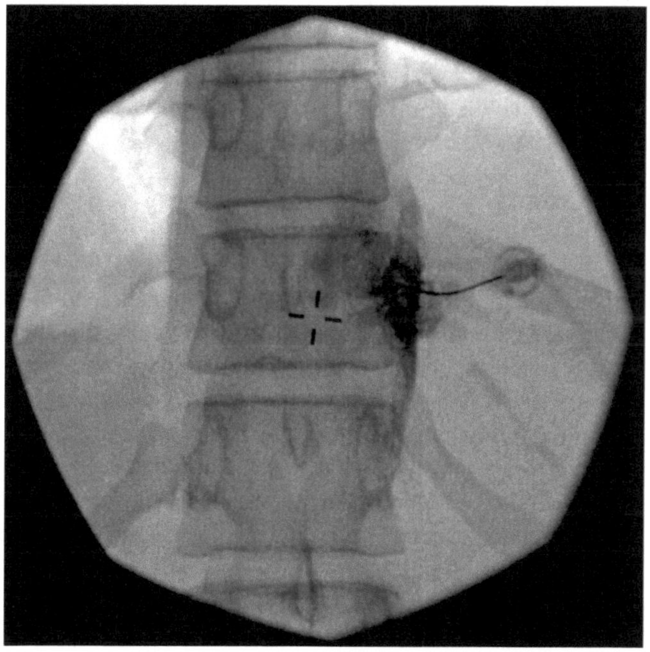

Fig. 15.5 Splanchnic nerve block. Contrast is seen hugging the vertebral body. A large volume injection at T11 can spread caudally to cover the least splanchnic nerve obviating the need to perform the block at the T12 level. A similar approach to the paired splanchnic nerves would be carried out on the contralateral side

When considering radiofrequency ablation, we advocate increasing specificity by performing at least one diagnostic block at both the T11 and T12 levels with a small volume of injectate which best approximates the size of the radiofrequency lesion.

Complications

Similar to the celiac plexus block, diarrhea, transient back pain, and hypotension may result from the splanchnic nerve block. Additional risks include subarachnoid or epidural injection, vascular or intradiscal injection, pneumothorax, chylothorax, and retrothoracic hematoma. Phenol neurolysis of the splanchnic nerves has generated reports of cardiac dysrhythmia and diaphagramic paralysis, thus favoring radiofrequency ablation as the preferred method of neurolysis.

Celiac Plexus Block

Anatomy and Related Structures

Surrounding the origin of the celiac and superior mesenteric arteries from the abdominal aorta lies the confluence of the splanchnic, parasympathetic, and sensory nerve fibers collectively termed the celiac plexus (Fig. 15.1). Neural blockade at the celiac plexus is not specific to sympathetic fibers, instead involving multiple types of fibers that course through the ganglion. For this reason, patient response to celiac block may vary from the response to the splanchnic nerve block.

Indications

The celiac plexus block, like the splanchnic nerve block, can be performed for a wide variety of malignant and nonmalignant pain conditions arising from the stomach, small bowel, pancreas, proximal large bowel, kidneys, and proximal ureters. It is perhaps best characterized as an accepted technique for malignant pain arising from pancreatic cancer. A large meta-analysis demonstrated that when celiac neurolysis is utilized for patients with pancreatic cancer, 90 % of patients report sustained pain relief at three-month follow-up [1, 2, 5, 10–12].

Technique

Nearly a dozen interventional techniques have been described by way of dorsal or ventral approaches, employing endoscopic ultrasound, fluoroscopy, CT, or MRI. Surgical approaches have also been described and include thoracoscopic and open approaches via laparotomy [13–23].

The celiac plexus can be approached with relative ease via a dorsal transaortic approach. Whereas once this technique was performed primarily with fluoroscopy, CT-guidance is gaining widespread acceptance due to the fact that this can identify anatomic variations, particularly those that develop with cancer progression [24].

After positioning the patient prone and visualizing the left L1 transverse process, a skin wheal is created approximately 6 cm to the left of midline. A 22 gauge spinal needle is then advanced ventrally avoiding the transverse process and coursing adjacent to the mid-portion of the L1 vertebral body. If the vertebral body is encountered, the needle should be redirected ventrally or "walked-off" employing rotation of the bevel. When the needle passes the anterior border of the body of L1, the stylete is removed and the needle is advanced under continuous aspiration until the appearance of pulsatile heme is evident suggesting an intra-aortic position. The needle is then advanced further until a change in resistance is again noted suggesting passage through the anterior wall of the aorta and the aspiration of heme is no longer possible. Some use a saline-filled syringe to appreciate the change in resistance as the anterior wall of the aorta is penetrated. Injection of 3 mL of radio-opaque contrast demonstrates cranio-caudal spread and excludes vascular uptake (Figs. 15.6 and 15.7).

Aortic puncture can be avoided via a similar dorsal approach. Upon identification of the L1 vertebral body, a

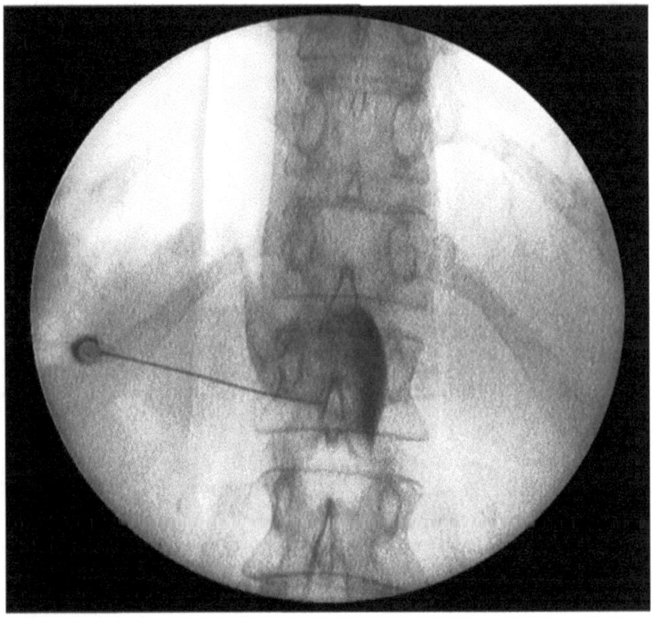

Fig. 15.6 Celiac plexus block. In the transaortic approach, the spinal needle is seen after having advanced through the aorta with contrast spreading anterior to the great vessels

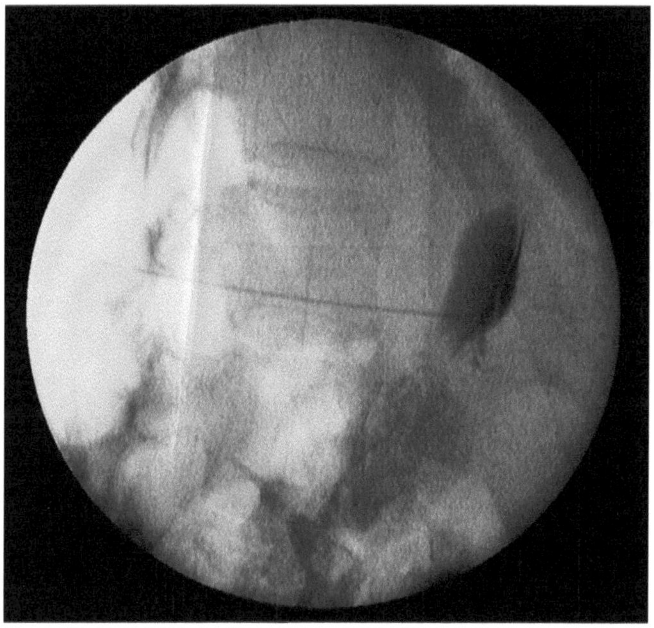

Fig. 15.7 Celiac plexus block. A cross-table lateral image demonstrates the cranio-caudal spread of contrast anterior to the great vessels when approaching the celiac plexus with a transaortic approach

skin wheal is introduced over an entry point corresponding to the lateral fluoroscopic aspect of both the left and right L1 transverse processes. A 22 gauge spinal needle is then introduced in a similar manner to the transaortic approach and advanced to terminate just beyond the anterior aspect of the L1 vertebral body behind the posterior wall of the aorta. 3 mL of radio-opaque contrast material is injected confirming precrural contrast spread and the injectate volume is increased to augment the ventral spread of local anesthetic by gravity. If the contrast spread is unilateral as visualized by an antero-posterior fluoroscopic image, the procedure often needs to be repeated via an identical approach on the contralateral side of the L1 vertebral body. Increasing the volume of injectate is one approach to account for anatomic variations in the take-off of the celiac artery along the cranio-caudal axis if the procedure is performed with fluoroscopy rather than CT.

Complications

Similar to the splanchnic nerve block, diarrhea is an expected outcome of the procedure; thus we recommend indicating this possibility as part of the informed consent. Reported complications include retroperitoneal abscess, bowel perforation, aortic wall dissection, aortic pseudoanyerism, and paraplegia. The transaortic route is not recommended in patients with aortic aneurisms or calcifications in proximity to the area of aortic puncture [25].

Superior Hypogastric Plexus Block

Anatomy and Related Structures and Indications

The superior hypogastric plexus contains sympathetic fibers that innervate the distal colon, bladder, uterus, vagina, cervix, prostate, testes, and urethra. The plexus is a paired structure located ventral to the L5–S1 disc space and medial to the bifurcation of the common iliac arteries (Fig. 15.8). Malignant and non-malignant pain arising from these organs is amenable to a superior hypogastic block if the pain condition is entirely, or in part, maintained by sympathetic transmission [1, 2, 4, 26–28].

Interventional Approaches

Several approaches to the superior hypogastric plexus have been identified. Plancarte first described a posterior approach that involves needle entry at the level of the L4–L5 disc space coursing caudally toward the anterior body of the L5 vertebral body. After positioning the patient prone on a procedure table, the L4–L5 disc space is identified via an anterior–posterior view. Following aseptic preparation of the injection site, a skin wheal is created over the needle entry site 5 cm from midline at the level of the L4–L5 disc. A 22 G spinal needle is then inserted through the skin wheal directed toward the midline with a 30-degree caudal and 45-degree medial angle to contact the lateral aspect of L5. Upon contact with the vertebral body, the needle should be walked off ventrally and a lateral fluoroscopic image will demonstrate the needle tip in front of the L5 vertebral body. A change in resistance will be noted with advancement of the needle 1 cm anterior to the vertebral body as the tip passes through the anterior aspect of the psoas muscle. Injection of 3 mL of radio-opaque contrast material will spread along the midline with anterior–posterior fluoroscopy and along the anterior psoas fascia via a lateral fluoroscopic view (Figs. 15.9 and 15.10). Radiofrequency neurotomy can be performed in a similar manner.

More recently, approaches using CT-guided needle placement from both dorsal and ventral approaches have been described. Anterior approaches can be performed with either fluoroscopy or CT and begin with identification of the L5 vertebral body. The ventral insertion site is located 6 cm below the umbilicus and a needle is advanced coaxially through the soft tissues of the abdomen to contact the anterior aspect of the L5 vertebral body. Though technically less challenging, the needle path can traverse bowel, bladder, and vascular structures predisposing patients to the complications associated with perforating these organs.

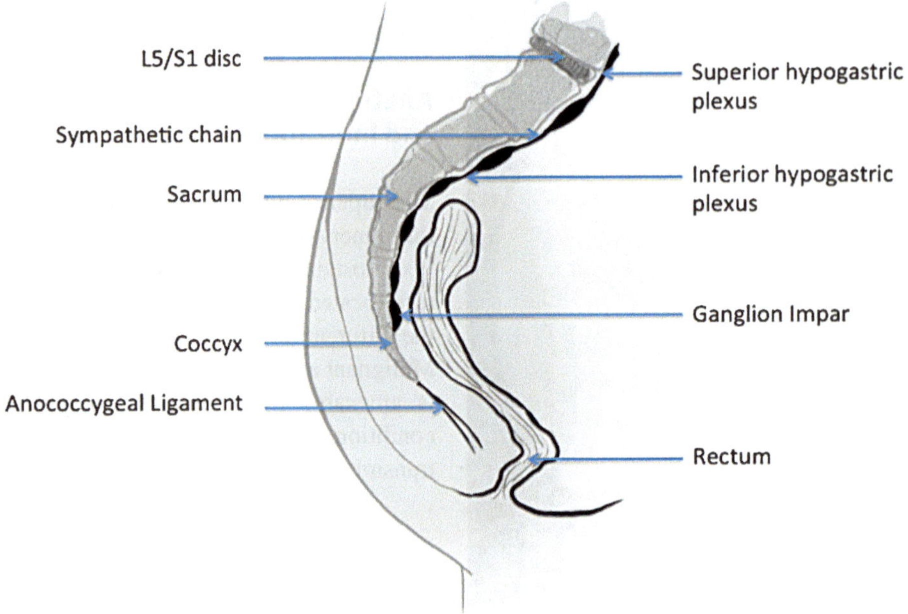

Fig. 15.8 The superior hypogastric plexus, inferior hypogastric plexus, and ganglion impar blocks can be performed via a dorsal approach. The paired sympathetic chains extend into the pelvis and coalesce at the ganglion impar. The superior hypogastric plexus lies anterior the L5–S1 disc space. The inferior hypogastric plexus is located along the ventral aspect of the sacrum at the level of the S2 foramen. The ganglion impar is ventral to the located at the sacrococcygeal junction

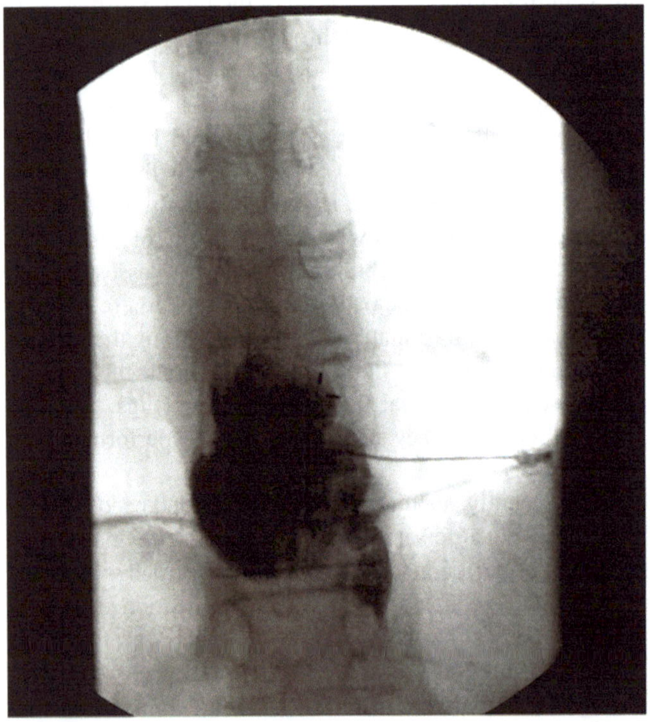

Fig. 15.9 Superior hypogastric nerve block. Radiographic contrast is visualized anterior to the vertebral body within the retroperitoneal space. A single large volume injection resulted in bilateral spread obviating the need to perform the injection on the contralateral side

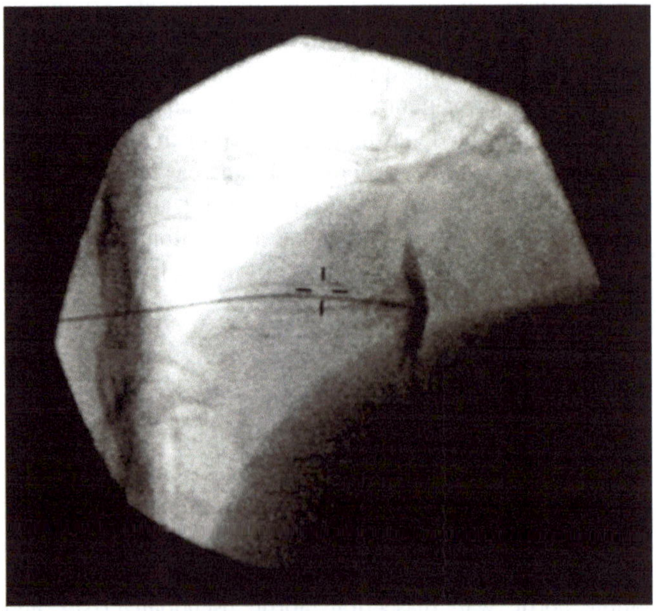

Fig. 15.10 Superior hypogastric plexus block. The spinal needle is advanced through the disc and the anterior longitudinal ligament to reveal contrast spreading within the retroperiotoneal space

A transdiscal approach using a 22 G spinal needle advanced through the L5–S1 disc space with either CT or fluoroscopic imaging combined with a loss of resistance technique is also described. The needle enters the skin taking an approach just lateral to the superior articular process and advanced through the disc. Upon advancement through the anterior aspect of the disc space, the injectate can be visualized within the retroperitoneal space. Though the median time required to perform this block was shorter in a head-to-head comparison with the classical posterior approach of Plancarte, one possible disadvantage includes the potential for discitis or disrupting the architecture of the disc required by this approach [29–32].

Choice of Injectate and Local Anesthetic Volume

Five to ten milliliters of local anesthetic with or without steroid is typically administered for therapeutic blocks. Neurolysis is accomplished with a similar volume of phenol. Radiofrequency ablation has been accomplished via these same approaches.

Complications

Vascular injury to the common iliac vessels may lead to retroperitoneal hematoma. When utilizing the classical approach of Plancarte, the needle path can be obstructed by the transverse process of L5 or by the iliac crest necessitating an entry site further cephalolateral to the L4–L5 disc space and employing further cranio-caudal fluoroscopic tilt to contact the anterior aspect of the L5 vertebral body. Modifications of the approach employing coaxial advancement of the needle have been suggested as an alternative and permit the practitioner to observe a clear needle tract before skin entry.

Inferior Hypogastric Plexus Block

Anatomy and Related Structures

The inferior hypogastic plexus is a paired paravertebral network of sympathetic fibers that lie presacrally bordered dorsally by the sacral bone and ventrally by the posterior wall of the rectum (Fig. 15.8).

Indications

Malignant and non-malignant sympathetically maintained pain conditions involving the distal colon, bladder, penis, vagina, rectum, anus, and perineum are amenable to treatment through disrupting transmission at the inferior hypogastric plexus.

Interventional Approaches

We advocate approaching the plexus using the trans-sacral approach. After positioning the patient prone, the site of needle insertion at the S2 foramen is identified using anterior–posterior fluoroscopic imaging. Following aseptic preparation of the injection site, a 3 mL skin wheal is created 1 cm lateral to the ipsilateral S2 foramen. A 22 G or 25 G spinal needle is advanced from the skin entry point toward the sacrum and contacts the sacral bone just lateral to the S2 foramen. The needle should then be walked into the lateral portion of the S2 foramen and advanced ventrally and medially through the foramen avoiding eliciting a paresthesia. The optimal trajectory will result in the needle exiting the foramen and coursing toward the midline, as seen on a lateral fluoroscopic radiograph. The needle tip is visualized by injecting 1 mL of radio-opaque contrast and demonstrating cranio-caudal spread within the presacral plane via an anterior–posterior radiograph. The procedure can be repeated on the contralateral side if unilateral contrast spread is observed [33, 34].

Choice of Injectate and Local Anesthetic Volume

Five to fifteen milliliters of local anesthetic with or without steroid is typically administered for therapeutic blocks. A single unilateral diagnostic injection may result in contralateral spread negating the need to perform bilateral blocks. In consideration for radiofrequency ablation, we recommend improving specificity by decreasing the volume to best approximate the size of the radiofrequency lesion.

Complications

The trans-sacral approach avoids penetrating the disc space and the possible resultant consequences. If the S2 foramen is not visible, or entry through the foramen is complicated by foraminal stenosis, the plexus can also be approached through either the S1 or S3 level. The presence of a paresthesia in the S2 nerve root is not uncommon and will necessitate withdrawing and redirecting the needle trajectory within the foramen. Because the rectum lies immediately ventral to the plexus, puncture of the hollow viscous is possible if the needle tip is advanced ventrally and may predispose the patient to an infection if the pelvis becomes contaminated with colonic microorganisms.

Ganglion Impar Block

Anatomy and Related Structures

The ganglion impar, also known as the Ganglion of Walther, is a single terminal structure containing cell bodies of the paired paravertebral sympathetic chain. It is located in the pre-coccygeal space, which is bordered dorsally by the anterior sacrococcygeal ligament and ventrally by the rectum (Fig. 15.8). The ganglion lies immediately anterior to the sacrococcygeal joint though anatomic variations along the craniocaudal axis of up to 2 cm have been described.

Indications

Ganglion impar blocks have been performed for both sympathetically mediated and nociceptive pain complaints of the perineum, distal rectum, urethra, vulva, vagina, and scrotum. A wide number of reported indications for benign and malignant pain complaints have been described including coccydynia, sacrococcygeal joint pain, vulvovaginitis, post-episiotomy pain, chronic proctitis, prostatitis, sacral postherpetic neuralgia, primary and metastatic malignancies of the pelvic organs.

Interventional Approaches

At least four interventional approaches to the ganglion impar have been described: (1) via the anococcygeal ligament in the midline approach, (2) via the anococcygeal ligament in the paramedian approach, (3) via a lateral approach, and (4) via the sacroccygeal joint space. These approaches have been performed blindly, with a loss of resistance technique and with imaging modalities including fluoroscopy, CT, or ultrasound.

Approaching the ganglion via the anococcygeal ligament from a midline caudal approach as the block was first described is technically challenging. The practitioner utilizes a 22 G spinal needle introduced through the anococcygeal ligament and directs the tip cephalad within the retroperitoneal space toward the sacrococcygeal junction. This necessitates introducing a bend in the spinal needle that mirrors the irregular anterior curvature of the patient's sacrum prior to insertion. Given the possibility of inadvertent rectal injury caused by this technique, some advocate the insertion of the nondominant finger into the rectum while advancing the needle cranially, though this raises similar concern for inadvertent digital needle stick injury to the operator, increased radiation exposure to the physician, and patient discomfort. Periosteal injection and needle fracture have also been noted to occur with this approach.

Similarly challenging is the paramedian approach via the anococcygeal ligament. Though this does not necessitate transversing the sacrum or coccyx, it requires that the physician introduce two 110° bends into a spinal needle and then simultaneously advance and corkscrew the needle so that the needle tip terminates in close proximity to the ganglion impar, again without inadvertently fracturing the needle or puncturing the rectum.

The trans-sacrococcygeal method is perhaps the fastest approach to the ganglion. Both fluoroscopic and ultrasound approaches have been described. Either imaging modality can be supplemented with a loss of resistance technique based upon the operator's preference. After positioning the patient in the prone position on a procedure table, the site of needle insertion at the sacrococcygeal ligament is identified using anterior–posterior fluoroscopic imaging. Following aseptic preparation of the injection site, a skin wheal is created over the midline insertion site. A 22 G or 25 G spinal needle is then inserted coaxially through the skin toward the sacrococcygeal ligament under intermittent fluoroscopic guidance. As the needle enters the disc space, a change in resistance is noted. At this interval, the C-arm is positioned laterally and the needle is advanced toward the ventral aspect of the disc. A second change in resistance is noted when exiting the anterior portion of the disc and passing through the anterior sacrococcygeal ligament (Fig. 15.11). The location of the needle tip is then confirmed by injection of 2 mL of radio-opaque contrast. A favorable distribution of contrast demonstrates a "reverse-comma" sign indicating cranio-caudal spread along the ventral wall of the sacrum (Fig. 15.12). After ensuring that vascular uptake is absent, the 5 mL injectate is introduced and the needle is removed. A similar approach is utilized for radiofrequency ablation of the ganglion [35–38].

Choice of Injectate and Local Anesthetic Volume

Local anesthetics, steroids, clonidine, Botulinum toxin, alcohol, and phenol have been utilized as single agents or combination in various reports. Given the variable cranio-caudal location of the ganglion, we advocate administration of 5 mL of injectate for therapeutic blocks, and a reduced volume for diagnostic blocks prior to radiofrequency ablation given the limited lesion size of the neurotomy probe.

Complications

Transversing the sacrococcygeal ligament in the midline approach may be challenging if the ligament is calcified or impossible if the coccyx is sacralized. Puncture of the rectum is possible if the needle is advanced ventrally and may result in peritonitis due to contamination of the pelvis with

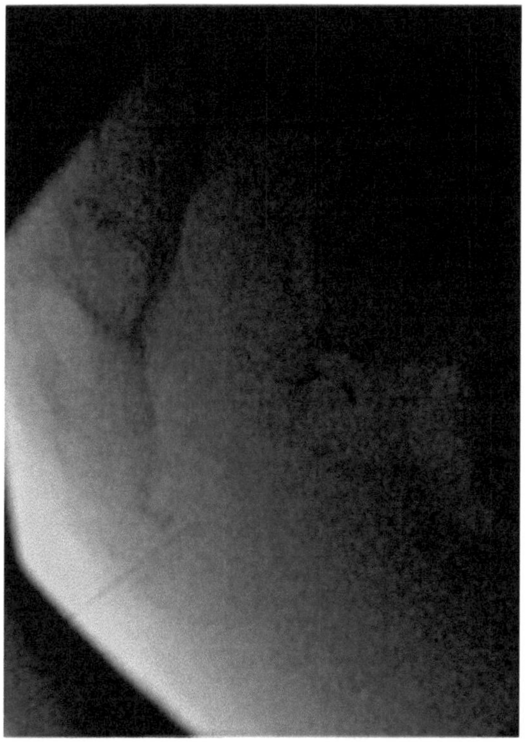

Fig. 15.11 Ganglion impar. A spinal needle courses through the sacrococcygeal disc space with the tip terminating just beyond the anterior aspect of the disc. A change in resistance is appreciable and the final position is confirmed by fluoroscopic imaging

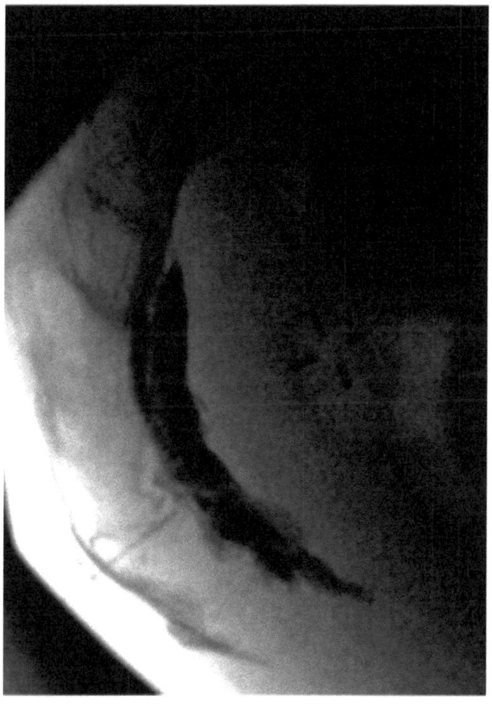

Fig. 15.12 Ganglion impar. Contrast spread along the ventral portion of the sacrum creates a "reverse comma" sign. Using a large volume for a therapeutic block ensures that anatomical variants along the craniocaudal axis are not missed. Note that contrast extravasated posteriorly prior to advancement through the rudimentary disc

fecal material. Similarly, discitis results from fecal or presacral skin contaminants when the ganglion is approached via the midline transdiscal approach.

Conclusion

Considering that the cost of treating patients with pharmacotherapy alone is substantial and often ineffective, the pain physician can offer interventional approaches to the patient suffering from sympathetically mediated abdominal or pelvic pain complaints. Several retrospective studies have demonstrated the decrease in analgesic intake and significant decrease in visual analog pain scores following sympathetic blocks for abdominal and pelvic pain complaints. These therapies warrant consideration when devising a multimodal analgesic strategy for patients suffering from chronic abdominal and pelvic pain complaints.

Acknowledgment This book chapter was funded by the Cleveland Clinic Anesthesiology Institute and Department of Pain Management.

References

1. Boas RA. Sympathetic blocks in clinical practice. Int Anesthesiol Clin. 1978;16:149.
2. de Leon-Casasola OA. Critical evaluation of chemical neurolysis of the sympathetic axis for cancer pain. Cancer Control. 2000;7(2):142–8.
3. de Oliveira R, dos Reis MP, Prado WA. The effects of early or late neurolytic sympathetic plexus block on the management of abdominal or pelvic cancer pain. Pain. 2004;110(1–2):400–8.
4. Nader A, Candido KD. Pelvic pain. Pain Pract. 2001;1(2):187–96.
5. Loukas M, Klaassen Z, Merbs W, Tubbs RS, Gielecki J, Zurada A. A review of the thoracic splanchnic nerves and celiac ganglia. Clin Anat. 2010;23(5):512–22.
6. Raj PP, Sahinler B, Lowe M. Radiofrequency lesioning of splanchnic nerves. Pain Pract. 2002;2(3):241–7.
7. Papadopoulos D, Kostopanagiotou G, Batistaki C. Bilateral thoracic splanchnic nerve radiofrequency thermocoagulation for the management of end-stage pancreatic abdominal cancer pain. Pain Physician. 2013;16(2):125–33.
8. Garcea G, Thomasset S, Berry DP, Tordoff S. Percutaneous splanchnic nerve radiofrequency ablation for chronic abdominal pain. ANZ J Surg. 2005;75(8):640–4.
9. Manchikanti KN, Pampati V, Damron KS, McManus CD. A double-blind, controlled evaluation of the value of sarapin in neural blockade. Pain Physician. 2004;7(1):59–62.
10. Puylaert M, Kapural L, Van Zundert J, Peek D, Lataster A, Mekhail N, van Kleef M, Keulemans YC. Pain in chronic pancreatitis. Pain Pract. 2011;11(5):492–505.
11. Sişu AM, Petrescu CI, Cebzan CC, Motoc A, Bolintineanu S, Vaida AM, Niculescu MC, Rusu MC. The adult coeliac ganglion: a morphologic study. Rom J Morphol Embryol. 2008;49(4):491–4.
12. Titton RL, Lucey BC, Gervais DA, Boland GW, Mueller PR. Celiac plexus block: a palliative tool underused by radiologists. AJR Am J Roentgenol. 2002;179(3):633–6.
13. De Cicco M, Matovic M, Bortolussi R, Coran F, Fantin D, Fabiani F, Caserta M, Santantonio C, Fracasso A. Celiac plexus block:

injectate spread and pain relief in patients with regional anatomic distortions. Anesthesiology. 2001;94(4):561–5.
14. Ischia S, Ischia A, Polati E, Finco G. Three posterior percutaneous celiac plexus block techniques. A prospective, randomized study in 61 patients with pancreatic cancer pain. Anesthesiology. 1992;76(4): 534–40.
15. Lieberman RP, Waldman SD. Celiac plexus neurolysis with the modified transaortic approach. Radiology. 1990;175(1):274–6.
16. Gress F, Schmitt C, Sherman S, Ikenberry S, Lehman G. A prospective randomized comparison of endoscopic ultrasound- and computed tomography-guided celiac plexus block for managing chronic pancreatitis pain. Am J Gastroenterol. 1999;94(4): 900–5.
17. Soweid AM, Azar C. Endoscopic ultrasound-guided celiac plexus neurolysis. World J Gastrointest Endosc. 2010;2(6):228–31.
18. Wyse JM, Carone M, Paquin SC, Usatii M, Sahai AV. Randomized, double-blind, controlled trial of early endoscopic ultrasound-guided celiac plexus neurolysis to prevent pain progression in patients with newly diagnosed, painful, inoperable pancreatic cancer. J Clin Oncol. 2011;29(26):3541–6.
19. Kambadakone A, Thabet A, Gervais DA, Mueller PR, Arellano RS. CT-guided celiac plexus neurolysis: a review of anatomy, indications, technique, and tips for successful treatment. Radiographics. 2011;31(6):1599–621.
20. Walsh N, Sarria JE. Management of chronic pain in a patient with autosomal dominant polycystic kidney disease by sequential celiac plexus blockade, radiofrequency ablation, and spinal cord stimulation. Am J Kidney Dis. 2012;59(6):858–61.
21. Erdek MA, Halpert DE, González Fernández M, Cohen SP. Assessment of celiac plexus block and neurolysis outcomes and technique in the management of refractory visceral cancer pain. Pain Med. 2010;11(1):92–100.
22. Eisenberg E, Carr DB, Chalmers TC. Neurolytic celiac plexus block for treatment of cancer pain: a meta-analysis. Anesth Analg. 1995;80(2):290–5.
23. Yan BM, Myers RP. Neurolytic celiac plexus block for pain control in unresectable pancreatic cancer. Am J Gastroenterol. 2007; 102(2):430–8.
24. Yang IY, Oraee S, Viejo C, Stern H. Computed tomography celiac trunk topography relating to celiac plexus block. Reg Anesth Pain Med. 2011;36(1):21–5.
25. Navarro-Martinez J, Montes A, Comps O, Sitges-Serra A. Retroperitoneal abscess after neurolytic celiac plexus block from the anterior approach. Reg Anesth Pain Med. 2003;28(6):528–30.
26. Plancarte R, Amescua C, Patt RB, Aldrete JA. Superior hypogastric plexus block for pelvic cancer pain. Anesthesiology. 1990;7 3(2):236–9.
27. Schmidt AP, Schmidt SR, Ribeiro SM. Is superior hypogastric plexus block effective for treatment of chronic pelvic pain? Rev Bras Anestesiol. 2005;55(6):669–79.
28. Bosscher H. Blockade of the superior hypogastric plexus block for visceral pelvic pain. Pain Pract. 2001;1(2):162–70.
29. Kanazi GE, Perkins FM, Thakur R, Dotson E. New technique for superior hypogastric plexus block. Reg Anesth Pain Med. 1999;24(5):473–6.
30. Stevens DS, Balatbat GR, Lee FMK. Coaxial imaging technique for superior hypogastric plexus block. Reg Anesth Pain Med. 2000;25(6):643–7.
31. Turker G, Basagan-Mogol E, Gurbet A, Ozturk C, Uckunkaya N, Sahin S. A new technique for superior hypogastric plexus block: the posteromedian transdiscal approach. Tohoku J Exp Med. 2005;206(3):277–81.
32. Gamal G, Helaly M, Labib YM. Superior hypogastric block: transdiscal versus classic posterior approach in pelvic cancer pain. Clin J Pain. 2006;22(6):544–7.
33. Schultz DM. Inferior hypogastric plexus blockade: a transsacral approach. Pain Physician. 2007;10(6):757–63.
34. Choi HS, Kim YH, Han JW, Moon DE. A new technique for inferior hypogastric plexus block: a coccygeal transverse approach. Korean J Pain. 2012;25(1):38–42.
35. Eker HE, Cok OY, Kocum A, Acil M, Turkoz A. Transsacrococcygeal approach to ganglion impar for pelvic cancer pain: a report of 3 cases. Reg Anesth Pain Med. 2008;33(4):381–2.
36. Khosla A, Adeyefa O, Nasir S. Successful treatment of radiation-induced proctitis pain by blockade of the ganglion impar in an elderly patient with prostate cancer: a case report. Pain Med. 2013;5.
37. Lin CS, Cheng JK, Hsu YW, Chen CC, Lao HC, Huang CJ, Cheng PH, Narouze S. Ultrasound-guided ganglion impar block: a technical report. Pain Med. 2010;11(3):390–4.
38. Johnston PJ, Michálek P. Blockade of the ganglion impar (walther), using ultrasound and a loss of resistance technique. Prague Med Rep. 2012;113(1):53–7.

Radiofrequency and Neurolysis for the Treatment of Chronic Visceral Abdominal Pain

16

Rajiv K. Shah and Amitabh Gulati

Introduction

When considering neurolytic techniques for visceral pain control, the practitioner has many options. Therapeutic options may rely on diagnostic blocks with local anesthetic, usually with image guidance of the target site. Once a target has been chosen for neurolysis, a neurolytic technique can be used for potential prolonged pain relief. This chapter describes the various neurolytic techniques available for pain control in patients with chronic visceral pain.

Neurolytic Techniques

Chemical Neurolysis

Chemical neurolysis provides long-term analgesia by administering chemical agents capable of destroying neural structures. These techniques originated in the 1930s when chemical agents were first used to treat severe malignant pain and other non-malignant chronic conditions [1]. Two common chemical agents used are phenol and alcohol.

Phenol was first used by Putnam and Hampton as a neurolytic agent in 1936 with the goal to destroy the nerves responsible for the patient's pain [1]. Today, phenol is prepared to a maximum concentration of 6–9 % solution in water and can be prepared with alcohol, glycerol, saline, and radiocontrast dye. When mixed with glycerol, the solution is localized to the targeted area as opposed to mixtures with water which are more permeable and cause larger areas of destruction [1].

Phenol causes nerve destruction by inducing protein precipitation, causing a loss of cellular fatty elements, separation of the myelin sheath from the axon, and axonal edema. Nerve arborization and neuroma formation can result at the site of nerve disruption, perhaps leading to deafferentation pain. The true effects of the block cannot be fully evaluated until 24–48 h after the treatment due to the use of local anesthetics during the procedure. As long as the nerve cell body is intact, nerve regeneration occurs at a rate of 1–3 mm/day. In addition to peripheral targets, phenol can be injected intrathecally or epidurally [1].

Ethyl alcohol was first reported to produce satisfactory analgesia by Labat and Greene in 1933 with an injection of 33.3 % alcohol. Ethyl alcohol is usually distilled to a 95 % solution and a 50–95 % alcohol can be used as a neurolytic agent. The exact minimum concentration for neurolysis has not been established. While its mechanism of nerve destruction may be similar to phenol, it can also lead to wallerian degeneration. The potential for the myelin sheath being intact may lead to less neuroma formation. Similar to phenol, the effects of neurolysis are usually seen after 12–24 h [1, 2]

Cryoablation

Throughout history, physicians have used "cold" techniques to treat pain with the earliest methods dating back to Hippocrates [3]. The advantage of cryoablation compared to other methods is that it may reduce the incidence of neuritis or neuralgia following treatment [3]. However, efficacy of cryoablation of peripheral nerve targets is limited in the current literature.

Cryoablation "freezes" specific nerve targets via contact with a cryoprobe, which extracts heat from surrounding tissue

R.K. Shah, M.D. (✉)
Department of Anesthesiology and Pain Medicine,
Weill Cornell School of Medicine, New York–Presbyterian Hospital,
525 E 68th Street, Box 124, New York, NY 10065, USA
e-mail: rks9004@nyp.org

A. Gulati, M.D.
Department of Anesthesiology, Memorial Sloan Kettering Cancer Center, 1275 York Ave, M 308, New York, NY 10065, USA
e-mail: gulatia@mskcc.org

L. Kapural (ed.), *Chronic Abdominal Pain: An Evidence-Based, Comprehensive Guide to Clinical Management*,
DOI 10.1007/978-1-4939-1992-5_16, © Springer Science+Business Media New York 2015

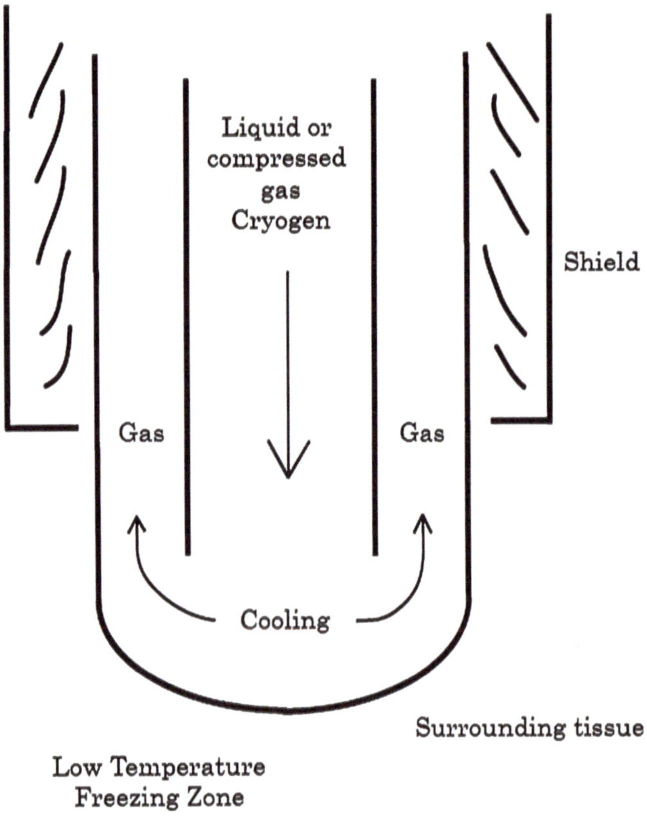

Fig. 16.1 The flow of cryogen in a cryoprobe. The majority of the cryoprobe is protected by a shield, allowing the tip to extract heat from the surrounding tissue. The choice of cryogen determines the lowest temperature of the cryoprobe

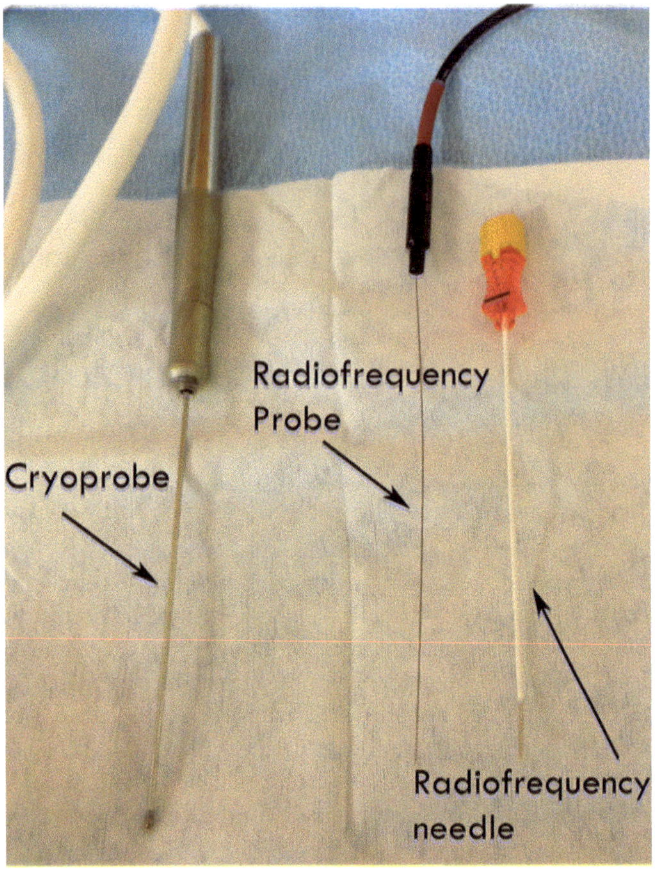

Fig. 16.2 Shown is a cryoprobe with a diameter of 1.7 mm, a RF probe with a much smaller relative diameter, and a 20-gauge RF needle that allows the probe to be introduced to the target

using supercooled liquid or compressed gas. Commonly, pressurized gas is forced through a small (0.002 mm) opening at the tip of the probe (Fig. 16.1). Heat is extracted from the tip of the cryoprobe, forming an ice ball with temperatures reaching −40 to −100 °C. Most probes are between 1.4 and 2.0 mm in size and have a thermistor to regulate temperature at the tip [3].

Using continuous freeze/defrost cycles and a larger probe can maximize disruptions within the nerve, causing both osmotic and mechanical damage. The repeat cycles decrease the temperature at sites away from the target nerve and maximize the size of the ice ball on the nerve target. At −20 °C, all nerve fibers stop conducting impulses through the nerve, and lower temperatures may lead to Wallerian degeneration, leaving the myelin sheath, endoneurium, and Schwann cell basal lamina intact allowing for regeneration. The extent of the block depends on the duration of exposure and temperature of the probe. However, long-term effects are thought to be caused by autoimmune responses via the release of sequestered proteins at the site of the treated nerve [3].

Radiofrequency Ablation

Radiofrequency (RF) ablation is a minimally invasive percutaneous procedure that is thought to reduce pain by altering the transmission of pain impulses [4, 5]. Conventional RF ablation (CRFA) causes controlled tissue destruction by irreversible coagulative necrosis [6]. However, recent evidence demonstrates that CRFA provides only a transient sensory loss in contrast to a much longer duration of pain relief. Electric fields produced by RF current may induce changes in the nerve cells and alter pain processing mechanisms at various sites, particularly at the molecular level [4].

All RF ablation techniques involve the transfer of alternating RF current (450–1,200 kHz) through insulated needle electrodes. The electrode is insulated, except for 2–10 mm at the tip. RF needles tend to be smaller than cryoprobes (Fig. 16.2). The electrode is positioned close to the target using nerve stimulators and image guidance. In conventional RF neurotomy, the electrode is positioned parallel to the target as the electrode coagulates transversely. A generator produces an electric field concentrated at the uninsulated tip of the electrode. The transfer of energy generates molecular oscillations that produce ionic friction and heat within the tissues. Once the cells are heated above a certain temperature, controlled tissue destruction occurs causing a lesion surrounding the uninsulated tip [6]. Early studies claimed that RF coagulation destroyed Aδ and C fibers preventing nociception. However, subsequent studies have shown that RF coagulation is non-selective and disrupts all nerves [4–7].

Pulsed Radiofrequency Ablation

Pulsed RF ablation (PRFA) was described by Slappendel et al. that compared the outcomes of cervical RF at the dorsal root ganglion (CRFA) in patients treated with lesions made at 40 °C to those at 67 °C [8]. While changing the temperature did not impact efficacy, it was believed that the RF current itself was therapeutic because of the overall electrical effects on the target nerve. The therapeutic effect of PRFA current is provided by applying brief bursts of RF energy and allowing the heat to dissipate at the target tissues, avoiding further damage to the nerve [9].

In PRFA, a current of 50,000 Hz is usually delivered in 20-msec pulses at a frequency of 2 per second (other protocols have been described). The electrode temperature is limited to 42 °C preventing any thermal lesion [8, 10]. The current is densest distal to the tip of the electrode and in contrast to CRFA, the electrode is applied perpendicular to the target nerve. Laboratory studies show that heating a nerve to a lower temperature (40–45 °C) causes reversible conduction blocks, but no pathologic lesion is produced [9, 11]. The distance of the electrode from the nerve target influences tissue damage in both CRFA and PRFA. Within 500 μm from the electrode, both CRFA and PRFA protocols produced tissue damage. Between 500 and 1,000 μm, tissue damage occurred with CRFA protocols, but not in PRFA protocols. Electron microscopy shows that ganglia treated with CRFA causes significant neuronal damage, whereas the ganglia treated with PRFA leaves nuclear membranes intact. However, studies demonstrating the efficacy of PRFA is limited as truly randomized clinical trials are lacking [8, 11].

Cooled RF Ablation

Cooled RF ablation (cooled RFA) is a newer RF technique used to treat various pain syndromes. The mechanism of pain relief is similar to CRFA. An electrode is placed close to the target nerve and conduction is disrupted relieving the pain. Cooled RFA utilizes a specialized electrode which is actively cooled by a continuous flow of water at ambient temperatures. This prevents the electrode from acquiring high surrounding tissue temperatures and increases the overall exposure to the RF current, heating larger tissue volumes with a higher thermal lesion. Similar to conventional RF, the lesion size depends on the size of the probe, the electrode temperature, and the duration of RF current that is applied [12]. Perhaps by delivering larger amounts of RF current to the target nerve, cooled RFA can be used in treating pain syndromes where conventional RF is unsuccessful [13, 14].

The Neuroanatomy of the Viscera

Most painful stimuli from abdominal viscera are transmitted by unmyelinated C fibers found in muscle, periosteum, mesentery, peritoneum, and viscera. The pain is characterized as dull, cramping, burning, gnawing, and gradual in onset. Secondary autonomic effects such as sweating, restlessness, nausea, vomiting, perspiration, and pallor can accompany the visceral pain. Abdominal visceral nociceptors respond to both mechanical and chemical stimuli. Visceral pain tends to be midline as sensory afferents are sent to both sides of the spinal cord. It is also poorly localized since the innervation is multi-segmental and the number of nerve endings is minimal [15]. The afferent fibers that mediate abdominal visceral pain usually follow the distribution of the autonomic nervous system, and consequently, the autonomic ganglia are the main targets for pain relief.

Sympathetic Nervous System

The sympathetic nervous system originates from the spinal cord in the thoracolumbar region, arising from the T1 to L3 levels. The preganglionic sympathetic fibers have cell bodies in the intermediolateral columns. From these cell bodies, nerve fibers continue to paired sympathetic chains, unpaired distal plexuses, or collateral ganglia near target organs. The paired sympathetic chains form 22 paired ganglia that lie on either side of the vertebral column. The preganglionic fibers leave the cord in the anterior nerve roots, join the spinal nerve trunks that connect the ganglia to each other, and enter the ganglion through the white ramus at their respective level. Additionally, the gray rami communicans connect the ganglia to the spinal nerves [15].

In the upper abdominal cavity, preganglionic fibers from T5 through T9 join together to form the greater splanchnic nerves serving the *celiac ganglia*. In the middle abdomen, nerve fibers from T10 and T11 form the lesser splanchnic nerves serving the *aorticorenal ganglia*. In the lower abdomen, nerve fibers from T12 form the least splanchnic nerves serving the *superior mesenteric ganglia* and nerve fibers from L1 through L3 form the lumbar splanchnic nerves serving the *inferior mesenteric ganglia*. The postganglionic fibers from the celiac, superior, and inferior mesenteric plexuses innervate the viscera of the abdomen and pelvis. Postganglionic fibers arising from synaptic links of the thoracic, lumbar, and pelvic sympathetic fibers form numerous plexuses, such as the cardiac, celiac, hypogastric, and plevic plexuses. Lastly, ganglia of the third type, the terminal or collateral ganglia, form near their target organs (e.g., adrenal medulla) [16, 17].

Parasympathetic Nervous System

The parasympathetic nervous system arises from cranial nerves III, VII, IX, and X and from the sacral spinal cord. The vagus nerve supplies the heart, tracheobronchial tree, liver, spleen, kidney, and entire gastrointestinal tract except for the distal part of the colon. Most vagal fibers do not synapse until they arrive at *small ganglia* on and about the thoracic and abdominal viscera. The preganglionic fibers are long, but the postganglionic fibers are short. The second through fourth sacral nerves form the pelvic splanchnic nerves. They synapse in terminal ganglia associated with the rectum and genitourinary organs also known as the *ganglion impar* [15].

Enteric Nervous System

The enteric nervous system (ENS) consists of a network of neurons within the walls of the gastrointestinal tract, the pancreas, and the gallbladder. The ENS functions independently from the sympathetic and parasympathetic systems as seen when digestion and peristalsis occur after spinal cord transection [18]. While not directly involved in pain sensation, secretory and neuronal mediators can cause signaling along the autonomic nervous system, which may be perceived as discomfort.

Referred Pain

Stimulation of the autonomic nervous system in the viscera may lead to referred pain, defined as a sensation perceived at a remote area from the site of the stimulus. Referred pain results from visceral and somatic afferent neurons converging on second-order neurons in the spinal cord. The best-known example is the pain experienced during a myocardial infarction. The damaged myocardium transmits pain signals via visceral afferent neurons to the T1–T4 levels of the spinal cord on the left side. These signals "converge" with somatic afferent neurons of the left chest and left arm at the same level. Thus, damaged myocardium is perceived as left chest and arm pain [19]. Similar pain is seen with pancreatic cancer (mid-back pain) and renal disease (groin and testicular pain) (Table 16.1).

Table 16.1 Common somatic referral patterns for chronic visceral pain [19]

Visceral pain location	Somatic referral pattern
Esophagus	Upper back and left chest
Pancreas and duodenum	Epigastric and mid-thoracic back
Liver disease and capsular pain	Right shoulder and right upper abdomen
Splenic disease and capsular pain	Left shoulder and left upper abdomen
Kidney and bladder	Flank pain, groin pain, testicular pain
Ovary	Groin and flank pain
Distal colon	Left lower abdominal quadrant
Testicular, prostate	Flank and groin pain

Autonomic Targets for Visceral Pain

Thoracic Sympathetic Block [20]

Indications: Pain related to lung and esophageal cancer, post-herpetic neuralgia, thoracic vertebral pain

Anatomy: Since the thoracic somatic nerves are close to the thoracic sympathetic chain, both neural pathways may be neurolyzed when approaching the thoracic sympathetic ganglion. The lower cervical ganglion and first thoracic ganglion are fused to make up the stellate ganglion at the level of the 7th cervical vertebrae. In moving caudad, each upper thoracic ganglia lie just beneath each rib. The lower thoracic ganglia are more anterior to the upper thoracic ganglia and lie along the posterolateral surface of each vertebral body. The pleural space is in close proximity and lies lateral and anterior to the thoracic sympathetic chain.

Technique: Usually the sympathetic chain is targeted using fluoroscopic or CT guidance. The needle is usually directed to the tip of the transverse process and redirected inferiorly to the inferior margin of the transverse process. After verifying the correct position, the needle is aspirated to ensure no blood or CSF and neurolysis is performed.

Complications: Pneumothorax, hemothorax, intrathecal neurolysis

Celiac Plexus Block [21, 22]

Indications: The celiac plexus block is used to treat pain related to pancreatic cancer, bile duct cancer, gastric cancer, or primary liver neoplasm, as well as chronic pancreatitis and chronic abdominal pain.

Anatomy: The plexus is located at the level of the upper part of the 1st lumbar vertebra and surrounds the celiac artery.

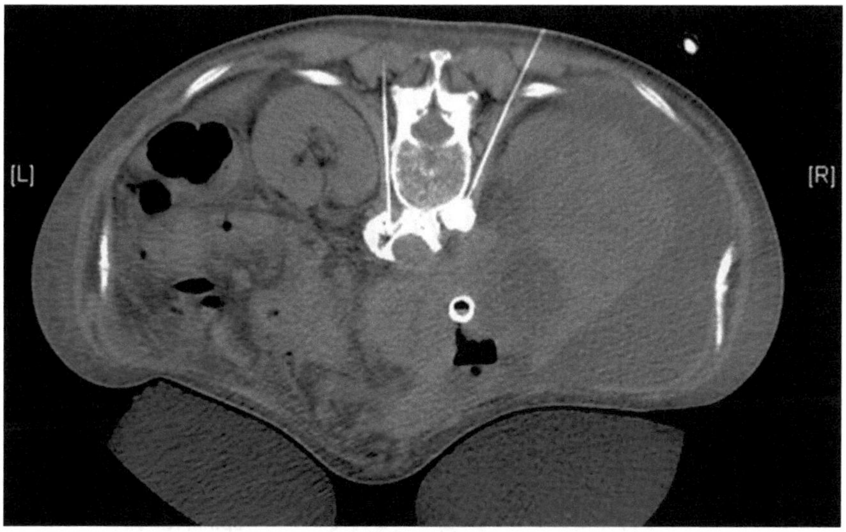

Fig. 16.3 Celiac plexus block performed under CT guidance. Alcohol neurolysis is performed after using contrast dye to determine the spread of the neurolytic. The right side approach is retrocrural and targets the lumbar splanchnics, while the left side approach is transcrural, targeting the left portion of the celiac plexus

It lies in between the suprarenal glands, in front of the crura of the diaphragm and abdominal aorta, and behind the stomach and omental bursa.

Technique: The approach has been described using CT, fluoroscopic, and ultrasound guidance. Given the various techniques, the celiac plexus and associated splanchnic nerves can be targeted independently for specific pain syndromes (Fig. 16.3).

Complications: Hypotension, diarrhea, intravascular/spinal/epidural injections of neurolytic substance, back/shoulder pain, leg weakness, sensory deficits, paresthesias, and paraplegia

Lumbar Sympathetic and Splanchnic Nerve Block [23]

Indications: The lumbar sympathetic block is used to treat claudication of the lower extremities, CRPS, herpetic neuralgia, or phantom limb syndrome; while the splanchnic nerves can be targeted for abdominal and pelvic visceral pain.

Anatomy: The lumbar sympathetic ganglion is located along the anterolateral surface of the lumbar vertebral bodies and anteromedial to the psoas muscle. The vena cava lies just anterior to the right sympathetic chain and the aorta lies anterior to the left sympathetic chain on the left. The splanchnic nerves and lumbar splanchnic chain are usually medial to the lumbar sympathetic chain and may travel around the large blood vessels of the abdomen (Fig. 16.4).

Technique: Either CT or fluoroscopic guidance is used to target the sympathetic and splanchnic chains. For specific targets, stimulation with a RF probe may allow the practitioner to locate either chain. RF may lead to more discreet lesions compared to chemical neurolysis.

Complications: Hypotension, Back/shoulder pain, Intravascular/spinal/epidural injections of neurolytic substance, genitofemoral neuralgia, psoas, and lumbar plexus damage.

Superior Hypogastric Plexus Block [24]

Indications: The superior hypogastric plexus block is used to treat lower abdominal and pelvic pain associated with cancer, bladder spasm/pain, or testicular pain

Anatomy: The plexus is located in the retroperitoneal space, starts at the lower part of the 5th lumbar vertebral body, and reaches the upper part of the 1st sacral vertebral body, close to the aortic bifurcation. It transfers visceral impulses from the upper vagina, cervix, uterus, fallopian tubes, bladder, and right colon to the dorsal horns of the spinal cord through sympathetic thoracic lumbar fibers

Technique: Approaching the hypogastric plexus has been described in many techniques using image guidance. Approaches include transdiscal, anterolateral vertebral, and trans-abdominal, under CT and fluoroscopic guidance. Because the plexus is diffuse, having imaging of the patient's abdomen and pelvis may help determine which

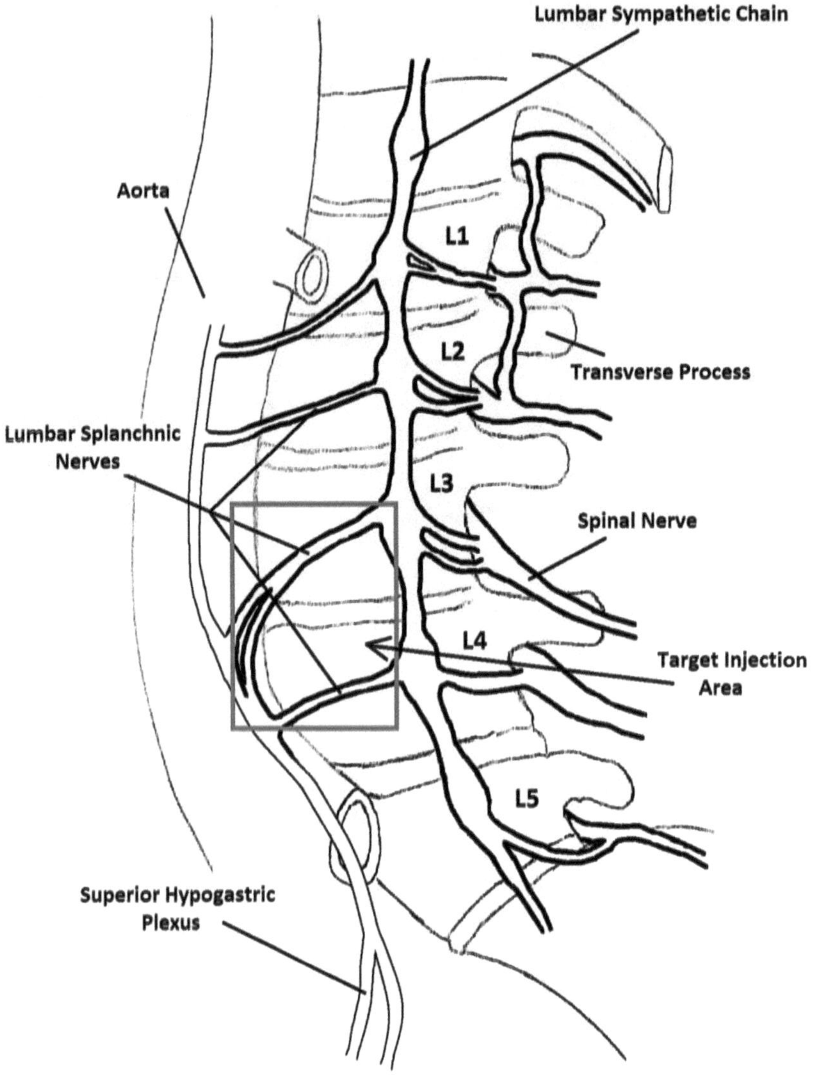

Fig. 16.4 The lumbar splanchnic nerves as they join the thoracic splanchnics and traverse around the abdominal vessels and merge to form the superior hypogastric plexus

approach may yield optimal results and reduce vascular trauma (Fig. 16.5).

Complications: Infection, damage to the aorta/iliac vein/ lumbar nerves, retroperitoneal bleeding.

Ganglion of Impar Block [25, 26]

Indications: The Ganglion of Impar block is used to treat rectal pain, perineal pain, rectal spasm, or coccydynia

Anatomy: The ganglion of Impar is located anterior to the sacrococcygeal junction where the two pelvic sympathetic trunks converge at the cranial base and travel retroperitoneal to form the solitary median ganglion.

Technique: The ganglion is usually targeted anterior to the sacrococcygeal or coccygeal ligament. Trans-ligament, lateral, or infracoccygeal technique has been described in the literature as effective approaches for neurolysis of the ganglion (Fig. 16.6).

Complications: Infection, accidental perforation of the rectum, impaired bladder/bowel/sexual/motor/sensory function, post-interventional neuralgia

Nociceptive Irritation and Somatic Pain

Nociceptive irritation results from mechanical, thermal, or chemical excitation of nociceptors. Nociceptors are located throughout the body, including the skin, subcutaneous tissue,

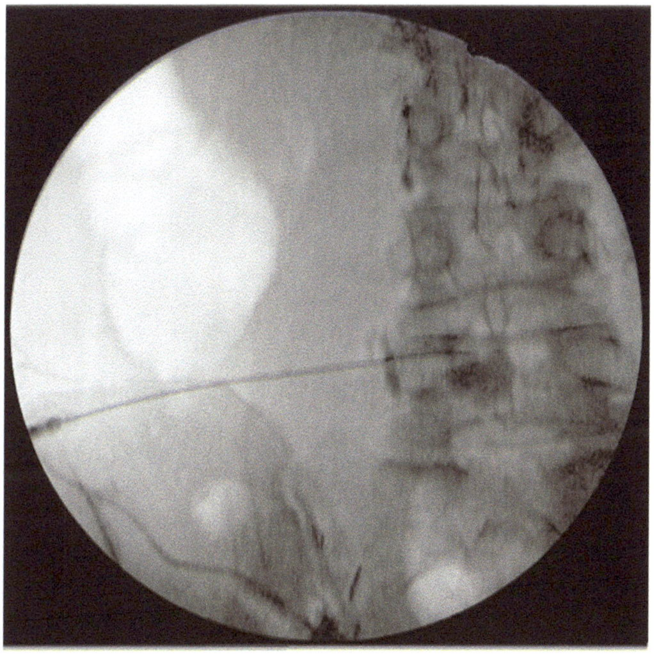

Fig. 16.5 Shown is an AP view of a fluoroscopic-guided L5 superior hypogastric block. The block can also target the sacral promontory below the L5–S1 discal junction

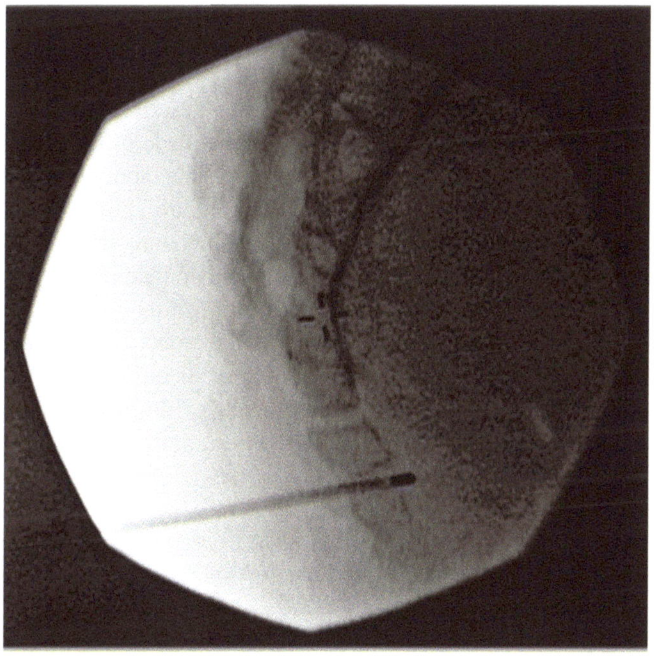

Fig. 16.6 An AP view under fluoroscopy showing a cryoprobe placed for neurolysis of the ganglion of impar. The probe is directed through the coccygeal ligament

bone, muscle, connective tissue, viscera, and blood vessels [27]. Somatic pain of the abdomen and pelvis may occur with visceral pain (such as in cancer pain syndromes) or may be difficult to distinguish from visceral pain (i.e., ilioinguinal neuralgia from surgery). As a result, diagnostic blocks and subsequent neurolysis may be performed for patients with concurrent somatic and visceral pain.

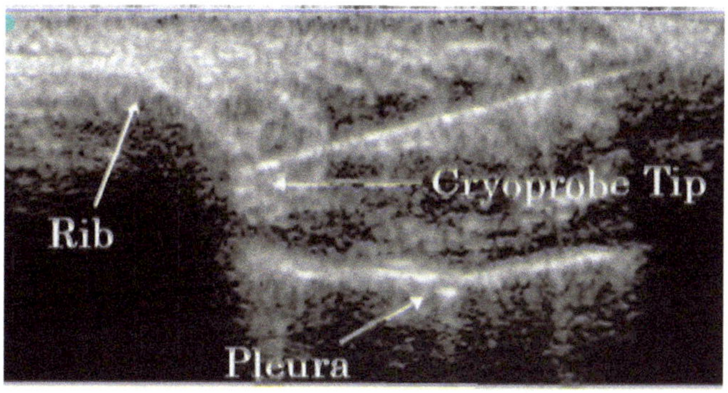

Fig. 16.7 A cryoprobe is placed to the inferior border of the rib under ultrasound guidance, targeting the ICN. Note the close proximity of the pleura to the location of the cyroprobe

Intercostal Nerves

The skin and muscles of the chest and abdominal are mostly innervated by the intercostal nerves (ICN). Acute use of local anesthetic around ICN can reduce pulmonary complications and narcotic requirements after upper abdominal surgery. As these procedures have advanced, ICN blocks are now used in a great variety of acute and chronic pain conditions involving the chest and upper abdomen (Fig. 16.7). The advantages of these blocks include superior analgesia, opioid-sparing, improved pulmonary mechanics, reduced CNS depression, and avoidance of urinary retention. The disadvantages include risks of pneumothorax and local anesthetic toxicity when blocking multiple levels [28, 29].

Iliohypogastric and Ilioinguinal Nerves

The iliohypogastric nerve courses the transverse abdominal muscles and the external oblique aponeurosis, while the ilioinguinal nerve travels between the second and third layers of abdominal muscles before coursing the inguinal canal. Both of these nerves are primarily derived from L1 spinal nerve with occasional contributions from T12, L2, or L3 nerves. These nerves are typically injured during hernia repair, trocar placement, Pfannenstiel incisions, needle suspensions of the bladder neck, and TVT procedures for correction of urinary complications [30]. Patients with low anterior pelvic and groin pain may benefit from neurolysis of these nerves.

Pudendal Nerves

Pudendal neuralgia is a neuropathic condition involving the dermatome of the pudendal nerve and is localized to the vulva, vagina, clitoris, perineum, and rectum in women and to the glans penis, scrotum (excluding testicles), perineum, and rectum in men. The incidence in the general population is around 1 % and affects women more than men. The nerve is derived from the sacral roots S2–S4 and forms the dorsal nerve of the penis/clitoris, the perineal nerve, and the inferior anal nerve. Since the pudendal nerve caries motor, sensory, and autonomic fibers, both afferent and efferent pathways are affected by nerve entrapment.

Patients often have associated symptoms of urinary frequency, urgency, dyspareunia, persistent sexual arousal, hyperalgesia, allodynia, and paresthesias. The three most common causes of pudendal nerve entrapment are surgical injury, pelvic trauma, and child birth. In terms of diagnosis, the "Nantes Criteria" are widely used and accepted to help diagnosis and treat pudendal neuralgia [31]. Because of motor and sensory function, neurolysis of the pudendal nerves is often seen as a "last resort", with possible consideration for PRFA techniques (Fig. 16.8).

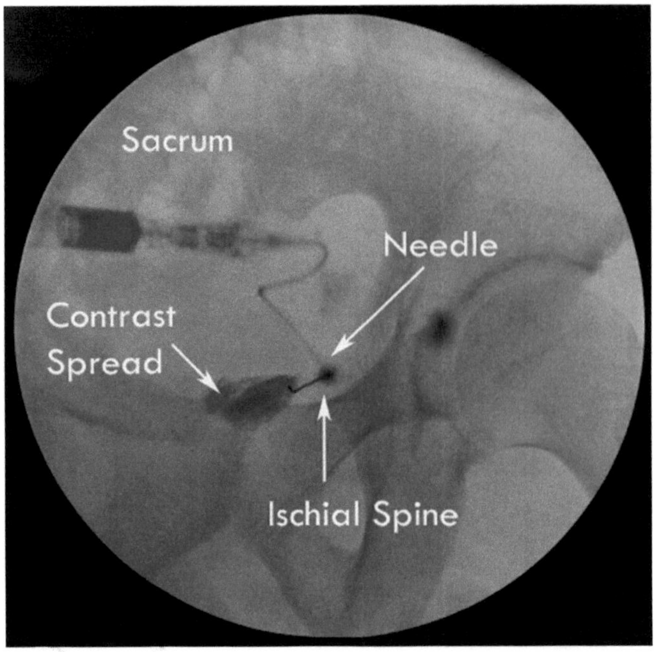

Fig. 16.8 Shown is a fluoroscopically guided technique in performing a pudendal nerve block. A needle is guided to the border of the ischial spine (*shown*), and contrast spread medial and above the ischial spine indicates the spread of the injectate and likely location of the pudendal nerve

Anterior Cutaneous Nerve Entrapment Syndrome

Anterior cutaneous nerve entrapment syndrome (ACNES) is a commonly misdiagnosed cause of abdominal pain. It generally occurs when a peripheral nerve is entrapped at specific anatomic sites, such as a fibrous or osseofibrous tunnel or when it passes over a fibrous or muscular band [32]. The pain at these sites is believed to be caused by mechanically induced irritation. The most common cause of abdominal wall pain is nerve entrapment at the lateral border of the rectus muscle [33]. It is believed that localized compression of the nerve at the ring of the rectus muscle bundle causes nerve ischemia, and subsequently pain. The acute pain is described as localized, dull, or burning. Often there is a sharp component radiating horizontally in the upper half of the abdomen and obliquely downward in the lower abdomen. The pain may exacerbate when the patient twists, bends, or sits up and is usually unilateral [32].

In terms of treatment, administration of local anesthetic can completely relieve the pain of ACNES [34]. The needle should be correctly positioned beneath the aponeurotic opening (using ultrasound guidance may improve the targeting of this area). The injection can relieve pain and reduce herniation of the neurovascular bundle through the fibrous ring. The patient usually will have immediate relief of pain if the treatment is effective [32]. When local anesthetic does not relieve the pain, phenol or alcohol can be used to treat ACNES.

Conclusion

Various neurolytic techniques have been described for treating chronic visceral pain. Diagnostic blockade of either the sympathetic or somatic innervation of the abdomen and pelvis will help determine which targets are amenable for neurolysis for prolonged pain control. Choosing which neurolytic technique involves careful understanding of the advantages and disadvantages when performing the neurolysis. Patient selection is also valuable in the decision whether to proceed to neurolysis for pain control. As our understanding of visceral neurophysiology improves, improved techniques and targets will allow for better pain control in our chronic visceral pain patient.

Acknowledgment This book chapter was funded by the Department of Anesthesiology and Critical Care at Memorial Sloan Kettering Cancer Center.

References

1. Leon-Casasola OA. Critical evaluation of chemical neurolysis of the sympathetic axis for cancer pain. Cancer Control. 2000;7(2):142–8.
2. Koyyalagunta D, Burton AW. The role of chemical neurolysis in cancer pain. Curr Pain Headache Rep. 2010;14(4):261–7.
3. Trescot AM. Cryoanalgesia in interventional pain management. Pain Physician. 2003;6(3):345–60.

4. Racz GB, Ruiz-Lopez R. Radiofrequency procedures. Pain Pract. 2006;6(1):46–50.
5. Lord SM, Bogduk N. Radiofrequency procedures in chronic pain. Best Pract Res Clin Anaesthesiol. 2002;16(4):597–617.
6. White WM, Kaouk JH. Ablative therapy for renal tumors (Chapter 56). In: Wein AJ, et al., editors. Campbell-Walsh urology. 10th ed. Elsevier Saunders; 2010. p. 1673–4.
7. Malik K, Benzon HT. Radiofrequency applications to dorsal root ganglia: a literature review. Anesthesiology. 2008;109(3):527–42.
8. Bogduk N. Pulsed radiofrequency. Pain Med. 2006;7(5):396–407.
9. van Boxem K, van Eerd M, et al. Radiofrequency and pulsed radiofrequency treatment of chronic pain syndromes: the available evidence. Pain Pract. 2008;8(5):385–93.
10. Chua NH, Vissers KC, Sluijter ME. Pulsed radiofrequency treatment in interventional pain management: mechanisms and potential indications. Acta Neurochir (Wien). 2011;153(4):763–71.
11. Cahana A, Van Zundert J, Macrea L, van Kleef M, Sluijter M. Pulsed radiofrequency: current clinical and biological literature available. Pain Med. 2006;7(5):411–23.
12. Malik K, Benzon H, Walega D. Water-cooled radiofrequency: a neuroablative or a neuromodulatory modality with broader applications? CaseRepAnesthesiol.2011;2011:263101.doi:10.1155/2011/263101.
13. Kapural L, Stojanovic M, Sessler DI, Bensitel T, Zovkic P. Cooled radiofrequency (RF) of L5 dorsal ramus for RF denervation of the sacroiliac joint: technical report. Pain Med. 2010;11(1):53–7.
14. Kapural L, Nageeb F, Kapural M, Cata JP, Narouze S, Mekhail N. Cooled radiofrequency system for the treatment of chronic pain from sacroiliitis: the first case-series. Pain Pract. 2008;8(5):348–54.
15. Glick D. The autonomic nervous system (Chapter 12). In: Miller RD et al., editors. Anesthesia. 7th ed. New York: Churchill Livingstone; 2009. p. 263–6.
16. The great plexuses of the sympathetic system. Gray's anatomy of the human body. http://education.yahoo.com/reference/gray/subjects/subject/220. Accessed 12 Feb 2012.
17. Chan CW. A case of sympathetically mediated headache treated with stellate ganglion blockade. Pain Med. 2010;11(8):1294–8.
18. Benarroch EE. Enteric nervous system: functional organization and neurologic implications. Neurology. 2007;69(20):1953–7.
19. Arendt-Nielsen L, Svensson P. Referred muscle pain: basic and clinical findings. Clin J Pain. 2001;17(1):9–1.
20. Waldman S. Pain review: Chapter 277-Thoracic sympathetic block. Philadelphia: Saunders–Elsevier; 2009. http://www.expertconsultbook.com/expertconsult/ob/book.do?method=display&type=book.
21. Lee JM. CT-guided celiac plexus block for intractable abdominal pain. J Korean Med Sci. 2000;15(2):173–8.
22. Michaels AJ, Draganov PV. Endoscopic ultrasonography guided celiac plexus neurolysis and celiac plexus block in the management of pain due to pancreatic cancer and chronic pancreatitis. World J Gastroenterol. 2007;13(26):3575–80.
23. Gulati A, Khelemsky Y, Loh J, Puttanniah V, Malhotra V, Cubert K. The use of lumbar sympathetic blockade at L4 for management of malignancy-related bladder spasms. Pain Physician. 2011;14(3): 305–10.
24. Cariati M, De Martini G, Pretolesi F, Roy MT. CT-guided superior hypogastric plexus block. J Comput Assist Tomogr. 2002;26(3): 428–31.
25. Agarwal-Kozlowski K, Lorke DE, Habermann CR, Am Esch JS, Beck H. CT-guided blocks and neuroablation of the ganglion impar (Walther) in perineal pain: anatomy, technique, safety, and efficacy. Clin J Pain. 2009;25(7):570–6.
26. Toshniwal GR, Dureja GP, Prashanth SM. Transsacrococcygeal approach to ganglion impar block for management of chronic perineal pain: a prospective observational study. Pain Physician. 2007;10(5):661–6.
27. Julius D, Basbaum AI. Molecular mechanisms of nociception. Nature. 2001;413(6852):203–10.
28. Byas-Smith MG, Gulati A. Ultrasound-guided intercostal nerve cryoablation. Anesth Analg. 2006;103(4):1033–5.
29. Detterbeck FC. Efficacy of methods of intercostal nerve blockade for pain relief after thoracotomy. Ann Thorac Surg. 2005;80(4): 1550–9. Review.
30. Klaassen Z, Marshall E, Tubbs RS, Louis Jr RG, Wartmann CT, Loukas M. Anatomy of the ilioinguinal and iliohypogastric nerves with observations of their spinal nerve contributions. Clin Anat. 2011;24(4):454–61.
31. Hibner M, Desai N, Robertson LJ, Nour M. Pudendal neuralgia. J Minim Invasive Gynecol. 2010;17(2):148–53.
32. Applegate WV. Abdominal cutaneous nerve entrapment syndrome ACNES: a commonly overlooked cause of abdominal pain. Perm J. 2002;6(3):20–7.
33. Srinivasan R, Greenbaum DS. Chronic abdominal wall pain: a frequently overlooked problem. Practical approach to diagnosis and management. Am J Gastroenterol. 2002;97(4):824–30.
34. Boelens OB, Scheltinga MR, Houterman S, Roumen RM. Management of anterior cutaneous nerve entrapment syndrome in a cohort of 139 patients. Ann Surg. 2011;254(6):1054–8.

Regional Anesthesia for Abdominal/ Truncal Pain

17

James Crews and Daryl Steven Henshaw

Introduction

Regional anesthetic neural blockade procedures are important diagnostic and therapeutic tools providing frequently useful information about the nature (somatic, visceral, or neuropathic) and location of abdominal pain. These procedures range in extent from relatively simple infiltration (local injection) of local anesthetic and analgesic adjuvant medications, to more complex procedures including peripheral nerve block techniques, or major visceral plexus block procedures. Some procedures may be appropriately performed in the office, clinic, or at the patient's bedside, while more complex subspecialty procedures may require ultrasound, fluoroscopic, or computed tomography guidance for needle placement, and observation of spread of the injectate. Many of the procedures discussed in this chapter should be performed only by physicians with subspecialty training and experience with complex regional anesthetic and neural blockade procedures. This chapter will cover the range of regional anesthetic neural blockade procedures, indications and contraindications, equipment required or commonly used, local anesthetic and non-local anesthetic analgesic drugs, as well as anticipated result of the commonly performed procedures.

J. Crews, M.D. (Deceased)

D.S. Henshaw, M.D. (✉)
Department of Anesthesiology, Wake Forest Baptist Health, Medical Center Boulevard, Winston Salem, NC 27157, USA
e-mail: dhenshaw@wakehealth.edu

Infiltration Techniques

Trigger Point Injections

The exact underlying cause of myofascial trigger points is unknown, but may result from entrapment of anterior cutaneous nerve branches in muscle or fascia. Such entrapment as in the case of the anterior rectus fascia, can lead to ischemia and pain, and has been termed abdominal cutaneous nerve entrapment syndrome (ACNES) [1]. However, there are other conditions associated with trigger points including peripheral nerve entrapment, or compression from surgical scars. Many patients with abdominal pain have no known underlying pathology and trigger point pain can arise from strenuous physical activity, overuse, hernias, hematomas, neuromas, weakened abdominal muscles, ascites, pregnancy, and obesity. Trigger points are also present in patients with fibromyalgia; however, they are not thought to have the same underlying pathology and are typically more diffuse.

Trigger points are usually no more than 2 cm in diameter, and depending on the causation, multiple trigger points may be present. Trigger point pain can often be elicited by direct palpation to the abdominal wall, and patients can oftentimes pinpoint the exact spot from where pain arises. As such, physical examination is instrumental when trigger points are suspected to be the cause of abdominal pain. In ACNES Carnett's test is often positive. To perform this test a supine patient is asked to raise his/her head and shoulders off the examination table in order to contract the abdominal musculature. The trigger point is palpated while the abdominal wall is tense [2]. If pain is more severe than while the abdominal muscles are relaxed, the test is considered positive. The "hover sign", described by Hershfield, is also often present [3]. This is present if the patient guards the painful area of their abdomen, often seizing the practitioner's hand when they get close to examining the trigger point.

L. Kapural (ed.), *Chronic Abdominal Pain: An Evidence-Based, Comprehensive Guide to Clinical Management*,
DOI 10.1007/978-1-4939-1992-5_17,

Procedure

The injection of an abdominal wall trigger point is a relatively straightforward and easily performed procedure. The underlying theory behind injecting an abdominal wall trigger point is that the injection of local anesthetic blocks the nerve, relaxes the muscle, and breaks the cycle of chronic pain within the trigger point. There is also some evidence that "dry needling" without injecting medications is equally effective short-term; however, the injection of local anesthetic can often decrease the discomfort that accompanies the needling of the trigger point both during and after the procedure [4].

Indications/Contraindications

Trigger point injections are indicated when the differential diagnosis includes ACNES or myofascial pain and is supported by the presence of one or more trigger points which can be directly identified on physical examination. Marked improvement, or complete resolution of pain following injection supports the diagnosis. Contraindications include systemic or local infection and allergies to local anesthetics or adjuvants in the injection mixture. Caution should be used in patients on anticoagulants.

Equipment

- Gloves, alcohol or chlorhexidine for skin preparation, and skin marker.
- Twenty-two to twenty-seven gauge 1.5 in. needle, although for obese patients a longer needle may be necessary.
- Ten mL three-ring syringe, multiple syringes may be needed if treating several trigger points.
- Local anesthetic and additives if desired.

Technique

With the patient supine to relax the abdominal musculature, the trigger point is confirmed by palpation. A skin marker may be used to mark the area. The skin is prepped, and two fingers, one on each side of the trigger point, are used to "hold down" the area to be injected (Fig. 17.1). The needle with syringe attached is then inserted into the trigger point. The needle can be inserted perpendicular to the abdominal wall with the trigger point pinched between the fingers; however, care must be taken not to penetrate too deep as the peritoneal cavity could be entered. The authors advocate a more parallel needle trajectory to minimize the chance of placing the needle through the abdominal wall. The patient may note a transient increase in pain as the trigger point is contacted. Fanning of the needle can be performed to assist in decisively locating the trigger point. The injection can then be made if dry needling is not being performed. Injection of 2–3 mL of local anesthetic is then completed at each trigger point. Ultrasound can be used for the procedure to visualize proper needle and injectate placement and to minimize the risk of placing the needle into the peritoneal cavity.

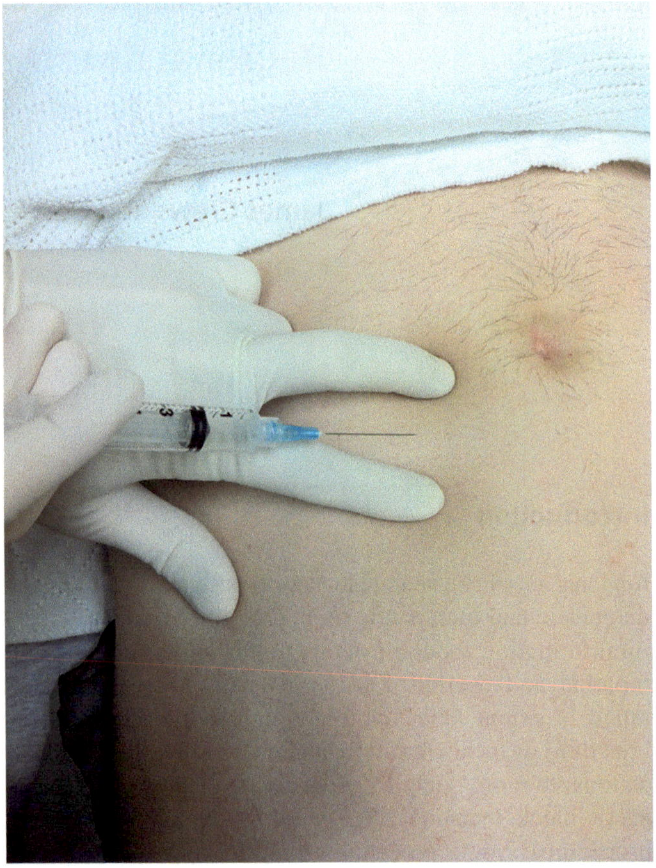

Fig. 17.1 Hand and needle position for performing a trigger point injection of the abdominal wall

Local Anesthetic

Multiple combinations of local anesthetics and adjuvants have been described for trigger point injection. There is little evidence that any one local anesthetic is superior to another, or that injectate of any kind is superior to dry needling. Local anesthetics that have been used include: lidocaine, procaine, mepivacaine, ropivacaine, and bupivacaine. Additives such as methylprednisolone or triamcinolone can be added. The authors suggest using equal parts 2 % lidocaine and 0.75 % bupivacaine with 40 mg of triamcinolone per 10 mL of solution. Avoidance of epinephrine may be important as vasodilation at the site is thought to be beneficial in the treatment of the trigger point.

Anticipated Results/Possible Complications

The anticipated result is that pain at the trigger point will dissipate over 5–10 min as the local anesthetic takes effect. Long-term or permanent pain relief may occur following single injections; however, repeat injection may be required. Complications include intraperitoneal needle placement, hematoma formation, needle breakage, infection, vasovagal syncope, intravascular injection and local anesthetic toxicity, scarring and failure of the procedure.

Scar Injections

Surgical scars can also lead to chronic abdominal pain secondary to entrapment of nerves in the fibrous scar or neuroma formation [5, 6]. The technique of scar injections is identical to that of trigger point injections.

Peripheral Nerve Block Techniques

Intercostal Nerve Block (ICNB)

The dorsal and ventral roots of the spinal cord unite to form 31 paired spinal nerves. The spinal nerves then exit through the intervertebral foramen and divide into several branches including the ventral and dorsal primary rami. The dorsal rami supply the paravertebral area with motor and sensory functions. The anterior primary rami of T1 through T11 make up the intercostal nerves. T12 is technically a subcostal nerve as it does not run between two ribs, but rather under the twelfth rib. The intercostal nerves run in the subcostal groove on the inferior side of each rib. The nerve is inferior to the posterior intercostal artery, which is itself inferior to the intercostal vein. The intercostal nerves travel between the internal intercostal and the innermost intercostal muscles (Fig. 17.2). Each nerve gives off a lateral cutaneous branch just anterior to the midaxillary line, which supplies the muscles and skin of the anterior and lateral torso. The intercostal nerves end as an anterior cutaneous branch. For the lower six intercostal nerves, which are applicable to abdominal pain, this branch terminates after piercing the rectus sheath to provide motor function and sensation to the anterior abdominal wall near the midline.

Procedure

An intercostal nerve block (ICNB) performed proximal to the take off of the lateral cutaneous branch will provide unilateral anesthesia/analgesia for the lateral and anterior thorax in a dermatomal pattern defined by the distribution of each specific nerve. Since spread to adjacent nerves is unlikely, each intercostal nerve must be blocked separately. This is a pure somatic block and no visceral coverage is provided.

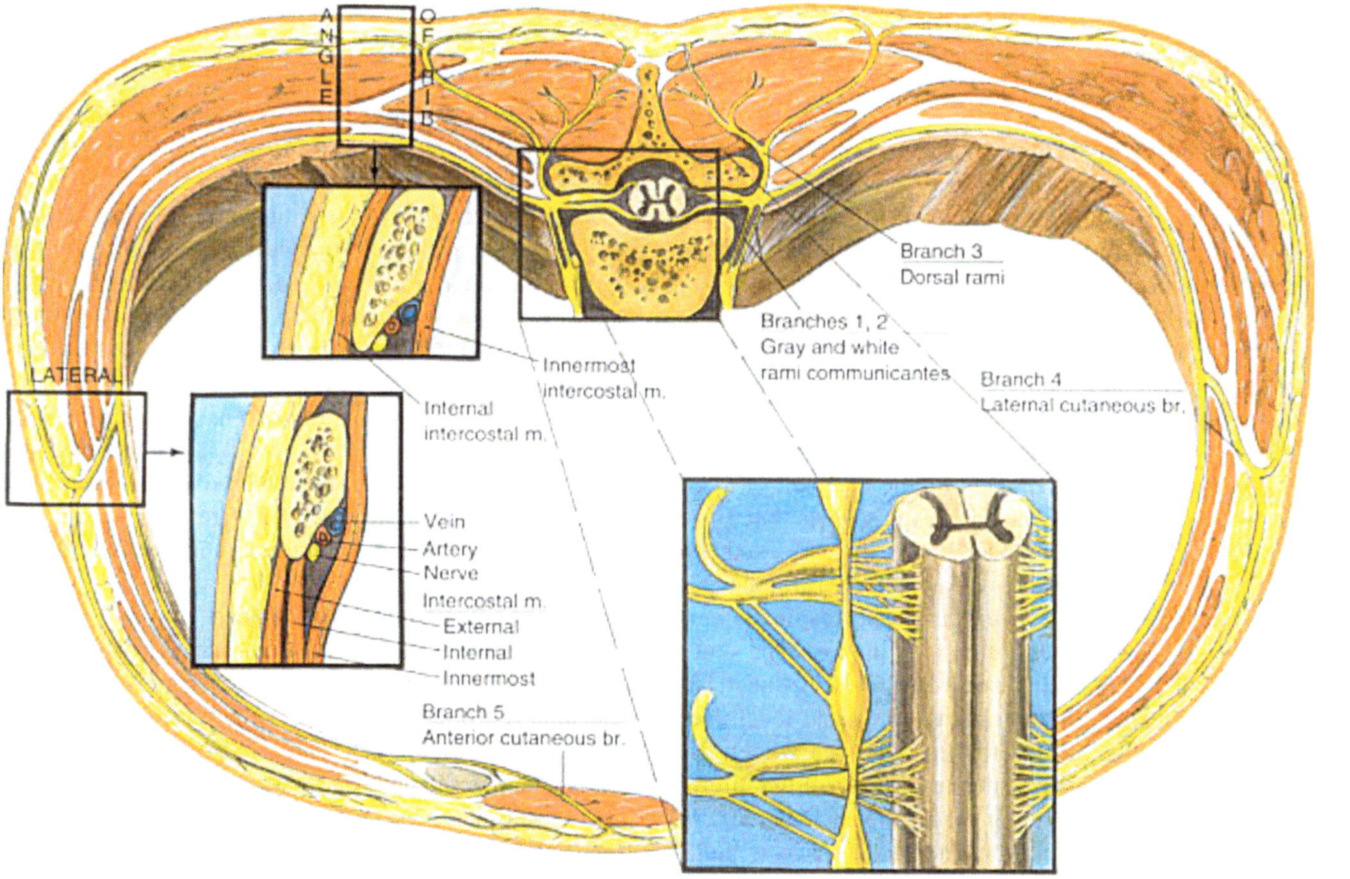

Fig. 17.2 Intercostal nerve block (ICNB): cross-sectional anatomy. Reprinted with permission from David L. Brown, Atlas of Regional Anesthesia, 4th edition, 2010, Elsevier

Indications/Contraindications

ICNBs for abdominal pain are indicated when somatic pain arises from the muscles, nerves, or connective tissues of the abdominal wall. Therefore, they can be used to treat trigger points, intercostal neuralgia and herpes zoster pain, rib fractures, post-thoracotomy pain, hernia pain, tumor-related pain, and musculoskeletal pain of the abdominal wall. They can also be used to distinguish somatic pain from visceral pain.

ICNB are contraindicated in patients who may not tolerate a pneumothorax, such as those with severe underlying pulmonary pathology or previous contralateral pneumonectomy. Local infection and allergies to local anesthetics or adjuvant medications are also contraindications. Caution should be used for patients on anticoagulation; however, this is not an absolute contraindication and clinical judgment should be exercised.

Equipment

- Skin marking pen, skin prep, gloves, and drape if desired.
- Twenty-five gauge needle, 1.5 % lidocaine, and 5 mL syringe for skin wheel.
- Twenty-one gauge short bevel 5 cm needle for block placement.
- Extension tubing.
- Syringe and local anesthetic for ICNB.

Techniques

Landmark technique: For performance of the block the patient may be sitting, lateral, or prone. Most commonly, the block is performed at the angle of the rib just lateral to the sacrospinalis muscles, which is approximately 6–8 cm from the vertebral spines. Blockade at this location should anesthetize the intercostal nerve prior to the takeoff of the lateral cutaneous branch. Additionally, at this point the ribs are superficial making palpation easier and the intercostal spaces are thicker so the risk of pleural puncture is decreased. The distance from the midline also minimizes the chance of proximal spread into the neuraxis. A marking pen is used to mark the spinous processes that correspond to the ribs and intercostal nerves to be blocked. The rib is palpated laterally and marked 6–8 cm from the midline. The mark should be lateral to the sacrospinalis muscles at the inferior border of the rib of interest. The skin is prepped and 1 % lidocaine is injected for skin wheel. The short bevel needle with extension tubing and local anesthetic connected is placed through the skin wheel. The goal of the block is to place the needle just off the inferior edge of the rib while angling the needle cephalad 20°. To do so, the skin overlying the intercostal space is pulled superior to lie over the rib, the needle is inserted to contact the rib, and then the skin is released as the needle is "walked" inferiorly until it just slips off the inferior margin of the rib. The cephalad angulation of the needle must be maintained to maximize block success (Fig. 17.3). The needle is advanced approximately 3–4 mm past the rib. Negative aspiration is ensured and 3–5 mL of local anesthetic is injected.

Ultrasound technique: An in-plane ultrasound-guided technique can also be used to perform an ICNB. Ultrasound has been shown to be as equally effective as fluoroscopy for this procedure [7]. A linear ultrasound probe is placed on the patient's back at a point that corresponds to the angle of the rib as described above. The probe is placed vertically over the ribs so that the ribs both above and below the intercostal space of interest can be seen (Fig. 17.4). Because of the caudal angulation of the ribs, the probe should be slightly oblique with the cephalad end of the probe being more lateral than the caudal end. In this fashion the probe will be perpendicular to the rib. The skin is anesthetized and the injection needle is inserted caudal to the probe and in-plane. It is advanced until it is 3–4 mm under the inferior border of the rib. Oftentimes, the pleura can be visualized as a bright white line deep to the ribs. As local anesthetic is injected, spread should be seen in the intercostal space posterior to the pleura. The major advantage of ultrasound for ICNB is a potential reduction in the risk of pneumothorax.

Local Anesthetics

An ICNB can be performed using a variety of local anesthetics. Bupivacaine 0.25–0.5 %, Ropivacaine 0.5–0.75 %, Lidocaine 1–2 %, and Mepivicaine 1–2 % can be used depending on the desired onset time and duration of block. For prolonged analgesia Bupivicaine or Ropivicaine are appropriate, while Lidocaine and Mepivicaine provide a shorter onset time. Epinephrine 1:200,000 or 1:400,000 should be added to the solution to reduce systemic absorption and as a marker for intravascular injection. It is important to note that systemic absorption following ICNB is higher than with any other nerve block, so maximal local anesthetic doses should not be exceeded and caution should be taken when multiple injections are being performed. Steroids may be added to the injectate solution if desired.

Anticipated Results/Complications

An ICNB results in loss of both motor function and sensation in the distribution of the individual intercostal nerve. Several sequential ICNB may be needed as the distribution of neighboring nerves may overlap.

Complications include pneumothorax, hemothorax, nerve injury, local anesthetic toxicity, and hematoma formation. There is also risk of penetrating the peritoneal cavity and injuring underlying abdominal viscera. Rarely, spread of local anesthetic to the spinal space may occur from injection into the dural sheath.

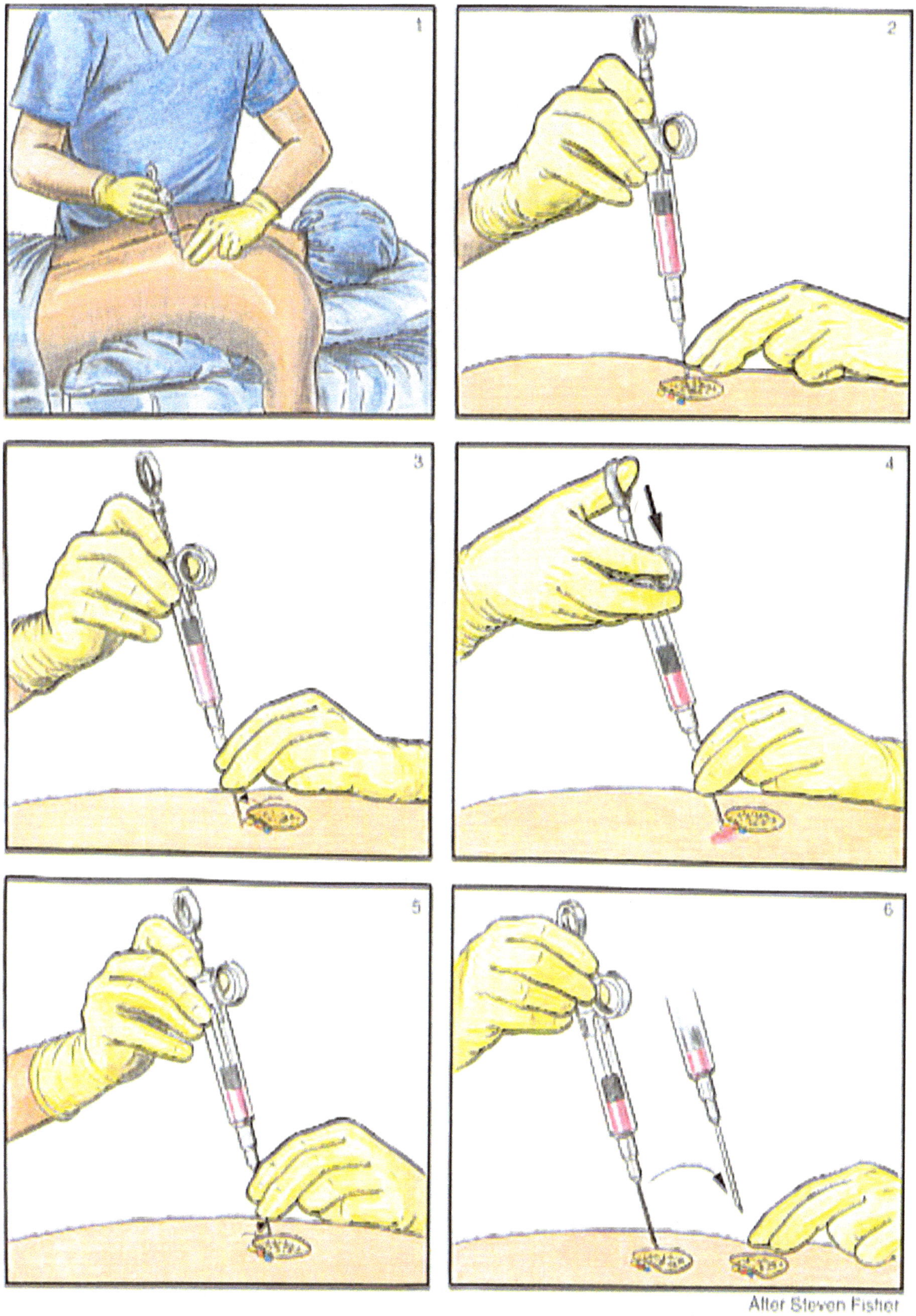

Fig. 17.3 ICNB: stepwise technique (1–6). Reprinted with permission from David L. Brown, Atlas of Regional Anesthesia, 4th edition, 2010, Elsevier

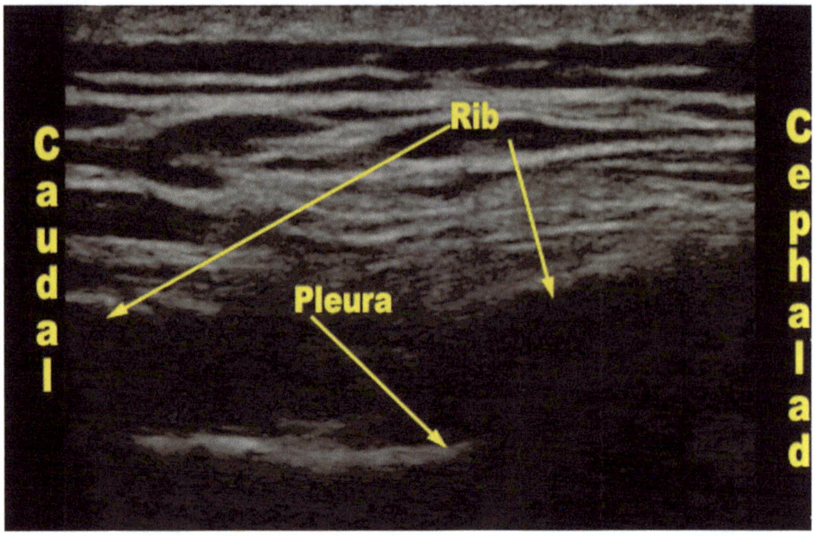

Fig. 17.4 Ultrasound view during intercostal nerve blockade with ultrasound transducer positioned vertically to show two adjacent ribs in cross-section

Thoracic Paravertebral Block (TPVB)

The thoracic paravertebral space is a triangular-shaped area adjacent to the spinal column on both sides. Within this space are the spinal nerves, which have exited the intervertebral foramen, and the sympathetic chain. It also contains fatty tissue, dorsal rami, rami communicantes, and the intercostal vessels. The paravertebral space is bordered by the parietal pleura anterolaterally, the vertebral body and disk medially, and the costotransverse ligament, ribs and transverse process posteriorly (Fig. 17.5). Continuing laterally, the paravertebral space is continuous with the intercostal space. Medially, it is continuous with the epidural space. The paravertebral space runs from the cervical region to the origin of the psoas muscle at the twelfth thoracic vertebrae.

Procedure

The thoracic paravertebral block (TPVB) can be thought of as unilateral epidural blockade as it results in unilateral blockade of sympathetic and somatic nerves and resultant ipsilateral dermatomal anesthesia. The injection of local anesthetic into this space results in analgesia/anesthesia of the spinal nerves. It has the potential advantage of minimizing autonomic dysfunction that accompanies neuraxial techniques, thus making it an attractive option for treatment of unilateral pain of the abdomen. Because the paravertebral space at one level may be continuous with the paravertebral spaces both above and below, a single injection of local anesthetic may result in blockade of multiple levels [8]. However, dependent on the patient's anatomy and the injection site, local anesthetic injection may be more confined [9]. Potential spread may also depend on the volume of local anesthetic injected.

Indications/Contraindications

The indications for a TPVB include rib fractures, upper abdominal wall pain both acute and chronic, herpetic neuralgia, acute herpes zoster, intercostal neuralgia, liver capsule pain, post-thoracotomy pain, and sympathetic blockade such as for the treatment of hyperhidrosis [10].

Contraindications include local and systemic infection and allergies to local anesthetic or adjuvants. Because there is also a risk of pneumothorax, this block should be avoided in patients who would not tolerate such complication. While anticoagulation is not an absolute contraindication for paravertebral blockade, the risk of bleeding and potential epidural hematoma formation are present, thus careful consideration should be given when performing this block in patients who are systemically anticoagulated.

Equipment

- Marking pen, skin prep, gloves, and drape. Equipment specific for the technique.
- 1.5 in. 25 gauge needle, 5 mL syringe, and 1.5 % lidocaine for local skin wheel.
- Needle: For single injections a 17–22 gauge Tuohy needle may be used. Alternatively, a 90–100 mm short bevel needle or a 22 gauge 3.5 in. Quincke needle may be used. For the placement of an indwelling catheter, a 17 gauge Touhy needle is needed depending on catheter size.
- Local anesthetic of choice, extension tubing and syringe.

Techniques

Landmark Technique: Patients can either be sitting or lateral during performance of a paravertebral block. As the block can be uncomfortable, especially if multiple injections are being performed, sedation is often necessary making the

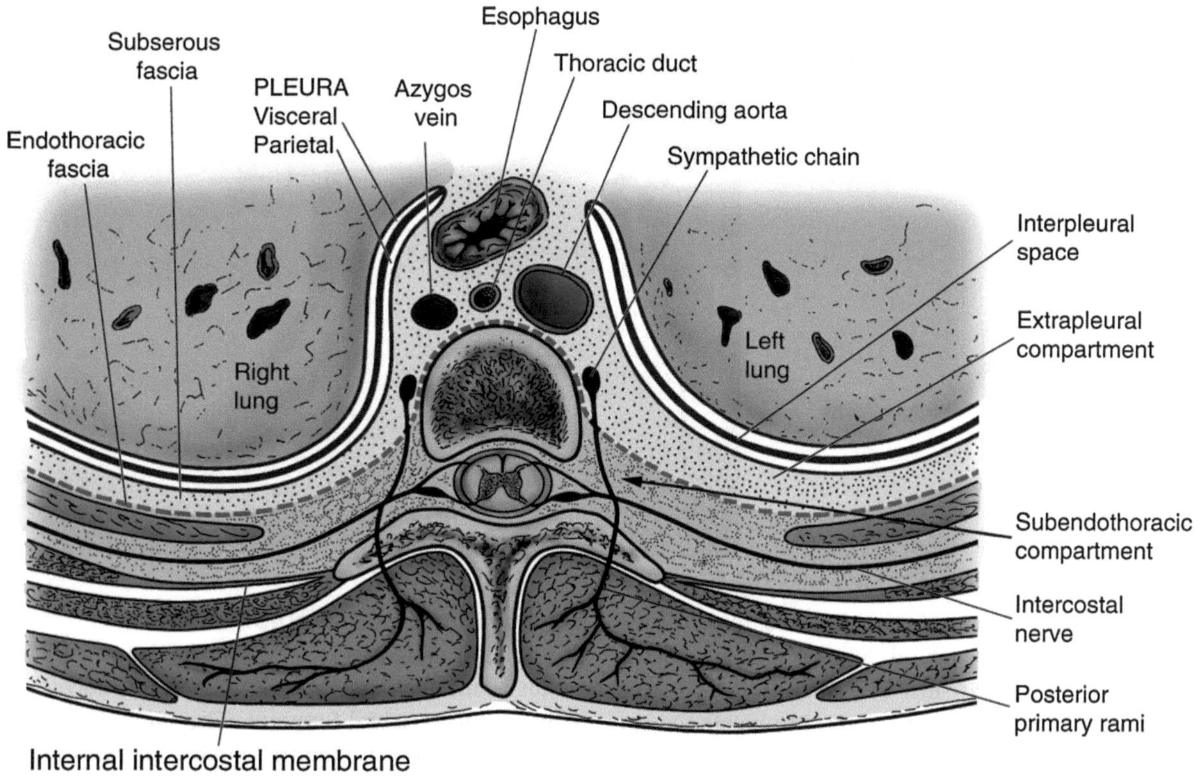

Fig. 17.5 Anatomy of the thoracic paravertebral space. Reprinted with permission from Samer N. Narouze, Atlas of Ultrasound-Guided Procedures in Interventional Pain Management, 2010, Springer

lateral position attractive. The spinous processes are identified and marked. Helpful landmarks for identification of specific spinous processes include the inferior angle of the scapula at T7. It is important to recall that because the spinous processes of the thoracic vertebrae are angled caudally, the spinous process of one vertebrae overlies the transverse process of the vertebrae below it. For example, the spinous process for T8 overlies the transverse process of T9. From the cephalad border of the spinous process of interest, a mark is made 2.5 cm laterally. This location is the needle insertion site for the block. After prepping the skin, and injecting local anesthetic for a skin wheel, the procedural needle is inserted perpendicular to the skin in all planes to contact the transverse process. The depth of this bone is variable depending on the thoracic level being blocked, but is deepest at the upper and lower thoracic spine (5–8 cm) and shallowest in the mid-thoracic spine (2–4 cm). After contacting the transverse process the needle is redirected caudally to walk off its inferior edge. The needle is advanced 1 cm past the depth at which the transverse process was located. Occasionally, a pop will be felt as the needle passes through the costotransverse ligament; however, this is unpredictable. After negative aspiration for cerebrospinal fluid (CSF), blood, and air, local anesthetic is injected. If a catheter is to be placed, it can be inserted through the Tuohy needle and threaded 3–4 cm.

Loss or Resistance Technique

The thoracic paravertebral space may also be located using a loss of resistance (LOR) technique. The procedure is identical to the landmark-based approach, except that it is routinely performed with a Tuohy needle and LOR syringe. After contacting the transverse process the needle is redirected inferior to the transverse process. The LOR syringe is then connected using either saline or air, depending on operator preference. The needle is slowly advanced while periodically checking for a LOR as the needle passes through the costotransverse ligament. This LOR can be subtle and can be difficult to appreciate, especially for those with less experience performing this technique. The LOR should occur within the first centimeter past the transverse process, and if it is not found within this distance the needle should be withdrawn, the landmarks checked, and the block procedure repeated. Once LOR is obtained, local anesthetic can be injected and a catheter can be placed if desired.

Ultrasound Technique

Various ultrasound-guided approaches to thoracic paravertebral blockade have been described [11–13]. Ultrasound can be used to locate and mark both the midline and transverse process in order to facilitate the landmark, or LOR

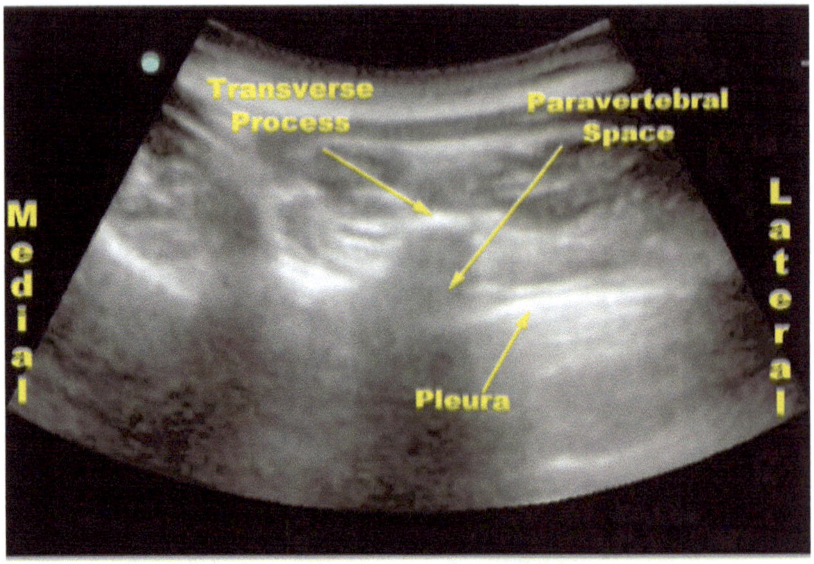

Fig. 17.6 Ultrasound view of the thoracic paravertebral space with the ultrasound probe oriented horizontally, and placed lateral to the midline

technique, as well as to perform blockade in real-time. With the patient supine or lateral, a linear probe, or curvilinear probe for obese patients, is placed on the patient's back in a horizontal orientation and used to locate the spinous process adjacent to the level of interest. The probe is then moved laterally to locate the transverse process as a hyperechoic structure with acoustic shadowing. The probe is then slid caudally into the intercostal space to obtain a view of the parietal pleura. This often requires a slightly oblique orientation of the ultrasound probe since the ribs angle inferiorly as they move laterally. With the transverse process and parietal pleura in view, the procedure needle is inserted in-plane from the lateral side of the ultrasound probe (Fig. 17.6). The needle is advanced into the paravertebral space just posterior to the parietal pleural and, following negative aspiration, local anesthetic is injected. Catheter placement can follow if desired.

Local Anesthetic

The choice of local anesthetic for thoracic paravertebral blockade will vary depending on the indication for nerve blockade. For diagnostic blocks a short acting local anesthetic such as lidocaine or mepivicaine is sufficient. However, in cases where longer duration of analgesia is desirable, 0.25–0.5 % bupivacaine or 0.5–0.75 % ropivicaine should be used. Epinephrine 1:200,000 or 1:400,000 should be added as a marker for intravascular injection and to potentially slow systemic absorption of the local anesthetic. When multiple levels are being blocked 4–5 mL per level should be used to minimize the overall dose of local anesthetic. When a single level injection is performed, a similar volume can be used if spread to adjacent nerve roots is not required. Volumes of 15–25 mL can be used for single level injections and may increase the spread of the block to multiple dermatomes.

Anticipated Results/Complications

Both motor and sensory blockade result from a TPVB as a result of anesthetizing the individual spinal nerves after they exit the intervertebral foramina. The number of dermatomes affected depends on both the number of levels blocked and also the spread of local anesthetic within the paravertebral space. Somatic coverage is expected, but some visceral coverage may result as well [14–16].

The complications of this block include: pneumothorax, bleeding, hemothorax, local anesthetic toxicity, autonomic dysfunction, and hypotension and nerve injury. There is also the possibility of local anesthetic spread to the epidural, subdural, and subarachnoid spaces.

Lumbar Paravertebral Block/Lumbar Somatic Block (LPVB)

The lumbar paravertebral space is a potential space formed by the vertebral body, intervertebral discs and intervertebral foramen medially, the psoas major muscle anterolaterally, and the transverse processes and intertransverse ligaments posteriorly. There is no costotransverse ligament in the lumbar region. The lumbar spinal nerve roots run through the paravertebral space and continue through the psoas major muscle where they form the lumbar plexus. Because the lumbar paravertebral space does not routinely communicate with the thoracic paravertebral space, local anesthetic injected into one region cannot be expected to spread to the other.

Procedure

Similar in concept to the TPVB, the lumbar paravertebral block (LPVB) aims to anesthetize the individual lumbar spinal nerve roots shortly after they exit the vertebral foramen. If small volumes of local anesthetic solution are injected, individual nerve roots may be blocked to aid in the diagnosis of pain originating from specific lumbar nerves. Upper lumbar nerves can often be blocked without resultant motor weakness. However, if spread occurs to the L2 nerve root or below, motor weakness may occur as a result of obturator or femoral nerve blockade. Just as with thoracic paravertebral blockade, epidural spread can occur in the lumbar region. Additionally, sympathetic blockade may occur as a result of local anesthetic injected into this area.

Indications/Contraindications

The indications for lumbar paravertebral nerve blockade include somatic lower abdominal and groin pain. Selective blockade of individual spinal nerves can help localize the origin of abdominal pain to specific lumbar nerve roots. This may be beneficial for the diagnosis and treatment of spinal and foraminal stenosis. Additionally, they may be used to assist in the diagnosis and treatment of chronic abdominal pain that results from nerve entrapment after inguinal herniorrhaphy.

Contraindications for this block are similar to those for TPVB. These include both local and systemic infection as well as an allergy to either local anesthetic or adjuvant medications. Because the needle is in close proximity to the epidural space and because this block is performed in a non-compressible area, caution should be used when performing this block on patients who are on anticoagulant medications.

Equipment

- Marking pen, skin prep, sterile drape, and gloves.
- 1.5 in. 25 gauge needle, 5 mL syringe, and 1.5 % lidocaine for local skin wheel.
- Eighteen to twenty-two gauge Tuohy needle or a 22 gauge Quincke needle for single injection.
- Local anesthetic of choice, extension tubing and syringe.

Technique

LPVB can be performed with the patient either sitting or lateral. If the block is to be performed in the lateral position, the side to be blocked should be in the up position. The spinous processes are identified and marked. Important landmarks in this region include the iliac crests, which correspond to L3–L4 interspace. From the superior edge of the spinous process of interest, a mark is made 2.5 cm lateral to the midline. If lower lumbar nerve roots are to be blocked, the distance from the midline can be 2 cm as the transverse processes in the lower lumber region are shorter. The skin is prepped and a skin wheel of local anesthetic is raised at this mark. The injection needle of choice is then inserted through the skin wheel and kept perpendicular to the skin in all planes. The needle is advanced to contact the transverse process, which will normally be located in lumbar region at the depth of 4–8 cm depending on the patient's body habitus, and the lumber level being approached. The needle is then withdrawn and redirected to walk caudally under the transverse process and advanced 1 cm past the depth at which the transverse process was located. After negative aspiration for CSF and blood, local anesthetic is injected.

Local Anesthetic

The choice of local anesthetics for lumbar paravertebral blockade is essentially the same as those for thoracic paravertebral blockade. Short acting fast onset local anesthetics can be used for diagnostic purposes, while local anesthetics with a longer duration can be used for the treatment of pain. Steroids may be added to the local anesthetic mixture if so desired and may be beneficial in the treatment of certain conditions, such as foraminal stenosis.

Anticipation/Complications

Lumbar paravertebral blockade will provide unilateral dermatomal anesthesia corresponding to the specific nerve root(s) that are blocked. With small volumes of local anesthetic (2–3 mL), local anesthetic may remain isolated to a single nerve root; however, larger volumes (>5 mL) may spread to adjacent nerve roots. For this reason small volumes should be used for diagnositic purposes to minimize the chance of spread if this is not desired.

Complications include hematoma formation, retroperitoneal bleeding, intra-abdominal needle placement and visceral injury, infection, local anesthetic toxicity, nerve injury and epidural, subdural or subarachnoid spread of local anesthetic. These last three could lead to undesired motor block, sympathectomy, and hypotension. Motor weakness may also occur as a result of nerve root blockade below L2 as a result of femoral or obuturator nerve anesthesia.

Ilioinguinal/Iliohypogastric Nerve Blocks

Both the ilioinguinal (II) and iliohypogastric (IH) nerves are components of the lumbar plexus. The L1 nerve root forms the II nerve, while the T12 and L1 nerve roots form the IH nerve. After these nerve roots exit the vertebral foramen they travel through the paravertebral space and contribute to the formation of the lumbar plexus posterior to the psoas major muscle. As the II and IH continue laterally, they pass through the transversus abdominis muscle to lie between it and the internal oblique muscle. As they continue around the abdominal wall towards the midline, they will at some point separately pass through the internal oblique muscle to lie between it and the external oblique. The points at which the nerves traverse the internal oblique muscle are variable. However, when the block is performed posterior to the anterior iliac spine as described below, the nerves lie between the internal oblique and transversus abdominis muscles in the vast majority of patients [17, 18]. The II nerve supplies cutaneous sensation to the upper and medial thigh and upper part of the genitalia. The IH nerve supplies some motor function to the lower abdominal musculature as well as cutaneous sensation to the suprapubic region and a small segment near the iliac crest.

Procedure

By placing local anesthetic between the internal oblique muscle and the transversus abdominis muscle, both the II and IH nerves can be anesthetized using ultrasound guidance. Although landmark-based techniques have previously been described, accurate placement of local anesthetic is inconsistent and success rates are low [19]. The result of successful blockade is ipsilateral anesthesia of the lower abdomen and inguinal region. No visceral coverage is supplied by blockade of these nerves.

Indications/Contraindications

For chronic pain, II and IH nerve blocks are commonly performed during the diagnosis and treatment of lower abdominal/inguinal pain/neuralgia. One major reason for chronic abdominal pain in this area is previous inguinal hernia repair [20].

The contraindications for II/IH nerve blockade are limited, but include local infection at the block location, implanted material in the area of the injection such as mesh and allergies to local anesthetics or adjuvants.

Equipment

- Gloves, marking pen, chlorhexidine, or alcohol for prep and sterile drape if desired.
- 1.5 % lidocaine and syringe for skin wheel.
- Twenty-one gauge short bevel needle or 22 gauge Quincke spinal needle.
- Extension tubing and 10–20 mL syringe for local anesthetic.
- Local anesthetic for procedure.
- Ultrasound machine and linear probe.

Technique-(Ultrasound)

With the patient supine, the anterior superior iliac spine (ASIS) is located and marked. The ultrasound transducer is placed horizontally on the patient's abdomen slightly superior and lateral to the ASIS. The lateral edge of the probe should be just over the iliac crest so that it can be seen on the edge of the ultrasound screen as a hypoechoic structure. The probe is then oriented obliquely so that the lateral edge is slightly caudal compared to the medial edge. This helps align the ultrasound beam perpendicular to the nerves. The ultrasound image will show three distinct muscles layers at this location, corresponding from superficial to deep to the external oblique muscle (EO), internal oblique muscle (IO), and transversus abdominis muscle (TA) (Fig. 17.7). Normally, the II and IH nerves can be seen between the IO and TA muscles as one or two hyperechoic structures. The nerves may be one structure or two separate structures depending on the patient's anatomy. The deep circumflex iliac artery, or a branch of it, usually accompanies the II nerve at this location and can be used as a landmark. Doppler can be used to assist in identifying the artery. Occassionaly, the nerves can't be identified between the two muscles layers, but local anesthetic can still be injected into the plane between the two muscles. Once the plane between the IO and TA muscles has been identified, the skin is prepped and a skin wheel is raised medial to the probe. The procedure needle is then brought in-plane from medial to lateral with a shallow angle and passed through the EO and IO muscles. Once the needle tip is between the IO and TA muscles and adjacent to the II and IH nerves, negative aspiration is verified and local anesthetic is injected.

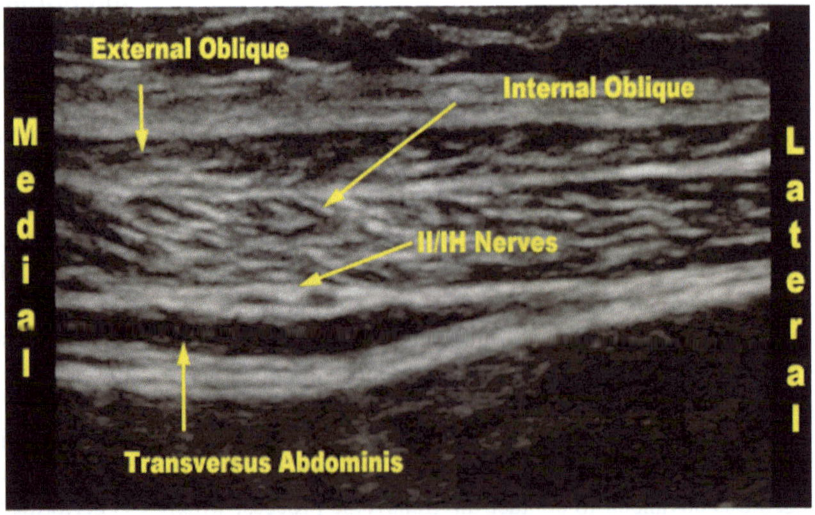

Fig. 17.7 Ultrasound view for ilioinguinal/iliohypogastric nerve blockade

Local Anesthetic

Because prolonged pain control is usually desired, long acting local anesthetics are primarily used for II and IH nerve blocks. Epinepherine should be routinely added to the local anesthetic as a marker for vascular injection secondary to the presence of the deep circumflex iliac artery and the possibility of intravascular injection. Bupivacaine 0.25 %, bupivacaine 0.5 %, or ropivacaine 0.5 % may all be used, with the latter two providing a denser block. 5–10 mL of local anesthetic can be used per side with lower volumes being adequate when accurate needle placement is obtained.

Anticipation/Complications

The II and IH nerves are usually blocked together as a method of selective blockade has not been proven. Blockade of these two nerves results in ipsilateral cutaneous anesthesia of the lower abdomen, genital region, and upper thigh. The distribution of sensory anesthesia for each nerve is variable, thus it is difficult to discern which nerve is the source of pain or relief following blockade [21].

Complications from II/IH nerve blocks include infection, nerve injury, intravascular injection and local anesthetic toxicity, abdominal hematoma formation, intra-abdominal needle placement and visceral injury, pelvic hematoma, femoral nerve block, and bowel puncture.

Rectus Sheath Block

The rectus abdominis (RA) muscles are paired anterior abdominal muscles, which run vertically on either side of the midline. They are separated at the midline by the linea alba and are bordered laterally on both sides by the linea semilunaris. The rectus sheath is comprised of the RA muscles and the aponeuroses of the internal oblique (IO), external oblique (EO), and transversus abdominis (TA) muscles. Which aponeuroses travel anterior to the RA muscle and which travel posterior to it depends on the location relative to the arcuate line. The aponeuroses of all three lateral abdominal wall muscles travel anterior to the RA muscle below the arcuate line. Above the arcuate line the aponeurosis of the IO muscle splits around the RA muscle. The RA muscle also normally contains three transverse tendinous intersections which compartmentalize the rectus muscle. However, these are usually incomplete and do not attach to the posterior rectus sheath [22].

Procedure

The anterior cutaneous branches of the lower intercostal nerves enter the rectus sheath from the posterior and lateral side to provide sensory innervation to the anterior abdominal wall and motor supply to the abdominal muscles [23]. There is no visceral coverage provided by this somatic block. The goal of a rectus sheath block is to place local anesthetic posterior to the rectus abdominis muscle and immediately anterior to the rectus sheath and transversalis fascia. In this location, the local anesthetic will be in close proximity to the anterior cutaneous branches of the intercostal nerves.

Indications/Contraindications

As the rectus sheath block only provides anesthesia or analgesia of the anterior abdominal wall with close proximity to the midline, its indications are limited to pain in this distribution. It may be utilized to treat and define pain that stems from nerve entrapment or pain following midline surgeries or hernia repairs. It can also be used to treat neuropathic pain such as that from post-herpetic neuralgia [24].

Contraindications are limited but include allergies to local anesthetics or adjuvants and local infection. Caution should be used when performing this block on anticoagulated patients; however, this is not an absolute contraindication.

Equipment

- Gloves, chlorhexidine or alcohol for skin prep, and drape if desired.
- Five mL syringe, 25 gauge 1.5 in. needle and 1.5 % lidocaine for skin wheel.
- Extension tubing and 20 mL syringe.
- Local anesthetic for procedure.
- Twenty-one gauge 9 cm short bevel needle or 22 gauge Quincke needle.
- Ultrasound machine and linear probe.

Technique

Although a landmark-based approach relying on tactile feedback has been described, the possibility of peritoneal needle placement and visceral injury are concerns [25]. As a result, ultrasound can be used to perform the block. The ultrasound probe is oriented horizontally on the abdomen in short-axis to the RA. Both in-plane and out-of-plane needle approaches are possible, but an in-plane approach allows for view of the entire needle. As the RA runs from the xiphoid process to the pubic symphasis, a rectus sheath block can be performed anywhere along this course depending on the location of pain and the underlying pathology. The block is most easily performed above the arcuate line, which is roughly 1/2 way between the umbilicus and pubic symphasis, as the posterior rectus sheath is more substantial above this point. With the patient supine, a linear ultrasound probe is placed adjacent to the location of the patient's pain so that the rectus muscle, posterior sheath, and underlying abdominal contents can be seen. Since branches of the intercostal nerves enter the RA muscle from the posterior and lateral side, this is the area that should be targeted. The RA muscle will appear as an oval muscular structure deep to the subcutaneous tissue. Deep to this a double hyperechoic line can be seen (Fig. 17.8). The more superficial

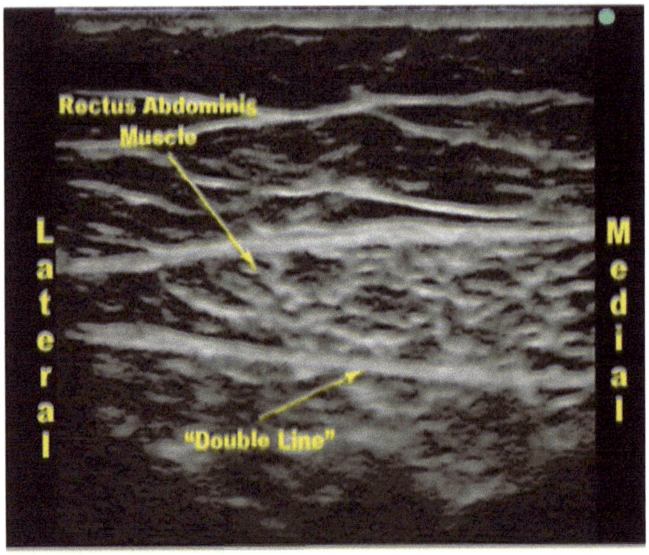

Fig. 17.8 Ultrasound guidance for rectus sheath block. The "double line" is formed by the aponeuroses of the lateral abdominal muscles (superficial) and transversalis fascia (deep)

of these two lines represents the aponeuroses of the lateral abdominal muscles. The second, deeper, line is the transversalis fascia. Under these structures are pre-peritoneal fat, the peritoneum, and viscera. After skin prep and injection of local anesthetic for skin wheel, the procedural needle is inserted in-plane. The objective of the block is to place the needle tip and local anesthetic between the posterior wall of the RA muscle and the aponeuroses of the lateral abdominal wall muscles. The local anesthetic should "lift" the RA muscle off the double hyperechoic line. Of note, if the block is being performed below the arcuate line, only a single hyperechoic line may be seen as the rectus muscle is lying directly on the transversalis fascia as all aponeuroses from the lateral abdominal wall muscles go anterior to the rectus muscle. The approach can be either from the medial or lateral side of the probe. Depending on the extent of desired abdominal coverage, multiple injection sites may be chosen.

Local Anesthetic

As there is little concern for significant motor blockade, high concentration local anesthetics can be used for rectus sheath blockade. Bupivacaine 0.5 % or Ropivacaine 0.5 % will both provide dense analgesia and a long duration of action. Bupivacaine 0.25 % can also be used when larger volumes of local anesthetic are to be used in order to reduce the risk of local anesthetic toxicity. Normally 10–20 mL of local anesthetic is used per side. If multiple injection sites are planned, local anesthetic doses can be divided between injections.

Anticipation/Complication

Rectus sheath blockade results in ipsilateral somatic analgesia/anesthesia of the anterior abdominal wall. This is a result of blockade of the anterior cutaneous branches of the intercostal nerves as they pierce the RA muscle. The area of coverage is from the lateral border of the RA muscle to the midline. Cephalad and caudal spread may be influenced by the volume of local anesthetic. No visceral coverage is provided.

Complications include intra-abdominal needle placement and possible visceral injury, bleeding, hematoma formation, nerve injury, intravascular injection, local anesthetic toxicity, and myonecrosis.

Transversus Abdominis Plane (TAP) Block

The transversus abdominis plane (TAP) lies between the internal oblique (IO) muscle and the transversus abdominis (TA) muscle. The lower six intercostal nerves, T6–T11, the subcostal nerve (T12), and the iliohypogastric and ilioinguinal nerves (T12-L1) all travel within this plane as they course distally. Within the plane the nerves branch and communicate extensively as they travel medially until they pierce the rectus sheath as the anterior cutaneous branches of the intercostal nerves [23]. The intercostal nerves innervate the thorax and abdomen in a dermatomal horizontal pattern. The intercostal nerve from T7 provides innervation to the epigastrium and the intercostal nerve from T10 provides innervation to the area around the umbilicus.

Procedure

The goal of a TAP block is to inject local anesthetic into the plan between the IO and TA muscles where the intercostal, subcostal, ilioinguinal, and iliohypogastric nerves are running. A single injection of local anesthetic placed into this plane can spread both cephalad and caudal to anesthetize multiple nerves unilaterally. If the block is performed sufficiently lateral, the intercostal nerves may be blocked prior to the takeoff of the lateral cutaneous branch.

Indications/Contraindications

TAP blockade can treat pain located in the anterior and/or lateral abdominal wall. Because local anesthetic injected into the TAP plane can spread to cover multiple intercostal nerves, pain that involves multiple adjacent dermatomes can potentially be treated using a single injection. Indications include neuropathic pain, including post-herpetic neuralgia, nerve entrapment, post-hernia pain, and musculoskeletal pain. It may also be useful for distinguishing somatic and visceral pain.

Contraindications for TAP blockade include allergies to local anesthetics or adjuvant medications, local infection at the procedure site, and patient refusal. Coagulopathy, or the use of anticoagulant medications are a relative contraindication.

Equipment

- Gloves, chlorhexidine or alcohol for skin prep, and sterile drape if desired.
- 1.5 % lidocaine, 25 gauge 1.5 in. needle, and 5 mL syringe for skin wheel.
- Eighteen gauge Tuohy needle or 22 gauge Quincke needle or 21 gauge short bevel 90–100 mm needle.
- Extension tubing and 30 mL syringe.
- Ultrasound machine and linear probe.

Technique

Landmark-based techniques using tactile feedback as the needle passes through tissue planes have been described, but given concerns for intra-abdominal needle placement, and accurate injection of local anesthetic, ultrasound-based techniques will be described [26, 27].

Posterior TAP Block Ultrasound Approach

The posterior approach to TAP blockade utilizing ultrasound relies on ultrasound visualization of the three muscular layers of the lateral abdominal wall. With the patient in the supine position, the ultrasound transducer is placed horizontally on the lateral abdominal wall, between the iliac crest and inferior costal margin at the mid-axillary line. At this location three abdominal muscle layers can be identified. They are, from superficial to deep, the EO muscle, the IO muscle, and the TA muscle (Fig. 17.9). The target for needle and local anesthetic placement is between the IO and TA muscles. The skin is prepped and local is injected for skin wheel anterior to the probe. The procedural needle is then inserted in-plane to the probe from the anterior side. The needle is directed posterior and medially to transverse the EO and IO muscles. Once the needle tip is through the IO muscle, and within the TAP, local anesthetic or saline can be injected to verify needle placement. When adequate needle location is obtained, the injectate, which will be hypoechoic, will spread along the plane between the IO and TA muscles.

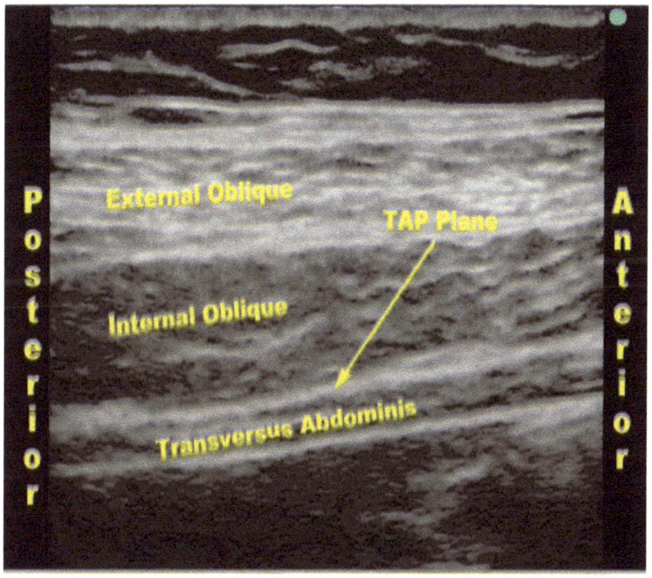

Fig. 17.9 Ultrasound image during a posterior TAP block showing three muscular layers of the abdominal wall. The needle should be placed at appropriate depth, so that local anesthetic is injected into the plane between the internal oblique muscle and the transversus abdominis muscle

Subcostal Ultrasound Approach

Because there is some controversy as to whether the posterior ultrasound approach provides adequate and reliable spread to the upper abdominal segments, the subcostal approach is often used when anesthesia/analgesia is needed above the umbilicus [28–30]. With the patient supine, the ultrasound transducer is placed on the upper abdomen adjacent to the midline and oriented obliquely, so that it is parallel to and immediately inferior to the costal margin. In this position, the ultrasound view will show the rectus sheath near the midline. The TA muscle can be seen lateral to the rectus sheath. In some individuals the two muscles will overlap each other for a short distance, with the TA underlying the RA; however, in other individuals the TA will become aponeurotic prior to reaching the RA. After skin prep and local anesthetic injection for skin wheel, the procedural needle is inserted in-plane to the probe from near the xiphoid process and directed inferiolateally. The needle tip should be placed between the RA and the TA muscles in the TAP or between the RA and the posterior rectus sheath if the TA is not present underneath the RA muscle. As local anesthetic is injected into the TAP, the needle can be advanced along the plane and incremental dosing can be performed to progressively distend the TAP parallel to the costal margin.

Local Anesthetic

The volumes of local anesthetic needed for adequate TAP blockade vary between studies, but clinically, moderately high volumes are needed [31]. If unilateral blockade is being performed then 30–40 mL of local anesthetic can be used. If bilateral TAP blocks are required then 20–25 mL per side can be used. Various local anesthetics can be used including 0.25 % bupivacaine, 0.5 % bupivacaine, and 0.5 % ropivacaine. Steroids can be added to the solution if desired depending on the suspected pathology involved.

Anticipation/Complications

TAP blocks provide ipsilateral somatic anesthesia/analgesia; visceral coverage is not expected. Although there is some disagreement on the coverage provided by the posterior ultrasound approach, T10-L1 coverage can be reliably expected. The subcostal approach may provide upper abdominal coverage more consistently, but may miss some of the lower thoracolumbar segments depending on the spread of local anesthetic.

Complications from TAP blockade include nerve injury, intravascular injection and local anesthetic toxicity, abdominal wall hematoma, intraperitoneal needle placement and visceral injury, intra-abdominal hematoma, infection, myonecrosis, and failure.

Neuraxial Techniques

Neuraxial regional anesthesia techniques for abdominal and truncal pain are primarily related to epidural anesthetic and analgesic techniques, as compared to intrathecal techniques due to the more selective result of thoracic epidural catheter placement and less risk of spinal cord injury as compared to either lumbar or thoracic intrathecal block procedures, respectively. An exception to this general rule would be the use of selective thoracic intrathecal neurolytic procedures which may most commonly be used for unilateral chest or abdominal wall pain associated with cancer-related pain.

Thoracic Intrathecal Neurolysis

Anatomy and Indications

Intrathecal neurolytic block procedures are valuable considerations in a very select group of patients. Most clinicians consider patients to be candidates for intrathecal neurolytic procedures only if they have: (1) a short life expectancy (less than 6–12 months); (2) have severe and well-localized pain (limited to two or three dermatomes) of a somatic source due to malignancy; (3) clear primary and secondary diagnoses; (4) pain unresponsive to less invasive techniques including maximum tolerated doses of opioid and non-opioid analgesics; (5) and a clearly positive response to a diagnostic/prognostic intrathecal block with small doses of local anesthetic. Patients should be advised of all possible complications including death and disability, as well as alternative procedures which may include subarachnoid catheter and drug delivery systems or surgical dorsal rhizotomy. Patients must also understand the possibility of continued tumor growth and the loss of effectiveness of the neurolytic procedure over time necessitating repeat procedures [32].

Technique

The effect of the chemical neurolytic agent, most commonly either alcohol or phenol, is to produce a dorsal rhizolysis through disruption of the sensory dorsal rootlets at their attachment to the spinal cord. The choice of neurolytic agent to be used may depend on a variety of factors including the patient's ability to be appropriately positioned for the procedure. Alcohol is hypobaric with respect to CSF, whereas phenol (usually in glycerin as the diluent) is hyperbaric with respect to CSF. Therefore phenol injection requires the patient to be positioned laterally with the painful side in the dependent (lowermost) position (Fig. 17.10), whereas alcohol injection allows the patient to be placed with the painful side in the nondependent (uppermost) position (Fig. 17.11). Phenol produces a local anesthetic effect and a sensation of warmth upon injection, whereas alcohol injection may produce a transient painful, burning sensation. These sensations noted upon injection of either agent are valuable in determining the appropriate localization of anticipated effect and should cover the selective dermatomal distribution of the patient's pain source.

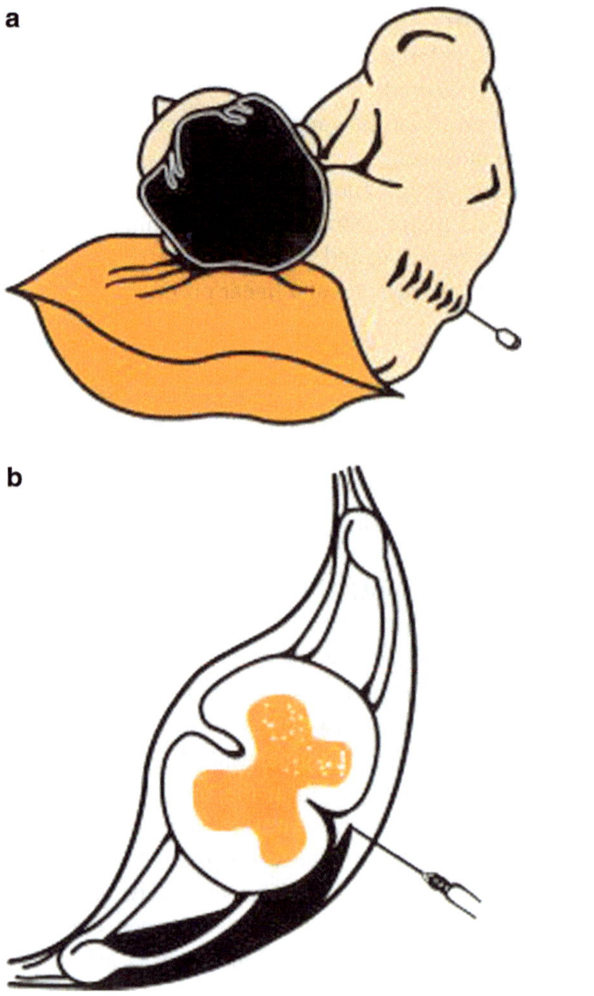

Fig. 17.10 Diagram showing patient positioning (**a**) and anticipated spread of injectate (**b**) with intrathecal phenol injection for neurolysis. Plenol being hyperbaric as compared to cerebrospinal fluid (CSF) will layer in the most dependent aspect of the intrathecal space

Drugs

Whether using alcohol (100 % ethyl alcohol) or phenol (6–12 % in glycerin) small doses of 0.1 mL volume are incrementally injected while assessing the distribution of sensations following injection to a total volume of ≤1 mL. Following completion of injection, the patient should remain in the same position as during the procedure for a period of at least

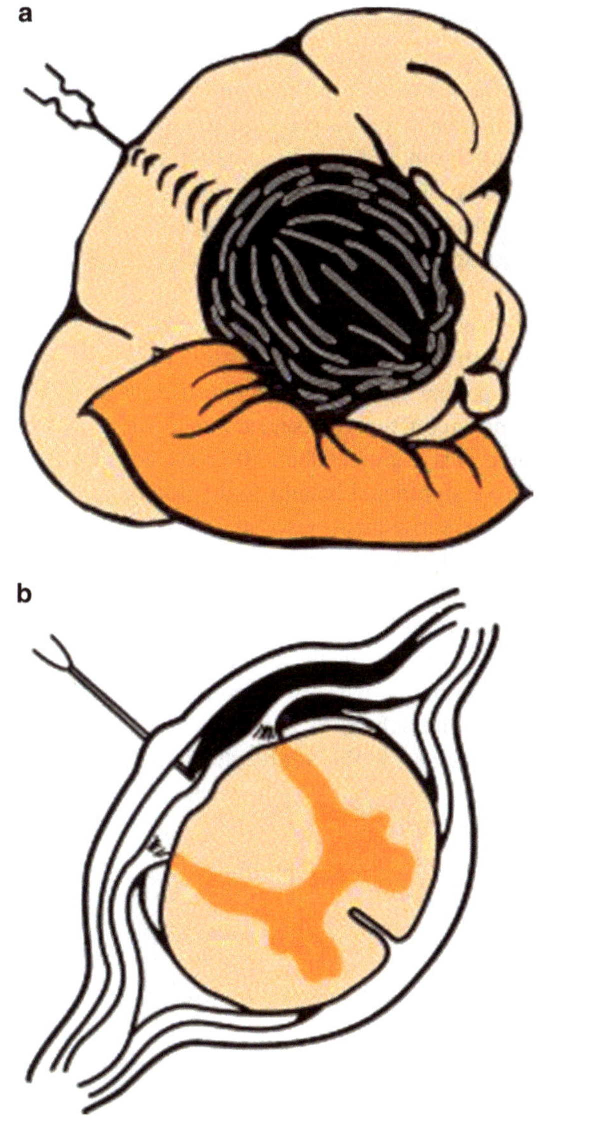

Fig. 17.11 Diagram showing patient positioning (**a**) and anticipated spread of injectate (**b**) with intrathecal alcohol injection for neurolysis. Absolute alcohol being hypobaric as compared to CSF will layer in the most superior aspect of the intrathecal space

15–20 min to maintain localization of the neurolytic agent until dilution/absorption has occurred. The needle should be "cleared" with a very small volume 0.1–0.2 mL of air or local anesthetic prior to withdrawal.

Epidural neurolysis via a needle or catheter placed into the thoracic epidural space with either alcohol or phenol has also been described [33].

Thoracic Epidural Analgesia

Anatomy and Indications

Thoracic epidural anesthetic and analgesic techniques which are commonly used for perioperative management of patients following major abdominal surgical procedures can also be a valuable technique for short-term analgesia for patients with a variety of more severe abdominal pain problems such as acute exacerbations of chronic pancreatitis, abdominal pain associated with sickle cell anemia, inflammatory bowel disease, or biliary or renal colic [34, 35]. The advantages of thoracic epidural analgesic techniques with local anesthetic and dilute concentrations of opioid analgesic include the prokinetic intestinal motility effects of the resulting visceral sympathetic nerve block and therefore unopposed parasympathetic system effects, as well as the decreased systemic dose of opioid analgesics required which therefore have less effect on inhibiting intestinal motility [36]. Visceral blood flow (perfusion) is increased [37, 38], and somatic, visceral, as well as potential neuropathic sources of pain are blocked [39–42].

Technique and Drugs

Placement of an epidural catheter in the region of T6-8 and dosing with Bupivaciane 0.25–0.5 % to achieve adequate dermatomal distribution for analgesia (4–8 mL) followed by a loading dose of epidural opioid such as Hydromorphone 0.2 mg, Morphine 1 mg, or Fentanyl 100 μg should be administered to establish analgesia. Then starting a continuous thoracic epidural infusion of Bupivacaine 0.25 % with a low dose opioid such as Hydromorphone 0.0005 %, Morphine 0.001 %, or Fentanyl 0.0005 %, at an infusion rate of 4–6 ml/h, with a patient-controlled bolus dose of 2–3 mL and a 30–60 min lockout interval is a reasonable starting point to provide adequate analgesia for a period of 3–5 days. The administration of non-opioid analgesics such as Ketorolac 15–30 mg IV q6h and Acetaminophen 650–1,000 mg IV q6h should also be considered in a multimodal analgesic approach. This period of time is usually sufficient to allow reinstitution of oral intake, increased physical activity including ambulation, and resolution of the acute pain exacerbation while minimizing systemic opioid dose requirement. At that point the epidural infusion can be reduced and discontinued, while converting the patient to an oral analgesic regimen.

Differential Neuraxial Blockade

Anatomy and Indications

Another common use of neuraxial regional anesthetic techniques in the patient with abdominal or truncal pain is as a differential neuraxial blockade in the evaluation of patients with chronic abdominal pain. Contemporary use in diagnostic purposes is detailed in Chap. 4 of this textbook. Differential neuraxial blockade is traditionally described using either an antegrade or retrograde approach, with a lumbar intrathecal dose of short-acting local anesthetic for lower extremity, or back pain syndromes. For patients with chronic abdominal pain, the more common approach is to utilize the thoracic epidural route for local anesthetic blockade. Following placement of a thoracic epidural catheter at the spinal level congruent with the spinal level of the patient's primary pain complaint, the epidural catheter can either be dosed with increasing concentrations and volumes of local anesthetic, while closely monitoring the patient's description of potential changes in the severity and location of their pain (anterograde approach), or after dosing the epidural catheter to produce an adequate sensory block and to produce complete loss of sensation in the distribution of the patient's pain, the patient is closely monitored for relief of pain if it occurs, and then return of pain while monitoring resolution of the sensory block (retrograde approach) [43, 44].

Technique

Utilizing the retrograde block approach, following careful assessment of the patient's pain complaint with regard to the location, character, and severity of the pain, an epidural catheter is placed at a spinal congruent level correlating with the patient's abdominal or truncal pain, using convention epidural placement techniques. The epidural catheter is secured and the patient is placed in the supine position and reassessed.

Local Anesthetic Drugs

A control dose of normal saline may be administered and the patient assessed for a placebo response. If the patient responds to the placebo injection of epidural saline, the patient may be brought back to the clinic and reassessed on another day. If the patient describes no pain relief with the saline injection, the epidural catheter is then incrementally dosed with a potent short-acting local anesthetic (e.g., 2 % 3-chloroprocaine) in 5 mL increments until the patient describes complete loss of sensation over the entire dermatomal distribution of the patients pain (approximately T4 to L2 distribution of surgical anesthesia). If the pain persists, despite adequate and complete segmental sensory block above and below the level of the pain, the pain may be interpreted as being supraspinal (central) in origin (or perhaps the pain is mediated by neural pathways outside the spinal level of block (vagal tracts)). If the patient has initial improvement in pain with the anesthetic block, the patient is carefully assessed with respect to pain intensity and dermatomal distribution of sensory block every 5 min. If the pain returns with resolution of the sensory block, the pain may be determined to be somatic in origin. If the patient has delayed return of pain beyond the period of resolution of the sensory block, the pain may be interpreted as visceral in origin. Not infrequently, patients may describe return of some component of the pain attributable to somatic pain and then later return of a component of pain attributable to visceral pain and would then be classified as having a mixed somatic and visceral pain source [39–42].

Anticipated Results

The results of a differential epidural block may be helpful in determining the source of pain and therefore directing subsequent therapy. Patients with a primary somatic pain source may benefit from anti-inflammatory analgesics, physical therapy, and somatic or infiltration type neural blockade procedures. Patients with a predominant visceral pain source, may benefit from additional evaluation to search for a diagnostic source of the pain or may be candidates for visceral nerve blocks such as splanchnic block, celiac plexus block, or hypogastric plexus block. In a study by Rizk and colleagues [44], differential epidural nerve block (DNB) was evaluated as a predictor of success of a visceral block in patients with chronic abdominal pain. These authors reported that if a patient has a DNB suggestive of a visceral pain source and a reduction in visual analog pain scores by ≥50 %, then there was a greater than 70 % chance that the patient would have significant benefit (reduction of pain by ≥50 %) following a subsequent visceral block (splanchnic block, celiac plexus block, hypogastric plexus block).

Interestingly a study by Conwell et al. [43] demonstrated that in patients with a firmly established diagnosis of chronic pancreatitis, 78 % of these patients demonstrated non-visceral pain on the basis of response to a differential neuraxial block, and only 22 % had results suggestive of visceral pain. Of 5 patients in this study who demonstrated a visceral pain source, 4 (80 %) went on to have successful response to medical (pancreatic enzyme therapy, endoscopic sphincterectomy, or celiac plexus blockade) or surgical therapy (Peustow procedure, distal pancreatectomy, sphincteroplasty, or biliary bypass) directed at pain control, whereas only 29 % of the patients in the group with non-visceral pain responded to medical or surgical therapy.

Visceral Sympathetic Blocks

The visceral sympathetic blocks which are primarily used for the management of abdominal and pelvic pain are the splanchnic nerve block, the celiac plexus block, and the superior hypogastric plexus block. These will be briefly reviewed here, as they were described in more details elsewhere in this book (Chaps. 17 and 18). The splanchnic and celiac plexus blocks are both highly effective for patients with visceral pain in the upper abdomen (stomach, duodenum, ascending and transverse colon, liver, and pancreas) whereas the superior hypogastric plexus block is utilized for pain in the lower abdomen and pelvis arising from the bladder, prostate, uterus, vagina, and rectum.

Splanchnic Nerve Block

Anatomy and Indications

The three splanchnic nerves of the thoracic sympathetic trunk arise from the lower eight ganglia and are formed by visceral preganglionic autonomic efferent fibers as well as pain conducting ascending visceral afferent fibers inervating the upper abdominal viscera. The greater splanchnic nerve is formed by branches of the T4-9 sympathetic ganglia, the lesser splanchnic nerve from T10 to T11 ganglia, and the least splanchnic nerve from the T12 ganglia. The splanchnic nerves penetrate the diaphragmatic crus and converge to form the celiac plexus located anteriolateral to the aorta and just caudal to the take-off of the celiac artery, usually at the level of the L1 vertebral body [45–47].

Therefore, the results following performance of a splanchnic nerve block would be anticipated to be the same as that obtained with a celiac plexus block. One potential advantage of the splancnic nerve block as compared to the celiac plexus block would be in the case of the patient with a significant upper abdominal mass which may alter the anatomy near the celiac plexus or prevent adequate spread of the injectate with the celiac plexus approach.

Celiac Plexus Block

Anatomy and Indications

The celiac plexus block is the most commonly performed sympathetic block for abdominal pain. As discussed above, the celiac plexus is the convergence of preganglionic sympathetic efferent fibers as well as visceral afferent fibers. Some parasympathetic preganglionic and afferent fibers originating from the vagus nerves also contribute to form the celiac plexus. The celiac plexus block can be utilized as a diagnostic procedure to determine the contribution of a visceral source of abdominal pain. As a treatment procedure it may be used to provide pain relief for a period of hours to days in patients with an acute exacerbation of chronic pancreatitis. It is highly effective and frequently used as a neurolytic technique to decrease the pain and opioid analgesic requirement for patients with pancreatic cancer for a period of weeks to months. It is a relatively safe and well-tolerated procedure and for these reasons, many advocate its use relatively early in the course of therapy for a patient with pain associated with pancreatic cancer.

Contraindications

The splanchnic nerve block and celiac plexus blocks will not be effective for somatic pain involving the parietal peritoneal or abdominal wall. Most patients will benefit from a diagnostic block performed with local anesthetic prior to proceeding with a neurolytic procedure. The block will result in a visceral sympathectomy which will likely produce some degree of hypotension that should be treated using peri-procedural intravascular volume expansion. The sympathetic block with relatively intact parasympathetic innervations of the gastrointestinal tract will likely result in transient diarrhea, especially in patients who may be having opioid-related constipation [48–50].

Equipment

Most splanchnic nerve block, or celiac plexus block procedures are currently performed under either fluoroscopic, computerized tomography (CT) [51, 52], or ultrasound guidance [53]. Endoscopic-guided procedures [53] and surgical sympathectomy, or direct application of neurolytic substances during open abdominal procedures, have been described.

Technique

Although the anterior approach has been described, the posterior approach remains most commonly utilized. The patient is placed prone on the operating table with appropriate padding and support for the head which is turned to one side. Intravenous access should be established and intravascular fluids can be given for volume expansion and the delivery of sedation or resuscitation drugs if needed. Routine monitoring including blood pressure, electrocardiograph, respiratory rate, and oxygen saturation should be maintained throughout the procedure. Supplemental oxygen may be required in some patients, especially if sedative medications are required.

Anatomic landmarks are identified including the spine, the inferior border of the 12th rib, and the iliac crests (Fig. 17.12). After sterile preparation of the skin and application of surgical drapes to the field, the vertebral body of T12 and L1 are identified by fluoroscopic or CT imaging. Approximately 4–8 cm lateral to the midline, depending on the size of the patient, and inferior to the border of the 11th or 12th rib, a 15 cm needle is advanced from the left side of the patient in a

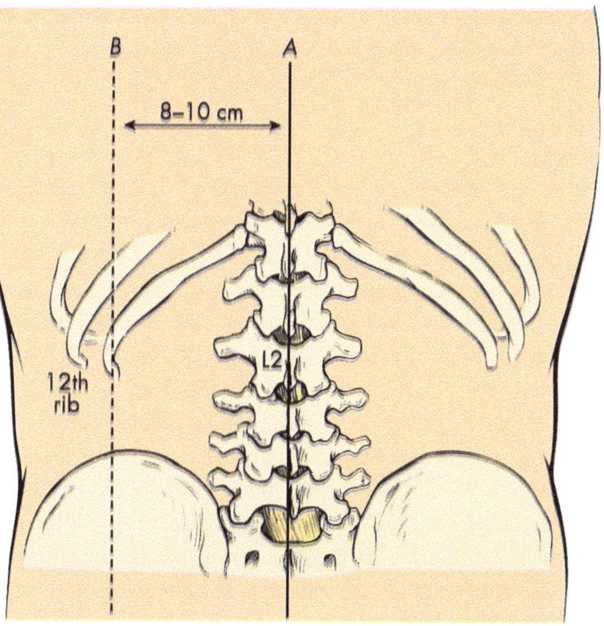

Fig. 17.12 Diagram of surface landmarks for splanchnic (retrocrural) and celiac plexus nerve block. Needle insertion is made 7.5–8 cm lateral to the mildline caudad to the 12th rib and directed superiorly, anteriorly, medially toward the anterolateral aspects of the T12 or L1 vertebral bodies. (A represents a line drawn down the spinous processes, B represents a parallel line through the inferior border of the 12th rib)

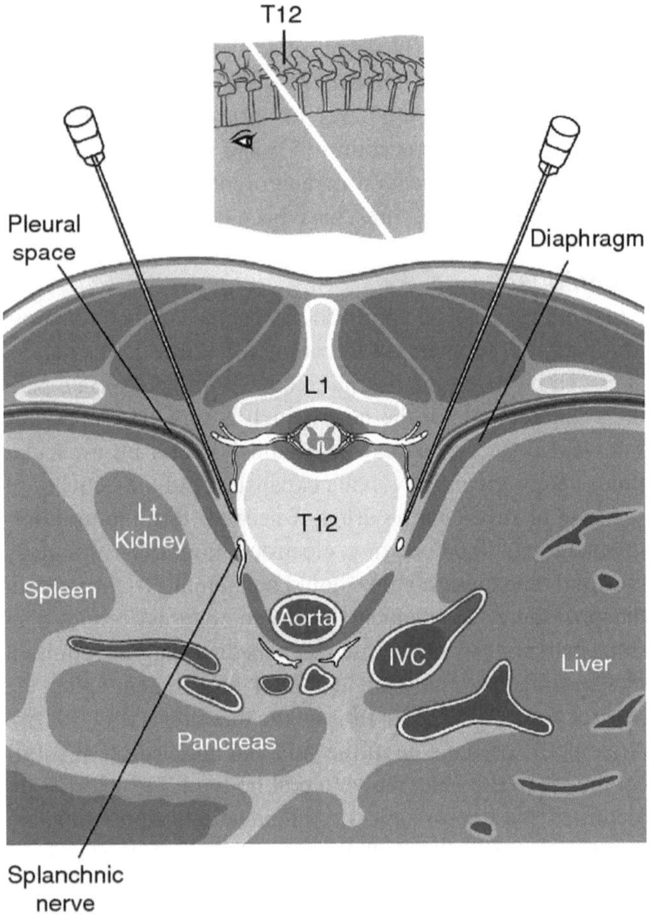

Fig. 17.13 Axial diagram of needle position for splanchnic nerve block. The needles are directed to the anterolateral aspects of the T12 vertebral body. At this level the splanchnic nerves and aorta lie above and posterior to the diaphragmatic crus

cephalomedial direction to a location just lateral to the T12 vertebral body. Avoidance of the transverse process or contact with the vertebral body is desirable as contact with these structures can be quite painful for the patient. The needle should be located just at the lateral aspect of the T11 or T12 vertebral body for a splanchnic nerve block (Figs. 17.13 and 17.14). The stylet is withdrawn from the needle and aspiration attempts are made for blood or other fluids. If blood is obtained, the needle may be within the aorta, and consideration may be given to performing a transaortic celiac plexus block, where the needle is placed through and just anterior the aorta. If no blood is obtained, the second needle is placed in similar fashion to the first on the contralateral side. At this point, 3–5 mL of a mixture of local anesthetic and water-soluble radiographic contrast medium can be injected through each needle to demonstrate the spread of the injectate anterior to the anterior edge of the T11 or T12 vertebral body and spreading cephalad and medial in the retrocrural fascial plane (Fig. 17.15). After confirmation of needle position, 10–25 mL of local anesthetic or neurolytic mixture is injected through each needle for a total volume of 20–40 mL.

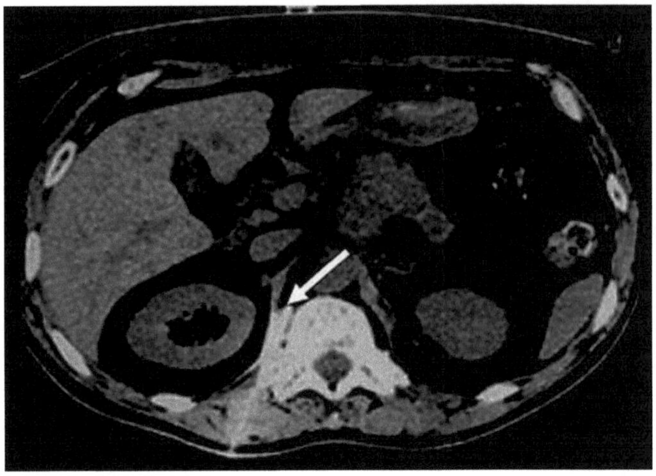

Fig. 17.14 Axial CT image of needle anterolocation lateral to the T12 vertebral body and posterior to the diagram in appropriate position for splanchnic nerve block injection

For a celiac plus block, the procedure is performed as above but the needles are advanced in a slightly less cephalad angle to approach the L1 vertebral body and then 1.5–2 cm anterior and through the crus of the diaphragm to reach the anterolateral edges of the aorta below the diaphragm. CT guidance is especially helpful here as the needle position can be adjusted to produce appropriate spread of radiographic contrast, surrounding the entire space anterior and lateral to the aorta or bilaterally (Figs. 17.16 and 17.17). With this needle position, the contrast will appear to spread medial and more caudad in direction, being anterior to the diaphragm (Fig. 17.18). If a neurolytic substance is injected the needles should be "cleared" with air or local anesthetic prior to withdrawal.

Local Anesthetic/Non-local Anesthetic Drugs

For a local anesthetic block a total volume of 30–40 mL of Bupivacaine 0.375 % (20–25 mL per side) is usually injected. For a neurolytic procedure, generally alcohol is preferred over phenol due to the better tissue spread characteristics of the less viscous ethanol. After confirmation of needle position with local anesthetic and radiographic contrast, a solution of 5–10 mL of Lidocaine 2 % or Bupivacaine 0.5 % is injected through each needle as a test dose. After several minutes the patient is questioned about any feeling of lower extremity numbness or weakness or other neurologic symptoms. If no such symptoms are present, then ethanol (60–100 %) or phenol 10 % diluted in radiographic contrast is injected. The total combined volume of local anesthetic agent and neurolytic agent will be between 30 and 50 mL.

Anticipated Results/Possible Complications

Following the splanchnic nerve block or celiac plexus block patients will usually describe a significant decrease in the visceral component of the abdominal pain. As noted above,

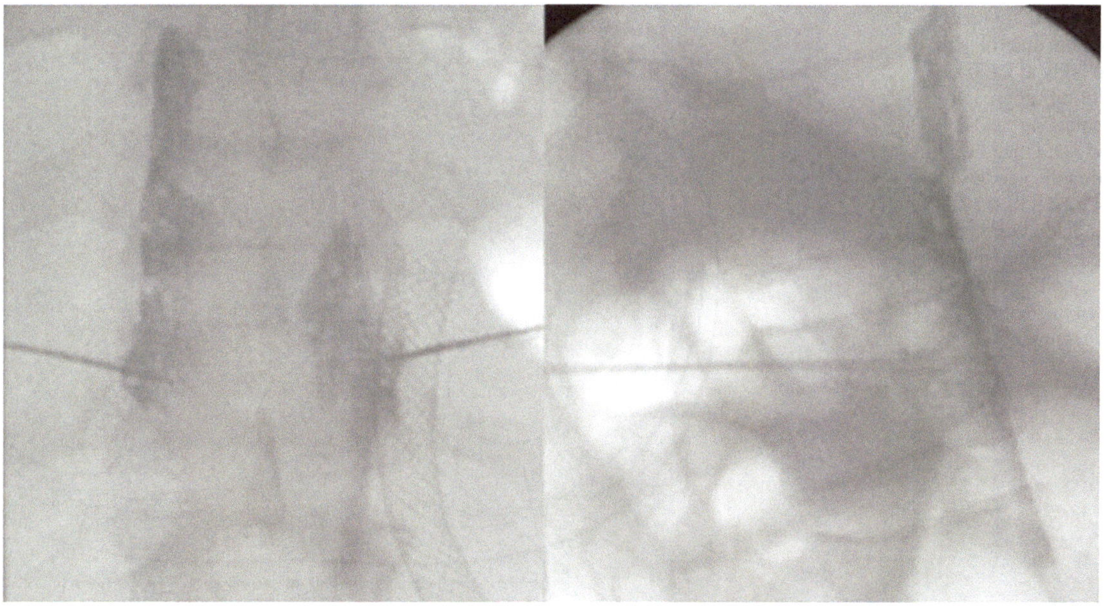

Fig. 17.15 Anterioposterior and lateral radiographs views showing radiopaque contrast spread for splanchnic (retrocrural) nerve block. Contrast can be seen spreading in a medial and cephalad direction with posterior spread impeded by the crus of the diaphragm with this retrocrural needle location

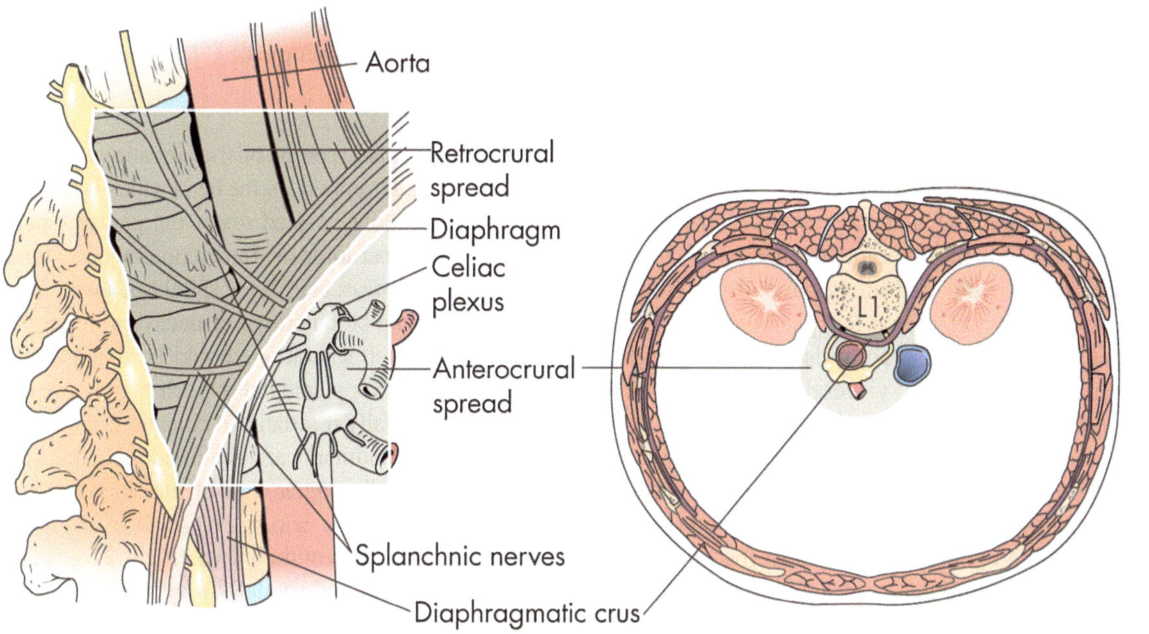

Fig. 17.16 Image of areas of spread of injectate for splanchnic (retrocrural) versus celiac (transcrural) nerve block. Injectate spread is cephalad of the diaphragm for the retrocrural splanchnic nerve block, whereas injectate spread is caudad of the diaphragm for the transcrural celiac plexus block

hypotension and diarrhea are not uncommon in the immediate post-block period. Serious complications could include transient, or permanent spinal cord injury if local anesthetic or neurolytic agents spread to the nerve roots, epidural, or intrathecal space. Another cause of permanent spinal cord injury may be related to spasm, injury, or direct injection into the anterior spinal artery (artery of Adamkiewicz), the vascular supply to the anterior two thirds of the spinal cord in the low thoracic region, resulting in spinal cord infarction. Other potential complications include pneumothorax, hematuria, and intravascular injection with possible local anesthetic systemic toxicity.

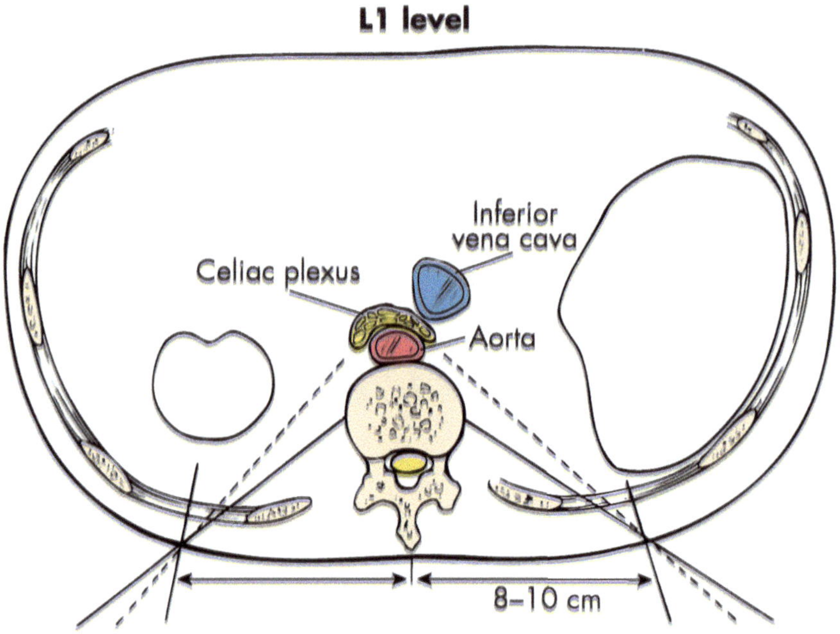

Fig. 17.17 Axial diagram of needle placement anterolateral to the vertebral body at L1. The traditional "blind" approach described contacting the L1 vertebral body and then "walking off" to position the needles at the anterolateral aspect of the L1 body. This contact with periosteum can be quite painful for the patient and can be avoided using a "down the beam" fluoroscopic technique

Superior Hypogastric Plexus Block

Anatomy and Indications

The superior hypogastric pleus is located anterior to the L4-5 and S1 vertebral bodies, medial and anterior to the psoas fascia, posterior to the peritoneum, and medial to the iliac vessels. The superior hypogastric plexus innervates the pelvic visceral organs (bladder, prostate, uterus, vagina, and rectum) through the hypogastric nerves and inferior hypogastric plexus and continues caudally to form the ganglion impar, which provides perineal innervation. The hypogastric plexus is comprised of preganglionic sympathetic efferent fibers, visceral afferent fibers, and parasympathetic fibers. This block is used to treat pain in the lower abdomen and pelvis, refractory to more conservative methods. Its primary indication may be diagnostic when used to identify a visceral pain component in chronic pelvic pain, or therapeutic when used in the management of advanced pelvic malignancy [45, 54–56].

Contraindications

Contraindications other than those associated with any neural blockade technique are few. Patients should be advised of possible loss of sensation and strength in the lower extremities, with spread of local anesthetic or neurolytic agents to the lumbar somatic plexus, and potential local anesthetic systemic toxicity if an intravascular injection occurs (plexus lies in close proximity to the iliac vessels).

Equipment

Superior hypogastric plexus blocks are most commonly performed using fluoroscopic or CT guidance.

Technique

The patient is placed prone on the operating table with appropriate padding and support for the head which is turned to one side. Additional padding placed under the patient's pelvis helps to flatten the lumbar lordosis. Intravenous access should be established, and intravascular fluids should be given. Routine monitoring including blood pressure, electrocardiograph, respiratory rate, and oxygen saturation should be maintained throughout the procedure. Supplemental oxygen may be required in some patients, especially if sedative medications are required.

Anatomic landmarks are identified including spine and iliac crest. After sterile preparation of the skin, and application of surgical drapes to the field, interspace between the spinous processes of L4-5 is identified. Two 15 cm 22 g needles are then directed medially and caudally from a point 5–7 cm lateral to the L4-5 interspace using fluoroscopic or CT guidance. The needle is advanced toward the anterolateral aspect of the L5 vertebral body (Fig. 17.19). The iliac crests and the L5 transverse processes are to be avoided with the use of radiographic guidance as contact with these structures can be painful for the patient. The needles are advanced to the point in front of the anterior psoas fascia and to the point where the needle tip is aligned with the anterior border, and just lateral to the L5 vertebral body on anterior–posterior and lateral fluoroscopic

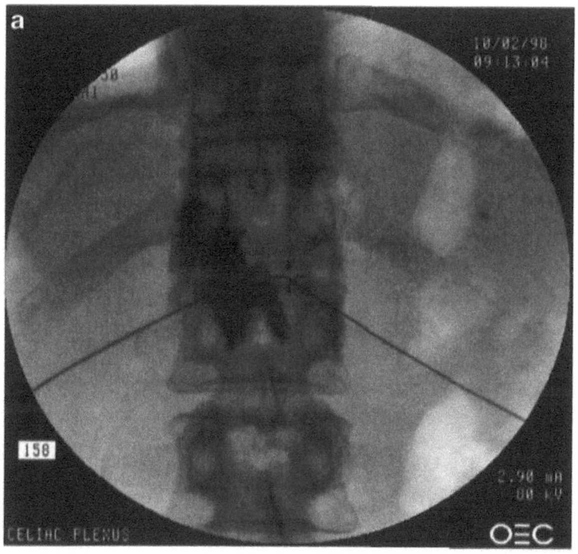

b

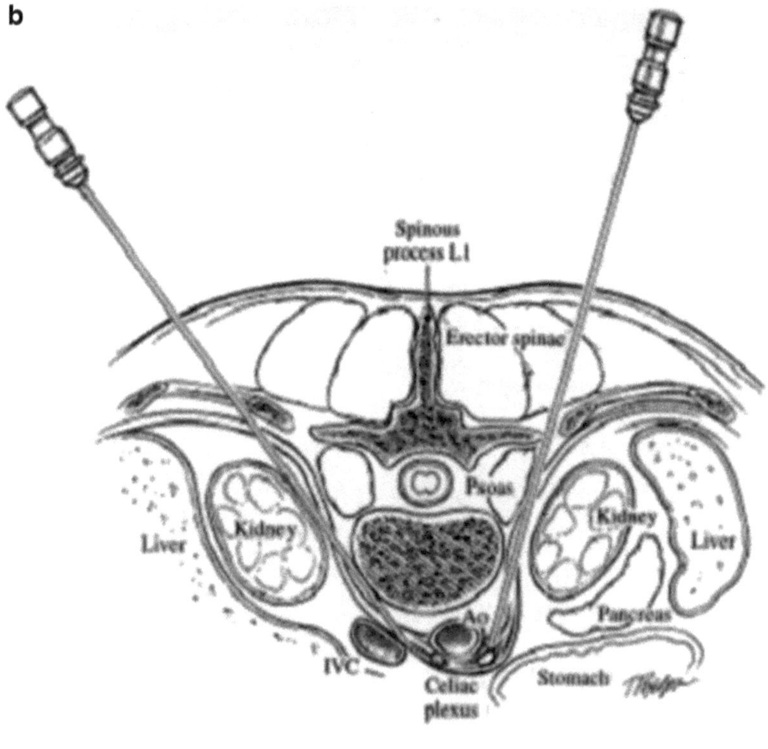

Fig. 17.18 Posteroanterior radiograph showing needle placement and contrast spread (**a**) and axial diagram demonstrating anatomy and needle placement for celiac (transcrural) plexus block. (**b**) Contrast is seen located anteromedial to the needle location at the anterolateral edge of the L1 vertebral body and anterocaudad to the diaphragm. (**a**)

projection or by CT (Figs. 17.20 and 17.21). Some authors advocate transdiscal approach through the L5-S1 intervertebral disc just anterior to the anterolateral surface of the vertebral column [57]. After radiographic confirmation of correct needle placement and careful aspiration for blood or other fluids, a test dose of 2–3 mL of water-soluble contrast should be injected and spread documented along the anterior surface of the lumbsacral space demonstrating proper needle placement (Fig. 17.22).

Local Anesthetic/Non-local Anesthetic Drugs

A diagnostic or temporary hypogastric plexus block is accomplished with a total volume of 20–30 mL of local anesthetic (10–15 mL per side), such as Bupivacaine 0.25–0.5 %. A neurolytic procedure is performed with a total dose of 20–25 mL (10–12 mL per side) of 10 % phenol in radiographic contrast, or alcohol.

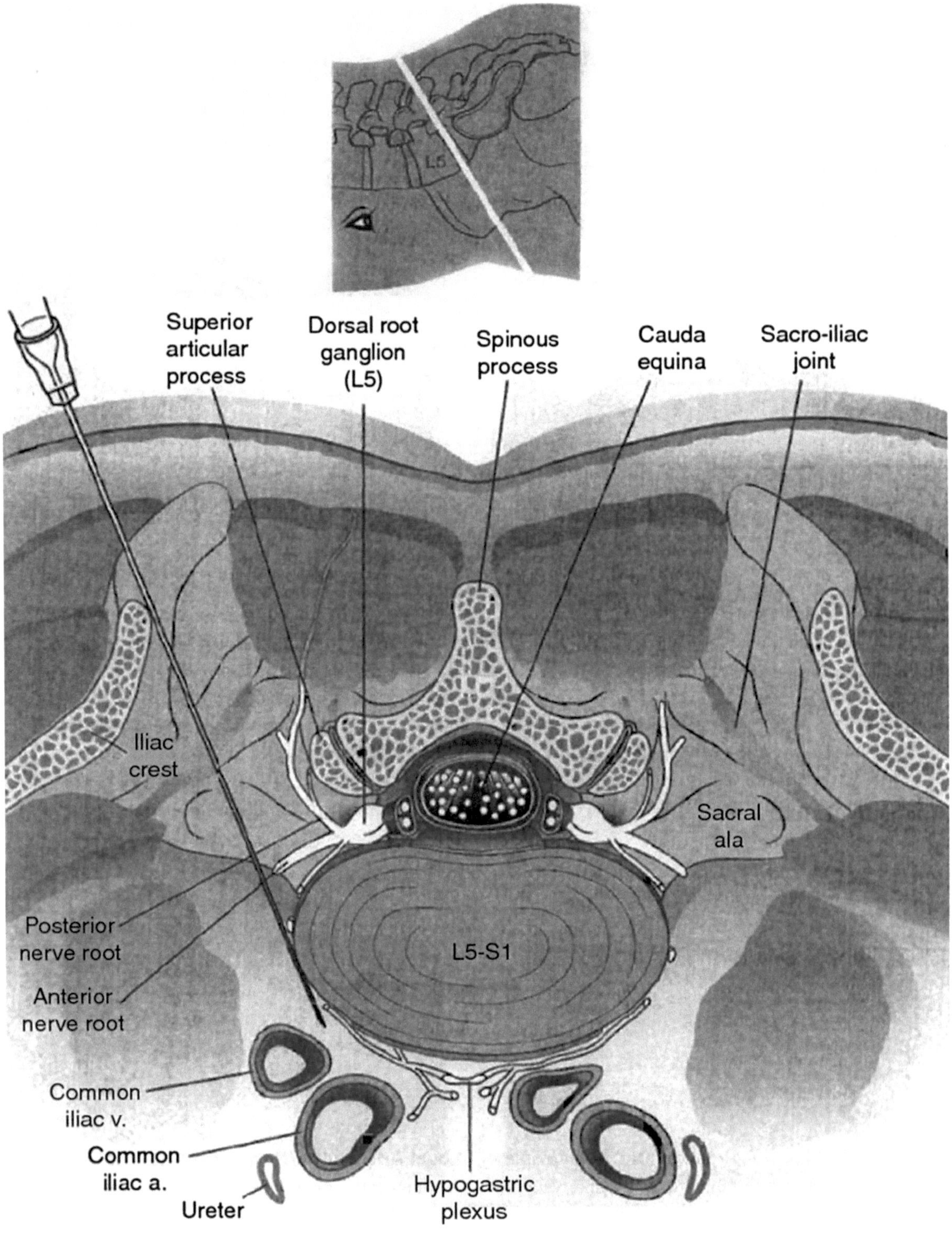

Fig. 17.19 Axial diagram showing anatomy and needle placement for superior hypogastric plexus block. In this axial orientation the needle can be seen to be placed at the anterolateral edge of the L5 vertebral body or the L5-S1 disc space, medial to the iliac vessels

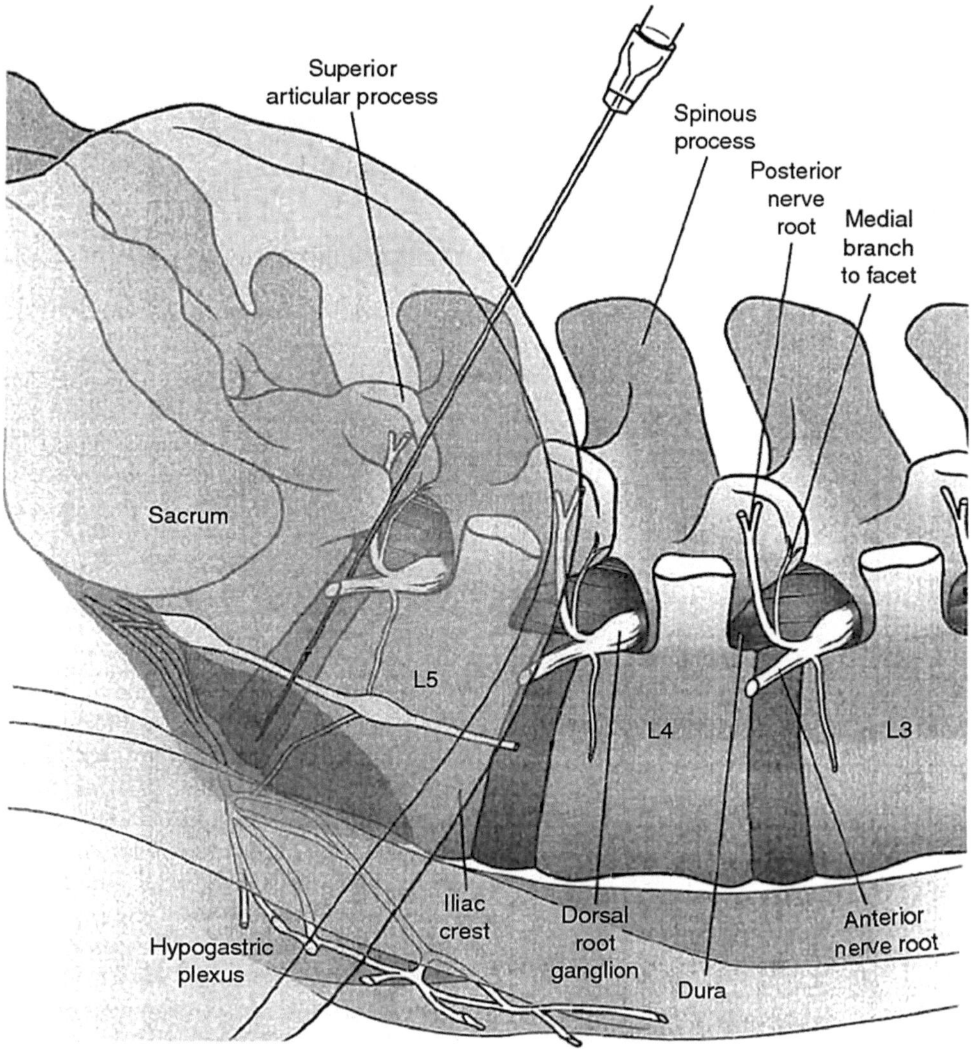

Fig. 17.20 Diagram of lateral view of anatomic structures and needle placement for superior hypogastric plexus block. Needle location can be seen at the anterolateral edge of the L5-S1 disc space

Anticipated Results/Possible Complications

It is anticipated that patients will report a ≥50 % reduction in visceral pelvic pain following the procedure. Hypotension is less likely with hypogastric plexus block as compared to celiac plexus block. Complications include intravascular injection, infection, and bleeding complications as with any invasive procedure. The risk of lower extremity sensory or motor block and intravascular injection is reduced with the monitoring of appropriate spread of the injectate with radiographic contrast.

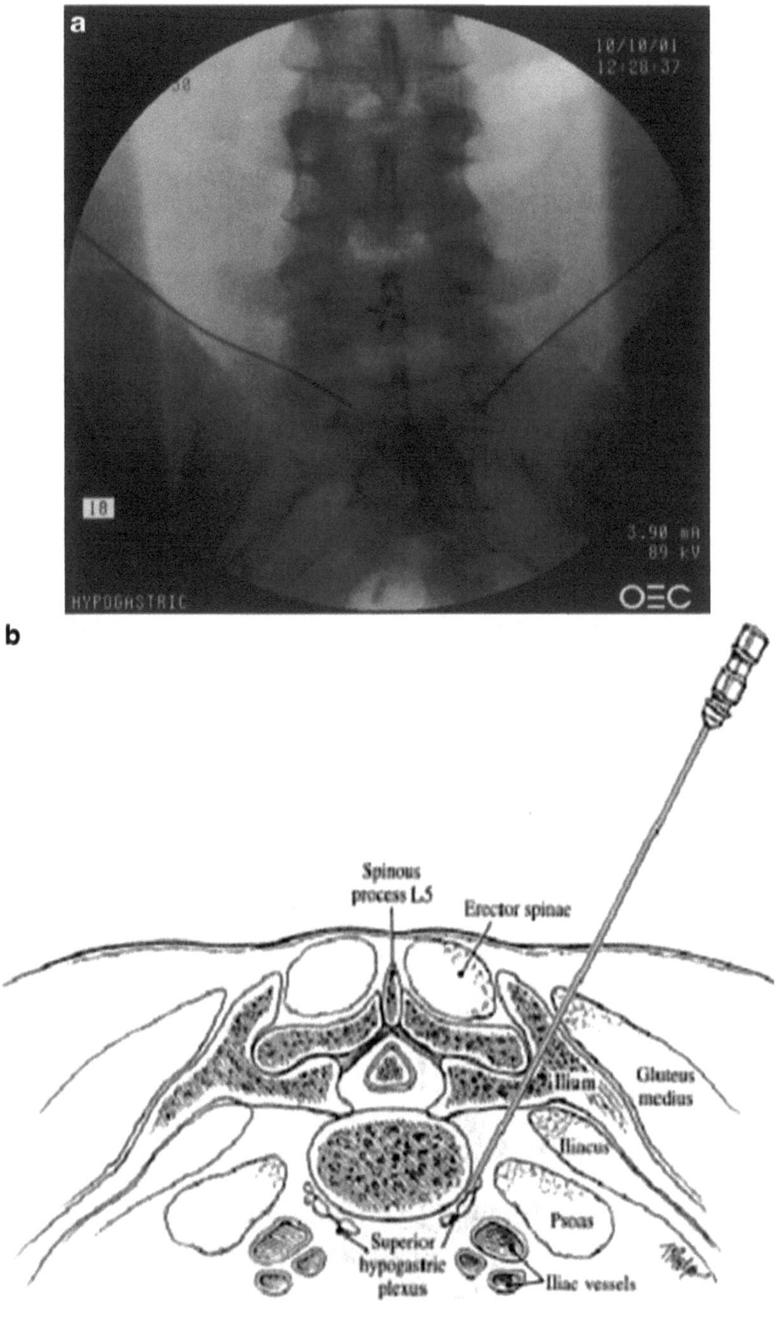

Fig. 17.21 (**a**) Posteroanterior radiograph showing needle position for superior hypogastric plexus block with needle tips lying at the anterolateral edge of the L5 vertebral body. (**b**) Corresponding axial diagram showing needle orientation and position

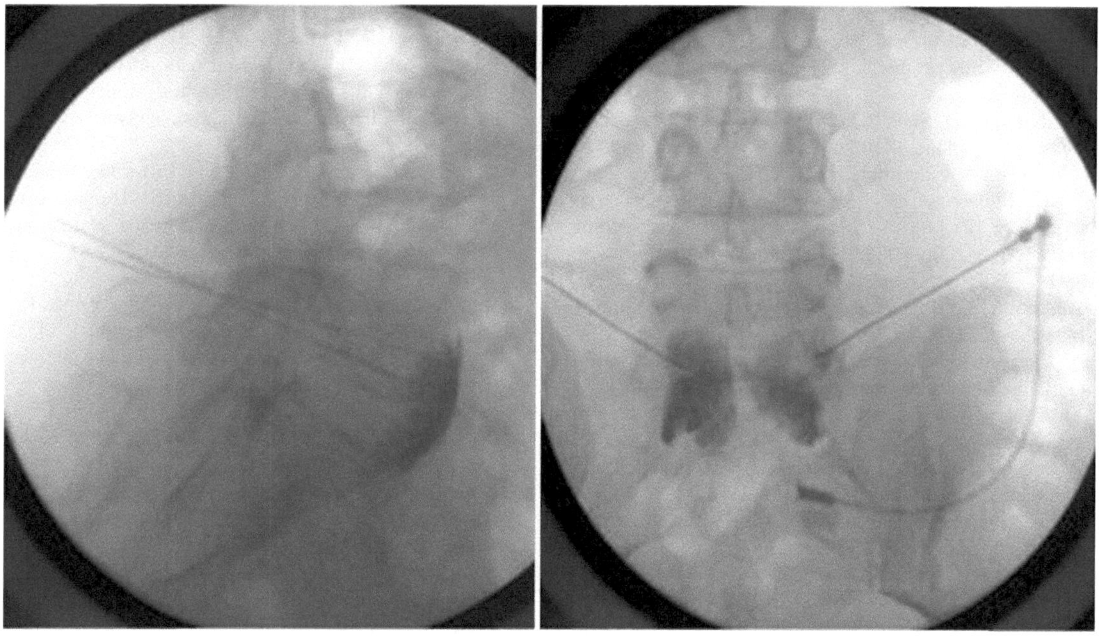

Fig. 17.22 Anteroposterior and lateral radiographic images showing needle placement and contrast spread for superior hypogastric plexus block. The needles are located at the anterolateral edge of the L5 vertebral body near the L5-S1 disc space. Radiopaque contrast can be seen to spread medially and cephalocaudally along the L5-S1 retroperitoneal space

References

1. Applegate MD. Abdominal cutaneous nerve entrapment syndrome (ACNES): a commonly overlooked cause of abdominal pain. Permanent J. 2002;6(3):20–7.
2. Carnett JB. Intercostal neuralgia as a cause of abdominal pain and tenderness. SurgGynecol Obstet. 1926;42:625–32.
3. Hershfield NB. The abdominal wall. A frequently overlooked source of abdominal pain. J Clin Gastroenterol. 1992;14(3):199–202.
4. Lewit K. The needle effect in the relief of myofascial pain. Pain. 1979;6(1):83–90.
5. Defalque RJ. Painful trigger points in surgical scars. Anesth Analg. 1982;61(6):518–20.
6. Rosson GD, Dellon AL. Abdominal wall neuroma pain after breast reconstruction with a transverse abdominal musculocutaneous flap: cause and treatment. Ann Plast Surg. 2005;55(3):330–44.
7. Shankar H, Eastwood D. Retrospective comparison of ultrasound and fluoroscopic image guidance for intercostal steroid injections. Pain Pract. 2010;10(4):312–7. Epub 2010 Mar 3.
8. Cheema SP, Ilsley D, Richardson J, Sabanathan S. A thermographic study of paravertebral analgesia. Anaesthesia. 1995;50(2):118–21.
9. Naja MZ, Ziade MF, El Rajab M, El Tayara K, Lönnqvist PA. Varying anatomical injection points within the thoracic paravertebral space: effect on spread of solution and nerve blockade. Anaesthesia. 2004;59(5):459–63.
10. Karmakar MK. Thoracic paravertebral block. Anesthesiology. 2001;95(3):771–80.
11. Ben-Ari A, Moreno M, Chelly JE, Bigeleisen PE. Ultrasound-guided paravertebral block using an intercostal approach. Anesth Analg. 2009;109(5):1691–4.
12. Marhofer P, Kettner SC, Hajbok L, Dubsky P, Fleischmann E. Lateral ultrasound-guided paravertebral blockade: an anatomical-based description of a new technique. Br J Anaesth. 2010;105(4):526–32. Epub 2010 Aug 3.
13. Renes SH, Bruhn J, Gielen MJ, Scheffer GJ, van Geffen GJ. In-plane ultrasound-guided thoracic paravertebral block: a preliminary report of 36 cases with radiologic confirmation of catheter position. Reg Anesth Pain Med. 2010;35(2):212–6.
14. Culp WC, Payne MN, Montgomery ML. Thoracic paravertebral block for analgesia following liver mass radiofrequency ablation. Br J Radiol. 2008;81(961):e23–5. Epub 2007 Dec 10.
15. Hall H, Leach A. Paravertebral block in the management of liver capsule pain after blunt trauma. Br J Anaesth. 1999;83(5):819–21.
16. Nikiforov S, Cronin AJ, Murray WB, Hall VE. Subcutaneous paravertebral block for renal colic. Anesthesiology. 2001;94(3):531–2.
17. Eichenberger U, Greher M, Kirchmair L, Curatolo M, Moriggl B. Ultrasound-guided blocks of the ilioinguinal and iliohypogastric nerve: accuracy of a selective new technique confirmed by anatomical dissection. Br J Anaesth. 2006;97(2):238–43. Epub 2006 May 12.
18. Jamieson RW, Swigart LL, Anson BJ. Points of parietal perforation of the ilioinguinal and iliohypogastric nerves in relation to optimal sites for local anaesthesia. Q Bull Northwest Univ Med Sch. 1952;26(1):22–6.
19. Weintraud M, Marhofer P, Bösenberg A, Kapral S, Willschke H, Felfernig M, Kettner S. Ilioinguinal/iliohypogastric blocks in children: where do we administer the local anesthetic without direct visualization? Anesth Analg. 2008;106(1):89–93.
20. Bay-Nielson M, Perkins F, Kehlet H, et al. Pain and functional impairment one year after inguinal herniorraphy—a nationwide questionnaire study. Ann Surg. 2001;233:1–7.
21. Rab M, Ebmer And J, Dellon AL. Anatomic variability of the ilioinguinal and genitofemoral nerve: implications for the treatment of groin pain. Plast Reconstr Surg. 2001;108(6):1618–23.
22. Willschke H, Bösenberg A, Marhofer P, Johnston S, Kettner SC, Wanzel O, Kapral S. Ultrasonography-guided rectus sheath block in paediatric anaesthesia—a new approach to an old technique. Br J Anaesth. 2006;97(2):244–9. Epub 2006 Jun 23.
23. Rozen WM, Tran TM, Ashton MW, Barrington MJ, Ivanusic JJ, Taylor GI. Refining the course of the thoracolumbar nerves: a new understanding of the innervation of the anterior abdominal wall. Clin Anat. 2008;21(4):325–33.

24. Kato J, Ueda K, Kondo Y, Aono M, Gokan D, Shimizu M, Ogawa S. Does ultrasound-guided rectus sheath block reduce abdominal pain in patients with postherpetic neuralgia? Anesth Analg. 2011;112(3):740–1.
25. Sviggum HP, Niesen AD, Sites BD, Dilger JA. Trunk blocks 101: transversus abdominis plane, ilioinguinal-iliohypogastric, and rectus sheath blocks. Int Anesthesiol Clin. 2012;50(1):74–92.
26. O'Donnell BD, McDonnell JG, McShane AJ. The transversus abdominis plane (TAP) block in open retropubic prostatectomy. Reg Anesth Pain Med. 2006;31(1):91. Erratum in: Reg Anesth Pain Med. 2006;31(3):286.
27. Rafi AN. Abdominal field block: a new approach via the lumbar triangle. Anaesthesia. 2001;56(10):1024–6.
28. Hebbard PD, Barrington MJ, Vasey C. Ultrasound-guided continuous oblique subcostal transversus abdominis plane blockade: description of anatomy and clinical technique. Reg Anesth Pain Med. 2010;35(5):436–41.
29. Hebbard P. Subcostal transversus abdominis plane block under ultrasound guidance. Anesth Analg. 2008;106(2):674–5.
30. Shibata Y, Sato Y, Fujiwara Y, Komatsu T. Transversus abdominis plane block. Anesth Analg. 2007;105(3):883.
31. Abdallah FW, Chan VW, Brull R. Transversus abdominis plane block: a systematic review. Reg Anesth Pain Med. 2012;37(2):193–209.
32. Canidido K, Stevens RA. Intrathecal neurolytic blocks for the relief of cancer pain. Best Pract Res Clin Anesthesiol. 2003;17(3): 407–28.
33. El-Sayed GG. A new catheter technique for thoracic subarachnoid neurolysis in advanced lung cancer patients. Pain Pract. 2007; 7(1):27–30.
34. Labat F, Dubousset AM, Baujard C, Waiser AP, Benhamou D, Cucchiaro G. Epidural analgesia in a child with sickle cell disease complicated by acute abdominal pain and priapism. Br J Anaesth. 2001;87:935–6.
35. Makharia GK. Understanding and treating abdominal pain and spasms in organic gastrointestinal diseases. J Clin Gastroenterol. 2011;45 Suppl 2:S89–93.
36. Agostini S, Eutamene H, Cartier C, Broccardo M, Improta G, Houdeau E, Petrella C, Ferrier L, Theodorou V, Bueno L. Evidence of central and peripheral sensitization in a rat model of narcotic bowel-like syndrome. Gastroenterology. 2010;139:553–63.
37. Freise H, Lauer S, Anthonsen S, Hlouschek V, Minn E, Fischer LG, Lerch MM, Van Aken HK, Sielenkämper AW. Thoracic epidural analgesia augments ileal mucosal capillary perfusion and improves survival in severe acute pancreatitis in rats. Anesthesiology. 2006;105(2):354–9.
38. Freise H, Lauer S, Konietzny E, Hinkelmann J, Minin E, Van Aken HK, Lerch MM, Sielenkäemper AW, Fischer LG. Hepatic effects of thoracic epidural analgesia in experimental severe acute pancreatitis. Anesthesiology. 2009;111(6):1249–56.
39. Anaparthy R, Pasricha PJ. Pain and chronic pancreatitis: is it the plumbing or the wiring? Curr Gastroenterol Rep. 2008;10:101–6.
40. Drewes AM, Krarup AL, Detlefsen S, Malstrom ML, Dimcevski G, Funch-Jensen P. Pain in chronic pancreatitis: the role of neuropathic pain mechanisms. Gut. 2008;57:1616–27.
41. Feng QX, Wang W, Feng XY, Mei XP, Zhu C, Liu ZC, Li YQ, Dou KF, Zhao QC. Astrocytic activation in thoracic spinal cord contributes to persistent pain in rat model of chronic pancreatitis. Neuroscience. 2010;167:501–9.
42. Kurucsai G, Joó I, Fejes R, Székely A, Székely I, Tihanyi Z, Altorjay A, Funch-Jensen P, Várkonyi T, Madácsy L. Somatosensory hypersensitivity in the referred pain area in patients with chronic biliary pain and a sphincter of Oddi dysfunction: new aspects of an almost forgotten phathogenic mechanism. Am J Gastroenterol. 2008;103(11):2717–25.
43. Conwell DL, Vargo JJ, Zuccaro G, Dews TE, Mekhail N, Scheman J, Walsh M, Grundfest-Broniatowski SF, Dumont JA, Shay SS. Role of differential neuroxial blockade in the evaluation and management of pain in chronic pancreatitis. Am J Gastroenterol. 2001;96(2):431–6.
44. Rizk MK, Tolba R, Kapural L, Mitchell J, Lopez R, Mahboobi R, Vrooman B, Mekhail N. Differential epidural block predicts the success of visceral block in patients with chronic visceral abdominal pain. Pain Pract. 2012;12(8):595–601.
45. Adolph MD, Benedetti C. Percutaneous-guided pain control: exploiting the neural basis of pain sensation. Gastroenterol Clin North Am. 2006;35:167–88.
46. Naidoo N, Partab P, Pather N, Moodley J, Singh B, Satyapal KS. Thoracic splanchnic nerves: implications for splanchnic denervation. J Anat. 2001;199:585–90.
47. Yang IY, Oraee S, Viejo C, Stern H. Transcrural coeliac plexus block simulated on 200 computed tomography images. Br J Anaesth. 2011;107(6):972–7.
48. Noble M, Gress FG. Techniques and results of neurolysis for chronic pancreatitis and pancreatic cancer pain. Curr Gastroenterol Rep. 2006;8:99–103.
49. de Oliveira R, dos Reis MP, Prado WA. The effects of early or late neurolytic sympathetic plexus block on the management of abdominal or pelvic cancer pain. Pain. 2004;110:400–8.
50. Yuen TST, Ng KFJ, Tsu SL. Neurolytic celiac plexus block for visceral abdominal malignancy: is prior diagnostic block warranted? Anaesth Intensive Care. 2002;30(4):442–8.
51. Avinash K, Ashraf T, Gervais DA, Mueller PR, Arellano RS. CT-guided celias plexus neurolysis: a review of anatomy, indications, technique, and tips for successful treatment. Radiographics. 2011;31(6):1599–621.
52. Kambadakone A, Thabet A, Gervais DA, Mueller PR, Arellano RS. CT-guided celiac plexus neurolysis: a review of anatomy, indications, technique, and tips for successful treatment. Radiographics. 2011;31:1599–621.
53. Bahatnagar S, Khanna S, Roshni S, Goyal GN, Mishra S, Rana SPS, Thulkar S. Early ultrasound-guided neurolysis for pain management in gastrointestinal and pelvic malignancies: an observational study in a tertiary care center of urban India. Pain Pract. 2012;12(1):23–32.
54. Ascunce G, Ribeiro A, Reis I, Rocha-Lima C, Steelman D, Merchan J. EUS visualization and direct celiac ganglia neurolysis predicts better pain relief in patients with pancreatic malignancy (with video). Gastrointest Endosc. 2011;73(2):267–74.
55. Kitoh T, Satoshi T, Ono K, Ohfusa Y, Ina H, Otagirl T. Combined neurolytic block of celiac, inferior mesenteric, and superior hypogastric plexus for incapacitating abdominal and/or pelvic cancer pain. J Anesth. 2005;19:328–32.
56. Plancarte R, Amescua C, Patt RB, Aldrete JA. Superior hypogastric plexus block for pelvic cancer pain. Anesthesiology. 1990;73:236–9.
57. Erdine S, Yucel A, Celik M, Talu GK. Transdiscal approach for hypogastric plexus block. Reg Anesth Pain Med. 2003;28(4):304–8.

Chronic Abdominal Wall Pain: Diagnosis and Interventional Treatment

18

Samer Narouze

Chronic pain stemming from the abdominal wall is frequently overlooked or misdiagnosed as visceral pain, often leading to extensive diagnostic testing and unnecessary treatments.

Chronic abdominal wall pain (CAWP) has been diagnosed in up to 10 % of patients with abdominal pain referred to gastroenterologists. The peak incidence of CAWP is between the ages of 30 and 50 years and women are more likely to be affected than men [1–3].

Differential Diagnosis of Abdominal Wall Pain

Although the list of differential diagnosis can be very extensive (Table 18.1), the most important cause of CAWP is entrapment of a cutaneous branch of the lower (T7–T12) intercostal nerves, the so-called anterior cutaneous nerve entrapment syndrome (ACNES) [4].

Clinical Presentation and Diagnosis

CAWP is best diagnosed based on patient's history and a physical examination. An important finding is that the pain is usually well localized with point tenderness on palpation. On the contrary, visceral pain is usually more dispersed and poorly localized [5].

Carnett's test is the hallmark of the physical examination for diagnosing abdominal wall pain [6]. The patient is placed in the prone position with slightly flexed knees and hips to relax the abdominal wall. The painful area is palpated while in this relaxed position, then the patient is asked to tighten his abdominal muscles by staining or lifting his head and shoulders off the bed. A positive test is demonstrated by increased tenderness as the patient tenses the abdominal wall indicating that the pain arises from the abdominal wall. On the other hand, when pain arises from an intra-abdominal source, the tensed abdominal wall muscles guard the underlying organs, thus reducing the pain.

A working clinical diagnosis of CAWP can be confirmed by a positive response to trigger point injections or nerve blocks. A successful injection after a positive Carnett sign was needed to be one of the most cost-effective procedures in gastroenterology [5]. Limitations to this approach are the high placebo response to injections especially in pain patients [7] and visceral abdominal disease with involvement of the peritoneum may give a false positive Carnett test as well [8].

Others advocated the use of differential epidural block to allow characterization of chronic abdominal pain into visceral and non-visceral pain.

Only few reports with small cohort of patients discussed the role of differential epidural block and showed weak evidence that it can predict treatment response [9–11].

Differential Epidural Block

Differential epidural block is a diagnostic nerve block that was initially described in 1964 for the evaluation of lower-back and lower-extremity pain [12]. Since then, several modifications of the procedure have been implemented using both subarachnoid and epidural approaches.

Differential epidural block involves the placement of a thoracic epidural catheter and the injection of saline (placebo) and different concentrations or incremental doses of local anesthetics. The procedure relies on the variable sensitivity of nerve fibers of various size, myelination, and function to local anesthetics. Sympathetic fibers and visceral afferent nerves are relatively more sensitive to local anesthetic blockade than large sensory or motor fibers (Table 18.2).

S. Narouze, M.D., Ph.D. (✉)
Center for Pain Medicine, Western Reserve Hospital,
1900 23rd Street, Cuyahoga Falls, OH 44333, USA
e-mail: narouzs@hotmail.com

L. Kapural (ed.), *Chronic Abdominal Pain: An Evidence-Based, Comprehensive Guide to Clinical Management*,
DOI 10.1007/978-1-4939-1992-5_18, © Springer Science+Business Media New York 2015

Table 18.1 Differential diagnosis of abdominal wall pain

Abdominal cutaneous nerve entrapment syndrome
Abdominal wall hernias and post-herniorrhaphy pain
Surgical scars and neuromas
Abdominal wall or rectus sheath hematoma
Thoracic disc degeneration and thoracic radiculopathy
Herpes Zoster infection and post-herpetic neuralgia
Chest wall pain etiologies, slipping rib, or ribs on pelvis syndrome
Abdominal wall endometriosis
Referred pain from an abdominal or thoracic source

Table 18.2 Interpretation of differential epidural block

1.	Visceral pain: No pain after surgical anesthesia of the relevant dermatome with persistent pain relief after the dermatomal somatic anesthesia level recedes
2.	Somatosensory: No pain after surgical anesthesia to the relevant dermatome with the return of pain as somatic dermatomal anesthesia level recedes
3.	Central: Persistent pain in spite of surgical anesthesia
4.	Mixed: mixed picture between the above three scenarios. Often encountered secondary to the subjective nature of pain
5.	Placebo responders: prolonged pain relief with saline injection

Pitfalls of the Differential Epidural Block Test

1. The interpretation of the differential block is non-standardized.
2. The interpretation of the differential block is very subjective.
3. The interaction between local anesthetic and nerve fibers is a dynamic and unpredictable phenomenon that may be influenced by a multitude of factors.
4. Overlap in the range of nerve fiber sizes makes it unlikely that any fiber type can be reliably isolated by this procedure.
5. As a result of the above, the interpretation of the test is often mixed (visceral/somatic/central), which defeats the purpose of the study!
6. The procedure takes between 4 and 8 h and requires continuous monitoring of the patient.
7. It has the limitations and disadvantages of neuraxial blocks.

Transversus Abdominis Plane Block for Chronic Abdominal Wall Pain

The transversus abdominis plane (TAP) block is very appealing as a valuable test in diagnosing pain stemming from the abdominal wall and thus helps differentiating somatosensory from visceral origin of pain [13].

The TAP block is a new regional anesthesia technique that provides analgesia to the abdominal wall. First described in 2001, the technique involves the injection of local anesthetic into the plane between the internal oblique and transversus abdominis muscles, the TAP [14, 15]. TAP block targets the entire anterolateral abdominal wall between the costal margin and inguinal ligament [16]. The introduction of ultrasound-guided TAP block allows the successful installation of local anesthetics around the anterior branches of the thoracolumbar ventral rami blocking "somatic sensations" from the anterior abdominal wall. As stated above, the limitations of differential epidural block are numerous and, contrary to TAP block (somatic) and celiac/hypogastric block (visceral), different nerve fibers cannot be reliably isolated. The author has found that transversus abdominus plane (TAP) block is very valuable in diagnosing pain originating from the abdominal wall and differentiating somatosensory from visceral pain [13]. Single injection as well as continuous infusions can be used for treatment of various abdominal wall pain syndromes [17].

Figure 18.1 offers a suggested algorithm with incorporation of TAP block in the diagnosis and management of chronic abdominal pain.

Anatomy and Innervation of the Anterior Abdominal Wall

The intercostal nerves run a very tortuous course through the abdominal wall muscle. After turning at a 90° angle, the nerve passes from the posterior sheath of the abdominal wall muscle (rectus abdominis) through a fibrous opening and then branches at right angles while passing through its anterior sheath. It has been thought that the underlying problem is nerve compression with resulting ischemia or lack of blood supply, explained by the nerve's course through the muscle. Applegate termed the condition as "anterior cutaneous nerve entrapment syndrome" and suggested the entrapped nerve may also be pushed by intra- or extra-abdominal pressure or pulled by a scar causing pain in the abdominal wall.

The abdominal wall consists of three muscle layers; the external oblique, the internal oblique, and the transversus abdominis and their associated fascial sheaths. These muscles are innervated via the ipsilateral ventral rami of T7-L1 thoraco-lumbar nerves.

After emerging through the intervertebral foramina, they follow a tortuous course through the abdominal wall muscles. They enter a fascial plane between the transversus abdominis and the internal oblique muscles what is known as the TAP accompanied by blood vessels. This neurovascular plan continues as far as the semilunar line. At the lateral border of the rectus, abdominis muscle, the external oblique and the anterior lamella of the internal oblique aponeuroses pass anterior to the muscle forming the anterior rectus sheath.

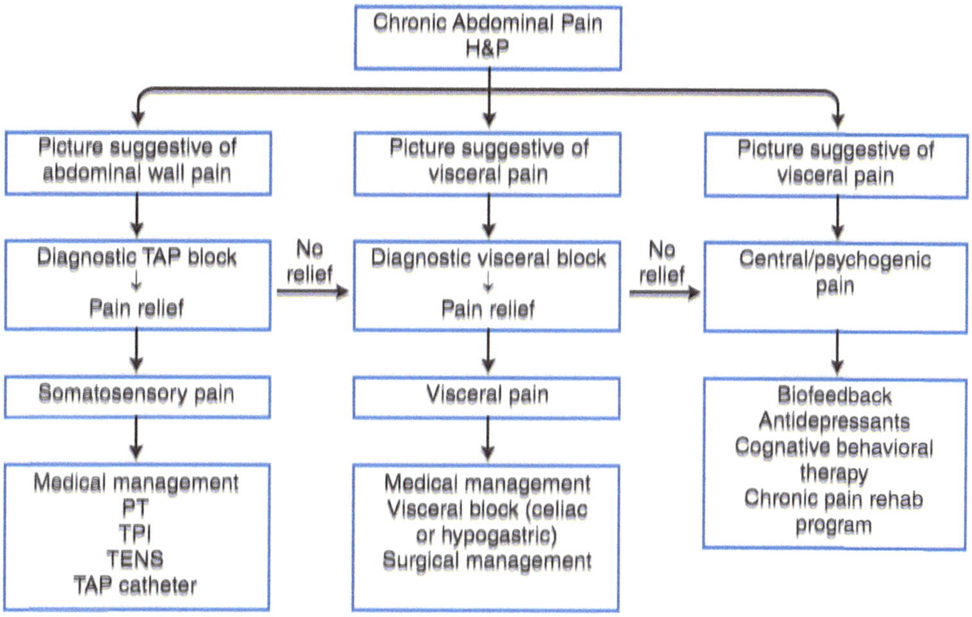

Fig. 18.1 Suggested algorithm for the management of chronic abdominal pain

The aponeuroses from the posterior lamella of the internal oblique muscle and the transversus abdominis muscle pass posterior to the rectus muscle forming the posterior layer of the sheath. At this point, the ventral rami of the thoracic spinal nerves are located between the posterior border of the rectus muscle and the posterior rectus sheath. They run medially within the sheath, through the rectus muscle, then branches at right angles while passing through its anterior sheath [16]. It has been postulated that the nerve's course through the muscle make it vulnerable to compression and entrapment. Applegate termed the condition ACNES and suggested that the entrapped nerve may also be pulled by a scar or pushed by an intra-abdominal or extra-abdominal pressure causing abdominal wall pain [4].

The Classic Approach for TAP Block

The TAP block was first described by Rafi and McDonell as a blind "double-pop" technique using a blunt needle introduced through the external and internal oblique muscles and fascia at the ilio-lumbar triangle of Petit [14, 15]. This triangle is bounded posteriorly by the latissimus dorsi muscle and anteriorly by the external oblique, with the iliac crest forming the base of the triangle. The introduction of ultrasound allows modification of this technique and the TAP can be accessed anywhere between the iliac crest and costal margin behind the anterior axillary line. A higher subcostal approach may block the upper thoraco-lumbar nerves more effectively than a lower approach immediately above the iliac crest.

Ultrasound-Guided Technique for TAP Block

The patient is positioned in the lateral decubitus position with the side to be blocked upward. A wedge can be placed underneath the patient in order to stretch the flank on the upper side. A high frequency or lower frequency transducers may be used according to body habitus. Pre-procedural scanning of the anterior abdominal wall along the midaxillary line is recommended to decide the best view of the three muscle layers. Care should be taken that scanning more medially may only show two layers of muscles since the external oblique muscle forms an aponeurosis that joins the rectus sheath. From superficial to deep the following structures are recognized: skin and subcutaneous fat, external oblique, internal oblique, and transversus abdominis muscles with their investing fascia (Figs. 18.2 and 18.3). Deeper to the transversus abdominis and its fascia, there is the pre-peritoneal fat separating it from the peritoneum and the bowels, which are often identified by its peristaltic movements. With ultrasound, the fascial layers appear as hyperechoic layers, and the muscles are identified by their relative hypoechoic structure with multiple striations.

The needle is usually inserted in-plane from the posterolateral side of the probe and is advanced in a medial and anterior direction. The needle is advanced through the different layers with a tactile feeling of a "pop" when crossing each fascial layer. Hydrolocalization is very helpful in identifying the tip of the needle while advancing under real-time sonography. Correct placement is identified by the solution separating the internal oblique muscle from the transversus abdominis muscle (Fig. 18.4).

For single shot block, a blunt 22G needle can be used while a Tuohy needle is used for continuous catheter technique. When a catheter is required, the space is dissected using 10 mL of saline followed by catheter insertion for about 5 cm beyond the tip of the needle.

Post-inguinal Herniorrhaphy Pain

Pain that persist after inguinal herniorrhaphy affecting daily activities is seen in 5–10 % of patients [18]. At least half the patient who suffers post-herniorrhaphy pain is thought to be due to entrapment or injury to the ilioinguinal, iliohypogastric, or genitofemoral nerves [19].

Aasvang et al. [20] conducted a large scaled multifactorial study reporting on the predictive risk factors for persistent post-herniotomy pain (PPP). The study showed that PPP is the result of both patient and surgical factors. Independent factors for PPP-related activity impairment are preoperative activity assessment scale (AAS) score, increased pain to preoperative heat stimulation, nerve injury, and early postoperative pain. Preoperative data on AAS score and response to heat stimulation can help clinicians in guiding high-risk patients to laparoscopic surgery with reduced risk for PPP.

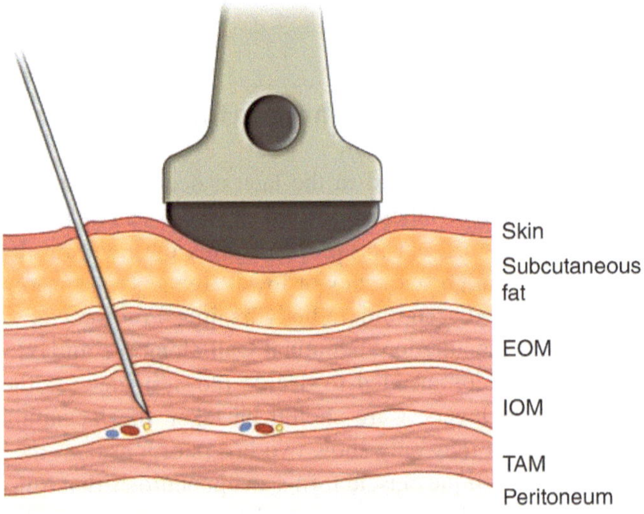

Fig. 18.2 Illustration showing the abdominal wall muscle layers and the needle in place for performing TAP block

Ilioinguinal and Iliohypogastric Nerve Blocks

Ilioinguinal and iliohypogastric nerve blocks can be used as a diagnostic, therapeutic, or preoperative block.

- Diagnostic nerve blocks: Ilioinguinal and iliohypogastric nerve blocks can help diagnosing injury or entrapment neuropathy of the specific nerves.
- Therapeutic nerve blocks: Few studies reported the effectiveness of ilioinguinal and iliohypogastric nerve blocks in the treatment of PPP [21–23]. However; one report showed that that ultrasound-guided ilioinguinal and iliohypogastric nerve blocks at the level of the ASIS were not useful in diagnosis and management of PPP [24].
- Pre-operative nerve blocks: Ilioinguinal and iliohypogastric nerve blocks may predict which patients will benefit from surgical neurectomy or neurolysis and have been used preoperatively in few studies [25, 26].

Ultrasound-Guided Technique for Ilioinguinal and Iliohypogastric Nerve Blocks

Traditionally these blocks have been performed with surface landmark technique at the ASIS, either blindly or with nerve stimulation.

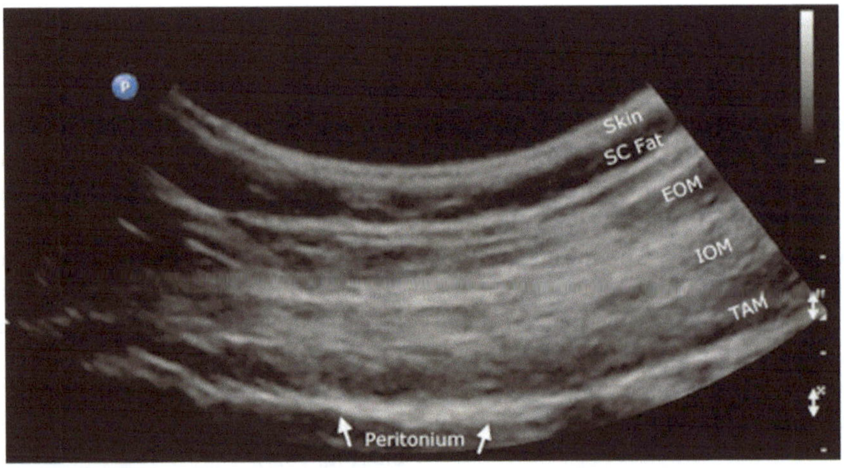

Fig. 18.3 Pre-injection short axis sonogram showing the abdominal wall muscle layers. *EOM* external oblique muscle, *IOM* internal oblique muscle, *TAM* transversus abdominis muscle. (Reprinted with permission from Ohio Pain and Headache Institute)

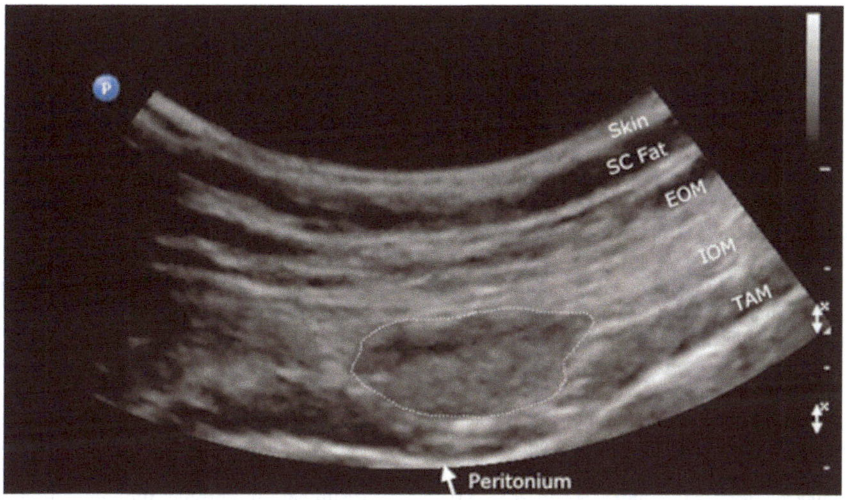

Fig. 18.4 Post-injection short axis sonogram showing the spread of the injectate in the plane between the internal oblique muscle (IOM) and the transversus abdominis muscle (TAM). Note that the TAM and the peritoneum were pushed away by the injectate. (Reprinted with permission from Ohio Pain and Headache Institute)

Recently, ultrasound-guided technique was described with the advantage of having a more precise block [27, 28]. However, a recent study showed that ultrasound was not superior to nerve stimulator-guided blocks [23].

The iliohypogastric and ilioinguinal nerves follow a similar course as the lower thoracic ventral rami (see above); however, they pierce the internal oblique muscle at different levels near the anterior superior iliac spine to supply the inguinal region. Accordingly the ultrasound-guided technique for ilioinguinal and iliohypogastric nerve blocks is basically a modified TAP block at the level of the ASIS. Even if individual nerves cannot be identified, the injectate can be administered at the fascial plane between internal oblique and transversus abdominis muscles. The spread of the injectate should be monitored under real-time sonography to ensure adequate spread to surround both nerves.

Interventional and Surgical Treatment

Cryablation, alcohol injection, radiofrequency ablation, mesh removal, or surgical neurectomy showed very good results in selected patients. However, all studies had a heterogeneous patient population and were either descriptive studies or case reports [29–34].

References

1. Srinivasan R, Greenbaum DS. Chronic abdominal wall pain: a frequently overlooked problem. Am J Gastroenterol. 2002;97:824–30.
2. Costanza CD, Longstreth GF, Liu AL. Chronic abdominal wall pain: clinical features, health care costs and long term outcome. Clin Gastroenterol Hepatol. 2004;2:394–9.
3. Langdon DE. Abdominal wall pain will be missed until examinations change! Am J Gastroenterol. 1998;2002:3207–8.
4. Applegate WV. Abdominal cutaneous nerve entrapment syndrome. Surgery. 1972;71:118–24.
5. Sharpstone D, Colin-Jones DG. Chronic, non-visceral abdominal pain. Gut. 1994;35:833–6.
6. Carnett JB. Intercostal neuralgia as a cause of abdominal pain and tenderness. Surg Gynecol Obstet. 1926;42:625–32.
7. Straus JL, von Ammon Cavanaugh S. Placebo effects. Issues for clinical practice in psychiatry and medicine. Psychosomatics. 1996;37:315–26.
8. Gray DWR, Seabrook G, Dixon JM, et al. Is abdominal wall tenderness a useful sign in the diagnosis of non-specific abdominal pain? Ann R Coll Surg Engl. 1988;70:233–4.
9. Conwell DL, Vargo JJ, Zuccaro G, et al. Role of differential neuroaxial blockade in the evaluation and management of pain in chronic pancreatitis. Am J Gastroenterol. 2001;96:431–6.
10. Bradley EL, Reynhout JA, Peer GL. Thorascopic splanchnicectomy for a small duct chronic pancreatitis: case selection by differential epidural anesthesia. J Gastrointest Surg. 1998;2:88–94.
11. Rizk MK, Tolba R, Kapural L, Mitchell J, Lopez R, Mahboobi R, Vrooman B, Mekhail N. Differential epidural block predicts the success of visceral block in patients with chronic visceral abdominal pain. Pain Pract. 2012;12:595–601.
12. McCollum DE, Stephen CR. The use of graduated spinal anesthesia in the differential diagnosis of pain of the back and lower extremities. South Med J. 1964;57:410–6.
13. Soilman LM, Narouze S. Ultrasound-guided transversus abdominus plan block for the management of abdominal pain: an alternative to differential epidural block. Tech Reg Anesth Pain Manag. 2009;13:117–20.
14. Rafi AN. Abdominal field block: a new approach via the lumbar triangle. Anaesthesia. 2001;56:1024–6.
15. McDonnell JG, O'Donnell BD, Curley G, et al. The analgesic efficacy of transverses abdominis plane block after abdominal surgery: a prospective randomized controlled trial. Anesth Analg. 2007;104:193–7.
16. Rozen WM, Tran TM, Ashton MW, et al. Refining the course of the thoracolumbar nerves: a new understanding of the innervation of the anterior abdominal wall. Clin Anat. 2008;21:325–33.
17. Guirguis MN, Abd-Elsayed AA, Girgis G, Soliman LM. Ultrasound-guided transversus abdominis plane catheter for chronic abdominal pain. Pain Pract. 2013;13:235–8.

18. Aasvang E, Kehlet H. Chronic postoperative pain: the case of inguinal herniorrhaphy. Br J Anaesth. 2005;95:69–76.
19. Aasvang EK, Brandsborg B, Christensen B, Jensen TS, Kehlet H. Neurophysiological characterization of postherniotomy pain. Pain. 2008;137:173–81.
20. Aasvang EK, Gmaehle E, Hansen JB, Gmaehle B, Forman JL, Schwarz J, Bittner R, Kehlet H. Predictive risk factors for persistent postherniotomy pain. Anesthesiology. 2010;112:957–69.
21. Deysine M, Deysine GR, Reed Jr WP. Groin pain in the absence of hernia: a new syndrome. Hernia. 2002;6:64–7.
22. Loos MJ, Verhagen T, Scheltinga MR, Roumen RM. A randomized controlled trial of injection therapy versus neurectomy for post-herniorrhaphy inguinal neuralgia: rationale and study design. Hernia. 2010;14:593–7.
23. Thomassen I, van Suijlekom JA, van de Gaag A, Ponten JE, Nienhuijs SW. Ultrasound-guided ilioinguinal/iliohypogastric nerve blocks for chronic pain after inguinal hernia repair. Hernia. 2013;17:329–32.
24. Bischoff JM, Koscielniak-Nielsen ZJ, Kehlet H, Werner MU. Ultrasound-guided ilioinguinal/iliohypogastric nerve blocks for persistent inguinal postherniorrhaphy pain: a randomized, double-blind, placebo-controlled, crossover trial. Anesth Analg. 2012;114(6):1323–9.
25. Loos MJ, Scheltinga MR, Roumen RM. Tailored neurectomy for treatment of postherniorrhaphy inguinal neuralgia. Surgery. 2010; 147:275–81.
26. Aasvang E, Kehlet H. Surgical management of chronic pain after inguinal hernia repair. Br J Surg. 2005;92:795–801.
27. Eichenberger U, Greher M, Kirchmair L, Curatolo M, Moriggl B. Ultrasound-guided blocks of the ilioinguinal and iliohypogastric nerve: accuracy of a selective new technique confirmed by anatomical dissection. Br J Anaesth. 2006;97:238–43.
28. Peng P, Narouze S. Ultrasound-guided interventional procedures in pain medicine: a review of anatomy, sonoanatomy, and procedures. Part I nonaxial structures. Reg Anesth Pain Med. 2009;34:458–74.
29. Fanelli RD, DiSiena MR, Lui FY, Gersin KS. Cryoanalgesic ablation for the treatment of chronic postherniorrhaphy neuropathic pain. Surg Endosc. 2003;17:196–200.
30. Bower S, Moore BB, Weiss SM. Neuralgia after inguinal hernia repair. Am Surg. 1996;62:664–7.
31. Cohen SP, Foster A. Pulsed radiofrequency as a treatment for groin pain and orchialgia. Urology. 2003;61:645.
32. Mitra R, Zeighami A, Mackey S. Pulsed radiofrequency for the treatment of chronic ilioinguinal neuropathy. Hernia. 2007;11: 369–71.
33. Amid PK. Causes, prevention, and surgical management of postherniorrhaphy neuropathic inguidynia: triple neurectomy with proximal end implantation. Hernia. 2004;8:343–9.
34. Aasvang EK, Kehlet H. The effect of mesh removal and selective neurectomy on persistent postherniotomy pain. Ann Surg. 2009;249: 327–34.

Spinal Cord Stimulation for Chronic Abdominal Pain

19

Leonardo Kapural

Introduction

Long-standing, chronic abdominal pain may significantly affect patients' quality of life [1], in addition, result in increased doctor visits, imaging, and surgery intervention [2, 3]. Poor localization, referral to somatic structures, and presence of visceral hyperalgesia sometimes makes diagnosis of such chronic pain illusive, or at least very difficult [4, 5].

Abdominal pain is a very frequent complaint in the primary care offices, and by far the most frequent reason for referral to gastroenterologist [6, 7]. Objective distinction between visceral, somatosensory, or possibly central cause of such pain was not possible until recently [8]. For long-standing chronic abdominal pain of visceral origin, spinal cord stimulation of the dorsal columns surfaced as an interesting new therapeutic option to provide non-pharmacologic control of severe chronic visceral pain and improve quality of life.

Proper integration of peripheral and central nervous system input to and from end organs is a requirement to maintain a normal gastrointestinal physiology. Imbalance of what is frequently referred as a "brain-gut axis" leads to many of functional GI disorders, and chronic visceral pain of various other sources [9–14]. Even more, chronic hyperalgesia may involve component of neuropathic pain [13, 14].

The mechanisms of chronic abdominal pain relief using electrical dorsal column stimulation or spinal cord stimulation (SCS) are not clearly identified, but has been speculated about, leading to several working theories that still needs to be studied and established [15, 16]. Activation of supraspinal pain modulatory pathways by SCS, release of inhibitory neuromodulators, such as GABA, blockade of nerve conduction by antidromic activation, direct stimulation of postsynaptic visceral dorsal column pathway, or downregulation of segmental or supraspinal sympathetic outflow may provide analgesic effect seen during SCS for chronic abdominal pain [15–20].

Although interruption of the above listed and recently described midline dorsal column pathway relieves visceral pelvic pain in cancer patients, at this time it is not clear if this visceral pathway can be modulated, excited, or suppressed by SCS [20].

Chemical or surgical neurectomy/sympathectomy involving the superior hypogastric or celiac plexus has been shown to suppress chronic abdominal pain [21, 22]. In addition, suppression of the sympathetic nervous system has been proposed as an important mechanism of pain control in intractable angina [16, 17]. It may be that the sympathectomy produced by SCS plays a role in the suppression of chronic visceral abdominal pain.

SCS has been studied in visceral hyperalgesia using well-established rat model of visceromotor reflex (VMR) elicited by colonic distention [23]. Such reflex was suppressed by SCS in both normal and sensitized rats when SCS electrodes were implanted in either cervical or lumbar regions. Hypothesis was then introduced that SCS may cause antidromic activation of peripheral sensory fibers negating the afferent input [23–25]. Those studies suggested possible mechanism of long-term analgesic effect in already seen when spinal cord stimulation was used for various causes of visceral abdominal pain [26–40].

Clinical Studies

Currently, compelling, but limited data on SCS for various chronic visceral pain syndromes are providing a good basis for further prospective and randomized, long-term trials to introduce SCS as a valid clinical therapeutic approach. Initially, numerous case reports and smaller case series have demonstrated significant clinical improvements in patients with chronic visceral syndromes suffering from: mesenteric

L. Kapural, M.D., Ph.D. (✉)
School of Medicine, Wake Forest University,
Kimel Park Drive 145, Winston-Salem, NC 27103, USA
e-mail: lkapural@ccrpain.com; lkapuralmd@gmail.com

L. Kapural (ed.), *Chronic Abdominal Pain: An Evidence-Based, Comprehensive Guide to Clinical Management*,
DOI 10.1007/978-1-4939-1992-5_19,

ischemic pain [26], esophageal dysmotility [27], gastroparesis [30, 35], IBS [29, 30], chronic pancreatitis [31, 32, 33, 35, 37, 41], familial Mediterranean fever [34], post-traumatic splenectomy [35], generalized chronic abdominal pain [35], and chronic visceral pelvic pain [36] (see Table 19.1). These case series provided evidence that such treatment can afford a long-term pain relief in patients with visceral hyperalgesia.

The first case report on SCS for the treatment of abdominal pain described a 78-year-old male with chronic, unrelieved severe postprandial pain caused by mesenteric ischemia. The patient experienced complete pain relief after SCS lead was placed in the epidural space at the T6 vertebral level and stimulation initiated [26]. A case report dealing with irritable bowel syndrome (IBS) described a female patient who suffered from 11 to 14 diarrheal episodes per day and extreme pain from IBS. After placement of a thoracic SCS system, the patient immediately became diarrhea free. However, her initial reduction in pain relief was not sustained [29, Table 1].

Khan et al. [32] described five successful cases of pain relief in patients with nonalcoholic pancreatitis who were trialed with both single and dual leads placed at the T5–T7 vertebral level in the posterior epidural space (Fig. 19.1). This report was followed by several others describing improvements in pain and function in patients with severe chronic pancreatitis [30–33] (see Table 19.1). Tiede and associates [28] described improvements in pain scores of patients with gastroparesis, and Jackson and Simpson [27] reported improved pain control and swallowing in a patient with a rather complicated history of esophageal problems.

Table 19.1 Shown below is a tabulated summary of SCS clinical outcomes when used for treatment of chronic nonmalignant abdominal pain

Causes of chronic abdominal pain	Study/case series/ case report	Number of patients studied	Outcome	Time interval follow-up	Complications
Mesenteric ischemia	Ceballos et al. [26]	1	8 to 0 VAS	1 year	None
Esophageal dysmotility	Jackson and Simpson [27]	1	9 to 3 VAS	>2 years	Two lead fractures
Irritable bowel syndrome	Krames and Mousad [29]	1	9-10 to 2-3 VAS	6 months	None
Chronic pancreatitis	Khan et al. [32]	5	Average 4.9 VAS decrease, >50 % opioid use decrease	>1 year	Lead migration
Familial Mediterranean fever	Kapur et al. [34]	2	8 to 5 VAS 10 to 1 VAS	1 year 3 months	None
Gastroparesis	Tiede et al. [28]	2	8 to 3 VAS 9 to 2 VAS	3 months	Lead migration
Chronic pancreatitis	Kapural and Rakic [31]	1	6 to 1 VAS 150 mg MSO4 to 0	3 months	None
Chronic pancreatitis	Kim et al. [33]	1	10 to 5 VAS	14 months	None
Chronic pancreatitis, abdominal adhesions, gastroparesis, mesenteric ischemia, postgastric bypass pain	Kapural et al. [30]	35	VAS from 8.2 to 3.8 MSO4 from 119 to 38 mg	1 year	Infection ($n=3$) Lead migration ($n=1$)
Chronic pancreatitis, postsurgical intraabdominal adhesion, gastroparesis	Kapural et al. [35]	70	VAS from 8 to 2.49 MSO4 from 158 to 36 mg	Average 84 weeks	Infection and Lead migration ($n=8$) Removed for headaches [1] and induced diarrhea [1]
Chronic pancreatitis	Kapural et al. [37]	30	VAS from 8 to 3.6 MSO4 from 165 to 48.6 mg	1 year	Infection ($n=2$) Lead migration ($n=1$)
Chronic pancreatitis	Al-Mahrouqi et al. [41]	1	680 mg to 510 mg MSO4 with "effective pain control"	9 months	None
Irritable bowel syndrome	Rana et al. [38]	1	VAS from 8-10 to 3	1 year	None
Mesenteric ischemia	Caruso et al. [39]	1	VAS from 8 to 2	1 month	Pocket infection
Bannayan–Riley–Ruvalcaba syndrome	Yakovlev and Resch[40]	1	VAS from 6 to 0	6 months	None

Established causes of the chronic pain, reference, number of patients studied, documented clinical outcome, follow-up time interval, and listed complications are shown

A more recently published report describes two cases of familial Mediterranean fever (FMF) in which intermittent painful abdominal attacks responded positively to SCS at the T8–T9 and T7–T8 vertebral levels respectively [34].

Despite initial enthusiasm for a novel modality of treatment for severe abdominal pain, interpreting such limited published experience was further complicated by the fact that there was considerable variability in patient selection, lead positioning, and type of hardware used in these reports. Consequently, it has remained unclear whether a reasonable fraction of patients may have long-term benefit from stimulation.

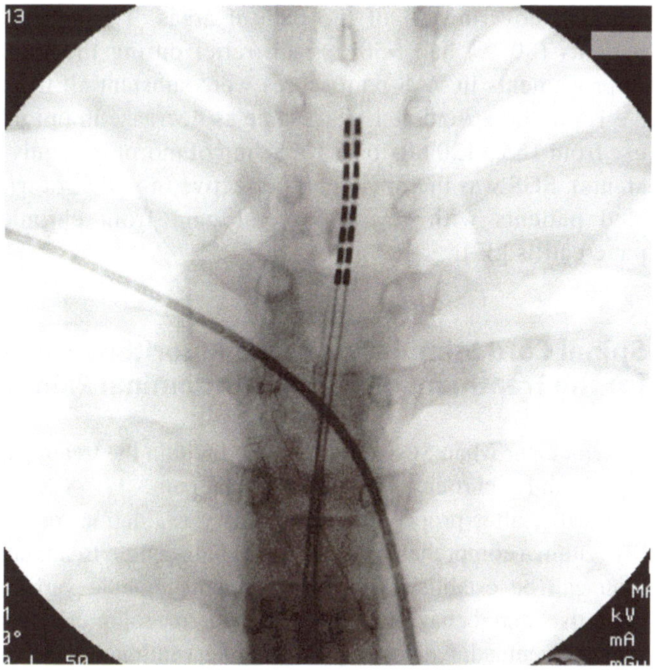

Fig. 19.1 Properly positioned two octrode leads shown here during fluoroscopic anterior–posterior (AP) radiograph of thoracic spine. Leads placed in posterior epidural space with tip of the leads reaching T5 is the most frequent lead positioning when used for the SCS in painful gastrointestinal disorders

More recently, a larger clinical retrospective review using SCS for treatment of chronic abdominal pain has been published [30] studied 35 patients and provided long-term (1 year) clinical follow-up data. Consistent with most technical descriptions in previous reports, SCS lead tips were positioned at T5 ($n = 11$; Figs. 19.1 and 19.2) or T6 ($n = 10$) height within the epidural space. Thirty patients (86 %) reported at least 50 % pain relief on completion of the trial (Fig. 19.3). Among 28 patients who received permanent implant, 19 were followed for at least 1 year. Their visual analog scale (VAS) pain scores remained low (3.8 ± 1.9 cm; $p < 0.001$) at 1 year, as did opioid use of 138.3 ± 134 to 38 ± 48 mg morphine equivalents (Fig. 19.4). This report for the first time suggested that SCS may provide consistent long-term improvements and be a useful therapeutic option for patients with severe visceral abdominal pain. A national survey was conducted to develop a consensus on patient selection and technical aspects of SCS for abdominal visceral pain, where 76 case reports were collected [35]. This brief survey confirmed that SCS for abdominal visceral pain is still rarely used, despite a possibility of high therapeutic success rates. Causes for this include very few studies describing the basic mechanisms of neuromodulation for longstanding visceral pain, comfort levels by the physicians, and issues with coverage of such treatment by payers. The technical aspects of SCS for the treatment of abdominal visceral pain seem to be uniform among physicians who use such technology across the United States and are also consistent with our larger retrospective study described previously [30, 35]. In most patients, the SCS leads were positioned with their tips at the level of the T5 (26 patients; Fig. 19.2) or T6 vertebral body (15 patients). Pain relief exceeded 50 % in 66 of 70 patients reported. VAS pain scores before an implant were 8 ± 1.9 cm, whereas after the implant they were

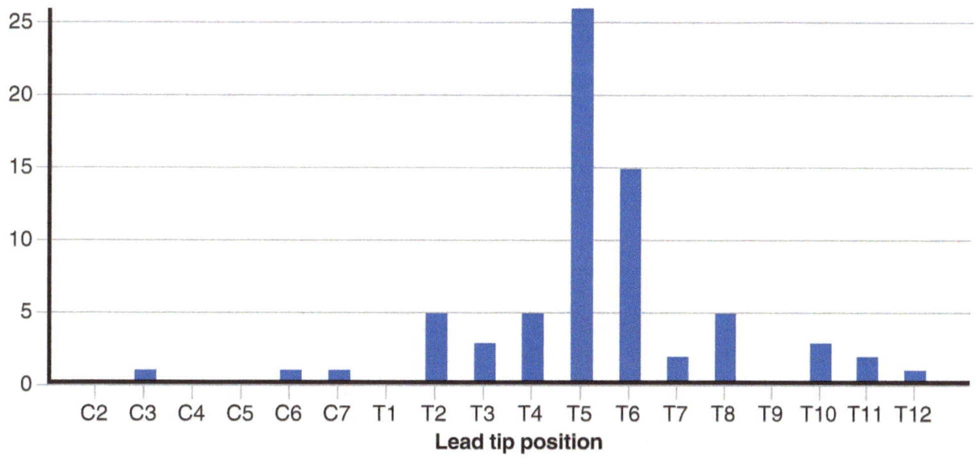

Fig. 19.2 A distribution of the lead tip positions when SCS used for various causes of chronic abdominal pain. The graph shown here is reproduced with permission from the survey study that collected data on 70 SCS cases. The optimal paresthesias within abdomen were achieved when leads were positioned midline up to T5 or T6 vertebral level. Modified from Pain Medicine [35]

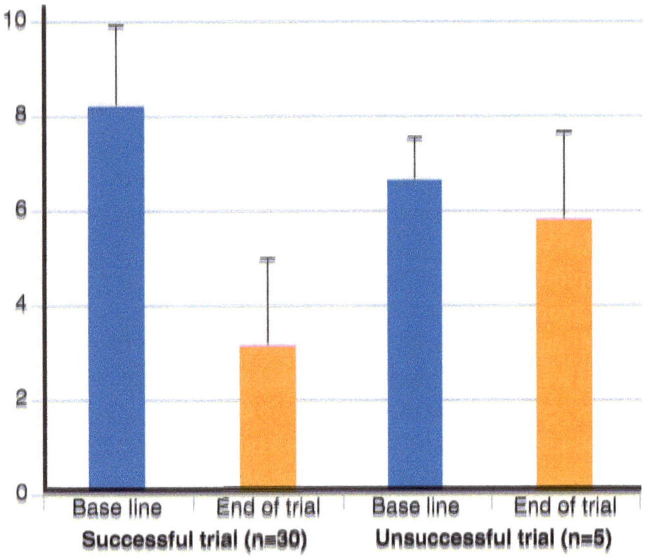

Fig. 19.3 Improvements in chronic abdominal pain, interpreted by patients as VAS score, during prolonged trial of SCS. Thirty out of 35 patients reported >50 % of pain relief, and went on for SCS implantation. Modified from Pain Medicine. 2010;11(3):347–355 [30]

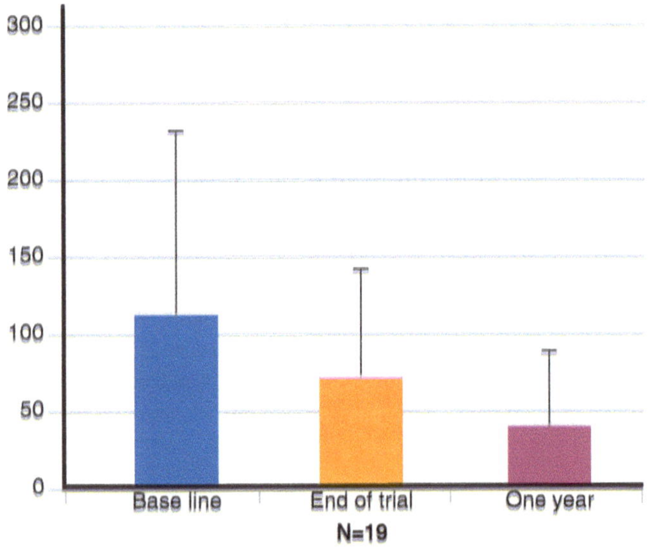

Fig. 19.4 A significant decrease in long-term opioid use when spinal cord stimulation is used for chronic abdominal pain. Nineteen patients followed for more than 1 year decreased their opioid use by 2/3 of baseline dose in average. Modified from Kapural et al. Spinal cord stimulation for visceral abdominal pain, Pain Med [30]

2.49 ± 1.9 cm. Opioid use before an implant was 158 ± 160 mg; at the last office visit after the implant, it was 36 ± 49 mg. The weakness of this survey was that the subgroup of responding physicians may not adequately represent the population of all of the physicians who trialed SCS for chronic visceral abdominal pain but rather those who had largely positive results. However, the goal of the survey was to examine technical aspects of SCS for chronic abdominal pain, and not the efficacy of spinal cord stimulation [35].

Our recent paper reported on acquired clinical experience using SCS in 30 consecutive with chronic pancreatitis [37]. Patient population was somewhat different than in our abdominal patient group [30], as there were 9 out of 30 patients with previous alcohol or opioid abuse. Patients selected for SCS trial suffered severe chronic pain from chronic pancreatitis, but had no exacerbations of acute pancreatitis with blood level of enzymes increased. Similarly, SCS leads were placed most frequently at T5 or T6 vertebral height and abdominal paresthesias were required covering all of the painful areas. Twenty-four patients (80 %) had >50 % pain relief during the trial. Improvements in VAS pain scores were substantial: from 8 ± 1.6 to 3.6 ± 2 cm at 1 year, same as decrease in opioid use from 165 ± 120 mg to 48.6 ± 58 mg of morphine equivalents). SCS was therapeutically effective in >70 % of trialed patients with severe visceral pain from chronic pancreatitis [37].

Spinal Cord Stimulation in the Algorithm for the Treatment of Chronic Abdominal Pain

It is not clear when SCS should be used within the treatment algorithm for chronic visceral abdominal pain.

Ideally, after proper multidisciplinary evaluation, possibly within a comprehensive abdominal pain center, treatment plan can be established. Such treatments include various cognitive and behavioral therapies, pharmacological pain management, adjuvant therapies, and interventional diagnostic and therapeutic nerve blocks (Fig. 19.5).

More recently, modified retrograde epidural differential block is used to clarify the underlying cause of abdominal pain, especially helpful in distinguishing between visceral, somatosensory or centralized pain causes.

For chronic visceral abdominal pain splanchnic or celiac plexus and hypogastric blocks can be used to acutely control the pain followed by radiofrequency ablation of splanchnic nerves to provide rather long-term relief.

Based on above detailed published clinical experience, SCS might be indicated when conservative therapies fail to improve analgesia and function. Psychological evaluation for implantable devices and case discussion within the interdisciplinary medical team should precede any attempt in neuromodulation. Only after SCS trial completed and results are largely positive with >50 % improvements in pain scores, and significant improvement in patients' function, decision can be made as to whether SCS implantation is an appropriate next step (Fig. 19.5). The data in support of SCS for visceral pain presently are encouraging. However, randomized controlled trials need to be initiated to support the role of SCS for long-term treatment of visceral pain.

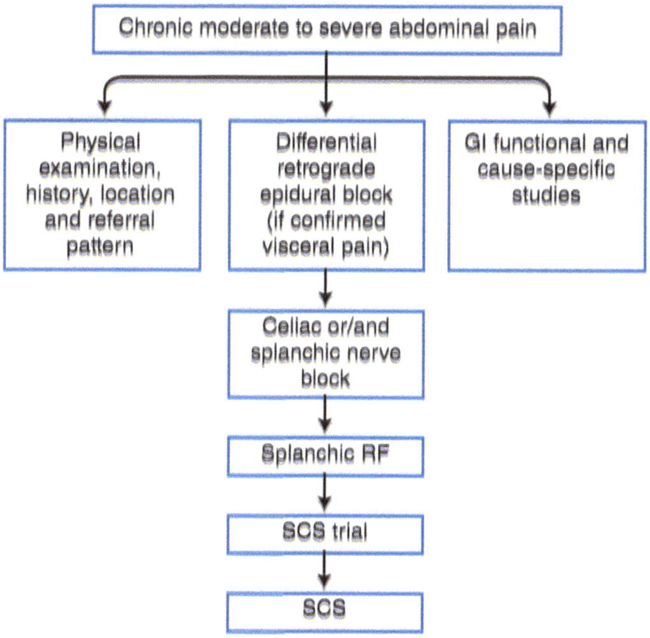

Fig. 19.5 Proposed interventional pain management arm of the algorithm for treatment of chronic, defined abdominal pain. Please note that none of the steps in this proposed algorithm was sufficiently studied to provide a definitive place in this treatment continuum. SCS as an implantable, invasive therapy is considered when patients failed to receive long-term improvements using conservative approaches and blocks/radiofrequency ablations. (*RF* radiofrequency ablation)

References

1. Pare P, et al. Health-related quality of life, work productivity, and health care resource utilization of subjects with irritable bowel syndrome: baseline results from LOGIC (Longitudinal Outcomes Study of Gastrointestinal Symptoms in Canada), a naturalistic study. Clin Ther. 2006;28(10):1726–35; discussion 1710–1.
2. Derbyshire SW. Imaging visceral pain. Curr Pain Headache Rep. 2007;11(3):178–82.
3. Russo MW, et al. Digestive and liver diseases statistics, 2004. Gastroenterology. 2004;126(5):1448–53.
4. Giamberardino MA, Vecchiet L. Pathophysiology of visceral pain. Curr Pain Headache Rep. 1997;1(1):23–33.
5. Grundy D. Neuroanatomy of visceral nociception: vagal and splanchnic afferent. Gut. 2002;51 Suppl 1:i2–5.
6. Everhart J. Digestive diseases in the United States: epidemiology and impact. P.H.S. US Department of Health and Human Services, National Institute of Diabetes and Digestive and Kidney Diseases. NIH publication no. 94-1447. p. 3–53.
7. Graney DO. General considerations of abdominal pain. In: Loesser JD, editor. Bonica's management of pain. Philadelphia: Lippincott, Williams & Williams; 2001. p. 1235–68.
8. Rizk M, Rocio L, Reda T, Yahia A, Justin M, Bell R, Shingo J, Mahoobi R, Kapural L. The Positive Predictive Value of Differential Nerve Block to determine success of visceral block in patients with visceral chronic abdominal pain. Pain Pract. 2012;12(8):595–601.
9. Puylaert M, Kapural L, van Zundert J, Peek D, Lataster A, Mekhail N, van Kleef M, Keulemans Y. Pain in chronic pancreatitis. Evidence-based Interventional Pain Medicine according to clinical diagnoses. Pain Pract. 2011;11(5):492–505.
10. Mertz H. Review article: visceral hypersensitivity. Aliment Pharmacol Ther. 2003;17(5):623–33.
11. Wood JD. Neuropathophysiology of irritable bowel syndrome. J Clin Gastroenterol. 2002;35(1 Suppl):S11–22.
12. Wood JD. Neuropathophysiology of functional gastrointestinal disorders. World J Gastroenterol. 2007;13(9):1313–32.
13. Gebhart GF. Visceral pain-peripheral sensitisation. Gut. 2000;47 Suppl 4:iv54–5; discussion iv58.
14. Gebhart GF. Pathobiology of visceral pain: molecular mechanisms and therapeutic implications IV. Visceral afferent contributions to the pathobiology of visceral pain. Am J Physiol Gastrointest Liver Physiol. 2000;278(6):G834–8.
15. Krames ES, Foreman R. Spinal cord stimulation modulates visceral nociception and hyperalgesia via the spinothalamic tracts and the postsynaptic dorsal column pathways: a literature review and hypothesis. Neuromodulation. 2007;10(3):224–37.
16. Linderoth B, Foreman R. Mechanisms of Spinal Cord Stimulation in painful syndromes: role of animal models. Pain Med. 2006;7 Suppl 1:S14–26.
17. Linderoth B, Meyerson BA. Physiology of spinal cord stimulation: mechanism of action. In: Burchiel K, editor. Surgical management of pain. Stuttgart: Thieme Medical Publisher; 2002. p. 505–26.
18. Cui JG, et al. Spinal cord stimulation attenuates augmented dorsal horn release of excitatory amino acids in mononeuropathy via a GABAergic mechanism. Pain. 1997;73(1):87–95.
19. Linderoth B, et al. Gamma-aminobutyric acid is released in the dorsal horn by electrical spinal cord stimulation: an in vivo microdialysis study in the rat. Neurosurgery. 1994;34(3):484–8; discussion 488–9.
20. Kapural L. Spinal cord stimulation for the chronic painful visceral abdominal disorders. In: Hayek SM, Levy R, editors. Neurostimulation for the treatment of chronic pain. Interventional and neuromodulatory techniques for pain management series, vol. 1. Philadelphia: Elsevier; 2011. p. 113–7.
21. Steege JF. Superior hypogastric block during microlaparoscopic pain mapping. J Am Assoc Gynecol Laparosc. 1998;5:265–7.
22. Rauck RL. Sympathetic nerve blocks. In: Raj PP, editor. Practical management of pain. 2nd ed. St Louis: Mosby; 1992. p. 778–812.
23. Van Meerveld B, et al. Attenuation by spinal cord stimulation of a nociceptive reflex generated by colorectal distention in a rat model. Auton Neurosci. 2003;104(1):17–24.
24. Linderoth B, Foreman R. Physiology of spinal cord stimulation: review and update. Neuromodulation. 1999;2:150–64.
25. Qin C, et al. Spinal cord stimulation modulates intraspinal colorectal visceroreceptive transmission in rats. Neurosci Res. 2007; 58(1):58–66.
26. Ceballos A, et al. Spinal cord stimulation: a possible therapeutic alternative for chronic mesenteric ischaemia. Pain. 2000;87(1): 99–101.
27. Jackson M, Simpson KH. Spinal cord stimulation in a patient with persistent oesophageal pain. Pain. 2004;112(3):406–8.
28. Tiede JM, et al. The use of spinal cord stimulation in refractory abdominal visceral pain: case reports and literature review. Pain Pract. 2006;6(3):197–202.
29. Krames ES, Mousad DG. Spinal cord stimulation reverses pain and diarrheal episodes of irritable bowel syndrome: a case report. Neuromodulation. 2005;8:82–8.
30. Kapural L, Sessler D, Tluczek H, Nagem H. Spinal Cord Stimulation for visceral abdominal pain. Pain Med. 2010;11(3):347–55.
31. Kapural L, Rakic M. Spinal cord stimulation for chronic visceral pain secondary to chronic non-alcoholic pancreatitis: a case report. Clin Gastroenterol Hepatol. 2008;42(6):750–1.
32. Khan Y, Raza S, Khan E. Application of spinal cord stimulation for the treatment of abdominal visceral pain syndromes: case reports. Neuromodulation. 2005;8:14–27.

33. Kim JK, et al. Spinal cord stimulation for intractable visceral pain due to chronic pancreatitis. J Korean Neurosurg Soc. 2009;46(2):165–7.
34. Kapur S, Mutagi H, Raphael J. Spinal cord stimulation for relief of abdominal pain in two patients with familial Mediterranean fever. Br J Anaesth. 2006;97(6):866–8.
35. Kapural L, Deer T, Yakovlev A, Bensitel T, Hayek S, Pyles S, Narouze S, Khan Y, Kapural A, Cooper D, Stearns L, Zovkic P. Spinal cord stimulation for visceral abdominal pain: results of the national survey. Pain Med. 2010;11(5):685–91.
36. Kapural L, Janicky T, Narouze S, Mekhail N. Spinal cord stimulation is an effective treatment for the severe pelvic visceral pain. Pain Med. 2006;7(5):440–3.
37. Kapural L, Cywinski J, Sparks D. Spinal cord stimulation for visceral pain from chronic pancreatitis. Neuromodulation. 2011;14(5):423–7.
38. Rana MV, Knezevic NN. Tripolar spinal cord stimulation for the treatment of abdominal pain associated with irritable bowel syndrome. Neuromodulation. 2013;16(1):73–7.
39. Caruso C, Lo Sapio D, Ragosa V, Lo Sapio S, Cafora C, Romano L. Abdominal angina due to obstruction of mesenteric artery treated with spinal cord stimulation: a clinical case. Neuromodulation. 2011;14:146–50.
40. Yakovlev AE, Resch BE. Treatment of intractable abdominal pain patient with Bannayan–Riley–Ruvalcaba syndrome using spinal cord stimulation. WMJ. 2009;108(6):323–6.
41. Al-Mahrouqi H, Munro Z, Acland RH, MacFarlane MR. Spinal cord stimulation for intractable chronic upper abdominal pain: a case report of the first patient in New Zealand. N Z Med J. 2012;125(1367):132–4.

Peripheral Nerve Stimulation for Chronic Abdominal Pain

20

David Pang and Teodor Goroszeniuk

Introduction

Visceral abdominal pain (VAP) intermittently affects over 25 % of the world population. Frequently, diagnosis is complex and difficult. In a large proportion of patients, no firm diagnosis is possible. This non-specific abdominal pain incurs over £100 million in expenses in the UK, which is spent investigating and treating this condition. The interventional and surgical treatments of VAP secondary to Irritable Bowel Syndrome (IBS), or chronic pancreatitis, are frequently unsatisfactory and only provide short duration pain relief. VAP is an enormous global problem causing great suffering for many, with additional social and economical implications. Therefore, the development of a new effective non-pharmacological, minimally invasive and reversible modality is extremely important. Neuromodulation, specifically conventional SCS and anterior SCS, is continuously gaining ground with increasing numbers of patients reported and encouraging results.

Peripheral neuromodulation in management of VAP is not fully explored, but it is attractive and emerging option tangential to the other modalities such as spinal cord stimulation. In the theory, there are multiple neural structures which can be considered as a possible stimulation target. Usually the sympathetic chain is used. The data available on the usefulness of this approach is limited, but encouraging. Targets, techniques, equipment, mode of action and the best programming-like frequencies are to be investigated further. Equipment will soon be designed for the specific requirements of peripheral stimulation and, specifically, for abdominal sympathetic chain stimulation.

The technique of peripheral nerve stimulation is amongst the oldest known treatments for chronic pain in man. The first written description was by Scribonius Largus in 15 AD [1]. He observed one of the Emperor's slaves who suffered from long-term foot pain from gout achieve pain relief when he accidentally stepped on a torpedo fish. The electric discharge from this fish was able to persistently alleviate his pain and this treatment was recommended for headaches as well as gout.

In 1887, Althaus described a case of abdominal pain and epigastric swelling in a 16-year-old girl that was treated with electrical stimulation with complete cessation of both pain and reduction in the size of the swelling, and after 33 5 min treatments the pain had completely resolved and the swelling had gone [2].

The use of electricity for the treatment of pain has continued to draw widespread interest, but the clues to its mechanism of action did not emerge until 1965, when the experimental work of Wall and Melzack led to their gate theory of pain [3]. This theory postulated that the transmission of pain could be inhibited by other, non-painful sensory impulses. Thus, nociception from pain transmitting C and Aδ nerve fibres at the dorsal horn of the spinal cord level could be inhibited by signals from Aβ fibres transmitting light touch and other non-nociceptive impulses. Thus, the gate to nociceptive sensations could be "closed" by other sensations such as light touch or the sensations elicited by peripheral nerve stimulation.

Following its publication, Wall and Sweet tested this gate theory by inserting insulated stimulating needles into their own infraorbital foramen [4]. They evoked paraesthesia and other sensations using stimulation at 100 Hz square waves and noted that this stimulation could induce analgesia to pinprick which stopped after cessation of stimulation.

D. Pang, M.B.Ch.B., F.R.C.A., F.F.P.M.R.C.A. (✉)
Pain Management and Neuromodulation Centre, St Thomas' Hospital, Guy's & St. Thomas' NHS Foundation Trust Hospital, Westminster Bridge Road, London SE1 7EH, UK
e-mail: dpang@doctors.org.uk

T. Goroszeniuk, F.F.A.R.C.S.I., F.F.P.M.C.A.I.
Interventional Pain Management and Neuromodulation Practice, London W1G 8GD, UK
e-mail: teogoroszeniuk@doctors.org.uk

L. Kapural (ed.), *Chronic Abdominal Pain: An Evidence-Based, Comprehensive Guide to Clinical Management*,
DOI 10.1007/978-1-4939-1992-5_20, © Springer Science+Business Media New York 2015

Subsequently, Sweet and Wepsic implanted a peripheral nerve stimulator in a 26-year-old woman with neuropathic hand pain. The implant consisted of a pair of platinum electrodes around the median and ulna nerve of the forearm. Stimulation at 100 Hz was able to produce paraesthesia in the hand and pain relief as a result [5]. It is of note that this occurred at the time of spinal cord stimulation by Shealy, who implanted the first spinal cord stimulator in 1967 [6].

The field of peripheral nerve stimulation for chronic painful conditions continued, but it was not without obstacles. These implants were by surgeons as the nerve that was to be stimulated required surgical exposure and nerve injury was reported from perineural fibrosis and direct injury from the insertion of the electrodes [7].

It was not until 1999 that percutaneous techniques of peripheral nerve stimulator implants were shown to be technically feasible. Weiner treated headache from occipital neuralgia with percutaneously inserted electrodes [8]. This avoided the necessity for surgical exploration of a nerve that was technically difficult and with significant associated morbidity. They demonstrated that using a percutaneously inserted electrode was less invasive, technically easier and with similar efficacy.

Further case reports of peripheral stimulation using percutaneous techniques began to emerge. With increasing awareness and knowledge of nerve and subcutaneous targets, there has been a continuous expansion in the use of the percutaneous technique applied in a number of new indications; supraorbital neuralgia [9], atypical facial pain [10], postherpetic neuralgia [11], ilioinguinal neuralgia [12], ulnar nerve neuralgia [13] or percutaneous implant targeted at sciatic nerve [14]. Imaginative techniques have been emerging, making the percutaneous peripheral approach even more attractive and successful. Plexus stimulation for brachial plexus neuropathy [15] has demonstrated that it can be used as an effective alternative to SCS. The development of Targeted Stimulation (STS)/Field Stimulation [16] has made it possible to reach many pain targets where SCS was not indicated or could be used as an alternative or in combination with SCS. Targeted/Field stimulation has been used to great effect in various diverse conditions such as back pain [17], costochondritis [16] and refractory angina [18, 19].

Although the use of peripheral neuromodulation in chronic abdominal pain has not received as much attention as other painful conditions, we describe later in this chapter how even this area of pain is starting to benefit from a peripheral approach to stimulation.

Anatomical Targets for Peripheral Neuromodulation

As the nerve pathways of nociception converge into well-described nerve plexuses and individual nerves these are suitable targets for peripheral neuromodulation. Nociceptive input from the entire GI tract has traditionally been via the celiac plexus and splanchnic nerves, entering the thoracic spinal cord at T5 to L2 [20–22]. The vagus nerve was thought not to play a part in nociception but only in transmission of functional and physiological information. Recent experiments have suggested that nociception can occur via the vagus nerve and this plays a role in the emotional-effective and autonomic aspects of gastric pain [23].

Another potential pathway of nociception that has been targeted in peripheral stimulation is the lumbar sympathetic chain. This pathway is commonly treated for sympathetically medicated pain from the lower limb, but a recent case report in its use for renal pain (described below) may suggest its role in renal nociception [24].

Thirdly, the role of peripheral stimulation at the subcutaneous epicentre target of pain has been successful in treating localised pain but its role is not confined to superficial pain [25]. It has been successful in treating angina [18, 19] and pelvic pain [26], suggesting that it may provide analgesia using both local and central mechanisms. This technique is described in more detail later.

Mechanism of Action in Peripheral Stimulation

The gate theory of pain remains the most important advance in our understanding of how electrical stimulation can alleviate pain [1]. Much of the work in understanding its mechanism of action has been done by investigators on spinal cord stimulation, but emerging research into occipital nerve stimulation and TENS have shed some light on how stimulation of the periphery works and on their differences with spinal cord stimulation.

As in spinal cord stimulation, TENS exerts its effect via local, spinal and supraspinal pathways. Its effect depends upon what frequency is used, as high (50–10 Hz) and low (<10 Hz) frequencies have different mechanisms of action. Low-frequency TENS can be antagonised by blockade of μ-opioid, $GABA_A$, serotonin and muscarinic M1 and M3 receptors [27–32]. Repeated stimulation can induce a tolerance and it suggests that this method of TENS exerts its effect via descending inhibitory pathways, utilising opioid, serotonin and cholinergic neurotransmitters [33]. High-frequency TENS also uses an opioid-related pathway as blockade of the δ opioid receptors can antagonise its analgesic effect [28]. In the CSF, high-frequency TENS has been shown to increase the levels of β Endorphins in the blood stream and methionine-enkephalin in the cerebrospinal fluid [34, 35]. GABA blockage at the spinal cord also reduces its efficacy, but unlike low-frequency TENS, serotonin blockade has no effect [30].

This suggests that peripheral stimulation may exert its effects at supraspinal levels and this is a phenomenon also seen in spinal cord stimulation. In a study of patients with chronic headaches that have been treated with occipital nerve implantation, PET scanning demonstrated an increase in blood flow to the dorsal rostral pons, anterior cingulated cortex and cuneus [36].

Direct stimulation of peripheral nerves results in decreased excitability, transient slowing of conduction velocity and an increase in electrical threshold [37]. The paraesthesia felt during peripheral nerve stimulation is mediated by Aβ fibres and this is consistent with the gate theory of pain. As these fibres travel in the dorsal columns, it is possible that peripheral stimulation shares its segmental mechanisms of action with spinal cord stimulation. In a rat model, low-frequency stimulation of Aδ fibres was shown to potentiate long-term depression of mono and polysynaptic excitatory post-synaptic potentials at the substantia gelatinosa. Notably, this effect lasted for several hours and in a few case reports peripheral stimulation has an effect that outlives the duration of stimulation [38, 39].

One exciting aspect of peripheral nerve stimulation is its possible effect on nociceptive pain. Spinal cord stimulation is extremely effective for pain of neuropathic or ischemic origin, but its effect on nociceptive pain has been disappointing. A number of case reports have suggested that peripheral nerve stimulation may be effective in pain of nociceptive origin such as low back and neck pain.

In a human volunteer study, Ellrich and Lamp used electric and laser infrared stimuli to test the effects of peripheral nerve stimulation on nociceptive pain [40]. Laser infrared stimuli are known to provoke Aδ and C fibre activation and the investigators used stimulation of the superficial radial nerve as it was easily accessible and was a pure sensory nerve. Cortical-evoked potentials were measured and in all volunteers the laser stimulation resulted in painful prickling sensations. During stimulation of the superficial radial nerve, the reduction in evoked potentials and latencies was significantly decreased compared with control stimulation. Again, this provides further evidence to support the gate theory and suggests that peripheral stimulation may have a role in treating nociceptive pain.

Caution must be advised in extrapolating these above mechanisms to peripheral neuromodulation in visceral pain. It has been suggested that mechanisms of action of spinal cord stimulation depends on the target organ and this may be the case in VAP. Many of the animal models used to investigate the role of SCS have used lesions of peripheral nerves such as sciatic nerve ligation and this provides a useful but limited model of neuropathic pain [41, 42].

One such animal model of visceral pain was described by Qin [43]. They noted that stimulation of somatic afferents in previous animal model studies could affect the cardiovascular response to gastric distension in rats. They used anaesthetised rats and used gastric distension as the visceral pain stimulus. Electrical stimulation of isolated forelimb median and hindlimb peroneal nerves was performed using bipolar platinum electrodes. Stimulation was 50 Hz at amplitude of 0.5–2.0 mA. This was deemed sufficient to activate both A and C fibres. To measure the modulating effect of this peripheral stimulation, microelectrodes were inserted in the thoracic spinal cord to measure extracellular potentials at T9 and T10.

The results showed that stimulation of somatic afferents at distant somatic sites could affect visceral nociceptive transmission. Median and peroneal nerve stimulation could alter activity of 63–67 % of spinal neurons at T9 and T10 in response to gastric distension. The mechanism of action has been suggested as via stimulation of propriospinal pathways and supraspinal inhibition. Another explanation is by the phenomenon of diffuse noxious inhibitory controls. This is inhibition of dorsal horn convergent neurons caused by noxious stimulation to widespread areas of the body and viscera.

Practical Application of Neuromodulation Techniques for Abdominal Visceral Pain

As the use of peripheral neuromodulation for abdominal visceral pain is still in its infancy, no randomised controlled trials of its efficacy exist. The evidence for the use of peripheral techniques have been presented as case reports, but this should not deter clinicians at the forefront of pain medicine from their use as patients with chronic abdominal visceral pains may have symptoms that are refractory to conventional pain management therapies [44]. We now describe techniques that have been used in patients with refractory chronic pain and how we can target peripheral neuromodulation techniques to abdominal visceral pains.

Stimulation at the Subcutaneous Target

Painful areas that are covered by a specific nerve distribution or dermatome are usually amenable to spinal or peripheral nerve stimulation. However, when the pain does not follow any of these patterns it can pose a challenge as there is no single nerve or dorsal column area that can be used as a target for nerve stimulation. This led Goroszeniuk in 2000 to develop an application of a peripheral neuromodulatory technique to stimulate the epicentre of pain itself in the subcutaneous tissue [25].

This technique started as a result of the encouraging clinical experience of using low frequency (2 Hz) stimulation of a patient suffering from neuropathic pain in an ulnar nerve distribution. Upon stimulation of the ulnar nerve with a stimulating monoelectrode at 2 Hz, the patient achieved 11 weeks of complete pain relief.

Subsequently, the application of a stimulating needle at the subcutaneous epicentre of pain rather than on individual nerves themselves resulted in successful pain relief in patients whom a non-dermatological distribution of pain exists. This new concept of stimulating the painful target itself rather than trying to find the neural pathways of pain has let to its use in an increasing number of painful conditions. The use of this technique is increasing due to its technical simplicity, low risk of complications and the ability to achieve pain relief in areas traditionally difficult to cover with spinal cord or peripheral nerve stimulation [16].

Use of this technique in visceral pain was reported in 2006 in three patients [45]. As abdominal visceral pain can be associated with secondary somatic hyperalgesia of the abdominal wall, it was thought that the use of subcutaneous stimulation over the painful target could result in pain relief. The three patients had refractory abdominal pain and due to its non-dermatomal pattern spinal cord stimulation was deemed not suitable as it would not cover adequately all the painful areas. Of note is that not all the pain was purely visceral and that abdominal wall pain existed in all three patients.

The first patient had chronic right lower quadrant pain secondary to inguinal hernia repair. After a successful trial of subcutaneous stimulation, a permanent peripheral stimulator implant was inserted and this had almost completely abolished her pain. The second patient had right-sided abdominal pain secondary to a liver transplant and repair of an incisional hernia. Placement of three subcutaneous leads over the scar and intercostal margin which was the site of greatest pain provided excellent pain relief and he was able to improve his level of function and reduce his oral analgesics. The third patient was a patient with chronic pancreatitis who had undergone a cholecystectomy and modified Whipple's procedure. His pain was localised in an area of 6 cm by 4 cm at the right upper quadrant. This pain was treated by placement of two octoelectrodes at the epicentre of pain. The initial placement was done using two multipolar electrodes placed in the subcutaneous area parallel to the spine in a vertical manner. Unfortunately, the stimulation was poor and upon revising the leads so that they were sitting in a horizontal pattern, parallel to the dermatomes resulted in excellent pain control. A third revision was required as the leads were too deep and coverage of pain was lost until the leads were moved more superficially.

Most interestingly, the orientation and depth of the leads are crucial to this technique. The common approach to electrode placement is to place the electrodes in a blind or image-guided manner via a modified 14-G Tuohy needle or a 14-G cannula and then test stimulation with the electrodes in situ. This has a major disadvantage in that if stimulation is not correct, the electrodes have to be reinserted from the beginning. These multiple passes can increase potential nearby tissue injury, discomfort and increase operating time. In subcutaneous stimulation, the exact position of the electrode at the epicentre of pain is crucial and the depth of electrode position can be difficult to determine accurately [46].

Initially, we have developed a novel stimulating needle based on the standard modified SCS Tuohy needle, which was a significant improvement over the existing practice [47].

However, the adaptation of the Coudé needle (Epimed, TX, USA) with the superior design, of the curvature and relatively bland end, that allows exact determination of stimulation at its tip and then electrodes can be passed through the needle to the position of stimulation has been a subsequent refinement. This is analogous to the technique of placing stimulating local anaesthetic catheters in regional anaesthesia. At 19-G diameter, current stimulating needles are too small to allow passage of implantable 1.2 mm electrodes which require a 14-G introducer; furthermore, they are straight so steerability under fluoroscopy is limited. Figure 20.1 shows the stimulating cannula and an implantable electrode.

This needle has been used in five patients undergoing peripheral neuromodulation implantation and its easy steerability and stimulation at the introducer tip allows passage of the electrode at the exact location and depth required to achieve optimal coverage of the painful area.

The terminology is still under debate as a number of different terms have been used to describe this technique. We use the term "subcutaneous targeted stimulation" as it uses the principle that the stimulation is targeted at a specific epicentre of pain and it is not designed to use the electrical field itself to cover the painful area [48]. The feeling of stimulation is much larger than the electrical field itself. Other terms used are peripheral nerve field stimulation and subcutaneous electrical neurostimulation [49].

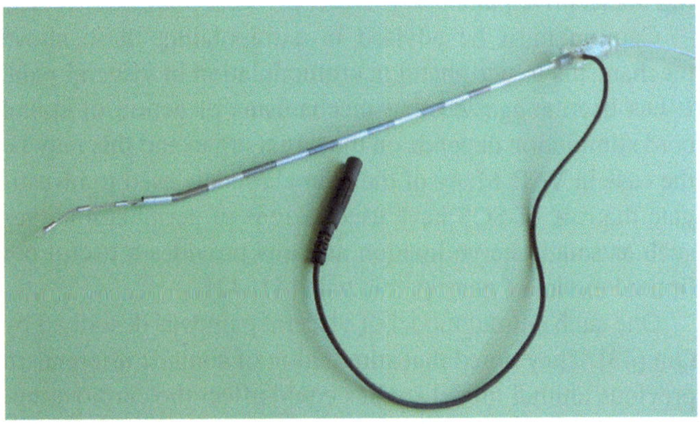

Fig. 20.1 Stimulating 14-G Coudé needle and multipolar electrode

The Use of Non-invasive External Neuromodulation

The application of low frequency non-invasive stimulation using an external nerve mapping probe is termed external neuromodulation and it allows mapping of potential peripheral targets for neuromodulation without needle insertion. The probe (Neuro-Trace, HDC Corporation, Milpitas, California, USA, Pajunk GmbH, Geisingen, Germany) was introduced to map nerve location for regional anaesthesia, but its adoption in using it to treat neuropathic pain states has led to its use as a method of stimulating nerves or the epicentre of pain in a non-invasive manner [16, 50]. This stimulation is at a frequency of 2 Hz and the ball-shaped probe is placed over the skin using known surface landmarks of nerve anatomy or at the point of greatest pain for non-nerve pain distributions. Figure 20.2 shows the probe being used over the inguinal region.

The amplitude of stimulation is increased so that paraesthesia is perceived. The procedure can be performed on an outpatient basis and it is important to emphasise that this technique should not be confused with TENS. The electrical field from the EN is deeper and more focused as it has a higher current density [16].

Its use can be both in screening for whether the patient's pain is suitable for other peripheral neuromodulation techniques or if analgesia is prolonged then this can be used as a single modality alone and the patient can be provided with their machine and be taught to self-administer [51].

With an increasing variety of techniques available, our practice is to use the various techniques in a logical manner from the least invasive at the beginning of treatment to invasive implanted therapies for the most refractory pain conditions [52].

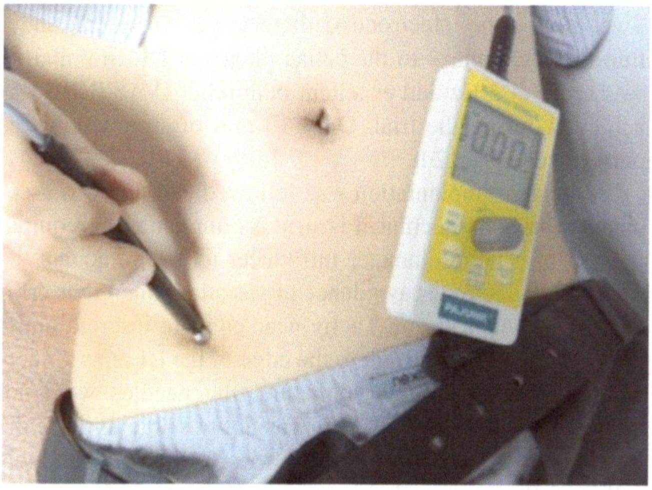

Fig. 20.2 Application of external neuromodulation over ilioinguinal nerve for groin pain

Stage I: The application of External Neuromodulation allows mapping of the nerves or target that may be amenable for stimulation. If analgesia is prolonged then this can be the sole therapy and patients can self-administer.

Stage II: Direct stimulation of the peripheral nerve or at the epicentre of pain using an insulated needle for approximately 5 min. Again, this provides further information on nerve mapping and what the best targets for stimulation would be for future treatment. Unlike external neuromodulation, this is usually not the sole therapy alone as repeat stimulation cannot be self-administered.

Stage III: Trial of stimulation. Once an appropriate area is mapped out with the above techniques, a trial of stimulation using either a monopolar stimulating catheter or standard spinal cord stimulator multipolar electrodes can be commenced. The monopolar lead (Stimulong, Pajunk) is inexpensive, but multipolar electrodes are useful if the area of stimulation is large or monopolar stimulation is inconclusive [53].

Stage IV: If pain relief from the trial is greater than 50 % and no adverse effects reported then we proceed to permanent insertion of multipolar leads to an internal implanted pulse generator.

A function of EX Stimulation in diagnosis and treatment of abdominal visceral pain has not been fully explored as yet; however, its success in pain states, such as refractory angina and neuropathic pain, may suggest a role in the future.

Lumbar Sympathetic Chain Stimulation: Another Target for Renal Pain

LPHS presents with flank pain and micro and macroscopic haematuria. Its aetiology is unknown and poorly understood, and diagnosis is one of exclusion. Renal function is normal and the aim is symptom control as there is no progression to chronic renal failure.

These patients have severe loin pain which is usually refractory to conventional treatments. One potential target of stimulation is the lumbar sympathetic chain as it was believed that many of the nociceptive afferents were autonomic. Four cases of loin pain haematuria syndrome have been successfully treated with stimulation of the lumbar sympathetic chain [24].

As with the case of splanchnic stimulation, the initial approach involved using low-frequency stimulation of the lumbar sympathetic chain at L3 with a stimulating needle at 2 Hz for 5 min. This allowed the authors to determine if sympathetic chain stimulation was effective in relieving the pain. All patients developed pleasant stimulation covering all of their pain and after 5 min of stimulation two of the patients

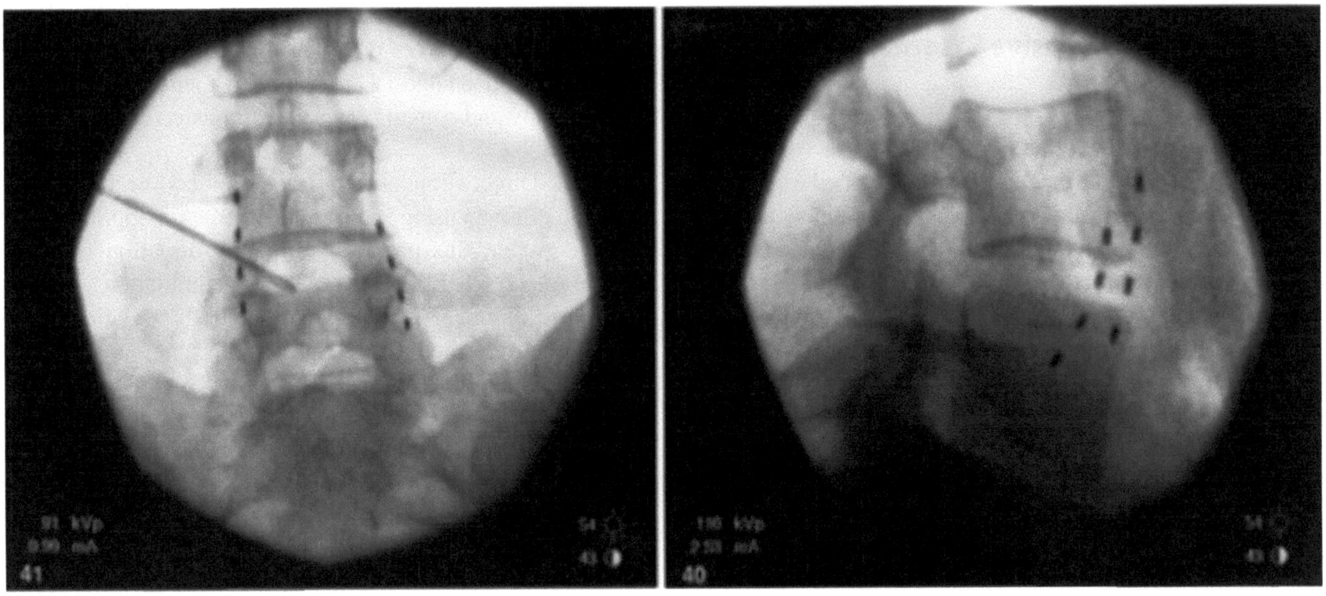

Fig. 20.3 Lead position for lumbar sympathetic chain stimulation

had 24 h of complete pain relief. After this, a trial of stimulation using monopolar electrodes was performed by inserting them at the level of the anterior border of the L3 vertebral body. In three patients the trial achieved full pain relief. The monopolar electrodes were chosen for their simplicity of insertion and economy. Full implantation in two patients with multipolar electrodes (Pisces quad, Medtronic Inc., Minnesota, USA) provided effective long-term pain relief. Insertion of these electrodes was performed by using a 14-G long Tuohy needle as an introducer inserted to the anterior border of the L3 and L4 vertebral body. The leads were connected to an implanted pulse generator in the anterior abdominal wall. Figure 20.3 shows the position of the leads. A role of this interesting and attractive new approach has to be evaluated further.

Following permanent implantation in these two patients, they continued with complete pain relief and stimulation was required only two or three times daily. The third patient had 7 months from a trial period of stimulation and full implantation was thus delayed and the fourth patient is awaiting implantation.

Stimulation of Splanchnic Nerves

The first case report of peripheral stimulation of the splanchnic nerves is a 36-year-old patient who had chronic pancreatitis and severe right upper quadrant abdominal pain [54]. This pain was exacerbated by food and she had undergone cholecystectomy, endoscopic incision of the sphincter of Oddi, and a Roux-en-Y procedure. She continued to have persistent pain and a celiac plexus block with alcohol only provided 3 months of pain relief. She required large doses of fentanyl daily and a Whipple's procedure was proposed as a last resort but she refused.

As external Neuromodulation would not be possible, it was decided that the first approach would to perform stimulation of the celiac plexus. This was performed by inserting two 150 mm stimulating needles at L1 to the celiac plexus and stimulation at 2 Hz for 5 min at amplitude that produced a pleasant sensation in the abdomen. Her pain scores dropped from 9 out of 10 on a VAS scale to 0 out of 10 for the next 48 h.

This encouraging result was followed by insertion of temporary monopolar electrodes as a trial of stimulation. Two monopolar electrode catheters were inserted under fluoroscopic control to the celiac plexus at L1. Stimulation was again at 2 Hz and excellent pain relief (VAS 0/10) was achieved during the trial. Only 10 min of stimulation was performed every 12 h.

Permanent implantation was followed and this was associated with some technical issues. As implantation required the use of a large diameter introducer to accommodate the multipolar electrodes avoidance of vascular puncture is critical. Two 14-G 150 mm Tuohy needles were inserted under fluoroscopy to lie adjacent to the edge of the L1 anterior border. To prevent vascular puncture, the needles were not placed beyond this point. As it was not possible to place the leads at the celiac plexus, a guide wire was inserted and this followed the anterior border of the T11 and T12 vertebral bodies. The Tuohy needle was removed and an introducer

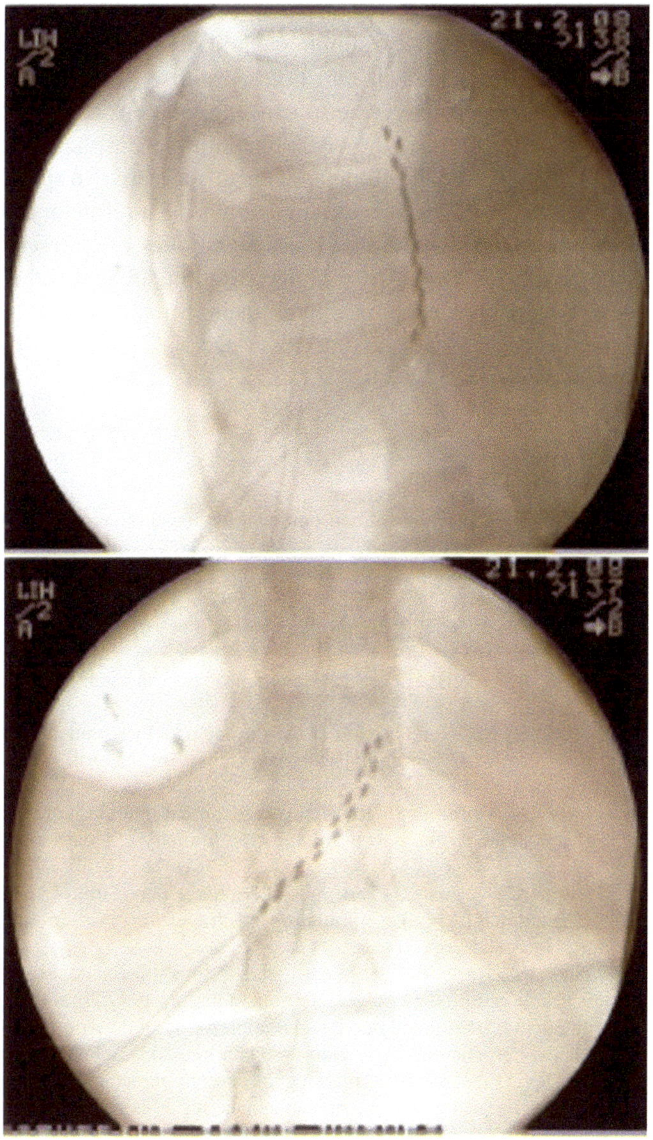

Fig. 20.4 Octopolar leads at the splanchnic nerves

railroaded over the guide wires. This allowed placement of two octopolar leads at the anterior border of the T11 and T12 bodies. Stimulation at 2 Hz produced pleasant abdominal stimulation covering the painful areas. The leads were connected to an extension and connected to an implanted pulse generator (EON, St Jude Medical Inc., USA) placed at the anterior abdominal wall.

The radiographic views are shown in Fig. 20.4.

Stimulation was only required for up to 6 h daily and she was able to reduce her fentanyl from 225 to 12.5 μg/h and stop the use of breakthrough analgesia. Her weight started to increase and she resumed employment.

This case illustrated the use of the potential of peripheral stimulation at the splanchnic nerves for VAP. As this treatment still requires validation with more patients and eventually randomised controlled trials we have to remain cautious but the long-term lead stability remains good and it is a possible alternative to spinal cord stimulation.

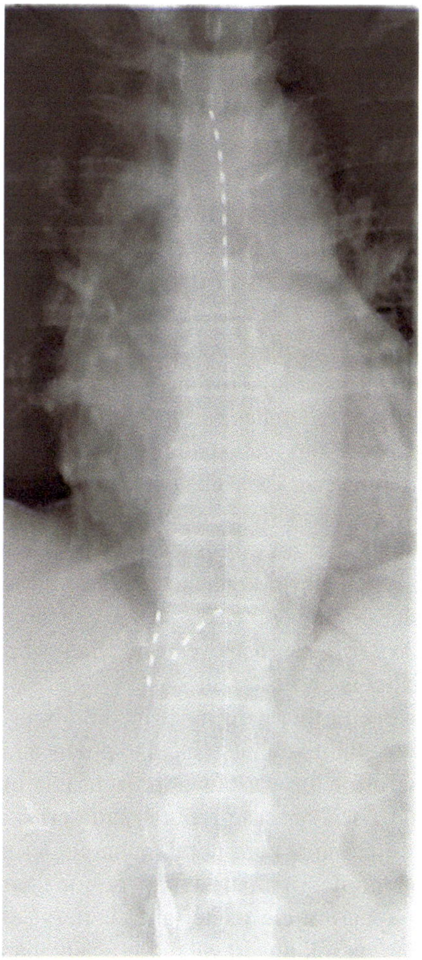

Fig. 20.5 Combined SCS and splanchnic leads in situ

Using a Combined Peripheral and Spinal Cord Stimulation Approach

The peripheral or spinal cord approaches do not have to be mutually exclusive and many cases of combined approaches have been described in the literature. In this case, we used this dual approach in a 63-year-old lady with severe abdominal and flank pains secondary to chronic non-alcoholic pancreatitis. She had a poor response to conventional medical management including opioids, celiac and splanchnic nerve blockade. A trial of splanchnic nerve neuromodulation was successful using two monopolar electrodes and she proceeded to full implantation. Two quadripolar electrodes were inserted at the level of T11 and T12 in a similar manner to the case above and in addition, a octopolar electrode was inserted in the epidural space to a level of T10 (Fig. 20.5.).

An EON mini was used as the implantable pulse generator. This provided excellent pain relief and she uses the stimulator three to four times a week for approximately 4 h at a time. The leads remained stable at 17-month follow-up (Fig. 20.5).

Current Indications for Using Peripheral Neuromodulation and Future Applications

The indications for using peripheral stimulation for abdominal visceral pain are similar to that of other Neuromodulation techniques. Pain relief has been demonstrated in chronic pancreatitis, post-surgical abdominal pain and loin pain haematuria syndrome. For pains with a specific nerve distribution or when no dermatological pattern is present the peripheral approach may be successful when spinal cord stimulation is not. However, the techniques can be combined and addition of peripheral electrodes to existing spinal cord stimulator systems can achieve pain relief when use of a single modality is unsuccessful.

Future work in this field must include both innovative approaches to stimulation itself and to produce high-quality evidence base for the various approaches as much of the literature is based on case reports and series at present. It is of note that most peripheral technologies are derived from spinal cord stimulation and regional anaesthesia. Peripheral specific electrode and battery design will improve stimulation coverage and efficacy, battery issues and morbidity.

Conclusion

Peripheral nerve stimulation is amongst the fastest growing aspects of Neuromodulation. Its promise in treating areas of pain that is not possible with spinal cord stimulation and the exciting possibility of treating pain of nociceptive origin justifies its attractions to an increasing number of clinical investigators. The previous restriction of requiring surgical access for placement of implanted electrodes has now been reduced with the development of percutaneous techniques alongside improved imaging technologies.

Despite the advances in peripheral nerve stimulation for painful chronic conditions, its mechanism of action has not been as well studied as for spinal cord stimulation and due to the heterogeneity in techniques and methods of implantation there remains a lack of high-quality evidence in its use. The challenges for the future of peripheral nerve stimulation are not only in simplifying and improving the technical aspects of implantation but also to show that the method itself has a high-quality evidence base to justify it becoming a standard and mainstream treatment for chronic pain.

For chronic abdominal pain, the novel use of subcutaneous stimulation allows a very simple method of controlling pain from the abdominal wall and holds promise in its use in deeper, visceral pain via peripheral modulation of central pathways. In contrast, directly stimulating the splanchnic nerves may provide the clinician with an alternative to spinal cord stimulation for visceral pain that has been refractory to other treatments. The use of these techniques of course is not mutually exclusive to spinal cord stimulation or other neuromodulatory techniques and their combination is a future topic for research and discussion.

References

1. Gildenberg PL. History of neuromodulation for chronic pain. Pain Med. 2006;7:S7–13.
2. Althaus J. Dilatation of the stomach. In: The value of electrical treatment. Paternoster Row: Longmans & Co.; 1895. p. 79–80.
3. Melzack R, Wall P. Pain mechanisms: a new theory. Science. 1965;150:971–8.
4. Wall P, Sweet WH. Temporary abolition of pain in man. Science. 1967;155:108–9.
5. Sweet WH, Wepsic JG. Treatment of chronic pain by stimulation of fibres of primary afferent neurons. Trans Am Neurol Assoc. 1968;93:103–7.
6. Shealy CN, Mortimer JT, Reswick JB. Electrical inhibition of pain by stimulation of the dorsal columns: preliminary clinical report. Anesth Analg. 1967;46(4):489–91.
7. Nielson KD, Watts C, Clark WK. Peripheral nerve injury from implantation of chronic stimulating electrodes for pain control. Surg Neurol. 1976;5:51–3.
8. Weiner RL, Reed KL. Peripheral neurostimulation for control of intractable occipital neuralgia. Neuromodulation. 1999;2:217–21.
9. Dunteman E. Peripheral nerve stimulation for unremitting ophthalmic post-herpetic neuralgia. Neuromodulation. 2002;5(1):32–7.
10. Yakovlev AE, Resch BE. Treatment of chronic intractable atypical facial pain using peripheral subcutaneous field stimulation. Neuromodulation. 2010;13:137–40.
11. Yakovlev AE, Peterson AT. Peripheral nerve stimulation in treatment of intractable post herpetic neuralgia. Neuromodulation. 2007;10(4):373–5.
12. Stinson LW, Roderer GT, Cross NE, David BE. Peripheral subcutaneous neurostimulation for control of intractable post-operative inguinal pain: a case report series. Neuromodulation. 2001;4(3): 99–104.
13. Kothari S, Goroszeniuk T. Percutaneous permanent electrode implantation to ulnar nerves for upper extremity chronic pain: 6 year follow up. Reg Anesth Pain Med. 2006;31(5 Suppl):16.
14. Siddalingaiah V, Goroszeniuk T, Palmisani S, Bagchi S, Sanderson K, Al-Kaisy A. Permanent percutaneous sciatic nerve stimulation for treatment of severe neuropathic pain. 13th World Congress on pain, Montreal, Canada, 29 August–2 September 2010.
15. Goroszeniuk T, Kothari S, Hamann W. Percutaneous implantation of a brachial plexus electrode for management of pain syndrome caused by a traction injury. Neuromodulation. 2007;10(2):148–55.
16. Goroszeniuk T, Kothari S, Hamann W. Subcutaneous neuromodulating implant targeted at the site of pain. Reg Anesth Pain Med. 2006;31(2):168–71.
17. Sator-Katzenschlager S, Fiala K, Kress HG, Kofler A, Neuhold J, Kloimstein H, Ilias W, Mozes-Balla EM, Pinter M, Loining N, Fuchs W, Heinze G, Likar R. Subcutaneous target stimulation

(STS) in chronic noncancer pain: a nationwide retrospective study. Pain Pract. 2010;10(4):265–377.
18. Goroszeniuk T, Pang D, Al-Kaisy A, Sanderson K. Subcutaneous target stimulation—peripheral subcutaneous field stimulation in the treatment of refractory angina: preliminary case reports. Pain Pract. 2011;11(2):1–9.
19. Buiten MS, DeJonste MJL, Beese U, Kliphuis C, Durenkamp A, Staal MJ. Subcutaneous electrical nerve stimulation: a feasible and new method for the treatment of patients with refractory angina. Neuromodulation. 2011;14:258–65.
20. Grundy D. Neuroanatomy of visceral nociception: vagal and splanchnic afferent. Gut. 2002;51 Suppl 1:2–5.
21. Naidoo N, Partaba P, Singh B, Moodley J, Satyapal KS. Thoracic splanchnic nerves: implications for splanchnic denervation. J Anat. 2001;199:585–90.
22. Al-Chaer ED, Willis WD. Neuroanatomy of visceral pain: pathways and processes. In: Pasricha PJ, Willis WD, Gebhart GF, editors. Chronic abdominal and visceral pain theory and practice. New York: Informa Healthcare; 2007. p. 33–43.
23. Qin C, Chen JD, Zhang J, Foreman RD. Modulatory effects and afferent pathways of gastric electrical stimulation on thoracic spinal neurons receiving input from stomach in rats. Neurosci Res. 2007;57:29–39.
24. Goroszeniuk T, Khan R, Kothari S. Lumbar sympathetic chain neuromodulation with implanted electrodes for long term pain relief in loin pain haematuria syndrome. Neuromodulation. 2009;12(4):284–91.
25. Goroszeniuk T. Short neuromodulation trial in neuropathic pain produces varying duration but reproducible pain relief. Proceedings of the 4th Congress of EFIC, September 2–6, Prague, Czech Republic; 2003. p. 326.
26. Al Tamimi M, Davids HR, Barolat G, Krutsch J, Ford T. Subcutaneous peripheral nerve stimulation treatment for chronic pelvic pain. Neuromodulation. 2008;11:277–81.
27. Sluka KA, Deacon M, Stibal A, Strissel S, Terpstra A. Spinal blockade of opioid receptors prevents the analgesia produced by TENS in arthritic rats. J Pharmacol Exp Ther. 1999;289:840–6.
28. Kalra A, Urban MO, Sluka KA. Blockade of opioid receptors in rostral ventral medulla prevents antihyperalgesia produced by transcutaneous electrical nerve stimulation (TENS). J Pharmacol Exp Ther. 2001;298:257–63.
29. Radhakrishnan R, Sluka KA. Spinal muscarinic receptors are activated during low or high frequency TENS-induced antihyperalgesia in rats. Neuropharmacology. 2003;45:1111–9.
30. Radhakrishnan R, King EW, Dickman J, Herold CA, Johnston NF, Spurgin ML, Sluka KA. Blockade of spinal 5-HT receptor subtypes prevents low, but not high, frequency TENS-induced antihyperalgesia in rats. Pain. 2003;105:205–13.
31. Maeda Y, Lisi TL, Vance CG, Sluka KA. Release of GABA and activation of $GABA_A$ receptors in the spinal cord mediates the effects of TENS in rats. Brain Res. 2007;1136:43–50.
32. Sluka KA, Lisi TL, Westlund KN. Increased release of serotonin in the spinal cord during low, but not high, frequency transcutaneous electric nerve stimulation in rats with joint inflammation. Arch Phys Med Rehabil. 2006;87:1137–40.
33. Chandran P, Sluka KA. Development of opioid tolerance with repeated transcutaneous electrical nerve stimulation administration. Pain. 2003;102:195–201.
34. Han JS, Chen XH, Sun SL, Xu XJ, Yuan Y, Yan SC, Hao JX, Terenius L. Effect of low and high frequency TENS on met-enkephalin-arg-phe and dynorphin A immunoreactivity in human lumbar CSF. Pain. 1991;47:295–8.
35. Salar G, Job I, Mingrino S, Bosio A, Trabucchi M. Effect of transcutaneous electrotherapy on CSF β-endorphin content in patients without pain problems. Pain. 1981;10:169–72.
36. Matharu M, Bartsch T, Ward N, Frackowiak R, Weiner R, Goadsby P. Central neuromodulation in chronic migraine patients with suboccipital stimulators: a PET study. Brain. 2004;127:220–30.
37. Ignelzi RJ, Nyquist JK. Direct effect of electrical stimulation on peripheral nerve evoked activity: implications in pain relief. J Neurosurg. 1976;45:159–65.
38. Sandkühler J, Chen JG, Cheng G, Randić M. Low-frequency stimulation of afferent Adelta-fibers induces long-term depression at primary afferent synapses with substantia gelatinosa neurons in the rat. J Neurosci. 1997;17(16):6483–91.
39. Randić M, Jiang MC, Cerne R. Long-term potentiation and long-term depression of primary afferent neurotransmission in the rat spinal cord. J Neurosci. 1993;13(12):5228–41.
40. Ellrich J, Lamp S. Peripheral nerve stimulation inhibits nociceptive processing: an electrophysiological study in healthy volunteers. Neuromodulation. 2005;8(4):225–32.
41. Linderoth B, Meyerson B. Spinal cord stimulation: exploration of the physiological basis of a widely used therapy. Anesthesiology. 2010;113(6):1265–7.
42. Linderoth B. Spinal cord stimulation: a brief update on mechanism of action. Eur J Pain Suppl. 2009;3(2):89–93.
43. Qin C, Chen JDZ, Zhang J, Foreman RD. Somatic afferent modulation of thoracic (T9-T10) spinal neurons receiving gastric mechanical input in rats. Neuromodulation. 2010;13:77–86.
44. Butterworth JF. Case reports: unstylish but still useful sources of clinical information. Reg Anesth Pain Med. 2009;34(3):187–8.
45. Paicius RM, Bernstein CA, Lempert-Cohen C. Peripheral nerve field stimulation in chronic abdominal pain. Pain Physician. 2006;9:261–6.
46. Abejón D, Deer T, Verrills P. Subcutaneous stimulation: how to assess optimal implantation depth. Neuromodulation. 2011;14:343–8.
47. Pang D, Goroszeniuk T, O'Keeffe D, Shetty A, Racz G. Novel stimulating needle for peripheral and subcutaneous neuromodulation. 6th World Congress-World Institute of Pain. Pain Practice, 2012;12:1–199. doi:10.1111/j.1533-2500.2011.00528.x.
48. Goroszeniuk T, Pang D, Kothari S. Targeting and fielding: not the same! Neuromodulation. 2011;14:569–70.
49. Abejón D, Krames ES. Peripheral nerve stimulation or is it peripheral subcutaneous field stimulation; what is in a moniker? Neuromodulation. 2009;12:1–4. doi:10.1111/j.1525-1403.2009.00192.x.
50. Goroszeniuk T, Kothari S. Targeted external area stimulation. Reg Anesth Pain Med. 2004;29(5):98.
51. Kothari S, Goroszeniuk T. External neuromodulation as a diagnostic and therapeutic procedure. Eur J Pain. 2006;10 Suppl 1:S158.
52. Goroszeniuk T, Pratap N, Kothari S, Sanderson K. An algorithm for peripheral neuromodulation in neuropathic pain. Proceedings of 8th World Congress of INS and 11th Annual Meeting of NANS, December 9–12, 2007b, Acapulco, Mexico. p. 25–26, p. 169–170.
53. Goroszeniuk T, Pratap N, Sanderson K, Kothari S. Therapeutic trial of peripheral neuromodulation using inexpensive monoelectrode temporary stimulating catheters. Proceedings of 8th World Congress of INS and 11th Annual Meeting of NANS, December 9–12, 2007c, Acapulco, Mexico. p. 150.
54. Goroszeniuk T, Khan R. Permanent percutaneous splanchnic nerve neuromodulation for management of pain due to chronic pancreatitis: a case report. Neuromodulation. 2011;14(3):253–7.

21 Gastric and Other Visceral Stimulation for Chronic Painful Gastrointestinal Motility Disorders

Yan Sun and Jiande Chen

Introduction

Gastrointestinal (GI) electrical stimulation has been introduced as a possible treatment for gastric motor dysfunction as early as in 1963. Bilgutay et al. firstly stimulated the gut of both dogs and humans using electrical stimuli consisting of pulses at a frequency of 50 Hz and observed augmented gastric contractions fluoroscopically and increased gastric emptying but did not record either electrical or mechanical activity [1]. In the late 1960s and early 1970s, experiments, primarily in the canine model, began to elucidate the nature of GI myoelectrical activity and its relation to contractile activity [2–4]. Since that time and before 1990s, there have been more reports on the applications of electrical stimulation to affect GI motility both acutely and chronically [5–7]. Over the past decade, a great progress has been made on the effects, mechanisms, and clinical applications of gastric electrical stimulation (GES). Electrical stimulation of various organs of the GI tract, such as stomach, small intestine, colon, and rectum for the treatment or therapeutic potentials of various conditions, such as gastroparesis, short bowel syndrome, intestinal pseudo-obstruction, and fecal incontinence have been reported in the literature. Recently, GES has also been introduced for the treatment of obesity. Clinical trials have reported significant weight loss in obese patients with GES using train of short pulses [8–10]. However, multicenter double-blinded placebo-controlled trials have failed to show significant weight loss in comparison with sham-GES [11].

Y. Sun, M.D., Ph.D.
Department of Internal Medicine, Texas Tech University Health Science Center, 4800 Alberta Avenue, El Paso, TX 79905, USA

J. Chen, Ph.D. (✉)
Division of Gastroenterology and Hepatology, Johns Hopkins Medicine, 4940 Eastern Ave, A-505, Baltimore, MD 21224, USA

Ningbo Pace Translational Research Center, Beilun, Zhejiang, Ningbo 315000, China
e-mail: Jiandedzchen@gmail.com

Although numerous studies have reported the treatment potentials of GI electrical stimulation for GI motor dysfunctions and obesity, much less is known about the potential of GI stimulation on visceral pain. Recently, a number of experiments have been performed to study the effects and mechanisms of GES on visceral pain in animal models.

Functional Gastrointestinal Disorders

Abdominal pain and/or discomfort is/are the major complaint(s) in patients with functional gastrointestinal disorders (FGIDs), such as functional dyspepsia (FD) and irritable bowel syndrome (IBS). FGIDs are clinical syndromes defined by chronic or recurrent abdominal symptoms without identifiable cause using conventional diagnostic measures [12]. In FD and IBS, symptoms are frequently related to meals and can also include bloating, early satiety, fullness, belching, and nausea, in addition to abdominal pain. Various pathophysiological mechanisms may account for chronic visceral pain, including visceral hypersensitivity to distension, gastrointestinal motor abnormalities and autonomic dysfunction, as well as disturbed central nervous function [13–16]. However, chronic visceral pain of FGIDs is still poorly understood and there is a lack of an effective therapy. Drugs used in treating visceral pain include opioid agents, low-dose tricyclic antidepressants and 5-HT reuptake inhibitors, etc. [17]. But these drugs are not satisfactory in all subtype patients and the long-term effects are still unclear. Other methods, such as hypnosis, cognitive-behavior therapy, electroacupuncture, and other biological agents need further investigation. Thus, it is important to explore novel therapeutic methods for visceral pain in patients with FGIDs.

Motor Function Abnormalities

Abnormal gastric motility may cause secondary abdominal discomfort or pain in FGIDs. For FD patients, main gastric

L. Kapural (ed.), *Chronic Abdominal Pain: An Evidence-Based, Comprehensive Guide to Clinical Management*,
DOI 10.1007/978-1-4939-1992-5_21,

motor dysfunctions include: (1) antral hypomotility and delayed gastric emptying. But the lack of a consistent correlation between symptoms and the detected abnormalities casts doubts on their primary role in symptom production [18]; (2) gastric dysrhythmias or abnormalities in gastric pacemaking activities. About 36 % of FD patients and 25 % of IBS patients display gastric dysrhythmias which have been linked to delayed gastric emptying, but not to symptoms [19]. However, gastric dysrhythmias are unlikely to have a primary pathophysiological role in FD and, in isolation, cannot account for symptoms in the majority of patients with FD; (3) abnormal gastric tone and accommodation. During eating, a vagally-mediated transient-receptive relaxation occurs in the stomach, followed by a more prolonged relaxation known as accommodation. There is evidence that gastric relaxation is impaired in FD [13]. A relationship between impaired gastric accommodation and dyspeptic symptoms has been reported in a subgroup of FD patients with meal-related symptoms.

For IBS, only 25–75 % patients exhibit the motility "abnormalities". In the ileum, colon, and rectum, IBS patients show an exaggerated response to a variety of provocative stimuli including meals, distension, stress, cholecystokinin, neostigmine, and corticotropin-releasing hormone injection [20]. No corresponding pattern of hyperreactivity has been shown in the proximal small intestine or stomach, where the response to stress (inhibition of contractions) differs from the response to meals (increase in contractions). Motility abnormalities may interact with low-sensory thresholds to produce symptoms: delayed transit of gas causes greater abdominal perception in IBS and IBS patients are more likely than healthy controls to perceive the occurrence of normal migrating motor complexes [15].

Visceral Hypersensitivity

A common feature of FGIDs is that patients often display a heightened sensitivity to experimental gut stimulation, termed visceral hypersensitivity. This may involve abnormal sensory receptor sensitization, abnormal central pain processing, autonomic nerve dysfunction, or a combination of these. A proportion of patients with FD and IBS have selective visceral hypersensitivity to mechanical distension and this has been correlated with the specific symptoms of belching, pain, and weight loss. However, a significant proportion of patients have no hypersensitivity to distension, suggesting that other factors must also be important in symptom production.

(1) *Peripheral and central sensitization.* Noxious stimuli may cause the peripheral release of several inflammatory mediators, such as adenosine triphosphate, 5-hydroxytryptamine 5-HT), bradykinins, prostaglandins, the transient receptor potential vallinoid (TRPV) receptors 1 and 4, the protease activated receptor 2 (PAR(2)), nitric oxide (NO), mast cells, etc. [21–26] induce peripheral sensitization of nociceptive afferent nerves by reducing their transduction thresholds and by inducing the expression and recruitment of previously silent nociceptors. A number of ion channels, such as voltage-gated sodium channels (VGSCs), neurotransmitter receptors, and trophic factors have also been implicated in the development of peripheral sensitization [27]. The main consequence of these inflammatory mediators is an increase in pain sensitivity at the site of injury known as primary hyperalgesia [28].

    Studies have shown that prostaglandin E2 (PGE2) and the *N*-methyl D-aspartate (NMDA) receptor are the most importance molecular factors in the development of central sensitization at the spinal dorsal horn [29, 30]. Human pharmacological studies have demonstrated that antagonism of the PGE2 or the NMDA receptor prevents the development of central sensitization and antagonism of the NMDA receptor with ketamine may even reverse established visceral hypersensitivity [30, 31]. Central sensitization may also occur after a noxious stimulus is applied to an anatomically distant site. For instance, esophageal sensitization may occur after a noxious stimulus is applied to the duodenum and balloon distension in the left colon may result in rectal sensitization [32]. In patients with IBS, following repetitive distension of the sigmoid colon, central sensitization may ensue as manifested by rectal hyperalgesia and increased viscerosomatic referral to experimental rectal distension [33].

(2) *Central abnormal processing of nociceptive stimuli.* Peripheral and central sensitizations are not exclusive entities in explaining visceral hypersensitivity in humans. Central processing of nociceptive input involves a number of cortical and subcortical brain structures. Functional neuroimaging, such as functional magnetic resonance imaging (fMRI), positron emission tomography (PET), magnetoencephalography (MEG), electroencephalography (EEG), and cortical-evoked potentials (CEPs) have facilitated the examination of the complete neuraxis implicated in central areas. For example, Mayer et al. found in an fMRI study, that in response to experimental rectosigmoid distension, IBS patients have inadequate activation in the subcortical brain regions involved with affective-emotional aspects of pain perception such as the limbic system, the periaqueductal gray (PAG) matter and thalamic regions [34]. Abnormal areas of activation have been observed in other areas such as the anterior cingulate cortex [35], amygdala and brainstem in IBS patients, suggesting that the aberrant visceral nociception observed in this group may be, in part, due to central mechanisms [36, 37]. Future studies using a combination of these functional imaging

techniques, in conjunction with improvements in study design, will no doubt further advance our understanding of the mechanisms involved.

(3) *Autonomic nerve dysfunction.* Central communication to the GI tract is via the parasympathetic nervous system (PNS) and sympathetic nervous system (SNS) pathways of the efferent autonomic nervous system (ANS). In a number of syndromes where chronic pain is a feature, such as IBS, it has been observed that autonomic dysfunction may coexist [38]. An important methodological consideration in the interpretation of results from ANS studies is heart rate variability (HRV), as a surrogate marker of ANS parameters truly reflects specific gut autonomic innervation. Most studies reported no difference in HRV when the IBS population was compared to healthy controls. However, dividing the IBS sample into subgroups—according to their predominant bowel symptoms, the severity of clinical course, the presence of depressive symptoms, or a history of abuse in the past—revealed changes in autonomic functioning [38].

Gastrointestinal Electrical Stimulation for FGIDs

GI electrical stimulation consists of a series of pulses, usually in a rectangular shape with a constant current or a constant voltage. Several stimulation parameters are involved in electrical stimulation, including frequency, pulse width, and amplitude. Various methods of GI electrical stimulation include long-pulse stimulation, short-pulse stimulation, and stimulation with trains of pulses. Based on the number of stimulation electrodes, GES can be classified into single-channel and multichannel GI electrical stimulation. Long-pulse stimulation is composed of repetitive single pulse with a pulse width in the order of milliseconds (10–600 ms), and a stimulation frequency in the vicinity of the physiological frequency of the gastric slow waves. The short-pulse stimulation is in the order of a few hundred microseconds (μs). The stimulation frequency is usually a few times higher than the physiological frequency of the gastric slow wave. In the trains of short-pulses stimulation, the stimulus is composed of repetitive trains of short pulses and is derived from the combination of two signals: (1) continuous short pulses with a high frequency (in the order of 5–100 Hz) and (2) a control signal to turn the pulses on and off [39].

Currently, there are a few new stimulation methods that have been developed, including synchronized electrical stimulation, and dual-pulse electrical stimulation [39]. In synchronized stimulation, stimuli are synchronized with the intrinsic slow waves. In the method of dual-pulse stimulation, the stimulus is composed of a short pulse (in the order of a few hundred microseconds), followed by a long pulse (in the order of a few hundred milliseconds), and stimulation was delivered at two different locations.

GI electrical stimulation has been applied in both humans and animals to modulate GI sensory-motor functions.

Gastric and Intestinal Electrical Stimulation for Motility Disorders

GI Electrical Stimulation Alters Myoelectrical Activity

In normal canine study, gastrointestinal pacing or entrainment with long-pulse electrical stimulation is achievable in the stomach and small intestine but not the colon, and the maximal entrainable frequency of the gastric and small intestinal slow waves is about 20 % higher than the intrinsic frequency [40]. The entrainment of slow waves with gastrointestinal pacing is not modulated by the vagal or sympathetic pathways, suggesting a purely peripheral or muscle effect [40]. Other studies showed that GES with long-pulse/low-frequency normalized vasopressin- or glucagon-induced gastric dysrthythmia in dogs and STZ-induced diabetic rats [41, 42]. Normalization of dysrhythmia was also reported in patients with gastroparesis and postsurgical patients using the same method of GES [43]. Cutaneous GES is also capable of altering gastric slow waves and inhibiting gastric motility, which may have therapeutic potential for treating eating disorders, such as obesity [44]. Intestinal electrical stimulation (IES) with long pulses has also been shown to entrain intestinal slow waves and normalize intestinal dysrhythmias [40, 45].

GI Electrical Stimulation Modulates GI Motor Dysfunction

A single-channel GES with long pulses has no effects on gastric emptying in healthy dogs but is capable of improving gastric emptying in animal and patients with gastroparesis [42, 43, 46]. Whereas, two- or four-channel GES, with long pulses, is able to improve gastric emptying in both healthy and diseased model of canines [47, 48]. Similar results were also observed with a multichannel microprocessor-controlled sequential GES with pulse-trains [49]. The novel methods of 2-channel dual-pulse GES or synchronized GES have been reported to improve antral contractions, and accelerated gastric emptying of liquid or solid, improve dysrhythmia and emetic responses in diabetics or vasopressin canine models, which is mediated via the cholinergic pathway [50, 51]. For gastric tone, with low-stimulation energy, GES with long pulses may change the gastric tone slightly, which may be beneficial to patients with an impaired gastric relaxation. With high-stimulation energy, GES could substantially inhibit gastric tone and result in a substantial distension of the stomach which may actually lead to an early satiety and be applied for treating obesity [52, 53].

IES may have multiple effects on gastrointestinal functions, including gastric emptying, small-bowel contractions and transit, nutrient absorption, and feedback signaling of satiety to the central nervous system. IES was reported to reduce gastric tone via the nitrergic pathway, inhibit antral contractions via the adrenergic pathway, and delay gastric emptying of liquid [54, 55]. Long-pulse IES with higher stimulation energy significantly inhibited intestinal contraction by 60–74 % (40–220 cm distal to the stimulation electrodes), mediated via sympathetic but not nitrergic, serotoninergic 5-HT_3, or opiate pathway [56]. The inhibitory effects were dependent on pulse width and amplitude. Whereas, short-pulse IES mainly reduces vomiting and nausea induced by vasopressin in dogs [57], improves duodenal distension-induced delayed gastric emptying, and prevents duodenal distension-induced vomiting and discomfort signs [58]. Synchronized IES (SIES) induces intestinal contractions and accelerates small intestinal transits delayed by glucagons via the cholinergic pathway [59]. In rodents, IES accelerates whole gut transit and promotes fat excrement, and these effects are mediated through the cholinergic nerves [60]. The location of IES also plays an important role on intestinal transit and absorption. IES via electrodes placed in the distal small intestine (backward IES) delays intestinal transit and increases absorption in a canine model of short bowel and dumping syndrome [61]; When IES is delivered via the electrodes placed in the proximal intestine (forward IES), it accelerates intestinal transit slowed by ileal brake [62].

Colon electrical stimulation (CES) has an excitatory effect on colonic transit and this excitatory effect may be mediated via the nitrergic pathway [63]. However, CES inhibited gastric emptying and small intestinal motility [64]. CES also inhibited gastric tone via sympathetic pathway [65, 66]; and inhibited rectal tone mediated by the nitrergic pathway [67].

Modulation Pathways

Vagal and sympathetic afferent and/or efferent pathways are involved in the regulation of GES on gastric motility. Studies using systemic administration of atropine, vagotomy, or spectral analysis of heart rate variability in dogs have suggested a possible involvement of the vagal pathway in the regulation of GES on gastric motility [68, 69]. A recent study in rats has shown that GES can activate single vagal afferent fibers [70]. Also, IES has been reported to alter intestinal slow waves, contractions, and transit mediated via both vagal and adrenergic pathways. In addition, nitrergic, cholinergic, sympathetic pathways are involved in the GI electrical stimulation effects on GI motor function (discussed above).

GI Electrical Stimulation and GI Sensory Function

There are few studies investigating the effect of GI electrical stimulation on visceral sensitivity. One of our recent studies found the GES with long pulses reduced visceral sensitivity to gastric distension in healthy canine, and visceral sensitivity to gastric distension was mediated via vagal and sympathetic pathways [71]. Another study showed that GES with high-frequency (100 Hz) pulse trains decreased the EMG response to gastric distension in gastric ulcer rats [72]. According to these findings, GES may have a therapeutic potential for visceral hypersensitivity. However, currently there is no clinical data.

Gastrointestinal Afferent in Central Nerve System

The gut is richly innervated with afferent fibers of the vagal (parasympathetic) and splanchnic nerves (sympathetic). The spinal cord plays an important role in chronic visceral pain as this structure is not only responsible for the communication between the peripheral system and the brain, but is also responsible for plastic changes that can modulate pain sensation. Visceral sensory axons that are comprised mostly of thinly myelinated Ab fibers and unmyelinated C-fibers inflow to the spinal cord dorsal column (specifically laminae I, II, V, and X where the sensory fibers terminate) in viscerosensory processing [14, 73–76]. Neurons excited by gastric distension were located in both superficial and deeper lamina. Acid applied either to the mucosa or serosa of the stomach causes a small stimulation of c-fos transcription in laminae I and II but not in the deeper laminae of the caudal thoracic spinal cord. A recent review showed that an area of the superficial dorsal horn, the substantia gelatinosa, might undergo plastic changes in the setting of chronic pain [77]. Not only local modulatory changes, but also the supraspinal structures, can modify the activity of the substantia gelatinosa under certain pathologic conditions and thus affect pain transmission in the spinal cord by activating "top-down" descending facilitatory systems [78].

The nociception input relay centrally to the somatosensory and limbic cortices in brain centers [79]. Projections to the hypothalamus are responsible for some autonomic reflexes and emotional response to pain [79]. Positron emission tomography studies have demonstrated that mechanical gastric distension in healthy volunteers causes activation of the brainstem periaqueductal area, thalami, caudate nuclei, and cerebellum, limbic and occipital cortices [80]. On the other hand, descending fibers from the cerebral cortex and hypothalamus can modulate afferent input at dorsal horn level. This modulation is mainly inhibitory [16]. An abnormality in central pain processing could interfere with the balance between descending inhibition and facilitation of nociceptive afferents, thereby leading to visceral hypersensitivity.

The nucleus of solitary tract (NTS) in the brain stem is an important central relay of vagal sensory afferents with cell bodies in nodose ganglia and neural endings in the stomach and/or intestine. Some NTS neurons receive afferent inputs from the stomach and respond to gastric and/or intestinal mechanical, thermal and chemical stimuli [81, 82]. Furthermore, gastric-afferent impulses arriving at NTS via the vagus nerve contribute to the visceral perception (e.g. satiety, nausea, and discomfort) and regulatory function (e.g. absorption, secretion, and motility) [83].

Central Nerve System Responses to GI Distension

Spinal cord T9–T10 segments received the sensory-afferent fibers from stomach. The neurons can be divided into different types according to their response to gastric distension (GD), such as Low-threshold neurons (LT, response to low volume GD), high-threshold neurons (HT, response to noxious GD). The HT spinal neurons are important for intraspinal processing of stomach noxious stimulation associated with visceral pain. LT neurons, on the other hand, might relate to nonpainful sensations (e.g., fullness, bloating, nausea) that arise from the stomach. In normal rats, 70 % spinal LT neurons and only 29 % HT neurons responded to GD [84]. However, in the gastric ulcer rats, less LT neurons (47.2 %) and more HT neurons (52.8 %) responded to GD [72]. This difference might be due to the different intraspinal signal processing mechanisms for gastric afferent information in gastric ulcer rats and the normal rats.

Spinal T9–T10 segments also received the sensory information from duodenum stimuli. Study showed 28 % T9–T10 spinal neurons responded to noxious duodenum distension (DD) (0.4 mL, 20 s) [85]. Of these, 6 % neurons had low-threshold responses to DD ($\leq$0.2 mL) and 22 % had high-threshold responses to DD ($\geq$0.4 mL). DD-responsive spinal neurons were encountered more frequently in deeper (depth: 0.3–1.2 mm) than in superficial laminae (depth: <0.3 mm) of the dorsal horn.

The neurons in spinal L6-S2 segments are activated by colon-rectal distension (CRD). Majority cells also received convergent somatic input from the scrotum, perianal region, hindlimb, and tail. Ness and Gebhart [86] Most cells were excited or excited/inhibited by CRD; and less (27 %) cells were inhibited or inhibited/excited by CRD. C1-2 cells activated with glutamate primarily produced inhibition of evoked responses to visceral stimulation of lumbosacral spinal cells. Inhibition resulting from activation of cells in C 6-7 segments required connections in the upper cervical segments. These results provide evidence that upper cervical cells integrate information that modulates activity of distant spinal neurons responding to visceral input [87].

In the brain central nerve system, a systematic c-Fos study showed low-intensity GD-induced c-Fos expression in the cranial part of nucleus of solitary tract (NTS), the nucleus ambiguus (NA), the lateral reticular area (LRt), and the ventrolateral medulla (RVL/CVL). High-intensity GD stimulation induced c-Fos expression in area postrema (AP), the lateral vestibular nucleus (LVe), and the caudal part of the NTS. Increasing the frequency of stimulation-induced c-Fos expression in further nuclei such as the parabrachial nucleus (PBN), the inferior olive subnuclei, the oral part of spinal trigeminal nucleus (Sp5O) and locus coeruleus, but decreased in NTS and LRt, disappeared in VLM and increased in NA. It was shown that 69 % NTS neurons responsive to GD responded to duodenal distension; effects of intestinal afferent input on NTS neurons are primarily excitatory [88].

Effects of GI Electrical Stimulation on Neurons in Central Nerve System

GES with different parameters have different effects on different central nerve neurons. In normal rat, GES was found to have excitatory effects on spinal cord neurons, and these effects were strengthened with an increased pulse width and/or pulse frequency (0.3–300 ms, 14–40 Hz). The modulatory effect of GES involved thoracic spinal-afferent fibers containing vanilloid receptor-1 [84, 89]. Similar results were also found in NTS (nucleus of solitary tract) neurons. GES was noted to have an excitatory effect on NTS neurons receiving input from the stomach; and the response to GES was enhanced with increased pulse width and/or amplitude (0.3–200 ms, 6–20 mA). The modulatory effect of GES on the central neurons receiving vagal inputs may contribute to the neural mechanisms of GES therapy for the treatment of patients with gastric motility disorders [89]. T9–T10 spinal neurons that process input from the duodenum might mediate the effects of GES on duodenal sensation and motility [85].

GES has been shown to have an analgesic effect on visceral pain. In a rodent model of gastric hypersensitivity attributed to gastric ulcers, GES with high-frequency (100 Hz) pulse trains inhibited spinal dorsal neuron activity. Moreover, the inhibitory effect was more potent on high threshold [5] neurons than low-threshold [90] neurons, suggesting a possible analgesic effect of GES since the high-threshold neurons are more responsive to pain.

Similarly, IES with different parameters activated 39–72 % of the NTS neurons responsive to gastric distension. The primary effects of IES on neuronal activity in the NTS were mostly excitatory and stimulation energy-dependent. The modulatory effect of IES on the central neurons receiving vagal inputs may contribute to the neural mechanisms of IES therapy for the treatment of patients with obesity and gastrointestinal motility disorders [91]. Another study using nonexcitatory electrical stimulation, which is used in cardiac contractility modulation, showed that nonexcitatory GES-activated distension-sensitive vagal afferents signaling to the central nervous system [70]. Therefore, it is conceivable to presume that IES might activate the duodenal distension-sensitive vagal primary afferents and directly enhance the activity of NTS neurons with convergent inputs from the stomach and the intestine.

Gastric Electrical Stimulation for Nausea and Vomiting

While various methods of GES have been explored, the only clinically available method is called Enterra Therapy, which has received approval from the US Food and Drug Administration (FDA) for the treatment of nausea and vomiting in gastroparesis patients. Enterra Therapy parameters, similar to the short-pulse train stimulation, are composed of two short pulses with an interval of 72 ms (or a frequency of 14 Hz), and the train repeated every 5 s. The pulse width is about 0.3 ms and amplitude is about 5 mA [92]. The FDA approval was given through a "humanitarian device exemption". This regulatory category was established in 1996 and only applies to devices intended to benefit <4,000 patients.

The Enterra gastric stimulation system consists of three main elements: an implantable pulse generator, an implantable lead with a pair of electrodes, and an external remote programming system. The electrode lead is implanted in the muscular layer of the body of the stomach, along the greater curvature, approximately 10 cm from the pylorus. The pulse generator is positioned in a subcutaneous pocket in the left or right upper quadrant. Battery life of the pulse generator is estimated to be at least 5 years, depending on the pulse parameters used. When the battery is depleted, the pulse generator is replaced by local intervention. Long-term studies show that the main complication associated with the device is infection on the subcutaneous pocket, occurring in up to 10 % of all subjects [93, 94].

Enterra Therapy and Gastroparesis

Gastroparesis is one of major GI motility diseases. It is a chronic disorder of diabetic (both type 1 and type 2 diabetes) or idiopathic etiology (approximately 25–30 % of cases are idiopathic). Symptoms of gastroparesis range from early satiety, fullness, bloating, nausea, and vomiting, to dehydration, nutritional deficiency, and poor glycemic control (in diabetics) in severe cases. Abnormalities in gastroparesis may include impaired gastric accommodation, visceral hypersensitivity, gastric dysrhythmia, antral hypomotility and delayed gastric emptying.

Improvement in nausea and vomiting in patients with gastroparesis is the major and most consistent finding reported with Enterra Therapy. Clinical studies have shown that the Enterra Therapy significantly decreases nausea and vomiting in about 60–75 % of patients with drug-refractory diabetic, idiopathic, or postsurgical gastroparesis [95–101]. The most interesting finding in these studies was the dramatic improvement in nausea and vomiting, but pain or bloating [102]. In addition, Enterra GES was found to improve quality of life [98, 100] and nutritional status [93, 103], and reduce healthcare costs and utilization of healthcare facilities [104, 105]. Studies seem to suggest that diabetic patients have a better outcome with the therapy than idiopathic patients [98, 105, 106, 107].

Enterra Therapy Mechanisms

Modulation GI Motor-Sensory Function

The Enterra Therapy has been reported to exhibit no demonstrable effects on the frequency of gastric electrical rhythm. In canine studies, GES with Enterra did not control vasopressin-induced gastric dysrhythmia, although it improved vomiting [108]. Likewise, gastric electrical activity in patients with gastroparesis was not affected by GES with Enterra [109]. The findings on the effect of the Enterra Therapy on gastric emptying are conflicting. Some studies reported an improvement in gastric emptying with chronic GES, whereas, others indicated no such improvement [98, 100, 110].

Gastric tone and accommodation are frequently impaired in patients with functional dyspepsia or gastroparesis [111]. GES with Enterra parameters was noted to slightly but significantly reduce fundic tone and enhance postprandial gastric accommodation in normal dogs [112, 113]. These findings suggest that enhanced gastric accommodation may be one mechanism for improving gastroparetic symptoms.

Not much information is available regarding the effect of the Enterra Therapy on visceral hypersensitivity. Two preliminary studies in humans seem to suggest a potential role of the therapy in altering visceral hypersensitivity. In patients with gastroparesis, GES with Enterra parameters was found to increase the perception threshold of the patients to gastric distension [114], reduced visceral perception to gastric distension via enhancement of gastric accommodation in normal dogs [112].

Neural Mechanisms

The Enterra parameters are similar to those used in nerve stimulation. Central and peripheral neural mechanisms have been explored in humans and animal modes. This antiemetic effect could be mediated via enteric, autonomic, and/or central neural mechanisms.

Vagal-mediated neural mechanisms are believed to be involved in the antiemetic effect of Enterra Therapy. A canine study showed that GES with the Enterra parameters reduced vasopressin-induced emesis in dogs, and that this effect was blocked by vagotomy [108]. However, GES was also shown to improve symptoms in postsurgical gastroparesis, some with vagal disruption [112, 115]. GES with Enterra parameters increased vagal activity in rats, which was abolished with vagotomy or denervation of vagal afferent fibers, suggesting a vagal–vagal reflex [116]. In humans, the Enterra Therapy was reported to alter sympathovagal activities, with the alterations associated with symptom improvement [113]. With the data for increased vagal activity, the researchers speculated that the symptomatic improvement with the Enterra Therapy may be mediated via activation of vagal afferent pathways, thereby influencing central nerve system control mechanisms for nausea and vomiting [113].

Central mechanisms involved in Enterra Therapy have been explored in both humans and animals. In humans, chronic GES with the Enterra device increased activity in the thalamus and caudate nuclei of patients with gastroparesis, as detected by positron emission tomography [21, 113]. In a preliminary study in seven patients with Enterra Therapy the scan of Fluoro-Deoxy Glucose PET revealed at least one-step increase in the color scale (10 %) in bilateral thalamic activity reflecting a substantial upregulation of metabolic activity [117]. In rats, GES with parameters similar to Enterra Therapy was found to activate neurons in the NTS [89] and inhibit the action potentials of those neurons in the paraventricular nucleus of the hypothalamus that received input from the stomach [118]. Recently, GES with Enterra parameters was reported to increase ghrelin mRNA, double the number of ghrelin-positive cells, and increase plasma ghrelin levels in rats; in the arcuate nucleus of the hypothalamus, GES increased c-Fos and agouti-related protein (AgRP) mRNA expression; reduced the number of c-Fos-positive cells throughout the NTS including catecholaminergic neurons [119]. These results suggest that GES with the Enterra parameters may improve appetite via stimulation of main orexigenic pathways, including ghrelin production in the stomach and AgRP in the hypothalamus, as well as by reducing the activity of catecholaminergic brainstem neurons.

In summary, the Enterra Therapy is an effective treatment for nausea and vomiting in patients with severe gastroparesis, especially diabetic gastroparesis. Various mechanisms are involved in the antiemetic effect of the Enterra Therapy, including gastric motor functions, autonomic and central nervous systems. However, these findings did not establish a direct link to nausea and vomiting associated with gastroparesis; therefore, the exact mechanisms involved in the antiemetic effect of Enterra Therapy need further investigations.

Spinal Cord Stimulation for Gastrointestinal Motility Disorders

Spinal cord stimulation (SCS) is an adjustable, nondestructive, neuromodulatory procedure that delivers therapeutic doses of electrical current to the spinal cord for the management of chronic severe pain [120, 121]. The main indication for using this treatment is not only neuropathic pain, but also nociceptive pain, such as in refractory angina, ischemic limb, and peripheral vascular diseases. Initially, there was lack of evidence for the application of SCS for visceral pain. However, recently, there have been a few studies reporting potential applications of SCS for the treatment of abdominal visceral pain.

Spinal Cord and Spinal Cord Stimulation for GI Motility

The neuronal regulation of GI motility involves intrinsic enteric nervous system (ENS), as well as extrinsic vagal and splanchnic nerves to the stomach and intestine. The extrinsic supply is divided into efferent and afferent categories with information carried in parasympathetic and sympathetic nerve tracts, provided by the vagus and the splanchnic nerves. Most efferent parasympathetic and sympathetic fibers terminate in the myenteric plexus and form connections in enteric ganglia, although some sympathetic axons terminate directly on sphincteric smooth muscle.

The GI tract is richly supplied with various sensory receptors, which relay information through afferent fibers. Information from activated sensory receptors is carried in vagal and spinal afferent nerves to the CNS. Most of the fibers in the vagus are afferent and synapse with neurons in the nodose ganglia. Spinal afferent fibers, carried in the splanchnic nerves, have cell bodies in the dorsal root ganglia and synapse in the dorsal horn of the spinal tract, where they activate second-order neurons, which relay information back to the gut or centrally trough ascending tracts. The spinal afferents in the lower six thoracic and the upper three lumbar spinal segments have been shown to transmit painful impulses from the viscera. Furthermore, most of the fibers in the splanchnic nerves are efferent. The predominant neural influence under basal conditions of sympathetic nerve is inhibitory. This has clinical implications, as demonstrated in humans with spinal cord injury. Patients with spinal cord injury frequently suffer from gut dysmotility, including delayed gastric emptying and intestinal transit, depending on the level of injury [122, 123].

The vagus nerves contain three groups of efferent fibers: preganglionic parasympathetic cholinergic nerves, preganglionic cholinergic nerves, and sympathetics from the cervical ganglia. Stimulation of efferent vagal cholinergic neurons principally activates nicotinic receptors within enteric ganglia, exiting motor activity. Whereas, sympathetic innervation from the splanchnic nerves is different from the vagal parasympathetic innervation in that neuronal cell bodies reside outside the wall within the prevertebral ganglia (i.e. celiac, superior, and inferior mesenteric ganglia). Preganglionic cholinergic neurons project from the spinal cord to the prevertebral ganglia, where they synapse through nicotinic receptors. The postganglionic neurons, which are noradrenergic, project to the enteric ganglia through the splanchnic nerves. Noradrenergic innervation from the splanchnic nerves generally inhibits excitatory cholinergic transmission within the myenteric plexus. The physiologic significance of these pathways is exemplified by the long inhibitory intestinal reflexes, which decrease motility through neural arcs involving the prevertebral ganglia [124].

So far, only few studies have investigated the effect of spinal cord stimulation on GI motility. One study has found SCS at T5 and T8 segments (15, 25, 50, 100, 200 Hz, 0.2 ms at 90 % motor threshold for 15 min) normalized gastric emptying and improve upper GI transit in a rodent model of postoperative ileus [125]. Geng-Qing Song et al. [126] systematically studied the SCS effects on GI motility. They found SCS at T9 and T10 segments intensity-dependently increased gastric tone; increased gastric emptying of liquids, and accelerated small intestinal transit in healthy rats; SCS accelerated gastric emptying of solids by about 24 % in healthy rats and by about 78 % in diabetic rats.

Spinal Cord Stimulation for Abdominal Visceral Pain

A number of studies have shown the SCS effects on severe visceral abdominal pain. In the clinical studies, SCS at T5-T7 was reported to reduce pain associated with abdominal angina [127], mesenteric ischemia [128], severe chronic pancreatitis [129–131], and other conditions [132–134]. In a case report, Kapur et al. described relief of abdominal pain associated with colchicine intolerance or resistance in patients with familial Mediterranean fever, with placement of the electrodes at the lower thoracic levels [135].

Recently, in rodent studies, SCS was found to induce a significant depression of the visceral motor reflex (VMR) produced by colorectal distension in both normal rats and those with acetic acid sensitized rats [136]. The suppressive effect of SCS on colonic sensitivity suggests that SCS may have therapeutic potential for the treatment of visceral pain of gastrointestinal origin associated with abdominal cramping and painful abdominal spasms, such as irritable bowel syndrome (IBS). In a case report, Krames and Mousad [137] described a patient treated for IBS who was developing escalating pain and diarrhea. The use of the tripolar SCS at T8 in this patient provided relief of abdominal and thoracic spine pain, regulated bowel habits, and improved the patient's quality of life. So they believed that the use of SCS should be considered as a treatment option in patients with IBS when all conservative treatments failed. Greenwood-Van Meerveld et al. [138] found SCS (50 Hz, 0.2 ms, amplitude 90 % of motor threshold for 30 min) reduced visceral-motor behavior response in a rodent model of post-inflammatory IBS, Yan Sun et al. [139] found that SCS at T9 and T10 decreased visceral pain in a rodent model of gastric hypersensitivity induced by gastric injection of acetic acid. An electrophysiological experiment in the same study showed 100 Hz SCS led to significant hyperpolarization of gastric-specific dorsal root ganglion (DRG) neurons and reduced the amplitude of evoked action potential (AP) in comparison with sham stimulation. They concluded that SCS ameliorates gastric hyperalgesia induced by gastric acetic acid and the inhibitory effect was probably mediated via its modulation on afferent sensory neurons in DRGs.

Mechanisms of SCS on GI Motor-Sensory Functions

Inhibition of Sympathetic Nerve System

The sympathetic nerves carry nociceptive information from the viscera to spinal nerve roots, which makes them a more viable target for SCS. Whereas, the parasympathetic nerves carry their afferents to anterior and posterior vagal trunks and are therefore not amenable to SCS. GI motility is known to be enhanced with the augment of vagal activity and inhibition of efferent sympathetic activity, and inhibited with the withdrawal of vagal activity and activation of sympathetic activity [140–142]. Extrinsically, GI motility is maintained by balancing the vagal and sympathetic activities. According to the spectral analysis of HRV, SCS was reported to decrease sympathetic activity and sympathovagal balance [126]. These results implicated that SCS-induced increase in GI motility might be attributed to the inhibition of sympathetic activity or reduction in sympathovagal balance. The effects of SCS on sympathetic activity have also been reported in other visceral organs studies. In patients with refractory angina, a similar decrease in sympathovagal balance was reported with SCS, also assessed by the spectral analysis of HRV [142]. In a few of animal studies, the peripheral vasodilatory effects of SCS were linked to the inhibition of efferent sympathetic activity [143–145].

Decrease the Spinal Cord Visceral Afferent

In addition to inhibitory effect of SCS to sympathetic activity, current study also showed that SCS inhibited spinal visceral afferent neurons in DRGs, suggesting that SCS may modulate GI motor-sensory function by inhibiting spinal cord visceral afferent. In a rat model of mononeuropathy, SCS was reported to suppress neuronal hyperexcitability of a wide dynamic range of spinal neurons after peripheral nerve lesions [146]. However, the mechanisms of SCS to modulate GI motor-sensory functions are still poorly understood. Further investigation is needed to explore the vagal central and other possible mechanisms.

Summary

In summary, this chapter reviewed: (1) the pathophysiological mechanisms of FGIDs; (2) the effects and mechanisms of GI electrical stimulation on motor dysfunction and visceral pain; (3) the antiemetic effects and mechanisms of the Enterra Therapy; (4) potentials of spinal cord stimulation for treating GI motility dysfunctions and visceral pain.

A number of the animal and clinical studies have suggested promising applications of GI electrical stimulation for painful GI motility disorders. However, clinical studies are still limited up till now and mechanisms are largely unknown, especially the central mechanisms. Currently, only Enterra Therapy has been approved by FDA to reduce nausea and vomit in gastroparesis patients. However, basic and clinical studies are needed to improve the efficacy of therapy by optimizing stimulation parameters and locations of delivery. In addition, less invasive methods are needed to reduce risks associated with potential therapies. The major hindrance in the advancement of GI electrical stimulation includes the invasive nature of the methodology and the lack of implantable device suitable for GI electrical stimulation. Accordingly, a less invasive method of placing stimulation electrodes would be of great significance, such as endoscopical placement of electrodes, and a new generation of implantable stimulator that is able to alter GI motility functions may have to be developed.

Although available data are limited, SCS has a great potential for treating functional GI motility disorders and visceral pain. Basic and clinical studies are needed to explore this unique application of SCS.

References

1. Bilgutay AM, et al. Gastro-intestinal pacing: a new concept in the treatment of ileus. Ann Surg. 1963;158:338–48.
2. Hinder RA, Kelly KA. Human gastric pacesetter potential. Site of origin, spread, and response to gastric transection and proximal gastric vagotomy. Am J Surg. 1977;133(1):29–33.
3. Hermon-Taylor J. The duodenal pacemaker and its role in the organisation of duodenal activity. J R Coll Surg Edinb. 1971;16(5): 268–70.
4. Szurszewski JH, Elveback LR, Code CF. Configuration and frequency gradient of electric slow wave over canine small bowel. Am J Physiol. 1970;218(5):1468–73.
5. Richter 3rd HM, Kelly KA. Effect of transection and pacing on human jejunal pacesetter potentials. Gastroenterology. 1986;91(6):1380–5.
6. Karlstrom L, Kelly KA. Ectopic jejunal pacemakers and gastric emptying after Roux gastrectomy: effect of intestinal pacing. Surgery. 1989;106(5):867–71.
7. Soper NJ, et al. Regulation of canine jejunal transit. Am J Physiol. 1990;259(6 Pt 1):G928–33.
8. Cigaina V. Long-term follow-up of gastric stimulation for obesity: the Mestre 8-year experience. Obes Surg. 2004;14 Suppl 1:S14–22.
9. Cigaina V, Hirschberg AL. Gastric pacing for morbid obesity: plasma levels of gastrointestinal peptides and leptin. Obes Res. 2003;11(12):1456–62.
10. De Luca M, et al. Progress in implantable gastric stimulation: summary of results of the European multi-center study. Obes Surg. 2004;14 Suppl 1:S33–9.
11. Shikora SA, et al. Implantable gastric stimulation for the treatment of clinically severe obesity: results of the SHAPE trial. Surg Obes Relat Dis. 2009;5(1):31–7.
12. Talley NJ, et al. Functional gastroduodenal disorders. Gut. 1999;45 Suppl 2:II37–42.
13. Tack J, et al. Role of impaired gastric accommodation to a meal in functional dyspepsia. Gastroenterology. 1998;115(6):1346–52.
14. Ness TJ, Metcalf AM, Gebhart GF. A psychophysiological study in humans using phasic colonic distension as a noxious visceral stimulus. Pain. 1990;43(3):377–86.
15. Mayer EA, Raybould HE. Role of visceral afferent mechanisms in functional bowel disorders. Gastroenterology. 1990;99(6): 1688–704.
16. Mayer EA, Gebhart GF. Basic and clinical aspects of visceral hyperalgesia. Gastroenterology. 1994;107(1):271–93.
17. Mayer EA, et al. Functional GI disorders: from animal models to drug development. Gut. 2008;57(3):384–404.
18. Timmons S, Liston R, Moriarty KJ. Functional dyspepsia: motor abnormalities, sensory dysfunction, and therapeutic options. Am J Gastroenterol. 2004;99(4):739–49.
19. Leahy A, et al. Abnormalities of the electrogastrogram in functional gastrointestinal disorders. Am J Gastroenterol. 1999;94(4): 1023–8.
20. Drossman DA, et al. AGA technical review on irritable bowel syndrome. Gastroenterology. 2002;123(6):2108–31.
21. Yu S, et al. TRPA1 in mast cell activation-induced long-lasting mechanical hypersensitivity of vagal afferent C-fibers in guinea pig esophagus. Am J Physiol Gastrointest Liver Physiol. 2009; 297(1):G34–42.
22. Jones 3rd RC, Xu L, Gebhart GF. The mechanosensitivity of mouse colon afferent fibers and their sensitization by inflammatory mediators require transient receptor potential vanilloid 1 and acid-sensing ion channel 3. J Neurosci. 2005;25(47):10981–9.
23. Sipe WE, et al. Transient receptor potential vanilloid 4 mediates protease activated receptor 2-induced sensitization of colonic afferent nerves and visceral hyperalgesia. Am J Physiol Gastrointest Liver Physiol. 2008;294(5):G1288–98.
24. Bueno L. Protease activated receptor 2: a new target for IBS treatment. Eur Rev Med Pharmacol Sci. 2008;12 Suppl 1:95–102.
25. Toda N, et al. Modulation of opioid actions by nitric oxide signaling. Anesthesiology. 2009;110(1):166–81.
26. Ohashi K, et al. Colonic mast cell infiltration in rats with TNBS-induced visceral hypersensitivity. J Vet Med Sci. 2007;69(12): 1223–8.

27. Liu M, Wood JN. The roles of sodium channels in nociception: implications for mechanisms of neuropathic pain. Pain Med. 2011;12 Suppl 3:S93–9.
28. Knowles CH, Aziz Q. Visceral hypersensitivity in non-erosive reflux disease. Gut. 2008;57(5):674–83.
29. Grundy D, et al. Fundamentals of neurogastroenterology: basic science. Gastroenterology. 2006;130(5):1391–411.
30. Sarkar S, et al. The prostaglandin E2 receptor-1 (EP-1) mediates acid-induced visceral pain hypersensitivity in humans. Gastroenterology. 2003;124(1):18–25.
31. Willert RP, et al. The development and maintenance of human visceral pain hypersensitivity is dependent on the *N*-methyl-D-aspartate receptor. Gastroenterology. 2004;126(3):683–92.
32. Hobson AR, et al. Development of esophageal hypersensitivity following experimental duodenal acidification. Am J Gastroenterol. 2004;99(5):813–20.
33. Munakata J, et al. Repetitive sigmoid stimulation induces rectal hyperalgesia in patients with irritable bowel syndrome. Gastroenterology. 1997;112(1):55–63.
34. Mayer EA. Spinal and supraspinal modulation of visceral sensation. Gut. 2000;47 Suppl 4:iv69–72; discussion iv76.
35. Sabbatini M, et al. The pattern of c-Fos immunoreactivity in the hindbrain of the rat following stomach distension. Exp Brain Res. 2004;157(3):315–23.
36. Chang L, et al. Brain responses to visceral and somatic stimuli in patients with irritable bowel syndrome with and without fibromyalgia. Am J Gastroenterol. 2003;98(6):1354–61.
37. Coen SJ, et al. Effects of attention on visceral stimulus intensity encoding in the male human brain. Gastroenterology. 2008; 135(6):2065–74, 2074 e1.
38. Mazurak N, et al. Heart rate variability in the irritable bowel syndrome: a review of the literature. Neurogastroenterol Motil. 2012;24(3):206–16.
39. Zhang J, Chen JD. Systematic review: applications and future of gastric electrical stimulation. Aliment Pharmacol Ther. 2006; 24(7):991–1002.
40. Sun Y, et al. Effects and mechanisms of gastrointestinal electrical stimulation on slow waves: a systematic canine study. Am J Physiol Regul Integr Comp Physiol. 2009;297(5):R1392–9.
41. Xu X, Brining DL, Chen JD. Effects of vasopressin and long pulse-low frequency gastric electrical stimulation on gastric emptying, gastric and intestinal myoelectrical activity and symptoms in dogs. Neurogastroenterol Motil. 2005;17(2):236–44.
42. Liu J, et al. Improvement of gastric motility with gastric electrical stimulation in STZ-induced diabetic rats. Digestion. 2004;70(3): 159–66.
43. McCallum RW, et al. Gastric pacing improves emptying and symptoms in patients with gastroparesis. Gastroenterology. 1998;114(3):456–61.
44. Yin J, et al. Cutaneous gastric electrical stimulation alters gastric motility in dogs: new option for gastric electrical stimulation? J Gastroenterol Hepatol. 2009;24(1):149–54.
45. Abo M, et al. Distension-induced myoelectrical dysrhythmia and effect of intestinal pacing in dogs. Dig Dis Sci. 2000;45(1): 129–35.
46. Bellahsene BE, et al. Acceleration of gastric emptying with electrical stimulation in a canine model of gastroparesis. Am J Physiol. 1992;262(5 Pt 1):G826–34.
47. Song G, et al. Two-channel gastric electrical stimulation accelerates delayed gastric emptying induced by vasopressin. Dig Dis Sci. 2005;50(4):662–8.
48. Chen JD, et al. Efficiency and efficacy of multi-channel gastric electrical stimulation. Neurogastroenterol Motil. 2005;17(6): 878–82.
49. Mintchev MP, et al. Microprocessor-controlled movement of solid gastric content using sequential neural electrical stimulation. Gastroenterology. 2000;118(2):258–63.
50. Zhu H, et al. Therapeutic potential of synchronized gastric electrical stimulation for gastroparesis: enhanced gastric motility in dogs. Am J Physiol Regul Integr Comp Physiol. 2007;293(5): R1875–81.
51. Song GQ, et al. A novel method of 2-channel dual-pulse gastric electrical stimulation improves solid gastric emptying in dogs. Surgery. 2008;143(1):72–8.
52. Sun Y, Chen JD. Gastric electrical stimulation reduces gastric tone energy dependently. Scand J Gastroenterol. 2005;40(2):154–9.
53. Xing JH, et al. Gastric electrical stimulation at proximal stomach induces gastric relaxation in dogs. Neurogastroenterol Motil. 2003;15(1):15–23.
54. Zhao X, et al. Inhibitory effects and mechanisms of intestinal electrical stimulation on gastric tone, antral contractions, pyloric tone, and gastric emptying in dogs. Am J Physiol Regul Integr Comp Physiol. 2009;296(1):R36–42.
55. Yin J, Ouyang H, Chen JD. Potential of intestinal electrical stimulation for obesity: a preliminary canine study. Obesity (Silver Spring). 2007;15(5):1133–8.
56. Liu S, Liu J, Chen JD. Neural mechanisms involved in the inhibition of intestinal motility induced by intestinal electrical stimulation in conscious dogs. Neurogastroenterol Motil. 2006;18(1): 62–8.
57. Liu J, Wang L, Chen JD. Effects of intestinal electrical stimulation on intestinal dysrhythmia and symptoms in dogs. Dig Dis Sci. 2004;49(5):720–8.
58. Xu J, Chen JD. Intestinal electrical stimulation improves delayed gastric emptying and vomiting induced by duodenal distension in dogs. Neurogastroenterol Motil. 2008;20(3):236–42.
59. Yin J, Chen J. Excitatory effects of synchronized intestinal electrical stimulation on small intestinal motility in dogs. Am J Physiol Gastrointest Liver Physiol. 2007;293(6):G1190–5.
60. Sun Y, Chen JD. Intestinal electric stimulation accelerates whole gut transit and promotes fat excrement in conscious rats. Int J Obes (Lond). 2009;33(8):817–23.
61. Yin J, Chen JD. Mechanisms and potential applications of intestinal electrical stimulation. Dig Dis Sci. 2010;55(5):1208–20.
62. Chen JD, Lin HC. Electrical pacing accelerates intestinal transit slowed by fat-induced ileal brake. Dig Dis Sci. 2003;48(2): 251–6.
63. Liu S, Chen JD. Colonic electrical stimulation regulates colonic transit via the nitrergic pathway in rats. Dig Dis Sci. 2006; 51(3):502–5.
64. Sallam HS, Chen JD. Colon electrical stimulation: potential use for treatment of obesity. Obesity (Silver Spring). 2011;19(9):1761–7.
65. Liu S, Lei Y, Chen JD. Inhibitory effects and mechanisms of colonic electric stimulation on gastric and rectal tone in conscious dogs. Dis Colon Rectum. 2006;49(11):1749–54.
66. Liu S, Wang L, Chen JD. Cross-talk along gastrointestinal tract during electrical stimulation: effects and mechanisms of gastric/colonic stimulation on rectal tone in dogs. Am J Physiol Gastrointest Liver Physiol. 2005;288(6):G1195–8.
67. Xu X, et al. Inhibitory effects of gastrointestinal electrical stimulation on rectal tone are both organ-specific and distance-related in dogs. Dis Colon Rectum. 2008;51(4):467–73.
68. Ouyang H, et al. Effects of gastric electrical field stimulation with long pulses on gastric emptying in dogs. Neurogastroenterol Motil. 2003;15(4):409–16.
69. Grundfest-Broniatowski S, et al. Electrical control of gastric emptying in denervated and reinnervated canine stomach: a pilot study. Artif Organs. 1990;14(4):254–9.
70. Peles S, et al. Enhancement of antral contractions and vagal afferent signaling with synchronized electrical stimulation. Am J Physiol Gastrointest Liver Physiol. 2003;285(3):G577–85.
71. Chen JH, et al. Gastric electrical stimulation reduces visceral sensitivity to gastric distention in healthy canines. Auton Neurosci. 2011;160(1–2):16–20.

72. Yan Sun CQ, Chen J. Effects and mechanisms of gastric electrical stimulation on visceral hypersensitivity in rats with gastric ulcers. Gastroenterology. 2010;138(5 Suppl 1):S314.
73. Al-Chaer ED, Traub RJ. Biological basis of visceral pain: recent developments. Pain. 2002;96(3):221–5.
74. Willis Jr WD. Dorsal root potentials and dorsal root reflexes: a double-edged sword. Exp Brain Res. 1999;124(4):395–421.
75. Saper CB. The central autonomic nervous system: conscious visceral perception and autonomic pattern generation. Annu Rev Neurosci. 2002;25:433–69.
76. Schicho R, et al. Nociceptive transmitter release in the dorsal spinal cord by capsaicin-sensitive fibers after noxious gastric stimulation. Brain Res. 2005;1039(1–2):108–15.
77. Furue H, Katafuchi T, Yoshimura M. Sensory processing and functional reorganization of sensory transmission under pathological conditions in the spinal dorsal horn. Neurosci Res. 2004; 48(4):361–8.
78. Zhuo M. Targeting central plasticity: a new direction of finding painkillers. Curr Pharm Des. 2005;11(21):2797–807.
79. Camilleri M, Saslow SB, Bharucha AE. Gastrointestinal sensation. Mechanisms and relation to functional gastrointestinal disorders. Gastroenterol Clin North Am. 1996;25(1):247–58.
80. Ladabaum U, et al. Gastric distention correlates with activation of multiple cortical and subcortical regions. Gastroenterology. 2001;120(2):369–76.
81. Emch GS, Hermann GE, Rogers RC. TNF-alpha activates solitary nucleus neurons responsive to gastric distension. Am J Physiol Gastrointest Liver Physiol. 2000;279(3):G582–6.
82. McCann MJ, Rogers RC. Oxytocin excites gastric-related neurones in rat dorsal vagal complex. J Physiol. 1990;428:95–108.
83. Berthoud HR, Neuhuber WL. Functional and chemical anatomy of the afferent vagal system. Auton Neurosci. 2000;85(1–3): 1–17.
84. Qin C, et al. Modulatory effects and afferent pathways of gastric electrical stimulation on rat thoracic spinal neurons receiving input from the stomach. Neurosci Res. 2007;57(1):29–39.
85. Qin C, et al. Duodenal afferent input converges onto T9-T10 spinal neurons responding to gastric distension in rats. Brain Res. 2007;1186:180–7.
86. Ness TJ, Gebhart GF. Characterization of neuronal responses to noxious visceral and somatic stimuli in the medial lumbosacral spinal cord of the rat. J Neurophysiol. 1987;57(6):1867–92.
87. Qin C, et al. Chemical activation of cervical cell bodies: effects on responses to colorectal distension in lumbosacral spinal cord of rats. J Neurophysiol. 1999;82(6):3423–33.
88. Zhang X, Fogel R, Renehan WE. Relationships between the morphology and function of gastric- and intestine-sensitive neurons in the nucleus of the solitary tract. J Comp Neurol. 1995;363(1): 37–52.
89. Qin C, et al. Gastric electrical stimulation modulates neuronal activity in nucleus tractus solitarii in rats. Auton Neurosci. 2005;119(1):1–8.
90. Holtmann G, et al. Differences in gastric mechanosensory function after repeated ramp distensions in non-consulters with dyspepsia and healthy controls. Gut. 2000;47(3):332–6.
91. Sun Y, et al. Intestinal electric stimulation modulates neuronal activity in the nucleus of the solitary tract in rats. Neurosci Lett. 2005;385(1):64–9.
92. Familoni BO, et al. Electrical stimulation at a frequency higher than basal rate in human stomach. Dig Dis Sci. 1997;42(5): 885–91.
93. Abell T, et al. Gastric electrical stimulation for gastroparesis improves nutritional parameters at short, intermediate, and long-term follow-up. JPEN J Parenter Enteral Nutr. 2003;27(4):277–81.
94. Anand C, et al. Gastric electrical stimulation is safe and effective: a long-term study in patients with drug-refractory gastroparesis in three regional centers. Digestion. 2007;75(2–3):83–9.
95. Abell TL, et al. Gastric electrical stimulation in intractable symptomatic gastroparesis. Digestion. 2002;66(4):204–12.
96. Lin Z, et al. Association between changes in symptoms and gastric emptying in gastroparetic patients treated with gastric electrical stimulation. Neurogastroenterol Motil. 2008;20(5):464–70.
97. van der Voort IR, et al. Gastric electrical stimulation results in improved metabolic control in diabetic patients suffering from gastroparesis. Exp Clin Endocrinol Diabetes. 2005;113(1):38–42.
98. Abell T, et al. Gastric electrical stimulation for medically refractory gastroparesis. Gastroenterology. 2003;125(2):421–8.
99. Lin Z, et al. Treatment of diabetic gastroparesis by high-frequency gastric electrical stimulation. Diabetes Care. 2004;27(5):1071–6.
100. Forster J, et al. Further experience with gastric stimulation to treat drug refractory gastroparesis. Am J Surg. 2003;186(6):690–5.
101. Oubre B, et al. Pilot study on gastric electrical stimulation on surgery-associated gastroparesis: long-term outcome. South Med J. 2005;98(7):693–7.
102. Forster J, et al. Absence of the interstitial cells of Cajal in patients with gastroparesis and correlation with clinical findings. J Gastrointest Surg. 2005;9(1):102–8.
103. Lin Z, et al. Symptom responses, long-term outcomes and adverse events beyond 3 years of high-frequency gastric electrical stimulation for gastroparesis. Neurogastroenterol Motil. 2006;18(1):18–27.
104. Lin Z, et al. Chronic gastric electrical stimulation for gastroparesis reduces the use of prokinetic and/or antiemetic medications and the need for hospitalizations. Dig Dis Sci. 2005;50(7):1328–34.
105. Maranki JL, et al. Predictive factors for clinical improvement with Enterra gastric electric stimulation treatment for refractory gastroparesis. Dig Dis Sci. 2008;53(8):2072–8.
106. Hou Q, et al. Is symptom relief associated with reduction in gastric retention after gastric electrical stimulation treatment in patients with gastroparesis? A sensitivity analysis with logistic regression models. Neurogastroenterol Motil. 2012;24(7):639–45, e274.
107. McCallum RW, et al. Gastric electrical stimulation improves outcomes of patients with gastroparesis for up to 10 years. Clin Gastroenterol Hepatol. 2011;9(4):314–9, e1.
108. Chen JD, et al. Gastric electrical stimulation with short pulses reduces vomiting but not dysrhythmias in dogs. Gastroenterology. 2003;124(2):401–9.
109. Lin Z, et al. Effect of high-frequency gastric electrical stimulation on gastric myoelectric activity in gastroparetic patients. Neurogastroenterol Motil. 2004;16(2):205–12.
110. Forster J, et al. Gastric pacing is a new surgical treatment for gastroparesis. Am J Surg. 2001;182(6):676–81.
111. Samsom M, et al. Proximal gastric motor activity in response to a liquid meal in type I diabetes mellitus with autonomic neuropathy. Dig Dis Sci. 1998;43(3):491–6.
112. Xing JH, Chen JD. Gastric electrical stimulation with parameters for gastroparesis enhances gastric accommodation and alleviates distention-induced symptoms in dogs. Dig Dis Sci. 2006;51(12): 2160–4.
113. McCallum RW, et al. Mechanisms of symptomatic improvement after gastric electrical stimulation in gastroparetic patients. Neurogastroenterol Motil. 2010;22(2):161–7, e50–1.
114. Tack J, Coulie B, Van Cutsem E, Ryden J, Janssens J. The influence of gastric electrical stimulation on proximal gastric motor and sensory function in severe idiopathic gastroparesis. Gastroenterology. 1999;116:A1090.
115. Undeland KA, et al. Lower postprandial gastric volume response in diabetic patients with vagal neuropathy. Neurogastroenterol Motil. 1997;9(1):19–24.
116. Liu J, Qiao X, Chen JD. Vagal afferent is involved in short-pulse gastric electrical stimulation in rats. Dig Dis Sci. 2004;49(5): 729–37.
117. McCallum RW, Dusing RW, McMillin C, et al. Fluro-Dexy Glucose (fdg) positron emission tomography (PET) in gastroparetic

patients before and during gastric electrical stimulation (GES). Gastroenterology. 2005;128:A622.
118. Tang M, et al. Implantable gastric stimulation alters expression of oxytocin- and orexin-containing neurons in the hypothalamus of rats. Obes Surg. 2006;16(6):762–9.
119. Gallas S, et al. Gastric electrical stimulation increases ghrelin production and inhibits catecholaminergic brainstem neurons in rats. Eur J Neurosci. 2011;33(2):276–84.
120. Devulder J, et al. Spinal cord stimulation in chronic pain therapy. Clin J Pain. 1990;6(1):51–6.
121. Meyerson BA, Linderoth B. Mechanisms of spinal cord stimulation in neuropathic pain. Neurol Res. 2000;22(3):285–92.
122. Gore RM, Mintzer RA, Calenoff L. Gastrointestinal complications of spinal cord injury. Spine (Phila Pa 1976). 1981;6(6):538–44.
123. Gondim FA, et al. Complete cervical or thoracic spinal cord transections delay gastric emptying and gastrointestinal transit of liquid in awake rats. Spinal Cord. 1999;37(11):793–9.
124. Hansen MB. Neurohumoral control of gastrointestinal motility. Physiol Res. 2003;52(1):1–30.
125. Maher J, et al. Effect of spinal cord stimulation in a rodent model of post-operative ileus. Neurogastroenterol Motil. 2009;21(6):672–7, e33–4.
126. Geng-Qing Song CQ, Sun Y, Jiande Chen J. Spinal cord stimulation improves gastrointestinal motility disorders by inhibiting sympathetic activities in rats. Gastroenterology. 2012;142(5 Suppl 1): S703–4.
127. Caruso C, et al. Abdominal angina due to obstruction of mesenteric artery treated with spinal cord stimulation: a clinical case. Neuromodulation. 2011;14(2):146–9; discussion 149–50.
128. Ceballos A, et al. Spinal cord stimulation: a possible therapeutic alternative for chronic mesenteric ischaemia. Pain. 2000;87(1): 99–101.
129. Kapural L, Rakic M. Spinal cord stimulation for chronic visceral pain secondary to chronic non-alcoholic pancreatitis. J Clin Gastroenterol. 2008;42(6):750–1.
130. Kim JK, et al. Spinal cord stimulation for intractable visceral pain due to chronic pancreatitis. J Korean Neurosurg Soc. 2009;46(2): 165–7.
131. Kapural L, Cywinski JB, Sparks DA. Spinal cord stimulation for visceral pain from chronic pancreatitis. Neuromodulation. 2011;14(5):423–6; discussion 426–7.
132. Tiede JM, et al. The use of spinal cord stimulation in refractory abdominal visceral pain: case reports and literature review. Pain Pract. 2006;6(3):197–202.
133. Khan YN, Raza SS, Khan EA. Application of spinal cord stimulation for the treatment of abdominal visceral pain syndromes: case reports. Neuromodulation. 2005;8(1):14–27.
134. Kapural L, et al. Technical aspects of spinal cord stimulation for managing chronic visceral abdominal pain: the results from the national survey. Pain Med. 2010;11(5):685–91.
135. Kapur S, Mutagi H, Raphael J. Spinal cord stimulation for relief of abdominal pain in two patients with familial Mediterranean fever. Br J Anaesth. 2006;97(6):866–8.
136. Greenwood-Van Meerveld B, et al. Attenuation by spinal cord stimulation of a nociceptive reflex generated by colorectal distention in a rat model. Auton Neurosci. 2003;104(1):17–24.
137. Krames E, Mousad DG. Spinal cord stimulation reverses pain and diarrheal episodes of irritable bowel syndrome: a case report. Neuromodulation. 2004;7(2):82–8.
138. Greenwood-Van Meerveld B, et al. Spinal cord stimulation attenuates visceromotor reflexes in a rat model of post-inflammatory colonic hypersensitivity. Auton Neurosci. 2005;122(1–2):69–76.
139. Yan Sun GX, Song G-Q, Chen J. Spinal cord electrical stimulation improved visceral pain by inhibiting gastric specific afferent neurons activity in rats. Gastroenterology. 2012;142(5 Suppl 1): S703–4.
140. Jänig W. The integrative action of the autonomic nervous system: neurobiology of homeostasis. Cambridge: Cambridge University Press; 2006.
141. de Jonge WJ, et al. Postoperative ileus is maintained by intestinal immune infiltrates that activate inhibitory neural pathways in mice. Gastroenterology. 2003;125(4):1137–47.
142. Anselmino M, et al. Spinal cord stimulation and 30-minute heart rate variability in refractory angina patients. Pacing Clin Electrophysiol. 2009;32(1):37–42.
143. Linderoth B, Fedorcsak I, Meyerson BA. Peripheral vasodilatation after spinal cord stimulation: animal studies of putative effector mechanisms. Neurosurgery. 1991;28(2):187–95.
144. Linderoth B, Herregodts P, Meyerson BA. Sympathetic mediation of peripheral vasodilation induced by spinal cord stimulation: animal studies of the role of cholinergic and adrenergic receptor subtypes. Neurosurgery. 1994;35(4):711–9.
145. Wu M, Linderoth B, Foreman RD. Putative mechanisms behind effects of spinal cord stimulation on vascular diseases: a review of experimental studies. Auton Neurosci. 2008;138(1–2):9–23.
146. Yakhnitsa V, Linderoth B, Meyerson BA. Spinal cord stimulation attenuates dorsal horn neuronal hyperexcitability in a rat model of mononeuropathy. Pain. 1999;79(2–3):223–33.

Intrathecal Therapy for Nonmalignant and Malignant Abdominal Pain

22

Oscar A. de Leon-Casasola

Background

Cancer pain is the result of cancer growth in human tissues, or the pain produced by any of the therapies implemented to treat it. Adequate pain control can be achieved in the great majority of patients with the implementation of an aggressive pharmacological treatment with the use of opioids and adjuvants [1, 2]. With the implementation of these strategies, 90–95 % of the patients may achieve adequate pain control [3]. Consequently, 5–10 % of patients will need some form of invasive therapy. For the successful management of these patients it is critically important to start with a thorough assessment via history and physical examination, and the judicious use of diagnostic testing to try to define the pathophysiological components involved in the expression of pain to implement optimal analgesic therapy. This is because intrathecal opioids are very effective for the treatment of somatic and visceral pain, but intrathecal bupivacaine and/or clonidine will be needed for the treatment of neuropathic pain. Thus, defining the specific pathophysiologic component(s) will be critical for the successful management of these patients. Consequently, when following specific guidelines, the great majority of patients with cancer-related pain may expect adequate pain control in the 21st century. Control of pain and related symptoms is a cornerstone of cancer treatment, as it promotes an enhanced quality of life, improved functioning, better compliance, and a means for patients to focus on those things that give meaning to life [4]. In addition to their salutary effects on quality of life, mounting evidence suggests that good pain control may positively influence survival [5, 6].

O.A. de Leon-Casasola, M.D. (✉)
Department of Anesthesiology, Roswell Park Cancer Institute, University at Buffalo, School of Medicine,
Elm and Carlton Streets, Buffalo, NY 14263, USA
e-mail: Oscar.deleon@roswellpark.org

Intraspinal Analgesia

Neuraxial analgesia is achieved by the epidural or intrathecal administration of an opioid alone (very rarely) or in combination with other agents such as bupivacaine, clonidine, or ziconotide. With the use of neuraxial analgesia, pain relief is obtained in a highly selective fashion with the absence of motor, and sympathetic blockade, making these modalities highly adaptable to the home care environment. When first introduced, the philosophy behind neuraxial opioid therapy was that administering small quantities of opioids in close proximity to their receptors in the substantia gelatinosa of the spinal cord, one could achieve high concentrations at these sites [7, 8]. Thus, analgesia is superior to that achieved when opioids are administered by other routes, and since the total amount of drug administered is reduced, side effects are minimized. Currently, the biggest advantage is the ability to use multiple agents to target multiple receptors resulting in better neuropathic, somatic, and visceral pain control while minimizing side effects.

In general, patients with survival expectancy greater than 3 months will be candidates for intrathecal therapy with a permanent intraspinal catheter and an implanted subcutaneous pump. Conversely, those patients with survival expectancy less than 3 months will require epidural therapy with an implanted system, such as the Du Pen's® epidural catheter [9], or the Sims® epidural port-a-cath which will be connected to an external pump with PCA capabilities. When considering a patient for intrathecal therapy with a permanent intrathecal catheter and a subcutaneous pump, a trial with an epidural catheter will be necessary to: (1) Assess the need for intrathecal multimodal therapy, (2) estimate the doses of the opioid to be used, (3) confirm the best site for catheter tip positioning.

Consequently, the tip of the epidural catheter will need to be placed at the site where nociception is being processed within the spinal cord. We conduct this trial on an outpatient basis to document a 50 % decrease in pain. If successful, we will proceed to implant the permanent device. For this purpose, we use the following protocol:

L. Kapural (ed.), *Chronic Abdominal Pain: An Evidence-Based, Comprehensive Guide to Clinical Management*,
DOI 10.1007/978-1-4939-1992-5_22, © Springer Science+Business Media New York 2015

Epidural Trial

1. Catheter position: dermatomal-specific for the area of nociception under fluoroscopy guidance
   (a) Opioids:
   - Morphine: 0.1 (60 mg)–0.2 (120 mg) mg/ml
   - Hydromorphone: 0.03 (20 mg)–0.12 (80 mg) mg/ml

   (b) Bupivacaine: 1–2 mg/ml (0.1–0.2 %)
   (c) Total volume 600 ml.
   (d) If the patient's source of nociception is in the lower lumbar or sacral areas, thus precluding the use of high concentrations of bupivacaine, then we use a more diluted solution of bupivacaine (0.05 %) to minimize the possibility of motor block and we compensate by adding clonidine: 3–5 mcg/ml
2. Determining epidural opioid doses:
   (a) If the patient is receiving > 300 mcg/h of Fentanyl or 1,200 mg/day of MS or 600 mg/day of Oxycodone or 160 mg/day of methadone, or > 300 mg/day of oxymorphone:
   - Hydromorphone: 0.12 mg/ml

   (b) If the patient is receiving between 100 and 300 mcg/h of fentanyl or an equivalent opioid dose:
   - Hydromorphone: 0.06 mg/ml

   (c) If the patient is receiving less than 100 mcg/h of fentanyl or equivalent dose:
   - Hydromorphone: 0.03 mg/ml
3. Basal infusion: 2 ml/h
4. No bolus during the first 72 h
   (a) Then 2 ml q 10 min
5. The goal is to determine patient requirements
6. Trial for 7–14 days as an outpatient.

If the patient had a successful trial, as defined above, we proceed to implant an intrathecal system. We suggest the following protocol to achieve more than 80 % success rate:

1. Conditions for success
   (a) Place the tip of the intrathecal catheter in the dermatome corresponding to the area of nociception under fluoroscopy guidance.
   (b) For severe somatic pain, combinations of local anesthetics and an opioid will be needed.
   (c) For neuropathic pain:
   - If the tip of the catheter is below L3–4: *Initial* therapy with opioid + clonidine
   - If the tip of the catheter is above L1–2: *Initial* therapy with opioid + bupivacaine

The doses and drugs that we use in our practice are (82).

Drug	Range of doses
Morphine	1.0–20 mg/day
Hydromorphone	0.5–25 mg/day
Sufentanil	10–100 μg/day
Bupivacaine	6–20 mg/day
Clonidine	250–2,000 μg/day

Thus, compounding by a trained pharmacist will be needed. The goal is to concentrate these drugs to twice the daily dose, so that the 20 ml programmable pumps may be programmed to deliver 0.5 ml/h. In this way, patients will need pump refills monthly and it will not be a burden to their quality of life by having too frequent visits to the pain specialist's office. The steps that we use to implement the therapy are:

1. Step 1:
   (a) Opioid + bupivacaine:
   - MS 3–25 mg/day or hydromorphone 0.5–15 mg/day
     **6 mg of MS/day = 1 mg of hydromorphone/day
   - Bupivacaine: 6–20 mg/day

   (b) Opioid + clonidine:
   - Clonidine: 250–2,000 μg/day
2. Step 2: Opioid + bupivacaine + clonidine
3. Step 3: Ziconotide:
   (a) Initiate therapy with ziconotide at a dose of 2.4 mcg/day (0.1 mcg/h) and titrate to patient response
   - Rinse the pump with 2 ml of the 25 mcg/ml solution three times and then fill the pump with the balance (16 ml)

   (b) Titration increments should not be more than 2.4 mcg/day or more frequent than once per week
   (c) Maximum recommended dose: 19.2 mcg/day (0.8 mcg/h)
4. In particular situations, the use of morphine + ziconotide may be an alternative [10]. However, the limitations include the following:
   (a) The benefit of a trial does not exist, as ziconotide may not be administered in the epidural space. Consequently, the patient will need progressive titration once the implanted system is in place.
   (b) Patients may not allow the practitioner to carry out a titration protocol over 4–6 weeks since:
   (c) Starting dose for ziconotide is 2.4 mcg/day with weekly increases of not more than 2.4 mcg/day
   - Therapeutic effects are not usually seen until a dose of 8–10 mcg/day is reached.

Recently, the option to coadminister ziconotide with morphine has emerged. A phase II, open-label, multicenter study of combined intrathecal morphine and ziconotide as add on therapy in 26 patients with noncancer pain showed that the mean improvement in pain, as judged by visual analog scale measurements was 14.5 % from baseline to week 5 [10]. Moreover, there was a mean decrease in opioid therapy of 14.3 % at week 5. Treatment-related side effects included mental confusion, dizziness, abnormal gait, hallucinations, and anxiety. Consequently, both the mean pain improvement and the mean opioid sparing effect are produced by the use of this agent where *clinically* insignificant. However, the maximum dose of ziconotide used in this study was 7.2 mcg/day and that may explain the marginal results.

If triple therapy with an opioid, bupivacaine, and clonidine at optimal doses is not working or one considers the need to implement therapy with ziconotine, then evaluation for catheter obstruction, disconnection, catheter migration, or pump malfunction is a must. In doing so, consider the following possibilities:

1. Pump: Computer program analysis for volume and the volume present within the pump needs to be within 10 % of each other, otherwise pump failure is suspected due to:
   (a) MRI Effects (Medtronic Medical Device Correction, August 2008).

   There is a potential for a delay in the return of proper drug infusion after a MRI affecting all SyncroMed pumps. Moreover, with SynchroMed II pumps, there is the potential for a delay in the logging of motor stall events after MRI. Although the reported incidence of these phenomena is very low (0.014 % and 0.11 % respectively) it is important to interrogate all the pumps after the MRI, to spare patients from not receiving medication. This is particularly important for SynchroMed pumps, as a "Pump Memory Error" may be generated and the pump will NOT restart infusing unless it is reprogrammed. In contrast, the SynchroMed II may continue infusing even though the interrogation may show a stall state. In either case, the pump will alarm in the face of a stall phenomenon.

   (b) Missing Propellant within the pump: Synchromed® II Missing Propellant. Models Affected: 8637–20, 8637–40 (Medtronic Medical Device Recall—May 2008)
   (c) Synchromed® EL pump motor stall due to gear shaft wear (Patient Management Information (Medtronic, August 2007)).
2. Catheter: A myelogram performed through the diagnostic port of the pump will be needed to determine if there is obstruction, disconnection (Medical Device Safety Alert—June 2008: Proper Connection of Sutureless Connector Intrathecal Catheters Models Affected: 8709SC, 8731SC, 8596SC, 8578), and the position of the tip of the catheter. When performing a myelogram through the diagnostic port of the pump, remember that this only accommodates a 25-gauge Huber needle. Moreover, consider:
   (a) *The dead space of the catheter when injecting the contrast medium*: 0.196 ml [89 cm total catheter length (81.4 cm for the spinal segment + 7.6 of the catheter interface with the sutureless connector) × 0.0022 ml/cm catheter volume for the model 8709 SC]
   (b) *The need for a bolus dose after the study is completed, as the catheter will be filled with contrast medium. Consequently, at a programmed rate of 0.5 ml/h it will take* ***9.4 h*** *for the pump to clear all this volume resulting in inadequate pain control and possibly opioid withdrawal symptoms.*

   When performing pump's diagnostic port injections, one needs:

   To withdraw enough amount of cerebrospinal fluid/therapeutic solution prior to injecting contrast medium to remove all the volume of the drug within the catheter and avoid giving the patient a bolus of the medications in use. If this was not performed, up to 0.196 ml of solution could be pushed alone with the contrast medium. Likewise, we suggest that one should aspirate the fluid with a 3 ml syringe at a very low negative pressure to avoid turbulent flow and the risk of leaving medication within the catheter (cavitations phenomenon). We usually aspirate a total of 3 ml of fluid, as this should contain all the medication left in the catheter's dead space and some CSF.

   A bolus dose should be programmed after the myelogram to clear the catheter's dead space containing contrast medium at this point. By doing so, one avoids leaving the patient without intrathecal treatment for periods of 16–20 h depending on how much catheter was implanted.

Clinical Studies

A recently published multicenter prospective randomized clinical trial by Smith, et al., compared intrathecal therapy to comprehensive medical management (CMM) after 1 month of therapy in 202 cancer patients with refractory pain [11]. The primary outcome measure was a 20 % improvement in analgesia, as measured via a 0–10 visual analog scale. Additionally, side effects change based on the National Cancer Institute's common toxicity criteria. There was a slight trend toward better analgesia in the intrathecal group but this difference did not achieve statistical significance. In contrast, there was a statistical difference in the side effect profile of those patients randomized to the intrathecal group. The two side effects where the therapy had its greatest impact were constipation and level of consciousness. After a 6-month analysis, there was also a trend towards an increased survival in the intrathecal group (54 % versus 37 %). Even though the number of patients who were alive at the end of the analysis was small, this difference is about a 25 % increase in survival in the patients randomized to the intrathecal group when compared to the CMM group.

A longitudinal prospective analysis of 30 crossover patients that received intrathecal therapy found significant decreases in pain scores and drug toxicity (27 % and 51 %, respectively) [12]. Median survival was 103 days after crossover to an IDDS, which was similar to that of patients in the randomized controlled trial [12].

The cost of implementing intrathecal therapy is initially high, because of equipment acquisition cost. In contrast, the cost of implementing long-term epidural therapy is low. Two studies evaluated the cost of implementing therapy with these two modalities. These analyses show a "break even" point at approximately 3 months [13, 14]. Thus, epidural therapy becomes very expensive after 3 months, and is one of the reasons to limit its use in patients with survival expectations of less than 3 months.

Clinical Guidelines

A consensus panel was recently published to update recommendations on the use of intrathecal medications in chronic *noncancer* pain [15]. Their goal was to:

1. Review the conclusions and guidelines of the Polyanalgesic Conference 2000 and Polyanalgesic Conference 2003.
2. Evaluate the current guidelines for IT drug infusion.
3. Review survey responses of fellow peers in the field of IT analgesics for pain management and use the findings to guide discussion during the conference.
4. Review preclinical and clinical data relevant to IT analgesics published since 2000.
5. Formulate consensus opinions on critical issues for IT polyanalgesic therapy.
6. Modify and update the IT analgesic drug selection algorithm, as appropriate, based on "best evidence" from published data and expert consensus opinion.
7. Identify areas, including promising under-researched and experimental analgesic agents, for future evidence based research that will advance the clinical practice of IT drug infusion therapy.
8. Disseminate the consensus opinions and primary conclusions of the expert panelists to the medical community through data-driven articles published in appropriate peer-reviewed biomedical journals.

Although the consensus limits its conclusions to the noncancer population, there are five issues that are important to discuss, in light of the recommendations given in this review:

1. Hydromorphone equianalgesic doses
2. Hydromorphone maximum dose
3. Bupivacaine spinal cord lesions
4. Ziconotide as first-line agent
5. The use of CT-Myelography for the diagnosis of granulomas at the tip of the intrathecal catheter.

Hydromorphone Equianalgesic Doses

The study by Johansen et al. [16] quoted in the consensus (reference 36) did not study equianalgesic doses between morphine and hydromorphone. Johansen et al. simply administered hydromorphone at "a dose equivalent to the minimum intrathecal morphine dose shown to produce inflammatory masses in our sheep model (12 mg/day)." Thus, there is no basis for the authors of the consensus to conclude that "intrathecal (IT) morphine and IT hydromorphone, in a dose 20 % of that of morphine, induce an equianalgesic response." Nonetheless, the discussion in the Johansen paper states that the morphine to hydromorphone conversion rate is 5–6:1: "No masses were observed at hydromorphone doses (3 and 6 mg/day) that were equianalgesic to morphine doses (18 mg/day and 36 mg/day, respectively) [16]." This is the conversion rate that we have used in our clinical practice but there has not been a trial to support the validity of this conversion figure.

Hydromorphone Maximum Doses

The consensus panel recommends a maximum hydromorphone concentration of 10 mg/cm^3 and a maximum dose of 4 mg/day for intrathecal use to prevent granuloma formation. Throughout the manuscript, there is not a single reference to support this recommendation and they acknowledge that "physicians are advised to titrate doses of these two opioids (morphine and hydromorphone) not beyond an *a priori* upper limit that has been determined from clinical practice [15]." To date, we have treated about 60 patients with IT hydromorphone in combination with bupivacaine and/or clonidine at concentrations and doses well beyond these recommended concentrations without a single incidence of granuloma. It is noteworthy that we survey these patients with magnetic resonance imaging on yearly basis to make an early diagnosis of this condition. Moreover, we ask patients on their monthly refill visits about symptoms that may be associated with the development of these masses (Table 22.1).

Bupivacaine Spinal Cord Lesions

The preclinical discussion on the use of IT bupivacaine in the Consensus [15] begins with the following statement: "Transient neurologic syndrome (TNS), defined as radicular irritation after spinal anesthesia with local anesthetics, is hypothesized to fall on the lower end of a spectrum of toxic effects caused by local anesthetics." It is noteworthy, that there is not a single report on TNS after bupivacaine spinal anesthesia. In contrast, it has been associated with the use of lidocaine and mepivacaine [17]. Consequently, the discussion of this syndrome in the bupivacaine section is out of contest and misleading. Additionally, there is the suggestion that bupivacaine/clonidine combinations could result in spinal cord lesions, based on a case report [18]. This appears as a footnote in the Recommendations section of the manuscript

Table 22.1 Reported patient's symptoms that led to the diagnosis of an inflammatory mass[a]

Symptoms[b]	Number of reports of symptoms	Percentage of cases with symptoms (%) ($n=448$)
Decreased therapeutic response/inadequate pain relief	150	33.5
Pain	146	32.6
Neurological deficit/dysfunction	78	17.4
Unknown (reports did not provide the patient's condition)	74	16.5
Paralysis/paraplegia/paresis	67	15.0
Weakness/muscle weakness	62	13.8
Numbness	43	9.6
Incontinence	32	7.1
Ambulation difficulties	12	2.7
Urinary retention	8	1.8
Tingling	8	1.8
Headache	7	1.6
Muscle spasm(s)	7	1.6
Burning sensation	6	1.3
Other[c]	68	15.2

[a]Medtronic updated information: inflammatory mass (granuloma) at or near the distal tip of intrathecal catheters—medical device correction (January 2008)
[b]There may be more than one symptom per report of inflammatory mass
[c]Multiple symptoms, each reported in less than 1 % of cases of inflammatory mass

that states: "a spinal cord lesion has been reported with the use of bupivacaine at a concentration of 20 mg/ml [18]." It is important to recognize that in the reported case, the neurologic deficit suddenly appeared 2 years after therapy with bupivacaine and clonidine at doses of 20 mg/day and 200 mcg/day respectively. The patient was a male individual who had been receiving a perfusion of this solution for about 2 years for a right sciatic cord compression neuropathy after a suicide attempt. The patient developed neurologic deficit 1 week after sustaining a fall and landing on his back. Neurologic examination 1 week after the fall revealed gait ataxia with impaired proprioception in the left leg. No vibration sensation up to the left knee and a left foot drop was noted. Three days after these findings, he was found to have complete loss of propioception bilaterally up to T11, hyperreflexia in the left lower extremity and bilateral hypoesthesia of all sacral segments. The MRI showed a round cavity within the spinal cord measuring 3 mm in diameter at the T9–11 level associated with edema that extended from the T5 level to the conus medullaris. The tip of the intrathecal catheter had migrated from the T12 to T10 level. The drug infusion was stopped and the patient's neurologic status improved over the following 3 months he experienced improvement of the corticospinal signs, but only moderate improvement in the propioception and the gait ataxia.

It is unclear if the spinal cord changes were related to the drug neurotoxicity, particularly because the rate of administration was 0.5 ml/h and the edema in the spinal cord extended from the conus medullaris to the T5 level. The CSF spread of the intrathecal solutions administered at a rate of 0.5 ml/h has been shown to be very limited both in the animal model [19], as well as in humans [20]. Consequently, it is hard to understand how the edema in the spinal cord was so extensive. Moreover, the tip of the catheter had migrated from the T12 to the T10 level, where the lesion was found raising the possibility that this could be the result of spinal cord catheter injury during the fall.

Ziconotide as First-Line Agent

The last polyanalgesic consensus recommended the use of ziconotide in chronic pain when all other options were exhausted [21]. At that time, the drug had not been FDA-approved and the only randomized clinical trial available was the study by Staats and collaborators [22]. In contrast, the panelists of the new recommendations have upgraded ziconotide to a first-line agent at the same level than morphine and hydromorphone [15]. Since it is acknowledged that "the medications in the current algorithm are arranged in a hierarchy based on evidence on safety, efficacy, and broad clinical parameters gleaned from previous and current consensus literature reviews, ratings of published studies, and expert opinion from three Polyanalgesic Consensus Conferences [21]," the question is whether there is enough new data on therapeutic efficacy and safety to support that recommendation.

In the study by Staats et al. [22], there are two concerns: First, the physiopathology of pain in cancer patients is disease- and site-specific, and may be multifactorial. Thus, treating patient without a clear description of the source of nociception (i.e., somatic, versus visceral, versus neuropathic) could be a problem.

Second, the 2-week follow-up may result in two problems. As previously discussed, ziconotide needs a significant titration window to reach a therapeutic effect and this is not normally achieved within a 2-week period. Thus, it is possible that the investigators were evaluating placebo effect at that time. Consequently, therapeutic responses beyond that time may have decreased and the success rate might have been lower if the follow-up would have been longer. Consequently, the results of this study do not fully support the use of this agent as a first-line agent.

Since the publication of the Staats and collaborators study, five other studies addressing the use of ziconotide in severe non-cancer chronic pain have been published [10, 23–26]. In the first study, 644 patients with severe chronic pain were studied in an open-label multicenter study with ziconotide [23]. In the end, 119 patients were treated for at least 1 year. Median duration of therapy was 2 months with a range of 1–1,215 days. Mean dose was 8.4 mcg/day (range 0.048–240 mcg/day). Pain scores decreased from 76 to 68 mm after 1 month of therapy and to 73 mm after 2 months of therapy. Virtually all patients experienced adverse events (99.7 %), of which 43.5 % were mild, 42.3 % moderate, and 14.2 % severe. Half of those adverse events were considered nontherapy-related. The most common side effects (≥25) were nausea, dizziness, headache, confusion, pain, somnolence, and memory impairment. The authors concluded that "long-term IT ziconotide is an option for patients with severe, refractory pain." However, the high incidence of side effects and the clinically insignificant pain reduction does not support therapeutic efficacy under the present protocol design.

In the second study, in what appears to be the need to address the lack of therapeutic effects reported in the first study, the safety and efficacy of adding IT ziconotide to intrathecal morphine in patients receiving a stable IT morphine dose [10]. Twenty-six patients receiving doses ranging between 2 and 20 mg/day of morphine received 0.6–7.2 mcg/day of IT ziconotide. The mean percentage improvement of pain in the visual analog scale was 14.5 % (95 % confidence interval of 9–38 %) from baseline to week 5. Mean percentage oral opioid doses change from baseline was 14 % at week 5. The investigators concluded that the coadministration of IT ziconotide and morphine may reduce pain and decrease systemic opioid use in patients receiving treatment with IT morphine alone [10]. However, both the mean decrease in pain intensity, as judged by the visual analog scale, and the amount of systemic opioid reduction are clinically insignificant and do not support these conclusions. Moreover, there is evidence of decreased ziconotide stability when coadministered with either morphine or hydromorphone [15]. Thus, at this point it is not clear what the clinical advantage of coadministering ziconotide with morphine is.

The third study [25], 255 patients were randomized to receive ziconotide ($n = 169$) or placebo ($n = 86$) during 6 days as inpatients. Patients received doses ranging from 9.6 to 168 mcg/day. But during the course of the study, doses were reduced to 2.4–57.6 mcg/day due to the high prevalence of side effects with the initial doses. The authors reported a 31 % pain reduction in the ziconotide group versus a 6 % reduction in the placebo group. Despite this significant pain reduction, it is noteworthy that of the 169 patients initially treated with ziconotide, only 54 patients (31 %) were considered responders and were eligible for outpatient 5-day treatment [25]. Treatment responders were defined as patients having (1) a ≥ 30 % pain improvement in the VASPI compared to baseline, (2) stable or decreased concomitant opioid analgesic use, and (3) no changes in type of opioid used during the study period.

The use of CT-Myelography for the Diagnosis of Granulomas at the Tip of the Intrathecal Catheter

The authors of the consensus suggest that MRI remains the gold standard for surveillance when evaluating the presence of a catheter-related inflammatory mass, although computed tomography/myelography through the pumpt offers a more cost-effective technique. This is true, provided that the practitioner is able to aspirate CSF from the diagnostic port, prior to performing a myelography study. As noted before, if CSF is not aspirated prior to injecting the contrast medium, the catheter dead space volume will be injected at once, and severe side effects may occur.

Summary

Acute and chronic pain is highly prevalent in cancer patients. Inadequate assessment and treatment of pain and other distressing symptoms may interfere with antitumor therapy and markedly detract from the quality of life. While a strong focus on pain control is important, independent of disease stage, it is a special priority in patients with advanced disease who are no longer candidates for potentially curative therapy.

Although rarely eliminated, pain can be controlled in the vast majority of patients, with the implementation of an aggressive comprehensive medical management. In the small but significant proportion of patients whose pain is not readily controlled with noninvasive analgesics, a variety of alternative invasive and noninvasive measures, when selected carefully, are also associated with a high degree of success. To this end, it is very reassuring to conclude that at this point, we have the appropriate tools to adequately treat cancer-related pain in close to 100 % of the patients.

References

1. American Pain Society. Principles of Analgesic use in the treatment of acute pain and chronic cancer pain. 3rd ed. Skokie: APS; 1992.
2. Jacox A, Carr DB, Payne R, et al. New clinical-practice guidelines for the management of pain in patients with cancer. N Engl J Med. 1994;330(9):651–5.
3. Zech DFG, Grong S, Lynch J, et al. Validation of the World Health Organization guidelines for cancer pain relief: a 10-year prospective study. Pain. 1996;63:65–76.
4. Ferrell BR, Wisdon C, Wenzl C. Quality of life as an outcome variable in management of cancer pain. Cancer. 1989;63:2321.
5. Liebeskind JC. Pain can kill. Pain. 1991;44:3–4.
6. Lillemoe KD, Cameron JL, Kaufman HS, et al. Chemical splanchnicectomy in patients with unresectable pancreatic cancer. Ann Surg. 1993;217:447–57.
7. Cousins MJ, Mather LE. Intrathecal and epidural administration of opioids. Anesthesiology. 1984;61:276–310.
8. Yaksh TL. Spinal opiates: a review of their effect on spinal function with an emphasis on pain processing. Acta Anaesthesiol Scand. 1987;31 Suppl 85:25.
9. Du Pen S, Kharasch ED, Williams A, et al. Chronic epidural bupivacaine-opioid infusion in intractable cancer pain. Pain. 1992;49:293–300.
10. Wallace MS, Kosek PS, Staats P, Fisher R, Schultz DM, Leong M. Phase II, open-label, multicenter study of combined intrathecal morphine and ziconotide: addition of ziconotide in patients receiving intrathecal morphine for severe chronic pain. Pain Med. 2008;9:271–81.
11. Smith TJ, Staats PS, Deer T, et al. Randomized clinical trial of an implantable drug delivery system compared with comprehensive medical management for refractory cancer pain: impact on pain, drug-related toxicity and survival. J Clin Oncol. 2002;20:4040–9.
12. Smith TJ, Coyne PJ. Implantable drug delivery systems (IDDS) after failure of comprehensive medical management(CMM) can palliate symptoms in the most refractory cancer pain patients. J Palliat Med. 2005;8:736–42.
13. Bedder MD, Burchiel KJ, Larson A. Cost analysis of two implantable narcotic delivery systems. J Pain Symptom Manage. 1991; 6(6):368–73.
14. Hassenbusch SJ, Bedder M, Patt RB, Bell GK. Current status of intrathecal therapy for nonmalignant pain management: clinical realities and economic unknowns. J Pain Symptom Manage. 1997;14:S36–48.
15. Deer T, Krames ES, Hassenbusch SJ, et al. Polyanalgesic consensus conference 2007: recommendations for the management of pain by intrathecal (intraspinal) drug delivery: report of an interdisciplinary expert panel. Neuromodulation. 2007;10:300–28.
16. Johansen MJ, Satterfield WC, Baze WB, Hildebrand KR, Gradert TL, Hassenbusch SJ. Continuous intrathecal infusion of hydromorphone: safety in the sheep model and clinical implications. Pain Med. 2004;5:14–25.
17. Liu S, Chiu AA, Neal JM, Carpenter RL, Bainton BG, Gerancher JC. Oral clonidine prolongs lidocaine spinal anesthesia in human volunteers. Anesthesiology. 1995;82(6):1353–9.
18. Perren F, Buchser E, Chedel D, Hirt L, Maeder P, Vingerhoets F. Spinal cord lesion after long-term intrathecal clonidine and bupivacaine treatment for the management of intractable pain. Pain. 2004;109:189–94.
19. Bernard CM. Cerebrospinal fluid and spinal cord distribution of Baclofen and bupivacaine during slow intrathecal infusion in pigs. Anesthesiology. 2006;105:169–78.
20. Kotob F, de Leon-Casasola OA, Lema M. Intrathecal infusion rates of 1 ml/day improve narrow analgesia with infusion rates of 0.5 ml/day. Anesthesiology. 2006;105:A347.
21. Hassenbusch S, Portenoy R, Cousins M, et al. Polyanalgesic consensus conference: an update on the management of pain by intraspinal drug delivery – report of an expert panel. J Pain Symptom Manage. 2004;27:540–63.
22. Staats PS, Yearwood T, Charapata SG, et al. Intrathecal ziconotide in the treatment of refractory pain in patients with cancer or AIDS. A randomized controlled trial. JAMA. 2004;291:63–70.
23. Wallace MS, Rauck R, Fisher R, et al. Intrathecal ziconotide for severe chronic pain: safety and tolerability results of an open-label, long-term trial. Anesth Analg. 2008;106:628–37.
24. Webster LR, Fakata KL, Fisher R, MineHart M. Open-label multicenter study of combined intrathecal morphine and ziconotide: addition of morphine in patients receiving ziconotide for severe chronic pain. Pain Med. 2008;9:282–90.
25. Wallace M, Charapata S, Fisher R, et al. Intrathecal ziconotide in the treatment of chronic nonmalignant pain: a randomized double-blind placebo-controlled trial. Neuromodulation. 2006;9:75–86.
26. Rauck RL, Wallace MS, Leong MS, et al. A Randomized, double-blind, placebo-controlled study of intrathecal ziconotide in adults with severe chronic pain. J Pain Symp Manage. 2006;5:393–406.

Surgical Options for Chronic Visceral Abdominal Pain

23

R. Matthew Walsh

Introduction

Chronic visceral abdominal pain is a complex and often misunderstood disease process. Contributing to the clinical challenge is the inability to often localize the source of the abdominal pain. The abdominal viscera are relatively insensate to many stimuli compared to other organ systems [1]. In addition to the relative paucity of sensory nerve endings, the same group of nerves may innervate several different viscera. This often makes it difficult to localize the exact source of the abdominal pain. As such, the focus of this chapter is to discuss the chronic abdominal pain syndromes where surgical intervention may be beneficial.

There are a few well-known nociceptive activators in the abdominal cavity. These include abnormal distention or contraction of hollow organs, traction or compression of ligaments, vessels or mesentery or ischemia of the visceral musculature. It may also be due to the direct action of chemical substances on the mucosa. Pain patterns are often not well differentiated as to their location or cause. Nevertheless, there are some recognizable pain patterns, and a careful history can often lead to the correct diagnosis.

Abdominal pain can arise from many organs in the abdominal cavity and can be caused both by functional and organic diseases. Figure 23.1 delineates the more common causes of chronic abdominal pain by organ system. Because of the nonspecific nature of the symptoms, it is imperative that a logical approach for localization of the organ be undertaken. The characteristics of some of the abdominal pain, such as pancreatic pain, biliary colic, acute intestinal colic, renal colic, and pain due to peptic ulcer have certain characteristics and a diagnostic algorithm can be followed (Table 23.1).

R.M. Walsh, M.D. (✉)
Department of General Surgery, Cleveland Clinic Foundation, 9500 Euclid Avenue, A100, Cleveland, OH 44195, USA
e-mail: Walshm@ccf.org

In other patients, the characteristics of abdominal pain may not be as specific. In addition, because of the increasing number of surgical procedures being performed, adhesions from prior abdominal surgery are an increasing problem. When no other diagnosis is readily apparent, abdominal exploration and lysis of adhesions may be the only alternative for these patients.

Pathophysiology

Visceral pain is transmitted from nociceptors found on the walls of the abdominal viscera via sympathetic and parasympathetic pathways [2]. This pain is nonspecific because of the wide divergence and relatively small number of afferent fibers innervating a large area with extensive ramification. Pain is often perceived in the midline since slow conducting afferent pain fibers from the viscera have bilateral entry into the spinal cord. The symptoms of abdominal pain are described as aching, cramping, or burning that fluctuates in intensity. In addition, others may complain of diffuse mid-abdominal discomfort described as paroxysmal, deep, squeezing, and diffuse. Referred pain to a different part of the body may also occur. The intensity of the stimulus is as essential as its quality. Severe visceral pain may also generate a secondary physiologic reaction mediated by the autonomic nervous system and manifested by nausea, vomiting, sweating, lightheadedness, and salivation further confounding the diagnosis.

Certain characteristics of the pain may facilitate the origin of the abdominal pain. Pain will typically present based on embryonic origin: foregut, midgut, and hindgut. Pain from the esophagus, stomach, pancreas, and hepatobiliary tree is usually referred to the epigastrium. Periumbilical localization occurs from the small bowel and right colon (midgut); the rest of the colon and the genitourinary organs (hindgut) cause pain that presents in the lower abdomen. It is also important to note its onset, whether acute or insidious, and its temporal profile. The circumstances that intensify or alleviate the pain are significant. Relief with eating or

L. Kapural (ed.), *Chronic Abdominal Pain: An Evidence-Based, Comprehensive Guide to Clinical Management*,
DOI 10.1007/978-1-4939-1992-5_23, © Springer Science+Business Media New York 2015

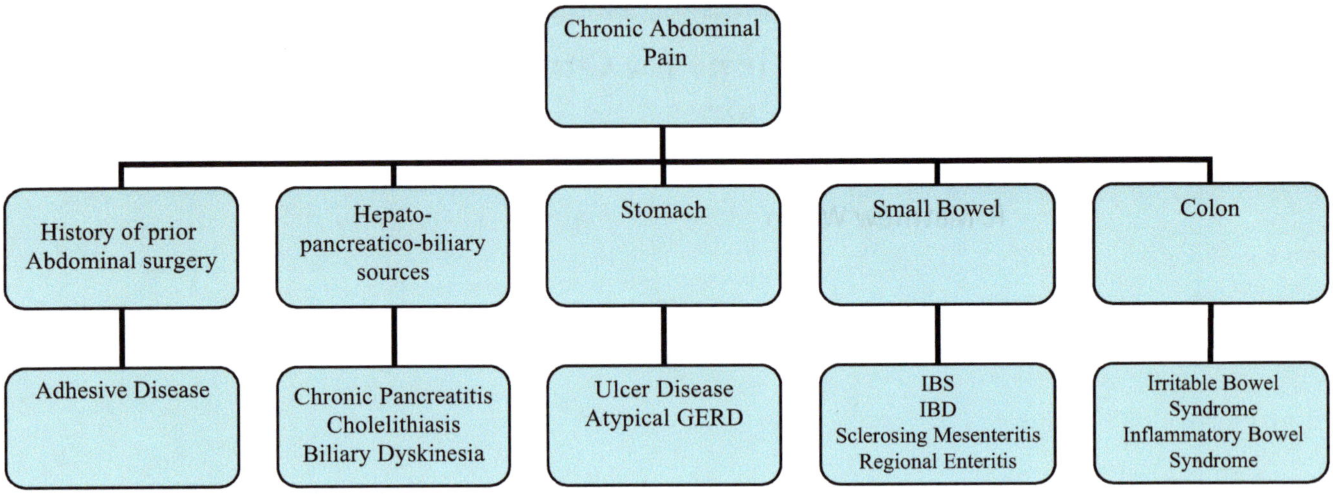

Fig. 23.1 Common causes and location of chronic abdominal pain

Table 23.1 Characteristics of abdominal pain

Epigastric pain
Referred pain from pancreas, stomach and hepatobiliary tree
Umbilical pain
Small bowel and right colon
Lower abdominal pain
Colon

antacids suggests ulcer disease or gastroesophageal reflux. Postprandial pain, depending on its location, character, and timing, could be biliary, ischemic, or associated with a more benign condition, such as lactose intolerance or irritable bowel syndrome. Seasonal patterns frequently are seen in ulcer disease and, occasionally, with regional enteritis. The pain of inflammatory bowel disease and irritable bowel syndrome may be relieved by defecation, whereas heat usually relieves pain of musculoskeletal origin. Posture, sudden movement, coughing, straining, and sneezing may worsen the pain from peritoneal irritation or of spinal origin. The abdomen is not exempt from psychogenic pain. This may be manifested as a component of irritable bowel syndrome. Although common, psychological pain should and does remain a diagnosis of exclusion.

Another important consideration in assessing abdominal pain is that it is an important manifestation of inflammation. Inflammatory cytokines and other inflammatory mediators sensitize primary afferent neurons in the intestinal lumen. Pain associated with pancreatitis is secondary to an inflammatory response to pancreatic parenchymal injury. Pancreatic inflammation, neural remodeling and injury, and changes in the central nervous system contribute to ongoing pain in this condition [3]. Pain management is multifactorial and may involve physiologic and psychologic factors.

Physiologic Causes of Chronic Abdominal Pain

Esophagus

Heartburn

Chronic abdominal pain related to gastroesophageal reflux disease (GERD) can be difficult to diagnose. The symptoms associated with GERD may be mimic other disease processes. Thus symptoms attributable to GERD may often go undiagnosed for long period of time. Patients with atypical GERD are often more challenging to treat than those with typical GERD because they often have unpredictable responses to antisecretory therapy. Aggressive acid reduction using proton pump inhibitors (PPIs) twice daily before meals for 3–4 months is the standard treatment for atypical GERD after other causative factors have been excluded including a normal EGD. An initial therapeutic trial of PPIs is more acceptable to most patients than pH testing, which can be uncomfortable, cumbersome, and unreliable in clinical correlation to pain. The advantage of using PPIs is that they can also demonstrate a cause-and-effect relationship. If atypical symptoms improve or resolve, expert opinion is to taper antisecretory therapy to once-daily PPIs or H2-receptor antagonists at the lowest effective dose over 3–6 months. This regimen may be the best way to demonstrate a causal relationship between GERD and extra-esophageal symptoms, with most treatment trials demonstrating a 50–70 % overall response rate [4]. To determine whether they have adequate acid control, patients who do not respond to empiric PPI therapy should undergo pH monitoring and esophageal manometry. At equivalent doses, PPIs are therapeutically equivalent for treatment of reflux symptoms.

Those patients with a known anatomic defect are best treated by surgical repair. Patients with intrathoracic stomachs require surgical intervention via reduction of their anatomic defect and Nissen fundoplication. These patients typically do not respond well to aggressive antisecretory therapy. There are several studies which demonstrate the efficacy and safety of surgical therapy for refractory GERD. One study found that 96.5 % of patients were satisfied with results of the procedure after 6.4 years, although 14 % were still taking continuous PPI therapy, and 27 % had GERD-related symptoms (e.g., regurgitation, dysphagia, bloating, noncardiac chest pain) [5]. Another study found that antireflux surgery was more effective than PPI therapy for long-term control of symptoms; however, obstructive symptoms (e.g., dysphagia, rectal flatulence, inability to belch or vomit) were more common in patients who underwent surgery. A total of 91.3 % of patients surveyed following surgery were satisfied with their operation, and 96 % commented they would have surgery again for their symptoms [6]. As such, patients with intrac GERD and an anatomic defect should be referred for surgical intervention.

Pain in Pancreatic Disease

Natural History of Chronic Pancreatitis

The most common cause of long-term abdominal pain due to pancreatic disease is chronic pancreatitis. Chronic pancreatitis is characterized by irreversible morphological and functional abnormalities due to longstanding inflammation and fibrosis of the pancreatic parenchyma. This is associated with intractable pain, malabsorption, and in some cases diabetes mellitus. Chronic abdominal pain is arguably the most important component of chronic pancreatitis and often leads to significant morbidity and disability. It places a tremendous burden on the medical care system as the cost of treating patients with chronic pancreatitis averages $17,000 per year. The most common cause of chronic pancreatitis is long-term alcohol use but other frequently observed causes include hereditary, autoimmune, and environmental factors. A large proportion of patients have no discernable cause for chronic pancreatic and are thus considered idiopathic. It commonly affects men more often than women with the onset of symptoms typically occurring in middle age, with the mean age of 48.9 (±15.4) years reported in one large North American survey [7].

The pathophysiology of pain from chronic pancreatitis remains poorly understood. Strictures in the main pancreatic duct, along with peri-pancreatic fibrosis, have been thought to result in increased pancreatic tissue pressure (ductal hypertension) and ischemia (compartment syndrome). Changes in neural density, hypertrophy, and both perineural and endoneural inflammatory infiltration (neuritis) have been described in patients with debilitating pancreatitis which have correlated with the intensity of the abdominal pain.

Most patients considered for surgical intervention for disabling abdominal pain should have a morphologic change in the pancreas. In those patients with no known discernable morphologic abnormality and chronic disabling pain, a total pancreatectomy with auto islet transfusion is one option. In a minority of patients, there are potentially treatable lesions such as pseudocysts or pancreatic ductal disruption. In other patients, strictures in the main pancreatic duct, along with peri-pancreatic fibrosis, have been thought to result in increased pancreatic tissue pressure (ductal hypertension) and ischemia (compartment syndrome). However, measurements of pancreatic ductal pressure in patients have not always correlated with pain or reliably predicted the success of ductal decompression procedures (e.g., lateral pancreaticojejunostomy or endoscopic intraductal stenting). In addition, the morphological changes are as common in patients with painless pancreatitis as in those with pain.

The major clinical features of chronic pancreatitis include functional failure (exocrine and endocrine) and pain. The glandular failure can usually be managed satisfactorily by replacement strategies (enzymes or insulin) to restore nutritional and metabolic stability. Pain though has remained a major clinical challenge; it is present in up to 90 % of cases and is the primary cause of hospitalization in most patients [8]. Unfortunately, pain in chronic pancreatitis remains difficult to treat. A lack of understanding about the underlying biology has led to various empirical approaches that are often highly invasive and based purely on anatomical grounds [9].

The workup for a patient with chronic pancreatitis requires computed axial tomography scan with a 2 mm slices through the pancreas. Endoscopic retrograde cholangiopancreatography (ERCP) with endoscopic ultrasound have proven to be extremely useful in defining the pancreatic anatomy. It not only provides fine anatomic details of the anatomy but also allows therapeutic interventions to be performed. Magnetic resonance cholangiopancreatography (MRCP) has also proven to be invaluable as an additional imaging modality and has replaced ERCP for studying of the ductal system. Finer details of ductal anatomy can be delineated with MRI that is not often demonstrated in other imaging modalities.

The use of ultrasound has become an increasingly common modality for diagnosis chronic pancreatitis at our institution. EUS detects structural abnormalities of the pancreatic duct and parenchyma indicative of fibrosis. To summarize the Rosemont criteria, the presence or absence of nine ductal and parenchymal criteria was routinely assessed in a prospective manner: hyperechoic foci, hyperechoic strands, cysts, lobularity, calcifications, hyperechoic duct margins, visible side branches, main pancreatic duct dilation, and main pancreatic duct irregularity (score 0–9) [10].

The presence of four or more features was considered abnormal. An adjunctive test is the use of secretin to stimulate pancreatic function. Pancreatic function testing is an independent, complementary marker for fibrosis, as they assess the degree of pancreatic exocrine insufficiency.

Selection of Surgical Procedure

General consensus exists that the initial therapy for chronic pancreatitis should be nonoperative management with an emphasis on eliminating etiologic factors, antioxidants and exocrine suppression with oral pancreatic enzymes. Surgery is considered in patients who fail conservative measures. In general, surgical intervention should only be performed in those patients with an expected low morbidity and mortality. An optimal surgical procedure should resolve intractable pain and concomitant complications of the chronic pancreatitis which may include duodenal and biliary obstruction. Part of the goal of the operation is to maximize of endocrine and exocrine function with sparing of the pancreatic parenchyma and improve quality of life.

The choice of surgical procedure to be undertaken depends upon the morphologic abnormality. In general, surgical procedures for chronic pancreatitis can be divided into operations that require resection, those that involve drainage of the pancreatic duct or a combination of both. In patients with a dilated pancreatic duct, decompression of the pancreatic duct is probably the best choice since pancreatic function can be preserved. Various decompressive techniques have been developed over the years which include the traditional lateral pancreaticojejunostomy (Partington–Rochelle or one of its variations) [11]. Those with an inflammatory head mass would most likely benefit from a Whipple, Beger, or Frey procedure. Patients with small-duct disease can be a challenging surgical dilemma. In these patients with no clear anatomic abnormality, a V-shaped excision or Izbicki procedure has been described with some success. Otherwise, a total pancreatectomy with or without auto islet transfusion may be considered. Unfortunately, no approach guarantees absolute success, although the data would suggest that surgical approaches have the most durable pain relief.

Pancreaticoduodenectomy

The most common operation for chronic pancreatitis in the United States is the Whipple procedure. The pancreaticoduodenectomy is performed classically (introduced by Whipple-Kausch) or as a modified pylorus-preserving pancreaticoduodenectomy (Longmire/Traverso). It offers an improvement of the quality of life and pain relief in short- and long-term follow up in up to 90 % of the patients. One of the major disadvantages of pancreaticoduodenectomy is the sacrifice of the surrounding nondiseased organs with loss of the natural bowel continuity. The pancreatic exocrine and endocrine function is significantly reduced. Contemporary series show the procedure can be performed with low mortality (0–5 %) in experienced centers, but the morbidity 20–40 % remains high [12].

Distal and Total Pancreatectomy

Near total and total pancreatectomy have been proposed in the treatment of chronic pancreatitis. In patients with complications after pancreatic surgery (pancreatic fistula or anastomotic leakage) or intractable pain after sufficient resection and drainage procedure, total pancreatectomy may be indicated as rescue procedure. Resections of the distal part of the pancreas are effective if there is an anatomic abnormality in the tail of the pancreas. This often relates to an enlarging pseudocyst or stricture of the main pancreatic duct.

Partington–Rochelle Procedure

The Peustow operation modified by Partington and Rochelle is a spleen-preserving longitudinal pancreaticojejunostomy without pancreatic tail resection. It is the most widely used and simple of all the drainage procedure and can be performed with relatively low mortality and morbidity (approximately 3 % and 20 %, respectively). One of the major advantages of a drainage procedure is that maximum pancreatic tissue is preserved. In most patients with a dilated main pancreatic duct, this operation will effectively decompress the pancreas providing all strictures of the duct are traversed. The short-term pain relief is found in approximately 75 % of all patients, but frequently it fails to provide long lasting pain relief. The reason for persisting or recurrent pain has been attributed to an incomplete decompression of the main pancreatic duct, especially in the head of the pancreas. An inflammatory mass in the pancreatic head including its strictures as well as the local intraductal hypertension of the ducts of second and third order are left behind. Additionally neuronal alteration, that are left behind and can cause pain, might be present in an inflammatory altered pancreatic head. Currently, the only suitable indication for a simple drainage procedure (Partington–Rochelle) with longitudinal pancreaticojejunostomy is an isolated dilatation of the pancreatic ductal (>7 mm) or "chain of lakes," without an inflammatory mass in the pancreatic head.

Beger and Frey Procedure

Initially, the first duodenum-preserving resection of the pancreatic head was introduced by Beger (Fig. 23.2).

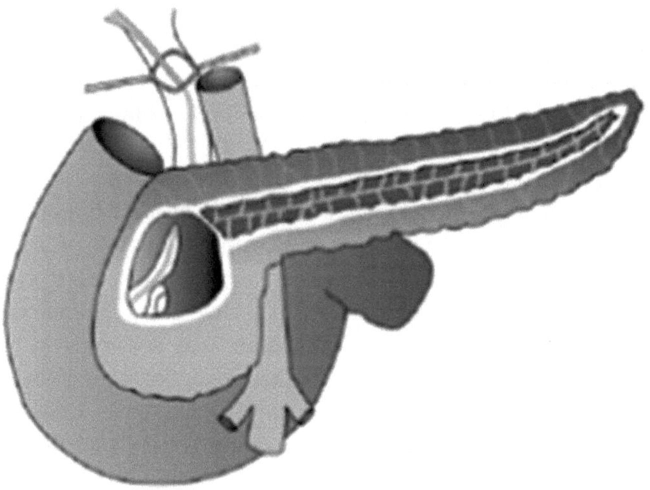

Fig. 23.2 Frey procedure

Fig. 23.3 Beger procedure

This procedure consists of a subtotal resection of the pancreatic head with transaction of the neck of the pancreas. The pancreas is then drained by an end-to-end or end-to-side pancreaticojejunostomy using a Roux-en-Y loop. Interestingly, this operation is almost exclusively performed in Europe and there may be a morphologic explanation for this variation. In a recent series from Germany, most patients with chronic pancreatitis presented with a pancreatic head mass as compared to patients in the United States. Because of this morphologic abnormality, the Frey procedure was performed more frequently as the inflammatory head mass was resected in addition to the longitudinal decompression (Fig. 23.3).

Total Pancreatectomy

In theory, a total pancreatectomy may be able to provide complete resolution of pain symptoms. This is generally not recommended as not only do a substantial number of patients continue to have pain, but also the metabolic consequences from removing the pancreas can be very severe. Autotransplantation of islet or pancreatic tissue is an intriguing option for these patients. Most auto islet transplantation currently being performed in combination with total pancreatectomy have encountered low success rates and technical challenges that make this option less feasible [13].

Abdominal Adhesive Disease

Adhesions are fibrous bands that form within the abdominal cavity. Complex abdominal and pelvic pain syndrome (CAPPS) is a disease process that often leads patients through an exhaustive course of tests and studies without a clear diagnosis or treatment plan and is often dismissed as drug seeking or malingering. As such, there has been a substantial commitment in the scientific and medical community in treating this ill-defined process. The role of adhesive disease in chronic abdominal pain is well established [14] but much of the current research underway is in the prevention of adhesions. The intra-abdominal adhesions that arise from any surgical procedure can cause complications decades later. Most studies report that 93 % of patients who undergo abdominal surgery will develop adhesions. A comprehensive study of inpatient care and expenditures associated with adhesiolysis procedures in the United States conducted in 1994 found that adhesions accounted for 303,836 hospitalizations (1 % of the hospitalizations in the United States), 846,415 days of inpatient care, and $1.33 billion in hospitalization and surgeon expenditures [15]. Though small bowel obstruction is the most common complication related to adhesions, chronic pelvic pain is another complication often cited to be caused by adhesive disease. Unfortunately, most surgeons will not perform operations for chronic pain unless it is related to chronic small bowel obstruction.

There are an increasing number of studies which have demonstrated the utility of laparoscopy in the treatment of chronic abdominal pain. Laparoscopic adhesiolysis was first described in the gynecologic literature for the treatment of chronic pelvic pain and infertility. Since then, this technique has been applied to the treatment of chronic abdominal pain in both adults and children [16, 17]. Intra-abdominal adhesions are often well-vascularized and innervated, which may explain the relationship to some chronic abdominal pain syndromes [18].

The diagnosis of chronic pain syndrome attributed to abdominal adhesions is difficult to determine. Despite the usefulness of contrast studies such as enteroclysis in determining the degree of obstruction, they seldom help in determining whether surgery is required for chronic pain symptoms [19]. Because of the limitations of radiographic imaging in identifying patients with abdominal pain related to adhesive disease, careful patient selection becomes the most important aspect of successful outcome. There are often no criteria to determine who will most benefit from adhesiolysis as no clinical objective data currently exist.

The most important predictive factor of adhesion formation is a history of previous abdominal surgery. In a prospective study by Menzies and Ellis 93 % of patients with prior laparotomy presented with intra-abdominal adhesions, whereas only 10 % of those without prior abdominal surgery had adhesions.

Inflammatory Bowel Disease

Abdominal pain is an important manifestation of inflammation. Inflammatory cytokines and other mediators that sensitize primary afferent neurons in the intestinal lumen can often manifest as pain in patients with inflammatory bowel disease [20]. It is believed that abdominal pain is present in approximately 50–70 % of patients with IBD [21]. In some patients, especially those with Crohn's disease, abdominal pain may be the only symptom of active disease.

Abdominal pain in IBD may be multifactorial. The exact cause of the abdominal pain is crucial to determine as this will ultimately determine the type of therapy required to treat these patients. The persistence of inflammatory activity in the intestine, intestinal stenosis, anorectal inflammation, enteric fistula, and abdominal abscesses may all be causes of abdominal pain in IBD patients. Although one expects abdominal pain to be marker of ongoing and/or inflammatory activity in the intestine, approximately one-third to one-half of patients with IBD complain of disabling abdominal pain while they are in clinical or endoscopic remission [22].

The use of laparoscopy in the treatment of IBD is gaining wider acceptance. The advantages of laparoscopy include shorter hospital stays, decrease in infectious wound complications, less narcotic use, and a faster return to an active life.

Chronic Appendicitis

Chronic pelvic pain may also prove to be a difficult and challenging disease process. Pelvic pain may arise from several origins including the reproductive organ, urological, musculoskeletal, neurological, gastrointestinal, or myofascial. In one study, 55 patients with chronic pelvic pain were treated by laparoscopy. The findings described include fibrous adhesions in 38, chronic appendicitis in 12, and endometriosis in 5. Of these patients, 44 had complete relief, 9 had satisfactory improvement, and 2 did not have any symptomatic relief [23].

In another study, 63 patients underwent laparoscopic appendectomy for chronic pelvic pain. Seventy-nine percent of these patients had symptoms of right lower quadrant pain. Histologically, 92 % of the appendices examined had pathologic abnormalities with 95 % experiencing no further pain following the procedure [24].

In an additional study published in the same year, 348 patients were treated laparoscopically for generalized chronic pelvic pain. Seventy-two percent of patients reported complete or significant relief of pain for at least 6 months. One hundred and three patients with symptoms of right lower quadrant pain underwent laparoscopy. Sixty-two (60 %) of these patients were noted to have gross appendiceal abnormalities. Histologic examination was abnormal in 30 (48 %) of these patients. In another cohort of patients who underwent pelvic reconstructive surgery and appendectomy, 60 (97 %) of 62 of these women reported complete relief of symptoms [25].

The resultant studies demonstrate that value of keeping the diagnosis of chronic appendicitis in the differential diagnosis. Gross morphologic abnormalities do not necessarily correlate with histologic pathology, but the clinical correlation of right lower quadrant pain and appendicitis cannot be overlooked. As such, it is recommended that an appendectomy be performed should there be any question on gross examination or the patient's clinical profile be consistent with possible chronic appendicitis. Such measures may be therapeutic and provide durable pain relief.

Conclusion

A detailed history and physical are crucial in focusing the source of the pain. A systematic approach is required as most patients present with vague, nonspecific abdominal symptoms. Individual patient factors must be taken into account in devising a treatment strategy but the advancement of laparoscopy has provided a powerful tool in the treatment of surgical causes of chronic pain. As such, by applying a systematic approach and selective use of laparoscopy, treatable causes of chronic pain may be alleviated in this difficult patient population.

References

1. Sandler RS, Stewart WF, Liberman JN, et al. Abdominal pain, bloating, and diarrhea in the United States: prevalence and impact. Dig Dis Sci. 2000;45:1166–71.
2. Bielefeldt K, Christianson JA, Davis BM. Basic and clinical aspects of visceral sensation: transmission in the CNS. Neurogastroenterol Motil. 2005;17(4):488–99.
3. Pasricha PJ. Unraveling the mystery of pain in chronic pancreatitis. Nat Rev Gastroenterol Hepatol. 2012;9(3):140–51.
4. Vakil N, van Zanten SV, Kahrilas P, Dent J, Jones R. The Montreal definition and classification of gastroesophageal reflux disease: a global evidence-based consensus. Z Gastroenterol. 2007;45(11): 1125–40.

5. DeVault KR, Castell DO. Updated guidelines for the diagnosis and treatment of gastroesophageal reflux disease. Am J Gastroenterol. 2005;100:190–200.
6. Zehetner J, Lipham JC, Ayazi S, Oezcelik A, Abate E, Chen W, Demeester SR, Sohn HJ, Banki F, Hagen JA, Dickey M, Demeester TR. A simplified technique for intrathoracic stomach repair: laparoscopic fundoplication with Vicryl mesh and BioGlue crural reinforcement. Surg Endosc. 2010;24(3):675–9.
7. Mullady DK, et al. Type of pain, pain-associated complications, quality of life, disability and resource utilisation in chronic pancreatitis: a prospective cohort study. Gut. 2011;60:77–84.
8. Ammann RW, Akovbiantz A, Largiader F, Schueler G. Course and outcome of chronic pancreatitis. Longitudinal study of a mixed medical-surgical series of 245 patients. Gastroenterology. 1984;86: 820–8.
9. DiMagno EP. Toward understanding (and management) of painful chronic pancreatitis. Gastroenterology. 1999;116:1252–7.
10. Sahai AV, Zimmerman M, Aabakken L, Tarnasky PR, Cunningham JT, van Velse A, Hawes RH, Hoffman BJ. Prospective assessment of the ability of endoscopic ultrasound to diagnose, exclude, or establish the severity of chronic pancreatitis found by endoscopic retrograde cholangiopancreatography. Gastrointest Endosc. 1998; 48(1):18–25.
11. Strate T, et al. Resection vs drainage in treatment of chronic pancreatitis: long-term results of a randomized trial. Gastroenterology. 2008;134:1406–11.
12. Hackert T, Buchler MW. Pancreatic cancer: advances in treatment, results and limitations. Dig Dis. 2013;31(1):51–6.
13. Mullhaupt B, Ammann RW. Total pancreatectomy for intractable pain in chronic pancreatitis? Pancreas. 2010;39:111–2.
14. Ward BC, Panitch A. Abdominal adhesions: current and novel therapies. J Surg Res. 2011;165(1):91–111. http://journals.ohiolink.edu/ejc/article.cgi?issn=00224804&issue=v165i0001&article=91_aacant
15. Ray NF, Denton WG, Thamer M, et al. Abdominal adhesiolysis: inpatient care and expenditures in the united states in 1994. J Am Coll Surg. 1998;186(1):1–9.
16. Di Lorenzo N. Impact of laparoscopic surgery in the treatment of chronic abdominal pain syndrome. Chir Ital. 2002;54:367–78.
17. Stringel G. Laparoscopy in the management of children with chronic recurrent abdominal pain. JSLS. 1999;3:215–9.
18. Herrick S. Human peritoneal adhesions are highly cellular, innervated, and vascularized. J Pathol. 2000;192:67–72.
19. Maglinte DDT, Peterson LA, Uahey TN, et al. Enteroclysis in partial small bowel obstruction. Am J Surg. 1984;147:325–9.
20. Bielefeldt K, Davis B, Binion DG. Pain and inflammatory bowel disease. Inflamm Bowel Dis. 2009;5:778–88.
21. Cross RK, Wilson KT, Binion DG. Narcotic use in patients with Crohn's disease. Am J Gastroenterol. 2005;100(10):2225–9.
22. Edwards JT, Radford-Smith GL, Florin TH. Chronic narcotic use in inflammatory bowel disease patients: prevalence and clinical characteristics. J Gastroenterol Hepatol. 2001;16(11):1235–8.
23. Bryson K. Laparoscopic appendectomy. J Gynecolog Surg. 1991; 7:93–5.
24. Fayez JA, Toy NJ, Flanagan TM. The appendix as the cause of chronic lower abdominal pain. Am J Obstet Gynecol. 1995;172(1 Pt 1): 122–3.
25. AlSalilli M, Vilos GA. Prospective evaluation of laparoscopic appendectomy in women with chronic right lower quadrant pain. J Am Assoc Gynecol Laparosc. 1995;2:139–42.

Neurosurgical Options for Control of Chronic and Malignant Abdominal Pain

24

Daniel M. Birk, Matthew K. Tobin, and Konstantin V. Slavin

The history of the neurosurgical treatment of pain dates to 1912 when Spiller described sectioning the anterolateral quadrant of the spinal cord to interrupt the ascending spinothalamic tract for unilateral body pain [1]. The twentieth century saw the development of an array of neurosurgical procedures that involve interruption of ascending pain pathways at different points in the central nervous system (CNS). These surgical strategies can be categorized as either destructive or nondestructive. Nondestructive procedures are also referred to as augmentative procedures and include electrical and chemical neuromodulation modalities such as peripheral nerve or spinal cord stimulation and the delivery of pharmacological agents directly into the thecal sac or ventricles. Because the nondestructive approaches to abdominal pain are discussed elsewhere in this book, this chapter will focus on ablative CNS techniques, which are performed exclusively by the neurological surgeon.

The advent of powerful opioids, the development of neuromodulation technology such as spinal cord and peripheral nerve stimulation, and programmable pumps for intrathecal or epidural delivery of medication all have a place in the management of abdominal pain. However, each of these modalities has its disadvantages. High dose opioids may be ineffective for some patients with malignant pain who become tolerant over time. Severe or malignant abdominal pain may require such high doses of opiates that side effects become unbearable or dangerous. Intrathecal morphine pumps require refills and maintenance and tachyphylaxis can reduce their effectiveness. Geographic or socioeconomic factors may also preclude the use of opioid analgesics or expensive technology such as morphine pumps or spinal cord stimulators. Some patients in socioeconomically poor areas may not have access to a continuous supply of medication or expensive technology. For such patients who do not get relief from medications or neuromodulation, develop intractable side effects, or for whom these medications and technology are not available, a neurosurgeon may be able to offer a surgical solution involving interruption of a specific ascending pain pathway through either an open or minimally invasive procedure.

Anatomy of Pain

Understanding the anatomy of the ascending spinal cord pathways has paved the way for the development of new procedures to alleviate abdominal pain. The lateral spinothalamic tract (Fig. 24.1a) conveys pain and temperature sensation from the contralateral side of the body [2]. Primary afferent fibers project from the dorsal root ganglion (DRG) to Rexed laminae I, II, and V where they synapse on second order neurons. Second order neurons decussate via the anterior white commissure where they ascend in the ventromedial spinal cord to the ventral posterolateral (VPL) nucleus of the thalamus where they synapse on third order neurons. Third order neurons then project via the posterior limb of the internal capsule to the primary somatosensory cortex.

The more recently discovered dorsal midline visceral nociceptive pathway (Fig. 24.1b) was further characterized in animal models and was shown to be distinct from the spinothalamic tracts. The cell bodies lie in the spinal gray matter dorsal to the central canal where they receive segmental primary afferent signals. The axons congregate near the central canal in the posterior midline before they terminate in the nucleus gracilis. This area is Rexed lamina X in the dorsal commissural region. Internal arcuate fibers then transmit the nociceptive signals to the visceroreceptive neurons of the VPL thalamus via the medial lemniscus. The dorsal nociceptive pathway has a viscerotopic organization. Fibers representing the pelvis region of the pathway are medial and the

D.M. Birk, M.D. • M.K. Tobin, B.S. • K.V. Slavin, M.D. (✉)
Department of Neurosurgery, University of Illinois at Chicago, 912 South Wood Street, M/C 799, Rm 451 N NPI, Chicago, IL 60612, USA
e-mail: dmbirk@uic.edu; mktobin2@uic.edu; kslavin@uic.edu

L. Kapural (ed.), *Chronic Abdominal Pain: An Evidence-Based, Comprehensive Guide to Clinical Management*, DOI 10.1007/978-1-4939-1992-5_24, © Springer Science+Business Media New York 2015

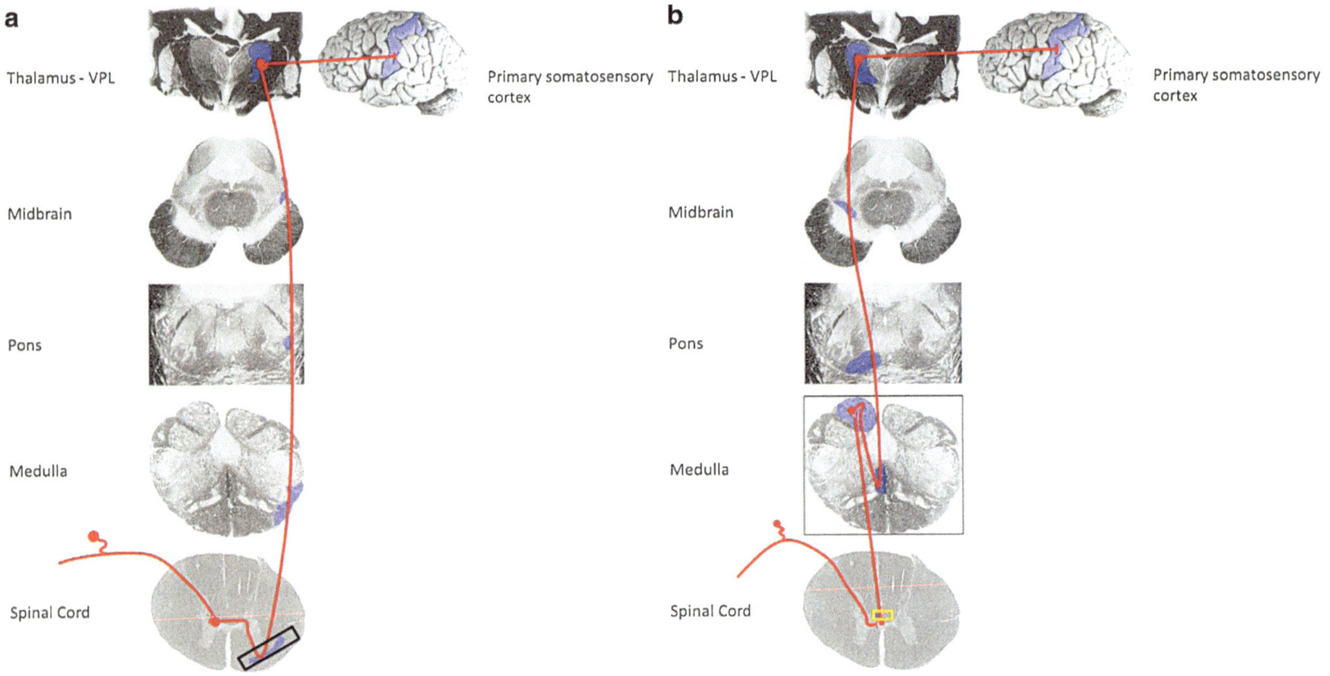

Fig. 24.1 Ascending sensory spinal cord pathways. (**a**) Spinothalamic tract. (**b**) Dorsal midline visceral nociceptive pathway. The black box indicates the location for cordotomy procedure and the yellow box indicates the location for a dorsal midline myelotomy (anatomical images courtesy of UIC Brainstem project by C H Anderson, RJ McAuley, J Unnerstall, http://tigger.uic.edu/classes/anat/anat403/Brainstem/publish/master.swf)

mid-thoracic tracts are lateral where the cuneatus and the fasciculi gracilis interface [3]. This may be the reason why punctate midline myelotomy is effective for chronic and malignant abdominal pain (Fig. 24.1).

Specific Procedures

The dearth of evidence-based data for ablative pain procedures can be attributed to the advent of intrathecal opioids. As intrathecal drug delivery became the procedure of choice for visceral cancer pain, the number of destructive procedures performed dwindled, eliminating training opportunities, and reducing industry investment in technology.

A review of the relevant neurosurgical literature is notable for various individual case reports of treating pain by making a targeted destructive lesion. In 1973, Andy described successful unilateral anterior thalamotomy with a bipolar electrode for a 37-year-old woman with hysterical pain and chronic severe visceral disturbances [4]. Targeting the thalamus for pain treatment was rather common in the 1970s as this approach was suggested and tried for treatment of all kinds of intractable pain syndromes [5–7]. Nevertheless, the majority of destructive neurosurgical procedures for abdominal pain have targeted spinal cord pathways. These techniques include anterolateral cordotomy, commissurotomy, and midline myelotomy.

Cordotomy

Cordotomies are generally effective for unilateral somatic or neuropathic pain while midline myelotomy is more useful for treating bilateral, diffuse visceral pain. Cordotomies aim to interrupt the ascending anterolateral spinothalamic spinal cord tracts and are most appropriate for unilateral nociceptive somatic pain below the neck as well as neuropathic pain. Cordotomy may reduce the severity of visceral pain but is not the best surgical option. Pain relief from cordotomy is unilateral (affecting the contralateral side of the body) but because visceral pain is frequently bilateral, it becomes necessary to perform cordotomies on both sides when trying to treat visceral pain patients. For this and other reasons mentioned below, midline myelotomy is currently the most effective ablative neurosurgical procedure for midline visceral pain and for deep, diffuse visceral abdominal pain.

There is more published evidence for cordotomy than any other procedure for cancer pain. Although none are Class I reports and none of the current prospective or retrospective cohorts qualified as Class II evidence, a recent meta-review identified 47 papers including 3,601 patients with the majority reporting excellent lasting relief from cancer pain [8]. This review clearly shows that most of the papers reported greater than 50 % pain reduction for more than 6 months with less than 1 % risk of postoperative weakness.

Moreover, hospital stays are brief and charges are low relative to implanting drug delivery or stimulation devices. For these reasons, the cordotomy procedure has a definite role for patients with terminal cancer with unilateral somatic abdominal pain.

Bilateral anterolateral cordotomies have seen success in treating some patients with midline visceral pain [9]. However, there are significant risks of complications such as incontinence and respiratory disorders. Malignant visceral pain cannot be eliminated with bilateral anterolateral cordotomy and there are risks of serious complications. Nevertheless, there are reports of successful bilateral cordotomies for pain related to pelvic cancer.

Midline Myelotomy

Midline myelotomy was originally designed to achieve bilateral spinothalamic lesions without damaging other functional fiber tract systems in the anterolateral quadrant of the spinal cord. Early midline myelotomy procedures were performed to interrupt the midline commissure over a retrocaudal length to sever the bilateral crossing fibers of the anterolateral ascending pain tracts. This commissural myelotomy for visceral pain was found to carry significant risk of loss of proprioception, dysesthesias, bowel and bladder dysfunction, and even death. When Armour in 1927 introduced open spinal midline myelotomy to achieve the same effect as bilateral cordotomy without the complications, he intended to interrupt the spinothalamic fibers as they decussate but the operation was abandoned because of unacceptable morbidity and mortality [10].

Hirshberg and colleagues postulated that the success of midline myelotomy in treating pain was due to coincidental lesioning of the midline dorsal columns. Using their own autopsies and clinical reports, they showed that some successful myelotomies were not deep enough or at incorrect levels [11]. This concept made more sense when in the 1990s Al-Chaer and colleagues discovered a new visceral pain pathway ascending in the posterior dorsal column (Fig. 24.1b) [12–14].

Hitchcock introduced the limited midline myelotomy with the intent of severing crossing spinothalamic tract fibers at only a single level. Even though the lesions were limited to the midline at a single level, lesions at C-1 and T-10 were reported to achieve widespread relief from chronic visceral abdominal pain yet spared proprioception and sensation [15, 16]. Schvarcz suggested that the pain relief resulted from destruction of a polysynaptic ascending pain tract in the central cord [17, 18] and it was soon discovered that the true culprit was in the midline of the posterior columns, an area incidentally damaged during myelotomy procedures.

With these developments in mind, Nauta et al. described a modified punctate midline myelotomy for chronic malignant lower abdominal pain in a 39-year-old woman with malignant abdominal pain from radiation damage to the bowel, bladder, and ureter in the setting of multiple abdominal surgeries. The lesion was made at the T8 level via open laminectomy. Prior to surgery, the patient reported a constant pressure in the right lower abdomen with a severe "ripping" pain after bowel movements. Afterwards, the patient reported 100 % resolution of her disabling lower abdominal pain and this effect persisted for at least 10 months postoperatively [19].

The first punctate midline myelotomies were performed via open laminectomy using an operative microscope to create a midline punctate incision with a needle inserted to a depth of 5 mm. The exact midline was determined by measuring and bisecting the distance between the two root entry zones since the dorsal vein and the septum posticum are not reliable markers of the true dorsal midline of the spinal cord.

Additional reports of successful punctate midline myelotomy were quick to follow Nauta's success with punctate midline myelotomy. Becker, Sure, and Bertalanffy reported success in treating a patient with severe visceral abdominal pain in the epi- and mesogastric regions of a 41-year-old man with multiple anaplastic carcinomas of the small intestine, peritoneal carcinosis, and retroperitoneal lymphomas. His pain was reduced from VAS 10 to 2–3 postoperatively [20]. Kim and Kwon reported performing high thoracic midline dorsal myelotomy for eight patients with severe visceral pain due to advance stomach cancer. All eight patients enjoyed relief from their preoperative abdominal pain and there were no reports of mortality. One patient suffered permanent paresthesias below the level of the myelotomy and two patients exhibited transient paresthesias that improved with corticosteroid treatment. They chose to lesion both the central gray area and the medial portion of the spinal cord and used a microdissector instead of a 16-gauge needle. They reiterate that the procedure is more effective for pain from diffuse abdominal metastases rather than patients with a large mass producing focalized pain [21].

The appropriate level for the punctate lesion is determined based on the level of the abdominal pain and the lesion level was usually made one segment above the level of the spinal cord that innervates the region causing the diffuse pain. Kim and Kwon made the lesion at T1-T4 for pain from stomach cancer [21] while Huang et al. did myelotomy at T4-5 for patients with hepatobiliary and pancreatic cancers [22]. However, based on a review of published cases, Hong et al. argue that the myelotomy should be performed several levels above the corresponding spinal cord level. For example, pain from cancers of the genitalia, rectum, or colon the dorsal midline pathway might be treated with a lesion at the T7-T8 spinal cord level [23].

Punctate midline myelotomy has been successfully performed for pain from pelvic cancer as well as pancreatic, and hepatobiliary cancer [22]. The procedure can be performed using local anesthesia with a 1–2 day hospital stay. Since the patient remains awake during the procedure, they can report sensory and motor changes, thereby reducing the risk of neurologic deficit [23].

Most case reports describe a lesion made to a depth of 5–6 mm when in reality the anterior–posterior diameter of the thoracic spinal cord varies along the length of the cord and between individuals [23]. This depth may not be appropriate based on several anatomic studies including Japanese postmortem data showing that the sagittal diameter of the spinal cord at the T4 and T10 segments is 7.59 ± 0.31 mm and 7.81 ± 0.25 mm, respectively [23–25]. For this reason, magnetic resonance imaging (MRI) must be used to calculate the depth of myelotomy. The surgeon must account for both individual anatomical variation as well as pathological changes.

Punctate midline myelotomy is the least destructive of the neuroablative procedures, however it still carries a risk of unintended injury to the dorsal column-medial lemniscus pathway. Therefore, the risks of this procedure include loss of sensation of touch, pressure, vibration, and proprioception. Bowl and bladder incontinence are also potential complications. Because the dorsal vein meanders along its course, using CT-guidance increases the risk of injuring the dorsal vein and causing a subarachnoid hemorrhage.

Despite having fewer neurological complications and proven efficacy, punctate midline myelotomy has not been standardized. Good surgical candidates have prominent visceral pain with poor pain control or intolerable side effects and a stable disease state with life expectancy greater than 3 months. They should also be in stable medical condition to minimize the risk of morbidity or mortality. Patients who have undergone radiation therapy and complain of diffuse, rather than local, visceral pain may be good candidates for neuroablation of the dorsal pain pathway [19].

Although outcome measures for myelotomy are generally less favorable then for cordotomy, myelotomy is consistently superior for relieving midline visceral cancer pain for cervical, pancreatic, and gastric cancers. While only Class III evidence exists for myelotomy for cancer pain relief, the anecdotal literature clearly supports its efficacy and highlights the potential for further development [8]. There are no prospective randomized studies of punctate midline myelotomy for abdominal pain and it has not yet become popular in the United States or Europe. Nevertheless, there is convincing empiric evidence that punctate midline myelotomy is an effective neurosurgical strategy for certain patients with diffuse visceral abdominal pain.

CT-Guided Procedures

The need to minimize invasiveness of pain relieving procedures became obvious long time ago. Back in 1963, Mullan introduced a percutaneous alternative to open cordotomy when he reported using a strontium needle to create the lesions during radiography-based percutaneous interventions [26]. Introduction of CT scanning further advanced the field enabling the practitioner to directly visualize the spinal cord and account for the differences in spinal cord diameter and shape between individual patients. CT guidance for stereotactic pain procedures dates to 1987 when first developed by Kanpolat and his colleagues [27]. They showed that CT measurements of the cervical cord are reliable and can be used to perform pain procedures such as percutaneous cordotomy.

In fact, Kanpolat et al. revolutionized and popularized the techniques employed in spinal cord tract leisoning by introducing minimally invasive methods using image guidance to achieve greater safety and efficacy in the 1990s [28]. These techniques have not yet become highly popular in the U.S. but they are becoming increasingly relevant. The costs of managing intrathecal systems are rising and there are risks of opioid hyperalgesia or dependence. Rapidly advancing intraoperative monitoring and neuronavigation technology have the potential to further increase the safety, accuracy and precision of spinal cordotomy, and midline myelotomy. With advances in medical and radiation oncology, patients with terminal cancer are living longer and also bearing a greater pain burden, further warranting the more widespread adoption of destructive spinal cord procedures for severe chronic abdominal pain.

The 2000s saw the successful application of CT-guided navigation to punctate midline myelotomy for the treatment of chronic abdominal pain. Based on the success of Nauta [29] and Kanpolat [27, 28, 30–32], Filho et al. [33] described using CT guidance to intentionally incise the dorsal midline ascending pain pathway. They reported operating on two women aged 58 and 54, with intractable pelvic visceral pain of uterine cancer. One patient remained pain free until she succumbed to her cancer 4 months later. The other patient remained pain free at the last follow up 2 months postoperatively.

Other Destructive Procedures

In addition to all of the above ablative procedures, there are numerous other destructive procedures that are currently in use for both neurosurgical and nonneurosurgical applications. These techniques include, but are not limited to, strontium-induced ablation for pain related to metastatic cancer [34], radiofrequency ablation (RFA) [35–38], cryodestruction [39–41], and chemical ablation [42, 43]. These procedures have a very broad spectrum of uses consequently making it difficult to predict the efficacy of these procedures for neurosurgical application. RFA, however, seems to offer the most promising data for successful treatment of neuropathic pain. Because of the extremely precise nature of RFA, it has very beneficial uses: primarily the ability to electrically stimulate the target before lesioning. By doing this, the surgeon has the ability to determine exactly what the effects of lesioning will be and what deficits may occur postoperatively. Additionally, it provides the ability to predict the size of the lesion and affords the possibility to do test lesions before the actual treatment. Overall, these techniques look to be promising options for the successful treatment of chronic abdominal pain. However, there has been limited research into the efficacy of these options. Therefore, more research into their applications and efficacies are needed to determine whether or not these provide long-term benefits as compared to more standard treatment options.

Conclusions

Neuroablative procedures offer the advantage of permanent pain relief and freeing the patient from the need for hospital care but this can be especially beneficial for patients with pain from advanced cancer who may not be able to tolerate frequent follow up, may not have access to a continuous supply of opioids, or have failed to find relief with any other modality.

Lack of prospective data in cases of neuroablative interventions should not stop interested clinicians in pursuing these established interventions in treatment of otherwise intractable pain syndromes. Decades of clinical experience and a large number of successful reports may prompt the pain management community to consider these procedures for their patients and consult experienced neurosurgery centers. At the same time, keeping the skill and maintaining training in neuroablative surgery is something that gives neurosurgeons a chance to stay involved in the multidisciplinary treatment process, supporting their century-long role of being the "last resort" and helping desperate patients when everything else fails.

References

1. Spiller W, Martin E. The treatment of persistent pain of organic origin in the lower part of the body by division of the anterolateral column of the spinal cord. JAMA. 1912;58:1489–90.
2. Gardner E, Cuneo HM. Lateral spinothalamic tract and associated tracts in man. Arch Neurol Psychiatr. 1945;53:423–30.
3. Wang CC, Willis WD, Westlund KN. Ascending projections form the area around the spinal cord central canal: a Phaseolus vulgaris leucoagglutinin study in rats. J Comp Neurol. 1999; 415:341–67.
4. Andy OJ. Successful treatment of long-standing hysterical pain and visceral disturbances by unilateral anterior thalamotomy. Case report. J Neurosurg. 1973;39:252–4.
5. Amano K, Kawamura H, Tanikawa T, et al. Relief of intractable pain from neurosurgical point of view with reference to present limits and clinical indications—a review of 100 consecutive cases. Neurol Med Chir (Tokyo). 1976;16:141–53.
6. Choi CR, Umbach W. Combined stereotaxic surgery for relief of intractable pain. Neurochirurgia (Stuttg). 1977;20:84–7.
7. Lende RA, Kirsch WM, Druckman R. Relief of facial pain after combined removal of precentral and postcentral cortex. J Neurosurg. 1971;34:537–43.
8. Raslan AM, Cetas JS, McCartney S, et al. Destructive procedures for control of cancer pain: the case for cordotomy. J Neurosurg. 2011;114:155–70.
9. Gybels JM, Sweet WH. Neurosurgical treatment of persistent pain. Physiological and pathological mechanisms of human pain. Pain Headache. 1989;11:1–402.
10. Armour D. Surgery of the spinal cord and its membranes. Lancet. 1927;1:691–7.
11. Gildenberg PL, Hirshberg RM. Limited myelotomy for the treatment of intractable cancer pain. J Neurol Neurosurg Psychiatry. 1984;47:94–6.
12. Al-Chaer ED, Lawand NB, Westlund KN, et al. Pelvic visceral input into the nucleus gracilis is largely mediated by the postsynaptic dorsal column pathway. J Neurophysiol. 1996;76: 2675–90.
13. Al-Chaer ED, Lawand NB, Westlund KN, et al. Visceral nociceptive input into the ventral posterolateral nucleus of the thalamus: a new function for the dorsal column pathway. J Neurophysiol. 1996;76:2661–74.
14. Hirshberg RM, Al-Chaer ED, Lawand NB, et al. Is there a pathway in the posterior funiculus that signals visceral pain? Pain. 1996; 67:291–305.
15. Hitchcock E. Stereotactic cervical myelotomy. J Neurol Neurosurg Psychiatry. 1970;33:224–30.
16. Hitchcock ER. Stereotaxis of the spinal cord. Confin Neurol. 1972;34:299–310.
17. Schvarcz JR. Spinal cord stereotactic techniques re trigeminal nucleotomy and extralemniscal myelotomy. Appl Neurophysiol. 1978;41:99–112.
18. Schvarcz JR. Stereotactic extralemniscal myelotomy. J Neurol Neurosurg Psychiatry. 1976;39:53–7.
19. Nauta HJ, Hewitt E, Westlund KN, et al. Surgical interruption of a midline dorsal column visceral pain pathway. Case report and review of the literature. J Neurosurg. 1997;86:538–42.
20. Becker R, Sure U, Bertalanffy H. Punctate midline myelotomy. A new approach in the management of visceral pain. Acta Neurochir (Wien). 1999;141:881–3.
21. Kim YS, Kwon SJ. High thoracic midline dorsal column myelotomy for severe visceral pain due to advanced stomach cancer. Neurosurgery. 2000;46:85–92.
22. Hwang SL, Lin CL, Lieu AS. Punctate midline myelotomy for intractable visceral pain caused by hepatobiliary or pancreatic cancer. J Pain Symptom Manage. 2004;27:79–84.
23. Hong D, Andren-Sandberg A. Punctate midline myelotomy: a minimally invasive procedure for the treatment of pain in inextirpable abdominal and pelvic cancer. J Pain Symptom Manage. 2007;33: 99–109.
24. Sherman JL, Nassaux PY, Citrin CM. Measurements of the normal cervical spinal cord on MR imaging. AJNR Am J Neuroradiol. 1990;11:369–72.
25. Nordquist L. The sagittal diameter of the spinal cord and subarachnoid space in different age groups. (A roentgenographic postmortem study). Acta Radiol Suppl. 1964;227:1–96.

26. Mullan S, Harper PV, Hekmatpanah J. Percutaneous interruption of spinal-pain tracts by means of a strontium 90 needle. J Neurosurg. 1963;20:931–9.
27. Kanpolat Y, Deda H, Akyar S, et al. CT-guided percutaneous cordotomy. Acta Neurochir Suppl (Wien). 1989;46:67–8.
28. Kanpolat Y, Caglar S, Akyar S, et al. CT-guided pain procedures for intractable pain in malignancy. Acta Neurochir Suppl. 1995;64:88–91.
29. Nauta HJ, Soukup VM, Fabian RH, et al. Punctate midline myelotomy for the relief of visceral cancer pain. J Neurosurg. 2000;92(2 Suppl):125–30.
30. Kanpolat Y. Percutaneous destructive pain procedures on the upper spinal cord and brain stem in cancer pain: CT-guided techniques, indications and results. Adv Tech Stand Neurosurg. 2007;32:147–73.
31. Kanpolat Y, Akyar S, Cağlar S, et al. CT-guided percutaneous selective cordotomy. Acta Neurochir Suppl. 1993;123:92–6.
32. Kanpolat Y, Savas A, Ucar T, et al. CT-guided percutaneous selective cordotomy for treatment of intractable pain in patients with malignant pleural mesothelioma. Acta Neurochir (Wien). 2002;144: 595–9.
33. Velela Filho O, Araujo MR, Florencio RS, et al. CT-guided percutaneous punctate midline myelotomy for the treatment of intractable visceral pain: a technical note. Stereotact Funct Neurosurg. 2001;77:177–82.
34. Charron M, Brown M, Rowland P, et al. Pain palliation with strontium-89 in children with metastatic disease. Med Pediatr Oncol. 1996;26:393–6.
35. Collighan N, Richardson J. Radiofrequency lesioning techniques in the management of chronic pain. Anaesth Intensive Care Med. 2005;6:56–8.
36. Van Zundert J, Patijn J, Kessels A, et al. Pulsed radiofrequency adjacent to the cervical dorsal root ganglion in chronic cervical radicular pain: a double blind sham controlled randomised clinical trial. Pain. 2007;127:173–82.
37. Boswell MV, Trescot AM, Datta S, et al. Interventional techniques: evidence-based guidelines in the management of chronic spinal pain. Pain Physician. 2007;10:7–11.
38. Lord SM, Barnsley L, Wallis BJ, et al. Percutaneous radiofrequency neurotomy for chronic cervical zygapophyseal joint pain. N Engl J Med. 1996;335:1721–6.
39. Yamauchi Y, Izumi Y, Yashiro Y, et al. Percutaneous cryoablation for pulmonary nodules in the residual lung after pneumonectomy: report of two cases. Chest. 2011;140:1633–7.
40. Kuflik EG, Gage AA. Recurrent basal cell carcinoma treated with cryosurgery. J Am Acad Dermatol. 1997;37:82–4.
41. Hanke CW, Bailin PL. Current trends in the practice of dermatologic surgery. J Dermatol Surg Oncol. 1990;16:130–1.
42. Ward S, Jenson M, Royal MA, et al. Fluoroscopy-guided sacroiliac joint injections with phenol ablation for persistent sacroiliitis: a case series. Pain Pract. 2002;2:332–5.
43. Kim DW, Baek JH, Kim YS, et al. Efficacy and safety of ethanol ablation for thyroglossal duct cysts. AJNR Am J Neuroradiol. 2011;32:306–9.

Psychological Determinants and Treatments for Chronic Abdominal Pain

25

Giries W. Sweis

Abbreviations

CBT	Cognitive-behavioral therapy
CR	Conditioned response
CS	Conditioned stimulus
EMG	Electromyographic
FGID	Functional gastrointestinal disorders
GI	Gastrointestinal
HPA	Hypothalamic–pituitary–adrenal
IBD	Inflammatory bowel disease
IBS	Irritable bowel syndrome
UR	Unconditioned response
US	Unconditioned stimulus

Introduction

Pain perceived within the abdomen may arise from a wide range of mechanisms, many of which can fall under a variety of diagnostic syndromes and categories. There are, however, frequent instances for which the cause of pain is unknown and where pain serves no warning signal function; such is the case with chronic abdominal pain. Chronic abdominal pain may begin with disorders such as chronic pancreatitis, inflammatory bowel disease (IBD), irritable bowel syndrome (IBS), and other functional gastrointestinal disorders (FGID) in which structural damage to organs contributes to pain. However, as these disorders persist or progress, the pain becomes more constant, affects daily functioning, impacts mood, and becomes less responsive to standard medical care.

In recent decades it has become increasingly evident that restrictive biomedical models are not sufficient to explain the development and maintenance of chronic pain problems [1]. For many persons with chronic pain, a single pathophysiologic mechanism that underlies all of their pain and suffering cannot be identified, and medical/pharmacologic treatments rarely, if ever, provide adequate pain control [2]. To address the limitations of biomedical models of chronic pain, more comprehensive biopsychosocial conceptualizations were developed to take into account a variety of factors (e.g., physical, social, cognitive, affective, behavioral, and sociocultural) that can contribute to chronic pain problems [3]. In this chapter, independent of the diagnosis, the reader will be presented with an overview of the different psychological understandings of pain and some of the current psychological therapies.

Psychophysiological Studies of Pain Perception

Our current knowledge related to pain arising from the internal organs of the body has most recently been extracted from psychophysical studies comparing visceral and nonvisceral stimuli [4]. These studies have been designed to determine whether uncontrolled clinical observations are in fact representative of responses evoked by visceral pain rather than nonspecific chronic pain. Various psychological factors such as depression, anxiety, and hypervigilence have been identified and differentiated, to some degree, between clinical population and health controls [5, 6].

Anatomical and physiological studies in animals, as well as functional imaging studies in humans, have shown that visceral pain perception is mediated at a cortical level [7] and influenced by cognitive mechanisms such as stress, attention, and anxiety. Increased vigilance to gastrointestinal (GI) stimuli has been shown to magnify symptoms [8], and there is evidence that psychological mechanisms such as anxiety play a role in modulating visceral sensory perception [9]. Research with controlled chemical, electrical, thermal, and mechanical stimuli [4] targeted on visceral and nonvisceral

G.W. Sweis, Psy.D., M.H.S. (✉)
Cleveland Clinic, Neurological Center for Pain,
9500 Euclid Avenue, C21, Cleveland, OH 44195, USA
e-mail: sweisg@ccf.org

L. Kapural (ed.), *Chronic Abdominal Pain: An Evidence-Based, Comprehensive Guide to Clinical Management*,
DOI 10.1007/978-1-4939-1992-5_25, © Springer Science+Business Media New York 2015

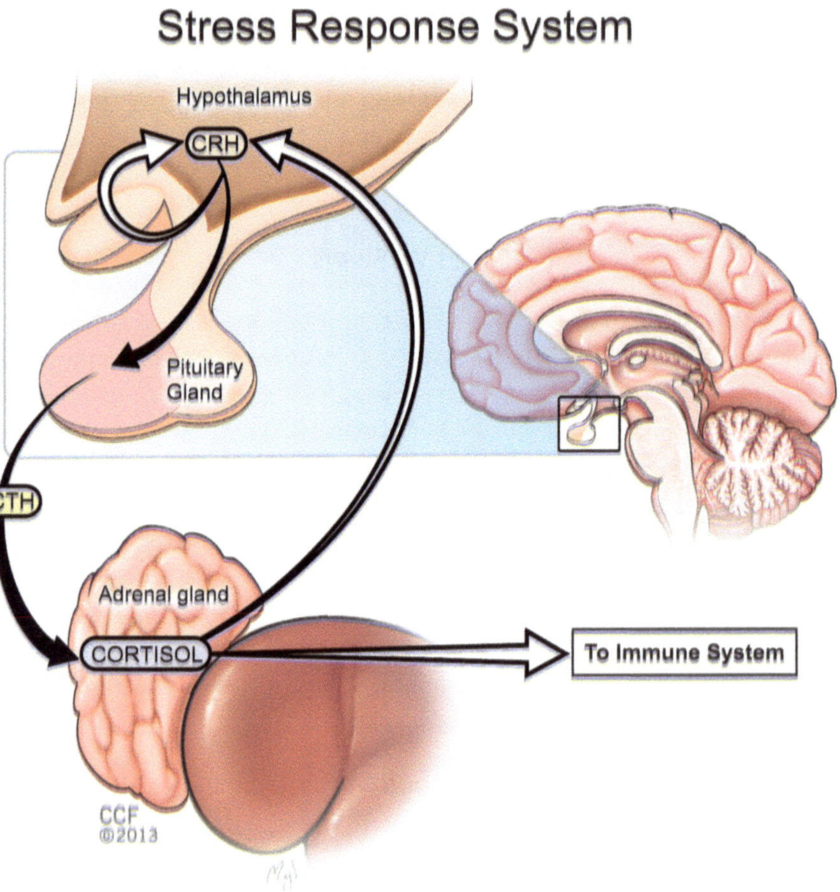

Fig. 25.1 The functional anatomy of the hypothalamic–pituitary–adrenal axis. Reprinted with permission, Cleveland Clinic Center for Medical Art & Photography © 2013. All Rights Reserved

areas has demonstrated evidence of hypersensitivity to visceral stimuli in most of the visceral pain disorders. This includes hypersensitivity to gastric distention in patients with functional dyspepsia [10], intestinal and rectal distention in patients with IBS [11, 12], biliary and/or pancreatic duct distention in patients with postcholecystectomy syndrome or chronic pancreatitis [13], and bladder distention in patients with interstitial cystitis [5].

Subpopulations within a certain diagnosis have also been identified. A number of IBS patients tested for rectal sensitivity were found to test rather consistently as hypersensitive, independent of the order of the stimulus or stimulus intensity, and others appear to be hypervigilant, with greater sensitivity associated with progressively increasing intensities of the stimulations [14]. It has also been shown that anticipatory knowledge, as compared to mental distraction, increases perception without modifying intestinal reflexes and that mental activity may modulate gut perception and override effects of somatic stimuli on gut perception [8].

These data raise the possibility that visceral pain patients are hypervigilant and pay more attention to gut events. Cognitive processes may selectively regulate the sensitivity to gut stimuli, while visceral reflexes operate independently. Symptoms of gastric distention may be modified by anxiety and to a degree induced by mental stress. Furthermore, it has been shown that psychological mechanisms also modulate gut perception.

Stress and Visceral Pain

Stressors can be thought of as being internal, such as infection or inflammation (physical), or external, such as a perceived threat (psychological) [15, 16]. These stressors evoke a complex network of adaptive responses, which serves to stabilize or balance the body's internal environment and ensure survival. The organism achieves stability, or homeostasis, through what is known as allostasis [17] which is essential in order to maintain internal viability amid changing conditions. Allostasis involves a number of neurobiological systems, such as the hypothalamic–pituitary–adrenal (HPA) axis and autonomic nervous system (Fig. 25.1). In healthy individuals, these neurobiological response systems actively adjust to both predictable and unpredictable events, in order to synchronize the stress response so the body is able to cope effectively with stressors [18].

Stress has been well documented in the pathophysiology, presentation, and treatment outcome in clinical pain states, in particular functional gastrointestinal disorders such as IBS [19–21]. Certain stressful life events have been linked to the onset and exacerbation of a number of GI tract disorders including functional gastrointestinal disorders (FGID) [22], post-infective-IBS [20], and IBD [23]. Stress has been shown to induce changes in gastrointestinal function and influences the development of visceral pain in IBS patients [21, 22, 24]. Psychopathology such as anxiety, somatization, neuroticism, hypochondriasis, and prior adverse life events have all been reported to increase the development of IBS [20, 25]. Childhood abuse and acute episodes of severe stress in adulthood such as sexual trauma have been reported to be significant risk factors in the development of FGID [26, 27]. There is also evidence of increased prevalence of GI symptoms and IBS in those who suffer from post-traumatic stress disorder [28–30]. Undoubtedly, stress and psychological factors play a significant role in GI disorders with convergent research demonstrating that childhood trauma (neglect, abuse, loss of caregiver, or life-threatening situation) impact the susceptibility to develop visceral pain and comorbidity with anxiety, depression, and emotional distress [31–33].

A number of animal studies have demonstrated evidence of stress-related visceral responses and susceptibility. In rodent and nonhuman primate studies, neonatal maternal separation and acute psychological stress were associated with increased stress reactivity. In adulthood, these animals showed greater activation of the HPA axis, sympathoadrenomedullary systems, and central monoaminergic systems, and thus, greater vulnerability for stress-induced illness [34, 35]. In another study using colonic distention in rodents, exposure to various stressors, including early-life stress, induced abdominal contractions [21, 36]. These investigations demonstrate that the responsiveness of these physiological systems can be altered by adverse life events, and appear to increase the organism's susceptibility to the negative effects of stress in later life.

Psychological Conceptualization of Pain

The attempt to understand pain in psychological terms has a very old history and there are a number of different ways to conceptualize pain. Although no single theoretical model can fully explain chronic pain, it is important to examine different conceptualizations, because the way individuals report pain and describe their symptoms will guide the methods to evaluate the individual and the type of treatment to initiate.

Psychogenic Pain

Of the earliest theories to explain pain, the psychoanalytic or psychodynamic view postulates that pain originates from psychological mechanisms. This notion has dominated the psychology of pain and has stirred controversy in the field of pain medicine, not only regarding its prevalence, but its very existence [37]. Psychogenic pain is essentially assumed in the absence of organic findings, if the pain is judged to be disproportionate to the physical pathology, or if the symptoms are refractory or unresponsive to standard treatment, then persistent pain is deemed to be primarily due to psychiatric illness.

A number of models of the psychogenic theory have been proposed. The predominant pain-related psychological model was the "pain-prone" personality which described predisposing psychological factors for chronic pain. The pain-prone disorder is characterized by denial of emotional and interpersonal problems, inactivity, depressed mood, guilt, inability to deal with anger and hostility, insomnia, craving for affection and dependency, lack of initiative, and a family history of depression, alcoholism, and chronic pain [38, 39]. This perspective assumes that pain may simply originate from psychological mechanisms, and once the psychic organization necessary for pain is established, the experience of pain no longer requires peripheral stimulation [39]. The model was later modified to suggest that inhibition of affect played a significant role in the etiology of chronic pain [40], and that difficulties in expressing anger and controlling intense emotions are predisposing factors linking chronic pain and negative affect [38, 39].

In the end, the psychoanalytic or psychodynamic theory forms the basis for the distinction of an underlying attempt to identify functional versus organic etiology. Put simply, a dichotomous reasoning is invoked. There is little research to support the etiological role of a pain-prone personality or inhibition of affect in chronic pain states [41]. However, personality attributes such as introversion/extraversion, optimism, perceived locus of control, and personality disorders appear to have a significant effect on patients' ability to cope with pain [42].

Respondent Conditioning in Pain

Certain types of stimuli typically elicit specific types of bodily responses. A person blinks when a puff of air is directed at the eye. The pupil of the eye constricts on exposure to bright light. Salivation occurs when food is in the mouth. These and other responses (Table 25.1) are called unconditioned responses (UR). These responses are elicited by antecedent stimuli even though no conditioning or learning has taken place. A UR is a natural reflexive action of the body that occurs when an unconditioned stimulus (US) is presented. URs are common to all people. Respondent conditioning (also called classical conditioning or Pavlovian conditioning) occurs when a previously neutral stimulus is paired with a US (the neutral stimulus and the US are presented together). As a result of this pairing,

Table 25.1 Respondent model

Unconditioned stimulus	Unconditioned response
Object touches infant's lips	Sucking reflex
Food in mouth	Salivation
Foreign object in throat	Gag reflex
Stimulation in throat	Coughing
Puff of air in the eye	Eye blink
Bright light in the eye	Pupil constriction
Painful stimulation to hand	Rapid withdrawal (e.g., hand from hot stove)

The occurrence of an unconditioned stimulus (US) with an unconditioned response (UR)

Table 25.2 Operant conditioning model of chronic pain

Response	Stimulus	
Complaining	Attention	Positive reinforcement of pain behavior
Medication intake	Pain reduction	Negative reinforcement of pain behavior
Inactivity/rest	Pain reduction	Negative reinforcement of pain behavior
Functional activity	Lack of positive reinforcement	Extinction of well behavior

the neutral stimulus becomes a conditioned stimulus (CS) and elicits a conditioned response (CR) similar to the UR [43].

In acute pain states it may be useful to reduce movement, and consequently avoid pain, to accelerate the healing process. However, in a respondent model of the development of chronic pain, frequent co-occurrence of harmless stimuli, such as activities, body positions, and environments (conditioned stimulus, CS) with acute pain (unconditioned stimulus, US) that may elicit motor, sympathetic, and endocrine responses (conditioned response, CR) may result in a direct relationship between the previously neutral stimuli (activities, body positions and environments) and the physiological responses (motor, sympathetic, and endocrine) [44]. This process of conditioning may lead to a pain-tension cycle that may maintain the chronic pain problem independently from the original tissue damage. Patients may learn to associate increased pain with all kinds of stimuli that were originally associated with nociceptive stimulation [45, 46]. Not only could it be possible for conditioned stimuli to contribute to the maintenance of pain indirectly (through motor, sympathetic, and endocrine systems), but it may also be possible for them to develop maladaptive responses to many stimuli and reduce the frequency of their performance of many activities other than those that initially elicited pain [47, 48].

Operant Conditioning in Pain

Operant conditioning (sometimes referred to as instrumental conditioning) is a method of learning that occurs through rewards and punishments for behavior [43]. Through operant conditioning, an association is made between a behavior and a consequence for that behavior. It is distinguished from respondent conditioning (or classical conditioning) in that operant conditioning deals with the modification of "voluntary behavior" or operant behavior. Operant behavior operates on the environment and is maintained by its consequences, while respondent conditioning deals with the conditioning of reflexive behaviors which are elicited by antecedent conditions.

The operant model of chronic pain hypothesizes that the perpetuation of pain behaviors after healing contributes to chronic pain problems and that these behaviors can contribute to suffering and disability [49, 50]. The model focuses on overt pain behaviors (e.g., limping, guarding, rubbing, pain medication use, inactivity) and well behaviors (i.e., adaptive behaviors, such as working and engaging in activities) exhibited by persons with pain (Table 25.2). It is theorized that pain behaviors are natural responses to acute pain that can persist after healing if they are reinforced and if competing well behaviors are not sufficiently reinforced. Over time, this can lead to pain behaviors occurring, at least in part, in response to environmental contingencies and discriminative stimuli (e.g., spouses, other family members, or healthcare providers who might reinforce pain behaviors) instead of only in response to nociception. For the most part, such behaviors involve limiting one's activity and functioning.

Although operant factors undoubtedly play a role in chronic pain, a fundamental problem with the operant approach is its emphasis on the communicative role of pain behaviors rather than on pain itself. According to this approach, observed behaviors are used as the basis to infer something about the internal state of the person. The observer may infer that certain behaviors are communications of pain, distress, and suffering. However, there is no way to determine from the behavior alone whether it results from pain, conditioning, structural abnormality, adherence to provider recommendation, pacing of activities, or whether it is a coping response. Unless observers ask patients why they are engaged in a certain behavior, they cannot know the motivation for the behavior.

Social Learning in Pain

Social learning theory is a perspective that states that people learn within a social context. It is facilitated through concepts such as observational learning, or modeling, and is an essential mechanism of learning new patterns of behaviors [51]. Individuals acquire perceptions and interpretations of symptoms and physiological processes from modeling and observations in their social environment. They also learn appropriate responses to injury and disease and consequently may be more or less likely to under or overrespond to the normal physical sensations they experience. A large number of studies have shown that physiological responses to pain stimuli may be vicariously conditioned during observations of others in pain [52] and that modeling can influence the expression and localization of pain and pain-coping behaviors, and pain perceptions in others activate both automatic and controlled processes [53].

Investigations of social learning have indicated that observational learning influences both observable expression of acute pain as well as the subjective experience [54]. In an experimental study, children observing their mothers' reactions during painful exposure of the hand to cold water subsequently displayed lower pain thresholds when their mothers had voluntarily exaggerated their pain [55]. Furthermore, children displayed reduced facial displays of pain when the mother had voluntarily suppressed her reaction. Instructive studies that examined observational learning of pain-related fear [56, 57], provided substantial evidence of the powerful impact that observing others in pain had on neurophysiological activity in observers. However, the broader mechanisms whereby observational learning plays a crucial role in establishing pain-related responses remain relatively uninvestigated; no systematic research is available examining cognitive and affective mechanisms (e.g., changes in beliefs about pain and attitudes toward pain) underlying the effects on behavior, including the moderators of these effects. Despite the great deal of data available on the modification of experimentally induced pain behaviors by means of modeling in normal (healthy) people, there are few experimental results concerning chronic pain patients.

Affective and Cognitive Factors in Pain

A plethora of psychological factors have been demonstrated to play important roles in chronic pain. Numerous studies have documented a strong association between chronic pain, affective factors, and the role of cognitive variables. They are each expressed within the context of chronic pain and are most often associated with depressive disorders and anxiety disorders.

Depression

A growing body of literature has focused on the interaction between depression and pain symptoms. This interaction has been labeled by some authors as the depression–pain syndrome [58] or depression–pain dyad, implying that the conditions often coexist, respond to similar treatments, exacerbate one another, and share biological pathways and neurotransmitters [59, 60]. Patients with depression often present with a complex set of overlapping symptoms, including emotional and physical complaints. Physical complaints typically include medically unexplained pain [61].

Although it is generally understood that depression and painful symptoms are common comorbidities, the extent to which depression and chronic pain are associated remains a contentious issue that empirical studies have failed to resolve it completely. Evaluation of the relevant literature provides increased support for an association between chronic pain and depression and suggests that coexisting syndromes may be a final common presentation [62]. However, in most cases, depression appears to be more of reaction to the condition [63].

Anxiety

Comorbidity between anxiety disorder and pain seems to be prevalent [64]. Studies conducted among selected groups of respondents have found the presence of pain to be associated with anxiety symptoms [65]. More specifically, there is ample evidence indicating that fear and avoidance are crucial factors not only in acute pain but in chronic pain [46]. Researchers have also found a close association between fear of pain and dysfunctional coping [66, 67] with muscle hyperactivity in response to stress being closely associated with the fear of pain [68]. Evidence from large epidemiologic studies suggests that the association of anxiety disorders with chronic pain is comparable to other mood disorders. The link between anxiety disorders and chronic pain, when they co-occur, is often complex and may not always lend itself to a simple cause-and-effect relationship, as both conditions may have a shared distal, rather than a proximal origin.

Anger

Anger has long been recognized as an integral part of pain experience [69, 70]. It has been discussed as an aversive emotional state ranging from mild irritation to fury [71], and comprising specific cognitive attributions and action tendencies [72, 73]. Anecdotal and empirical data suggest that anger is commonplace among chronic pain sufferers [74] and that inhibiting anger expression may amplify acute and chronic pain at a later time [75]. Studies have consistently indicated that greater trait of anger is associated with elevated acute and chronic pain responsiveness [76–78]. In other studies, increased sensitivity to acute pain was found, in roughly equal degrees, among participants who claim to become easily angered, those who claim that they bottle up their anger, and those who claim that they vent their anger more explosively [40, 79, 80].

Beliefs About Pain

Beliefs have been defined as personally or culturally shared cognitive configurations [81] that may be generalized or specific to certain contexts, mold the individual's perception of the environment, and shape the meaning of their experiences [82]. It has been suggested that beliefs about persistent pain have two dimensions. These dimensions are organic pain beliefs (the physiological experience of pain) and psychological pain beliefs (the internal influences and feelings of the experience of pain) [83]. Both dimensions can potentially influence viewpoints about pain control either positively (having personal control over the pain experience) or negatively (feeling helpless to manage the potential threat to their well-being).

Research shows that negative pain beliefs have a detrimental impact on patients' overall health, self-efficacy, and function [84]. A number of studies reported that negative pain beliefs can contribute to the transition from acute pain to persistent pain [81, 85]. Beliefs of physical capabilities and not the experience of self-reported pain, appears to affect physical functioning and contributes to disability [84–86]. Conviction of personal control was shown to ameliorate the experience of experimentally induced nociception in a study that used experimental pain stimuli [87]. Specific pain beliefs that contribute to poor compliance, motivation, and misunderstanding about pain have been identified. These include catastrophizing, limited perception of control over the pain experience, and emotional distress [85] with evidence that addressing negative pain beliefs in the management of persistent pain can affect treatment outcomes [88–90].

Self-Efficacy

Based on the theory of social learning, self-efficacy describes the confidence the person has in his or her own ability to achieve a desired outcome [91]. Higher levels of self-efficacy have been found to be associated with lower levels of pain and disability in patients with chronic pain [92–94]. In a study by Dolce et al. [95], self-efficacy was found to be related to exercise performance in chronic pain patients. In their study, they found that beliefs regarding ability to exercise predicted improvements in work status and exercise level 6–12 months after a physical reactivation program. In another study, researchers demonstrated that pain-related self-efficacy ratings are likely to change following cognitive-behavioral management and that these changes were associated with better outcomes such as reduced disability [96, 97]. In a study that asked patients to rate their self-efficacy and expectancy on performance of movement tasks, performance levels were highly related to their self-efficacy expectations, which in turn appeared to be determined by their expectancy of pain levels [98]. A number of other studies have reported that success in response to rehabilitation was predicted by perceived self-efficacy [99, 100].

Catastrophizing

The term catastrophizing was formally introduced by Albert Ellis [101] and subsequently adapted by Aaron Beck [102] to describe a maladaptive cognitive style employed by patients with anxiety and depressive disorders. At the core of their definitions of catastrophizing was the concept of an irrationally negative forecast of future events. Similarly, pain-related catastrophizing is broadly conceived as a set of exaggerated and negative cognitive and emotional schema brought to bear during actual or anticipated painful stimulation [103, 104].

The catastrophizing literature to date provides rather demonstrative evidence for the influence of catastrophizing in shaping emotional, functional, and physiological responses to pain, and was found to be the strongest pain predictor, followed by pain-related fear and bodily vigilance [105]. Catastrophizing has been shown to be associated with persistent pain and to be a predictor of poor outcomes in pain management interventions [89, 106]. The literature also points to consistent and generally robust associations between pain catastrophizing and an array of clinical pain-related outcomes, including measures of clinical pain severity, pain-related activity interference, disability, depression (and other negative mood indices), and alterations in social support networks [107, 108]. Moreover, catastrophizing has been linked to increased behavioral expressions of pain, as well as a variety of illness behaviors (e.g., more frequent visits to healthcare professionals). It is important to note that the magnitude of these relationships is variable, with catastrophizing accounting for minimal variance in pain severity in some studies, and up to 31 % of the variance in pain severity in others [107].

Acceptance

Acceptance is emerging as a potentially valuable concept in contemporary theories of how patients react and adapt to chronic pain. It denotes a state of mind where a person is aware of the pain but does not try to actively change or avoid it [109, 110]. In a study examining patients awaiting interdisciplinary treatment, researchers found that acceptance of chronic pain was more successful in predicting pain, depression, disability, pain-related anxiety, and patient physical and vocational functioning than were measures of coping [111]. A host of studies demonstrated that acceptance is associated with better physical, social, and emotional functioning, in work-related functioning, and in analgesic and healthcare use [112–115]. Other studies have shown convincingly that acceptance of chronic pain is associated with reports of less pain, psychological distress, and physical and psychological disability, as well as more daily up time [112, 116].

Psychological Interventions

A significant literature has developed, primarily on IBS, examining psychological variables in functional gastrointestinal disorders (FGID). Psychological treatments can help improve FGID outcomes, and such treatment should be considered for patients who have moderate or severe symptoms, who show inadequate response to usual medical care, or whose pain is clearly exacerbated by stress or emotional symptoms [117, 118]. Although such studies have been criticized for small sample sizes, generalizability of diagnoses, use of waiting list control groups, and various methodological flaws, the existing literature on psychological treatments found an overall effectiveness in reducing symptoms [119].

Biofeedback

Three professional biofeedback organizations, the Association for Applied Psychophysiology and Biofeedback, Biofeedback Certification International Alliance, and the International Society for Neurofeedback and Research, arrived at a consensus definition of biofeedback in 2008 [120].

> Biofeedback is a process that enables an individual to learn how to change physiological activity for the purposes of improving health and performance. Precise instruments measure physiological activity such as brainwaves, heart function, breathing, muscle activity, and skin temperature. These instruments rapidly and accurately 'feedback' information to the user. The presentation of this information—often in conjunction with changes in thinking, emotions, and behavior—supports desired physiological changes. Over time, these changes can endure without continued use of an instrument.

Simply put, biofeedback is a means for gaining control of our physiological functioning (e.g., skin temperature, muscle tone, skin conductance, heart rate) primarily using instruments that provide information (e.g., visual) on the activity of those same systems with a goal of being able to manipulate them at will. Although not always the case, biofeedback may typically include training in relaxation procedures.

The biofeedback literature on chronic abdominal pain is limited and primarily based on IBS studies. Some studies have been promising, but the most current review indicated that there was insufficient evidence to support the use of biofeedback for some GI conditions [121]. The review concluded that the evidence is insufficient to support the efficacy of biofeedback for constipation, encopresis, fecal incontinence, and abdominal pain. However, a study using thermal biofeedback and diet (increased fiber) was shown to be effective and efficient as a treatment modality for recurrent abdominal pain [122]. Forehead Electromyographic (EMG) biofeedback and thermal biofeedback have shown some benefits to counteract the effects of stress in patients with IBS [123]. EMG biofeedback demonstrated benefits on reduction of constipation [124–126], but a multicomponent behavioral intervention for IBS that included thermal biofeedback was not more effective than an attention-control intervention [127]. Relaxation training, involving progressive muscle relaxation, thermal biofeedback, cognitive therapy, and education, implemented through biofeedback training in patients with IBS was found to be more effective than the control group (waiting list) in reductions in GI symptoms (e.g., abdominal pain, constipation, and diarrhea) [128–130]. However, wait-list control groups do not adequately control for expectancy (placebo), and such experimental designs may overestimate treatment effects. Biofeedback therapy using breathing exercises on autonomic imbalance in patients with functional dyspepsia improved tolerance to water intake and improved quality-of-life scores, but clinical outcome and control treatment were both poorly defined [131]. Ability to significantly influence gastric contraction was demonstrated after 4 h of heart beat biofeedback training [132]. Although it is possible to teach individuals to alter their gastric motility, the clinical utility of research protocols aimed at influencing gastric motility has yet to be established.

Hypnosis

According to the American Society of Clinical Hypnosis, "Hypnosis is a state of inner absorption, concentration and focused attention." [133]. While there is general agreement that certain effects of hypnosis exist, there are differences of opinion among research and clinical communities about how analgesic hypnosis works. Some researchers generally describe hypnotic analgesic interventions as including the induction of relaxed states of focused attention and inner absorption, with a relative suspension of peripheral awareness, combined with suggestions for analgesia [134].

Hypnotherapy has been shown to be a strongly supported psychological intervention for IBS with at least three separate randomized controlled trials supporting its efficacy in reducing symptomatology and improving quality of life [135]. In a systematic review, hypnotherapeutic interventions showed long-lasting symptom relief with influences on colorectal sensitivity, colorectal motility, and mental strain (anxiety, depression, maladaptive cognitions) [136]. In another review, the authors concluded that hypnosis consistently produced significant results and improved the cardinal symptoms of IBS in the majority of patients, as well as positively affecting non-colonic symptoms [137]. Hypnotherapy also normalized visceral pain thresholds in IBS patients; threshold-associated changes were highly correlated with improvement in clinical symptoms [138]. In addition, hypnotherapy demonstrated reduced sensory and motor components of gastrocolonic responses in patients with IBS [139]. Hypnotherapy studies with IBS patients have yielded promising results, however, the underlying mechanisms of action are not well understood. Hypnotherapy appears to maintain its long-term benefits quite well, with 81 % maintaining improvements in IBS symptoms for up to 5 years [140].

Cognitive Behavior Therapy

Cognitive behavior therapy (CBT) is a type of psychotherapeutic treatment that helps patients understand thoughts and feelings that influence behaviors. The underlying concept behind CBT is that one's thoughts and feelings play a fundamental role in behavior. Cognitive behavior therapy teaches patients a blend of behavioral skills (e.g., relaxation and pain-coping skills training) and cognitive therapy (e.g., restructuring negative cognitions such as catastrophizing) with the goal of reducing pain, pain-related disability and distress, as well as increasing self-efficacy [141].

There is a significant body of research indicating that CBT is effective in reducing IBS symptoms of abdominal pain, diarrhea, and constipation [142, 143]. CBT for IBS usually involves teaching the patient-specific strategies for calming the body, coping with unpleasant symptoms, and facing difficult situations. In one study, seven IBS patients received 8 sessions of cognitive-behavioral therapy. Before initiating the treatment, all patients were assessed for psychological performance and severity of GI signs. After the treatment, it was observed that five did not have any IBS signs. Although the frequency of expression of symptoms by patients did not decrease, the frequency of depression and anxiety decreased significantly. As a whole, the results indicated that cognitive-behavioral therapy reduces the disability caused by IBS; however, it did not affect the expression of the symptoms by patients [144]. A study that reviewed all psychological treatments applied to persons suffering from IBS found that cognitive therapy was significantly efficient [145] and that cognitive therapy was demonstrated to be quite effective compared to placebo, with effects lasting nearly 4 years [146]. In IBS, both progressive muscle relaxation training [147] and cognitive therapy [148] reduced GI symptoms more than self-help interventions. In a well-designed multisite trial of CBT for severe functional bowel disorder, CBT was reported to be more effective than education alone across IBS subtypes [149].

Multidisciplinary Treatment

Research has continually demonstrated that patients with chronic nonmalignant pain disorders attain significant benefits from treatment in a multidisciplinary chronic pain rehabilitation program [150]. One study has shown that repeated sessions of patient education, cognitive therapy, thermal biofeedback, and progressive muscle relaxation resulted in both short-term and long-term benefits [151]. Multidisciplinary treatment has shown significant reductions in abdominal pain and diarrhea for irritable bowel syndrome [129] with improvements lasting up to 2 years following program completion [152]. It has also shown to be superior to standard care in randomized clinical trials in the treatment of chronic pelvic pain [153]. However, despite positive reviews highlighting the effectiveness of these programs [154], randomized clinical trials have not determined the necessary components of multidisciplinary treatment.

Conclusion

Chronic abdominal pain is a complex physical and psychological phenomenon; this review has touched only briefly on some of its facets. The care of patients with chronic abdominal pain requires understanding the psychological mechanism patients utilize to conceptualize their pain; appreciating the clinical, physical, and psychosocial features that characterize and amplify patients' pain-related symptoms and behaviors; and implementing a variety of treatment options tailored to the needs of patients. An integrated biopsychosocial treatment approach to chronic pain is intuitively logical, and when properly implemented, offers promising outcomes.

References

1. Novy DM, Nelson DV, Francis DJ, Turk DC. Perspectives of chronic pain: an evaluative comparison of restrictive and comprehensive models. Psychol Bull. 1995;118(2):238–47.
2. Loeser JD. What is chronic pain? Theor Med. 1991;12:213–25.
3. Turk DC, Monarch ES. Biopsychosocial perspective on chronic pain. In: Turk DC, Gatchel RJ, editors. Psychological approaches to pain management: a practitioner's handbook. 2nd ed. New York: Guilford Press; 2002. p. 3–29.
4. Drews AM, Gregersen H, Arendt-Mielsen L. Experimental pain in gastroenterology: a reappraisal of human studies. Scand J Gastroenterol. 2003;38:1115–30.
5. Ness TJ, Powell-Boone T, Cannon R, Lloyd LK, Fillingim RB. Psychophysical evidence of hypersensitivity in subjects with interstitial cystitis. J Urol. 2005;173(6):1983–7.
6. Verne GN, Robinson ME, Price DD. Hypersensitivity to visceral and cutaneous pain in the irritable bowel syndrome. Pain. 2001; 93(1):7–14.
7. Dunckley P, Wise RG, Aziz Q, Painter D, Brooks J, Tracey I, et al. Cortical processing of visceral and somatic stimulation: differentiating pain intensity from unpleasantness. Neuroscience. 2005;133(2): 533–42.

8. Accarino AM, Azpiroz F, Malagelda JR. Attention and distraction: effects on gut perception. Gastroenterology. 1997;113(2):415–22.
9. Ford MJ, Camilleri M, Zinsmeister AR, Hanson RB. Psychosensory modulation of colonic sensation in the human transverse and sigmoid colon. Gastroenterology. 1995;109(6):1772–80.
10. Salet GA, Samsom M, Roelofs JM, VanBerge-Henegouwen GP, Smout AJ, Akkermans LM. Responses to gastric distension in functional dyspepsia. Gut. 1998;42(6):823–9.
11. Mertz H. Visceral hypersensitivity. Aliment Pharmacol Ther. 2003;17(5):623–33.
12. Ritchie J. Pain from distension of the pelvic colon by inflating a balloon in the irritable colon syndrome. Gut. 1973;14(2):125–32.
13. Corazziari E, Shaffer EA, Hogan WJ, Sherman S, Toouli J. Functional disorders of the biliary tract and pancreas. Gut. 1999;45 Suppl 2:48–54.
14. Naliboff BD, Munakata J, Fullerton S, Gracely RH, Kodner A, Harraf F, et al. Evidence for two distinct perceptual alterations in irritable bowel syndrome. Gut. 1997;41(4):505–12.
15. Selye H. Stress and the general adaptation syndrome. Br Med J. 1950;1(4667):1383–92.
16. Chrousos GP, Gold PW. The concepts of stress and stress system disorders. Overview of physical and behavioral homeostasis. JAMA. 1992;267(9):1244–52.
17. Carlson NR. Emotions and stress. In: Carlson NR, editor. Physiology of behavior. 5th ed. Boston: Allyn and Bacon; 1994. p. 359–69.
18. Mayer EA. The neurobiology of stress and gastrointestinal disease. Gut. 2000;47(6):861–9.
19. Bennett EJ, Tennant CC, Piesse C, Badcock CA, Kellow JE. Level of chronic life stress predicts clinical outcome in irritable bowel syndrome. Gut. 1998;43(2):256–61.
20. Gwee KA, Leong YL, Graham C, McKendrick MW, Collins SM, Walters SJ, et al. The role of psychological and biological factors in postinfective gut dysfunction. Gut. 1999;44(3):400–6.
21. Mönnikes H, Tebbe JJ, Hildebrandt M, Arck P, Osmanoglou E, Rose M, et al. Role of stress in functional gastrointestinal disorders. Evidence for stress-induced alterations in gastrointestinal motility and sensitivity. Dig Dis. 2001;19(3):201–11.
22. Whitehead WE, Crowell MD, Robinson JC, Heller BR, Schuster MM. Effects of stressful life events on bowel symptoms: subjects with irritable bowel syndrome compared with subjects without bowel dysfunction. Gut. 1992;33(6):825–30.
23. Garrett VD, Brantley PJ, Jones GN, McKnight GT. The relation between daily stress and Crohns disease. J Behav Med. 1991;14(1): 87–96.
24. Mayer EA, Naliboff BD, Chang L, Coutinho SV. Stress and the gastrointestinal tract. V. Stress and irritable bowel syndrome. Am J Physiol Gastrointest Liver Physiol. 2001;280(4):G519–24.
25. Gwee KA, Graham JC, McKendrick MW, Collins SM, Marshall JS, Walters SJ, et al. Psychometric scores and persistence of irritable bowel after infectious diarrhea. Lancet. 1996;347(8995):150–3.
26. Stam R, Akkermans LM, Wiegant VM. Trauma and the gut: interactions between stressful experience and intestinal function. Gut. 1997;40(6):704–9.
27. Drossman DA, Talley NJ, Leserman J, Olden KW, Barreiro MA. Sexual and physical abuse and gastrointestinal illness. Review and recommendations. Ann Intern Med. 1995;123(10):782–94.
28. Cohen H, Jotkowitz A, Buskila D, Pelles-Avraham S, Kaplan Z, Neumann L, et al. Post-traumatic stress disorder and other co-morbidities in a sample population of patients with irritable bowel syndrome. Eur J Intern Med. 2006;17(8):567–71.
29. Irwin C, Falsetti SA, Lydiard RB, Ballenger JC, Brock CD, Brener W. Comorbidity of posttraumatic stress disorder and irritable bowel syndrome. J Clin Psychiatry. 1996;57(12):576–8.
30. Savas LS, White DL, Wieman M, Daci K, Fitzgerald S, Laday Smith S, et al. Irritable bowel syndrome and dyspepsia among women veterans: prevalence and association with psychological distress. Aliment Pharmacol Ther. 2009;29(1):115–25.
31. Halpert A, Drossman D. Biopsychosocial issues in irritable bowel syndrome. J Clin Gastroenterol. 2005;39(8):665–9.
32. Kamm MA. The role of psychosocial factors in functional gut disease. Eur J Surg Suppl. 1998;583:37–40.
33. Nemeroff CB. Neurobiological consequences of childhood trauma. J Clin Psychiatry. 2004;65 Suppl 1:18–28.
34. Meaney MJ, Diorio J, Francis D, Widdowson J, LaPlante P, Caldji C, et al. Early environmental regulation of forebrain glucocorticoid receptor gene expression: implications for adrenocortical responses to stress. Dev Neurosci. 1996;18(1–2):49–72.
35. Coplan JD, Andrews MW, Rosenblum LA, Owens MJ, Friedman S, Gorman JM, et al. Persistent elevations of cerebrospinal fluid concentrations of corticotropin-releasing factor in adult nonhuman primates exposed to early-life stressors: implications for the pathophysiology of mood and anxiety disorders. Proc Natl Acad Sci U S A. 1996;93(4):1619–23.
36. Ness TJ, Gebhart GF. Colorectal distension as a noxious visceral stimulus: physiologic and pharmacologic characterization of pseudoaffective reflexes in the rat. Brain Res. 1988;450(1–2): 153–69.
37. Covington EC. Psychogenic pain: what it means, why it does not exist, and how to diagnose it. Pain Med. 2000;1:287–94.
38. Blumer D, Heilbronn M. Chronic pain as a variant of depressive disease. The pain-prone disorder. J Nerv Ment Dis. 1982;170: 381–406.
39. Engel G. Psychogenic pain and the pain-prone patient. Am J Med. 1959;26:899–18.
40. Beutler LE, Engle D, Oro'-Beutler ME, Daldrup R, Meredith KJ. Inability to express intense affect: a common link between depression and pain? J Consult Clin Psychol. 1986;54(6):752–9.
41. Turk DC, Salovey PJ. Chronic pain as a variant of depressive disease a critical reappraisal. Nerv Ment Dis. 1984;172(7):398–404.
42. Gatchel RJ, Weisberg JN. Personality characteristics of patients with pain. Washington, DC: American Psychological Association; 2000.
43. Miltonberger RG. Respondent conditioning. In: Miltonberger RG, editor. Behavior modification: principles and procedures. 2nd ed. Belmont: Wadsworth-Thompson; 2001. p. 137–53.
44. Gentry WD, Bernal GA. Chronic pain. In: Williams R, Gentry WD, editors. Behavioral approaches to medical treatment. Cambridge: Ballinger; 1977. p. 173–82.
45. Linton SJ, Melin L, Gotedtam KG. Behavioral analysis of chronic pain and its management. In: Herson M, Bellack A, Eisler M, editors. Progress in behavior modification. New York: Academic; 1985. p. 1–42.
46. Leeuw M, Goossens ME, Linton SJ, Crombez G, Boersma K, Vlaeyen JW. The fear-avoidance model of musculoskeletal pain: current state of scientific evidence. J Behav Med. 2007;30(1): 77–94.
47. Vlaeyen JW, Linton SJ. Fear-avoidance and its consequences in chronic musculoskeletal pain: a state of the art. Pain. 2000;85(3): 317–32.
48. Rhudy JL, Meagher MW. Fear and anxiety: divergent effects on human pain thresholds. Pain. 2000;84(1):65–75.
49. Fordyce WE. Behavioral methods for chronic pain and illness. St. Louis: Mosby; 1976.
50. Fordyce WE, Fowler RS, Lehmann JF, DeLateur BJ. Some implications of learning in problems of chronic pain. J Chronic Dis. 1968;21(3):179–90.
51. Ormrod JE. Educational psychology: developing learners. 5th ed. Upper Saddle River: Prentice Hall; 2006.
52. Rickard K. The occurrence of maladaptive health-related behaviors and teacher-rated conduct problems in children of chronic low back pain patients. J Behav Med. 1988;11(2):107–16.

53. Armitage KJ, Schneiderman LJ, Bass RA. Response of physicians to medical complaints in men and women. JAMA. 1979;241(20):2186–7.
54. Craig KD, Prkachin KM. Social modeling influences on sensory decision theory and psychophysiological indexes of pain. J Pers Soc Psychol. 1978;36:805–15.
55. Goodman JE, McGrath PJ. Mothers' modeling influences children's pain during a cold pressor task. Pain. 2003;104:559–65.
56. Craig KD, Lowery HJ. Heart-rate components of conditioned vicarious autonomic responses. J Pers Soc Psychol. 1969;11:381–7.
57. Jackson PL, Meltzoff AN, Decety J. How do we perceive the pain of others? A window into the neural processes involved in empathy. Neuroimage. 2005;24:771–9.
58. Lindsay PG, Wyckoff M. The depression-pain syndrome and its response to antidepressants. Psychosomatics. 1981;22:571–3.
59. Gallagher RM, Verma S. Managing pain and comorbid depression: a public health challenge. Semin Clin Neuropsychiatry. 1999;4(3):203–20.
60. Blier PA, Abbott FV. Putative mechanisms of action of antidepressant drugs in affective and anxiety disorders and pain. J Psychiatry Neurosci. 2001;26:37–43.
61. Katon W, Sullivan M, Walker E. Medical symptoms without identified pathology: relationship to psychiatric disorders, childhood and adult trauma, and personality traits. Ann Intern Med. 2001;134:917–25.
62. Romano JM, Turner JA. Chronic pain and depression: does the evidence support a relationship? Psychol Bull. 1985;97(1):18–34.
63. Tunks ER, Crook J, Weir R. Epidemiology of chronic pain with psychological comorbidity: prevalence, risk, course, and prognosis. Can J Psychiatry. 2008;53(4):224–34.
64. Leo RJ. Chronic pain and comorbid depression. Curr Treat Options Neurol. 2005;7(5):403–12.
65. Dersh J, Polatin PB, Gatchel RJ. Chronic pain and psychopathology: research findings and theoretical considerations. Psychosom Med. 2002;64:773–86.
66. Asmundson GJ, Norton GR, Allerdings MD. Fear and avoidance in dysfunctional chronic back pain patients. Pain. 1997;69(3):231–6.
67. Keefe FJ, Rumble ME, Scipio CD, Giordano LA, Perri LM. Psychological aspects of persistent pain: current state of the science. J Pain. 2004;5(4):195–211.
68. Vlaeyen JW, Seelen HA, Peters M, DeJong P, Aretz E, Beisiegel E, et al. Fear of movement/reinjury and muscular reactivity in chronic low back pain patients: an experimental investigation. Pain. 1999;82(3):297–304.
69. Berkowitz L. On the formation and regulation of anger and aggression. A cognitive-neoassociationistic analysis. Am Psychol. 1990;45:494–503.
70. Fernandez E, Wasan A. The anger of pain sufferers: attributions to agents and appraisals of wrongdoing. In: Potegal M, Stemmler G, Spielberger C, editors. International handbook of anger: constituent and concomitant biological, psychological, and social processes. New York: Springer; 2009. p. 449–64.
71. Iyer P, Rom Korin M, Higginbotham L, Davidson KW. Anger, anger expression, and health. In: Iyer P, Rom Korin M, editors. Handbook of health psychology and behavioral medicine. New York: Guilford Press; 2010. p. 120–33.
72. Frijda NH. The emotions. Cambridge: Cambridge University Press; 1986.
73. Scherer KR, Schorr A, Johnstone T. Appraisal processes in emotion: theory, methods, research. New York: Oxford University Press; 2001.
74. Okifuji A, Turk DC, Curran SL. Anger in chronic pain: investigations of anger targets and intensity. J Psychosom Res. 1999;47:1–12.
75. Burns JW, Holly A, Quartana P, Wolff B, Gray E, Bruehl S. Trait anger management style moderates effects of actual state anger regulation on symptom-specific reactivity and recovery among chronic low back pain patients. Psychosom Med. 2008;70(8):898–905.
76. Bruehl S, Chung OY, Burns JW. Differential effects of expressive anger regulation on chronic pain intensity in CRPS and non-CRPS limb pain patients. Pain. 2003;104:647–54.
77. Bruehl S, Chung OY, Burns JW, Biridepalli S. The association between anger expression and chronic pain intensity: evidence for partial mediation by endogenous opioid dysfunction. Pain. 2003;106:317–24.
78. Bruehl S, Chung OY, Burns JW, Diedrich L. Trait anger expressiveness and pain induced beta-endorphin release: support for the opioid dysfunction hypothesis. Pain. 2007;130:208–15.
79. Pilowsky I, Spence ND. Pain, anger and illness behavior. J Psychosom Res. 1976;20(5):411–6.
80. Kerns RD, Rosenberg R, Jacob MC. Anger expression and chronic pain. J Behav Med. 1994;17(1):57–67.
81. DeGood DE, Tait R. Assessment of pain beliefs and pain coping. In: Turk DC, Melzack R, editors. Handbook of pain assessment. New York: Guilford Press; 2001. p. 213–46.
82. Iles RA, Davidson M, Taylor NF. Psychosocial predictors of failure to return to work in non-chronic non-specific low back pain: a systematic review. Occup Environ Med. 2008;65(8):507–17.
83. Edwards LC, Pearce SA, Turner-Stokes L, Jones A. The pain beliefs questionnaire: an investigation of beliefs in the causes and consequences of pain. Pain. 1992;51(3):267–72.
84. Main CJ, Foster N, Buchbinder R. How important are back pain beliefs and expectations for satisfactory recovery from back pain? Best Pract Res Clin Rheumatol. 2010;24(2):205–17.
85. Ramond A, Bouton C, Richard I, Roquelaure Y, Baufreton C, Legrand E, et al. Psychosocial risk factors for chronic low back pain in primary care—a systematic review. Fam Pract. 2011;28(1):12–21.
86. Elfering A, Mannion AF, Jacobshagen N, Tamcan O, Muller U. Beliefs about back pain predict the recovery rate over 52 consecutive weeks. Scand J Work Environ Health. 2009;35(6):437–45.
87. Jensen MP, Turner JA, Romano JM. Self-efficacy and outcome expectancies: relationship to chronic pain coping strategies and adjustment. Pain. 1994;44:263–9.
88. Somers TJ, Shelby RA, Keefe FJ, Godiwala N, Lumley MA, Mosley-Williams A, et al. Disease severity and domain-specific arthritis self-efficacy: relationships to pain and functioning in patients with rheumatoid arthritis. Arthritis Care Res. 2010;62(6):848–56.
89. Truchon M, Cote D, Fillion L, Arsenault B, Dionne C. Low-back-pain related disability: an integration of psychological risk factors into the stress process model. Pain. 2008;137(3):564–73.
90. Mitchell LA, MacDonald RA. Qualitative research on pain. Curr Opin Support Palliat Care. 2009;3(2):131–5.
91. Bandura A. Self-efficacy: toward a unifying theory of behavioral change. Psychol Rev. 1977;84:191–215.
92. Reid MC, Williams CS, Gill TM. The relationship between psychological factors and disabling musculoskeletal pain in community-dwelling older persons. J Am Geriatr Soc. 2003;51:1092–8.
93. Denison E, Asenlof P, Lindberg P, Denison E. Self-efficacy, fear avoidance, and pain intensity as predictors of disability in subacute and chronic musculoskeletal pain patients in primary health care. Pain. 2004;111:245–52.
94. Dohnke B, Knäuper B, Müller-Fahrnow W. Perceived self-efficacy gained from, and health effects of a rehabilitation program after hip joint replacement. Arthritis Care Res. 2005;53:585–92.
95. Dolce JJ, Cracker MF, Doleys DM. Prediction of outcome among chronic pain patients. Behav Res Ther. 1986;24:313–9.

96. Nicholas MK, Wilson PH, Goyen J. Comparison of cognitive-behavioral group treatment and an alternative non-psychological treatment for chronic low back pain. Pain. 1992;48:339–47.
97. Altmaier EM, Russel DW, Kao CF, Lehmann TR, Weinstein JN. Role of self-efficacy in rehabilitation outcome among chronic low back pain patients. J Counsel Psychol. 1993;40:335–9.
98. Council JR, Ahern DK, Follick MJ, Kline CL. Expectencies and functional impairment in chronic low backpain. Pain. 1988;33: 323–31.
99. Dobkin PL, Liu A, Abrahamowicz M, Ionescu-Ittu R, Bernatsky S, Goldberger A, et al. Predictors of disability and pain six months after the end of treatment for fibromyalgia. Clin J Pain. 2010; 26(1):23–9.
100. O'Leary A, Brown S. Self-efficacy and the physiological stress response. In: Maddux JE, editor. Self-efficacy, adaptation, and adjustment: theory, research, and application. New York: Plenum Press; 1995. p. 227–46.
101. Ellis A. Reason and emotion in psychotherapy. New York: Lyle Stuart; 1962.
102. Beck AT, Rush AJ, Shaw BF, Emery G. Cognitive therapy of depression. New York: Guilford Press; 1979.
103. Chaves JF, Brown JM. Spontaneous cognitive strategies for the control of clinical pain and stress. J Behav Med. 1987;10:263–76.
104. Spanos NP, Radtke-Bodorik HL, Ferguson JD, Jones B. The effects of hypnotic suggestibility, suggestions for analgesia, and utilization of cognitive strategies on the reduction of pain. J Abnorm Psychol. 1979;88:282–92.
105. Sorbi MJ, Peters ML, Kruise DA, Maas CJ, Kerssens JJ, Verhaak PF, et al. Electronic momentary assessment in chronic pain: pain and psychological pain responses as predictors of pain disability. Clin J Pain. 2006;22(1):67–81.
106. Abbott AD, Tyni-Lenné R, Hedlund R. The influence of psychological factors on pre-operative levels of pain intensity, disability and health-related quality of life in lumbar spinal fusion surgery patients. Physiotherapy. 2010;96(3):213–21.
107. Wolfe F. Clinical, laboratory, and radiographic assessments. Curr Opin Rheumatol. 1993;5:138–45.
108. Stucki G. International classification of functioning, disability, and health: a promising framework and classification for rehabilitation medicine. Am J Phys Med Rehabil. 2005;84(10):733–40.
109. McCracken LM. Contextual cognitive-behavioral therapy for chronic pain. Seattle: IASP Press; 2005.
110. McCracken LM, Vowles KE. Acceptance of chronic pain. Curr Pain Headache Rep. 2006;10(2):90–4.
111. McCracken LM, Eccleston C. Coping or acceptance: what to do about chronic pain? Pain. 2003;105(1):197–204.
112. McCracken LM. Learning to live with the pain: acceptance of pain predicts adjustment in persons with chronic pain. Pain. 1998;74: 21–7.
113. Viane I, Crombez G, Eccleston C, Poppe C, Devulder J, Van Houdenhove B, et al. Acceptance of pain is an independent predictor of mental well-being in patients with chronic pain: empirical evidence and reappraisal. Pain. 2003;106:65–72.
114. McCracken LM, Eccleston C, Bell C. Clinical assessment of behavioral coping responses: preliminary results from a brief inventory. Eur J Pain. 2005;9:69–78.
115. McCracken LM, Eccleston C. A comparison of the relative utility of coping and acceptance-based measures in a sample of chronic pain sufferers. Eur J Pain. 2006;10:23–9.
116. McCracken LM. Behavioral constituents of chronic pain acceptance: results from factor analysis of the Chronic Pain Acceptance Questionnaire. J Back Musculoskelet Rehabil. 1999;13:93–100.
117. Drossman DA, Camilleri M, Mayer EA, Whitehead WE. AGA technical review on Irritale bowel syndrome. Gastroenterology. 2002;123:2108–31.
118. Palsson OS, Whitehead WE. Psychological treatments in functional gastrointestinal disorders: a primer for the gastroenterologist. Clin Gastroenterol Hepatol. 2013;11(3):208–16.
119. Lackner JM, Mesmer C, Morley S, Dowzer C, Hamilton S. Psychological treatments for irritable bowel syndrome: a systematic review and meta-analysis. J Consult Clin Psychol. 2004;72(6): 1100–13.
120. What is biofeedback [Internet]. Association for Applied Psychophysiology and Biofeedback. [cited 2008 May 18]. http://www.aapb.org/i4a/pages/index.cfm?pageid=1.
121. Coulter ID, Favreau JT, Hardy ML, Morton SC, Roth EA, Shekelle P. Biofeedback interventions for gastrointestinal conditions: a systematic review. Altern Ther Health Med. 2002;8(3):76–83.
122. Humphreys P, Gevirtz R. Treatment of recurrent abdominal pain: components analysis of four treatment protocols. J Pediatr Gastroenterol Nutr. 2000;31:47–51.
123. Whitehead WE, Schuster MM. Gastrointestinal disorders: behavioral and physiological basis for treatment. Orlando: Academic; 1985.
124. Bassotti G, Chistolini F, Sietchiping-Nzepa F, De Roberto G, Morelli A, Chiarioni G. Biofeedback for pelvic floor dysfunction in constipation. Br Med J. 2004;328(7436):393–6.
125. Chiarioni G, Salandini L, Whitehead WE. Biofeedback benefits only patients with outlet dysfunction, not patients with isolated slow transit constipation. Gastroenterology. 2005;129(1): 86–97.
126. Battaglia E, Serra AM, Buonafede G, Dughera L, Chistolini F, Morelli A, et al. Long-term study on the effects of visual biofeedback and muscle training as a therapeutic modality in pelvic floor dyssynergia and slow-transit constipation. Dis Colon Rectum. 2004;47(1):90–5.
127. Blanchard EB, Schwarz SP, Suls JM, Gerardi MA, Scharff L, Greene B, et al. Two controlled evaluations of multicomponent psychological treatment of irritable bowel syndrome. Behav Res Ther. 1992;30(2):175–89.
128. Neff DF, Blanchard EB. A multicomponent treatment for the irritable bowel syndrome. Behav Ther. 1987;18:70–3.
129. Blanchard EB, Schwarz SP. Adaptation of a multicomponent treatment program for irritable bowel syndrome to a small group format. Biofeedback Self Regul. 1987;12:63–9.
130. Schwarz SP, Blanchard EB, Neff DF. Behavioral treatment of irritable bowel syndrome: a 1 year follow-up study. Biofeedback Self Regul. 1986;11(3):189–98.
131. Hjelland IE, Svebak S, Berstad A, Flatabo G, Hausken T. Breathing exercises with vagal biofeedback may benefit patients with functional dyspepsia. Scand J Gastroenterol. 2007;42:1054–62.
132. Whitehead WE, Drescher VM. Perception of gastric contractions and self-control of gastric motility. Psychophys. 1980;17:522–58.
133. What is hypnosis? [Internet]. American Society of Clinical Hypnosis. [cited 2013 May 1]. http://www.asch.net/.
134. Kupers R, Faymonville ME, Laureys S. The cognitive modulation of pain: hypnosis and placebo induced analgesia. Prog Brain Res. 2005;150:251–69.
135. Blanchard EB, Scharff L. Psychosocial aspects of assessment and treatment of irritable bowel syndrome in adults and recurrent abdominal pain in children. J Consult Clin Psychol. 2002;70:725–38.
136. Hefner J, Rilk A, Herbert BM, Zipfel S, Enck P, Martens U. Hypnotherapy for irritable bowel syndrome: a systematic review. Gastroenterology. 2009;47(11):1153–9.
137. Tan G, Hammond DC, Joseph G. Hypnosis and irritable bowel syndrome: a review of efficacy and mechanism of action. Am J Clin Hypn. 2005;47(3):161–78.
138. Lea R, Houghton LA, Calvert EL, Larder S, Gonsalkorale WM, Whelan V, et al. Gut-focused hypnotherapy normalizes disordered rectal sensitivity in patients with irritable bowel syndrome. Aliment Pharmacol Ther. 2003;17:635–42.

139. Simren M, Ringstrom G, Bjornsson ES, Abrahamsson H. Treatment with hypnotherapy reduces the sensory and motor component of the gastrocolonic response in irritable bowel syndrome. Psychosom Med. 2004;66:233–8.
140. Gonsalkorale WM, Miller V, Afzal A, Whorwell PJ. Long term benefits of hypnotherapy for irritable bowel syndrome. Gut. 2003;52(11):1623–31.
141. Sullivan MJ, Thorn B, Haythornthwaite JA. Theoretical perspectives on the relation between catastrophizing and pain. Clin J Pain. 2001;17:52–64.
142. Blanchard E. Irritable bowel syndrome: psychosocial assessment and treatment. Washington, DC: American Psychological Association; 2001.
143. Toner BB, Segal ZV, Emmott SD, Myran D. Cognitive-behavioral treatment of irritable bowel syndrome: the brain-gut connection. New York: Guilford Press; 2000.
144. Blanchard EB, Keefer L, Galovski TE, Taylor AE, Turner SM. Gender differences in psychological distress among patients with irritable bowel syndrome. J Psychosom Res. 2001;50:271–5.
145. Edward B, Bharucha L. Psychological aspects of assessment and treatment of IBS, stress and anxiety disorder. New York: Center of an American University; 2001.
146. Edward B, Louis K. Psychological treatment in IBS patient, best practice and research. Clin Gastroenterol. 1999;13:473–87.
147. Blanchard EB, Greene B, Scharff L, Schwarz-McMorris SP. Relaxation training as a treatment for irritable bowel syndrome. Biofeedback Self Regul. 1993;8:125–32.
148. Greene B, Blanchard EB. Cognitive therapy for irritable bowel syndrome. J Consult Clin Psychol. 1994;62:576–82.
149. Drossman DA, Toner BB, Whitehead WE. Cognitive-behavioral therapy versus education and desipramine versu placebo for moderate to severe functional disorders. Gastroenterology. 2003;125: 19–31.
150. Chou R, Loeser J, Owens D, Rosenquist R, Atlas S, Baisden J, et al. Interventional therapies, surgery, & interdisciplinary rehabilitation for low back pain: an evidence-based clinical practice guideline from the APS. Spine. 2009;34(10):1066–77.
151. Schwarz SP, Taylor AE, Scharff L, Blanchard EB. Behaviorally treated irritable bowel syndrome patients: a four-year follow-up. Behav Res Ther. 1990;28:331–5.
152. Blanchard EB, Schwarz SP, Neff DF. Two-year follow-up of behavioral treatment of irritable bowel syndrome. Behav Ther. 1988;19(1):67–73.
153. Rapkins AJ, Kames LD. The pain management approach to chronic pelvic pain. J Reprod Med. 1987;32:323–37.
154. Geco CD. Management of adolescent chronic pelvic from endometriosis: a pain center perspective. J Pediatr Adolesc Gynecol. 2003;16:17–9.

Index

L. Kapural (ed.), *Chronic Abdominal Pain: An Evidence-Based, Comprehensive Guide to Clinical Management*, DOI 10.1007/978-1-4939-1992-5, © Springer Science+Business Media New York 2015